ITS Sensors and Architectures for Traffic Management and Connected Vehicles

T0133788

ITS Sensors and Architectures for Traffic Management and Connected Vehicles

Lawrence A. Klein

CRC Press
Taylor & Francis Group
Boca Raton London New York

CRC Press is an imprint of the
Taylor & Francis Group, an **informa** business

CRC Press
Taylor & Francis Group
6000 Broken Sound Parkway NW, Suite 300
Boca Raton, FL 33487-2742

© 2018 by Taylor & Francis Group, LLC
CRC Press is an imprint of Taylor & Francis Group, an Informa business

No claim to original U.S. Government works

Printed on acid-free paper

International Standard Book Number-13: 978-1-138-74737-1 (Paperback)
978-1-138-63407-7 (Hardback)

Library of Congress Cataloging-in-Publication Data

Names: Klein, Lawrence A., author.
Title: ITS sensors and architectures for traffic management and connected vehicles / Lawrence A. Klein.
Description: Boca Raton : Taylor & Francis, CRC Press, 2017. | Includes bibliographical references.
Identifiers: LCCN 2017007856| ISBN 9781138634077 | ISBN 9781315206905 (ebook)
Subjects: LCSH: Vehicular ad hoc networks (Computer networks) | Intelligent transportation systems--Equipment and supplies. | Sensor networks.
Classification: LCC TE228.37 .K445 2017 | DDC 388.3/120284--dc23
LC record available at https://lccn.loc.gov/2017007856

Visit the Taylor & Francis Web site at
http://www.taylorandfrancis.com

and the CRC Press Web site at
http://www.crcpress.com

Printed and bound in the United States of America by Sheridan

To Jonathan, Amy, Gregory, and the Magnificent Seven:
Maya, Theo, Coco, Casper, Cassie, Tessa, and Johnny

Contents

Preface

ITS Sensors and Architectures for Traffic Management and Connected Vehicles is based on a semester-length course for undergraduate and graduate students taught for several years at the Harbin Institute of Technology (HIT) in Harbin, China. As such, a section of the introductory chapter includes a description of typical U.S. city street types and their functions. This serves to familiarize foreign readers with street configurations commonly found in U.S. cities and provides some motivation for the need for traffic monitoring and traffic control devices.

The subject matter is also suited for novice and more experienced practitioners from transportation institutes and agencies, and contracting companies who wish to obtain the knowledge required to develop requirements and design concepts for intelligent transportation systems. This book is responsive to the needs of personnel in agencies serving local, regional, state, and multistate or multinational jurisdictions desiring knowledge of modern traffic management systems, traffic flow data acquisition methods, specification of data requirements, automated vehicles and vehicle systems, connected (cooperative) vehicles, the systems engineering process, and National Intelligent Transportation System (ITS) Architectures. Readers gain insights into sensor (detector) operation and selection for effective gathering of street and controlled-access highway data and information needed for enhancing the safety of the travelling public, increasing their mobility on freeways and tollways, and adding predictability to travel times. Intelligent transportation systems address these goals through strategies that include automatic incident detection, active transportation and demand management, traffic-adaptive signal control, and efficient dispatching of emergency response providers.

The growth in the number and types of driver assist and automated features appearing in automobiles and the rapid acceleration of autonomous or self-driving vehicle technologies and configurations is influencing not only technical matters, but also security, policy, planning, legal, and institutional interests. Vehicle automation is evolving rapidly and, while the material presented in this area was up-to-date as of the time of publication, the reader is advised to consult trade publications and government notices and press releases for the latest regulations and policies affecting these vehicles.

The systems engineering process and the National ITS Architecture frameworks are ideal for originating concepts and architectures that meet the needs of stakeholders that own, operate, and rely on multimodal transportation systems for commuting and their livelihood. The latter chapters of this book introduce the reader to sensor and data fusion and its application to traffic management. This subject is gaining relevance as traffic data acquisition devices proliferate and the need for more accurate and timely traffic flow information increases.

The types of sensors examined include inductive loops, magnetometers, magnetic sensors, video detection systems (machine vision sensors), presence-detecting microwave radar sensors, microwave Doppler sensors, passive infrared sensors, lidars, ultrasonic sensors, and acoustic sensors. The strengths and limitations of each are explored so that an informed choice can be made for a particular application. Data utilization is discussed to illustrate how it influences the corresponding accuracy specification for the sensor. In addition, alternative sources of traffic flow data are described. These include license-plate, media access control (MAC) address, and toll-tag readers that exploit mobile devices to gather travel time, speed, and origin–destination pair data. In the future, cooperative vehicle or connected vehicle data will also be available via vehicle-to-vehicle (V2V), vehicle-to-infrastructure (V2I), and infrastructure-to-vehicle (I2V) wireless communications.

Several people contributed valuable suggestions that were incorporated into this edition. Professor Hua Wang of HIT, Professor Yinhai Wang of the University of Washington, and Carol Jacoby, a former colleague at Hughes Aircraft Company, reviewed several of the chapters. Their insightful suggestions improved upon the presentation of the concepts and other material in the chapters. Lisa Burgess of Kimley-Horn along with the Freeway Operations Committee of the Transportation Research Board provided a draft of "Chapter 15: Traffic Management Centers" from the *Freeway Management and Operations Handbook* that proved invaluable in preparing Chapter 2 of this book. Steven Shladover of the University of California, Berkeley, suggested several references that greatly assisted in the preparation of the book. Tony Moore, the Taylor and Francis acquisitions editor who I first met at the annual Transportation Research Board meeting several years ago, Ariel Crockett, Editorial Assistant, Scott Oakley, Editorial Assistant, Cynthia Klivecka, Project Editor, Joette Lynch, Project Editor, and Karthick Parthasarathy, Assistant Manager at NovaTechset proved indispensable in improving the quality of the text and subject matter and expediting the printing of the book. Appreciation is also acknowledged to the many companies and organizations that graciously provided permission to reprint photographs and illustrations that appear in this book.

Lawrence A. Klein
April 2017

Author

Lawrence A. Klein has more than 40 years of aerospace and traffic management experience as a specialist in systems engineering and the development of sensors and data fusion concepts. He is a consultant and has been a research engineer and program manager, including Program Manager and Principal Investigator of the Detection Technology for IVHS Program while at Hughes Aircraft Company, director of Advanced Technology Programs at Waveband Corporation, and research scientist at the French Institute of Science and Technology for Transport, Development and Networks (IFSTTAR). He is a member of Transportation Research Board's Highway Traffic Monitoring Committee and past member of the Freeway Operations Committee, and led the ASTM E17 Group V-ITS development of worldwide standards to evaluate traffic sensors. Klein is also a visiting professor at Harbin Institute of Technology in China, and is the principal author of the third edition of the FHWA's *Traffic Detector Handbook*. His other books of note are *Sensor Technologies and Data Requirements for ITS* (Artech House) and *Sensor and Data Fusion: A Tool for Information Assessment and Decision Making*, Second Edition (SPIE).

Chapter 1

Introduction

Modern traffic and transportation management systems concern themselves with the safety of travelers, efficiency in moving travelers from one point to another (often referred to as mobility), and the environmental impacts of the transportation modes. Measures of effectiveness (MOEs) are utilized to quantify the success of the system in meeting these goals. MOEs in turn require data for their evaluation. Historically, data were obtained from traffic flow sensors, such as inductive loop detectors (ILDs) in the roadway, which were installed by the traffic management agency. ILDs are able to provide vehicle counts, presence, passage, a measure of lane occupancy, and local estimates of vehicle speed. As the variety of sensor technologies increased and matured, additional types of sensors became available. These include video detection systems, microwave radar sensors, Doppler microwave sensors, acoustic sensors, ultrasonic sensors, magnetometers and magnetic sensors, passive infrared sensors, lidar sensors, and sensors that employ combinations of these technologies. Many are capable of multilane coverage.

In addition to roadway sensors, transportation agencies gather data such as travel times and origin–destination pairs from personal and mobile communication devices via Bluetooth® readers. Similar types of data may also be obtained from toll-tag readers and license plate readers installed along roadways. Global positioning system (GPS) and inertial navigation system (INS) information is available on mobile devices to motorists, other travelers, and system operators. These data provide travel route alternatives and travel time information, and can track commercial, transit, and traffic management agency vehicles to improve safety and operational efficiency. More recently, initiatives such as the Connected Vehicle Program in the United States, Cooperative Intelligent Transportation Systems initiatives in Europe, and similar programs elsewhere are enabling vehicle-to-vehicle and vehicle-to-infrastructure data transfer that promises to further increase safety and mobility, and reduce the environmental impacts of the automobile and other types of vehicles.

The ever-increasing availability of automation systems in vehicles is one more example of how technology is improving the safety of motorists by reducing the potential for accidents. Standard and optional equipment offer blind spot detection, parking assist, rearview camera imagery, adaptive cruise control and emergency braking, lane keeping warnings, cross traffic alerts and avoidance maneuvers, and automated parking in parking structures. Eventually, the automation will lead to completely automated vehicle operation. But before this occurs, several issues will require resolution. These include technology, security, policy (technical, legal, and implementation), and institutional issues being addressed by vehicle manufacturers, governmental agencies, and professional transportation organizations as they develop automated vehicle functions and systems and the infrastructure that lead to a fully autonomous or self-driving vehicle.

Another key concept that affects the design and effectiveness of transportation management systems, whether for limited-access highways or arterials, is systems engineering.

Applications of its principles are essential to ensure that the concerns of all stakeholders, for example, owners of the system, operators, maintenance personnel, and users of all types (motorists, pedestrians, cyclists, emergency service providers, commercial vehicle and transit system operators and drivers, and law enforcement agencies) are addressed as the concept for the design and operation of the system is developed.

Since there are many sources of traffic data, it is only logical to consider how estimates of traffic flow parameters such as vehicle speed, count, flow rate, travel times, congestion, and queue length can be improved by combining information from more than one data source. Hence, the book concludes by describing sensor and data fusion concepts that have been or can be applied to traffic management.

1.1 SENSOR APPLICATIONS TO TRAFFIC MANAGEMENT

Figure 1.1 depicts a few of the many applications of sensors to traffic management. Those discussed in this book are related to traffic signal control, ramp metering, travel time estimation and forecasting, wrong-way vehicle detection, freeway incident detection and congestion monitoring, and active transportation and demand management. Other sensor types and applications not specifically addressed are weigh-in-motion sensors and road-weather sensors. A description of weigh-in-motion sensors is found in McCall and Vodrazka [1] and Klein [2]. A review of best practices for road-weather management is available in Murphy et al. [3].

The selection of a traffic sensor depends on many factors such as the use of the data (e.g., incident detection through traffic flow parameter measurement, traffic signal actuation by means of lane-by-lane vehicle detection, or toll collection through vehicle classification and weight measurement), accuracy required of the data, weather conditions in which the sensor will operate, sensor and installation costs, road geometry, road condition, vendor support, agency preferences, and availability of appropriate communications media. Not all sensors output the same sorts of data and information. For instance, sensors can provide some, but generally not all, of the data listed below [2,4]:

- Flow rate (volume), lane occupancy, and density.
- Count, presence, and passage.
- Speed of individual vehicles and vehicle platoons.
- Vehicle class.
- Queue lengths.
- Approach flow profile.

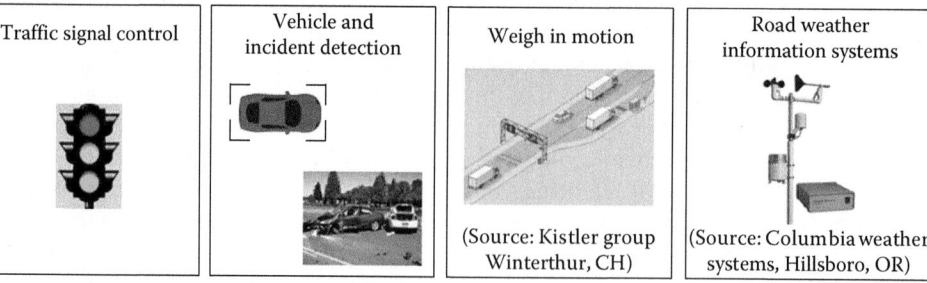

Figure 1.1 Applications of sensors to traffic management.

- Approach stops.
- Link travel time.
- Origin–destination pairs.

1.2 STREET TYPES AND FUNCTIONS

Several categories of streets and arterials are encountered in the United States. Figure 1.2 illustrates those normally found in a single-family housing area in a typical U.S. city. Usually, there are not any control devices such as stop signs, yield-to-oncoming-traffic signs, or traffic signals at the intersections of these roadways. It is the responsibility of the drivers to slow down at the intersections and proceed only when no other vehicles are entering the intersection and the intersection is clear. If two vehicles approach the intersection at the same time, the driver on the right is presumed to have the right-of-way.

When a residential-area street intersects a major street or an arterial, the minor street may have a stop sign, as shown on the left of Figure 1.3, or a yield sign as in the middle or on the right. The yield sign on the right warns drivers to give way to pedestrians who may be

(a) (b)

Figure 1.2 Streets commonly found in a single-family housing area in a typical U.S. city. These particular intersections do not have traffic control devices at the intersections. (a) Four-way intersection and (b) three-way intersection.

Yield sign

Figure 1.3 Stop sign and yield sign configurations. In the photograph on the left, the stop sign inset shows the octagonal shape of the sign.

crossing the street. If none of these control devices are present, it is the responsibility of the driver on the minor street to proceed only when safe, that is, there is no oncoming traffic on the major street.

Figure 1.4 depicts a collector street that channels traffic from the minor residential streets onto an arterial. In some states in the United States such as California, it is common to see four-way stop signs at the intersection of two collector streets. This sign configuration requires drivers on all intersecting streets to come to a full stop before proceeding through the intersection. In these photographs, opposite-direction traffic is separated by a median or a set of painted lines. The painted lines in the photograph on the right indicate that vehicles traveling in either direction may enter the middle lane to make left turns or U-turns.

When major streets and arterials intersect as represented in the left photograph of Figure 1.5, traffic is usually controlled by a traffic signal. The right photograph shows video cameras mounted at the top of the luminaries. They are part of a video detection system (at times referred to as a machine vision sensor or video image processor) that controls the phase and timing of the traffic signals at this intersection. Specialized traffic control devices for bicycles, such as those in Figure 1.6, may also be present at the intersection.

Sometimes, arterials lead to a freeway or controlled-access highway as illustrated in Figure 1.7. Entrance onto the controlled-access highway is via a ramp that may be metered in larger urban areas to regulate the number of vehicles entering the highway in a given time

Figure 1.4 Collector street.

Figure 1.5 Residential and commercial area with traffic signal controlled intersection.

Figure 1.6 Traffic signal controls for requesting and notifying a cyclist of a green signal phase. The bike signals are highlighted with a yellow housing in the photograph on the right.

period. Vehicle presence on the ramp is detected by sensors, in this case ILDs installed in the roadbed at the traffic signal stopline to measure vehicle presence and upstream of it to measure vehicle queue. Additional sensors are installed beyond the stopline to indicate passage of the vehicle.

Figure 1.7 Arterial leading to a freeway or controlled-access highway.

1.3 MANUAL ON UNIFORM TRAFFIC CONTROL DEVICES

Approved types of traffic control devices suitable for installation on streets, highways, bikeways, and private roads open to public travel are found in the *Manual on Uniform Traffic Control Devices* (MUTCD) published by the U.S. Federal Highway Administration (FHWA). It provides the transportation professional with the standards, guidance, options, and support materials needed to select the proper device for the location [5,6].

Traffic control devices are defined as the signs, signals, markings, and other apparatus used to regulate, warn, or guide traffic. They can be placed on, over, or adjacent to a street, highway, pedestrian facility, bikeway, or private road open to public travel by authority of a public agency or official having jurisdiction. For a private road, the authority is the private owner or private official having jurisdiction.

The MUTCD is recognized as the U.S. national standard for all traffic control devices installed on any street, highway, bikeway, or private road open to public travel. It assists the FHWA attain policies and procedures to implement basic uniformity of traffic control devices. Conformance with these recommendations usually limits the liability of local and state agencies in the event of an accident.

Any traffic control device design or application provision contained in the MUTCD is considered to be in the public domain. Traffic control devices found in the MUTCD are not protected by a patent, trademark, or copyright, except for the Interstate Shield and any items owned by FHWA.

1.4 CHAPTER SUMMARY

This book is divided into four broad sections. Chapters 1 and 2 contain information about the types of roadways and traffic management centers (TMCs) that are typical in the United States. Chapters 3–10 discuss traffic flow sensor applications to intelligent transportation systems (ITSs), data requirements of several ITS strategies, sensor technologies, sensor installation and initialization procedures, field testing of sensors, and alternate sources of traffic flow data. Chapters 11–15 address automated, connected, and cooperative vehicles; systems engineering principles; National ITS Architectures; and other architectures and applications that exploit the data available from connected and cooperative vehicles. Chapters 16 and 17 explore sensor and data fusion and the benefits it can bring to ITS. Bayesian inference and Dempster–Shafer evidential reasoning, two of the more ubiquitous data fusion algorithms, are discussed in detail. The following paragraphs describe the specific chapter-by-chapter contents.

Chapter 2 contains descriptions of TMCs, also referred to as traffic operations centers or transportation management and operations centers (TMOCs). They facilitate day-to-day traffic management and operations for limited-access highway, integrated corridor management systems, and modern traffic signal control systems. Here, real-time data and situational awareness information are monitored, processed, and acted upon to improve the operational efficiency of highway and arterial networks. The role of a TMC may extend beyond the limited-access highway network and the particular responsible agency, functioning as the key technical and institutional hub to bring together multiple jurisdictions, transportation modal interests, and service providers to focus on the common goal of optimizing the performance of the entire surface transportation system. Often, TMCs are part of a larger emergency operations center where they monitor situational awareness on the road network and communicate this information to other local, state, and federal agencies and commercial interests.

Chapter 3 discusses various applications of sensors to ITS for collecting accurate and timely traffic flow data. These include local isolated intersection signal control, interconnected intersection signal control, ramp and freeway metering, travel time estimation, wrong-way vehicle detection, freeway incident detection and congestion monitoring, active transportation and demand management, and traffic data collection for planning and archival or historical purposes. Sensor data also support vehicle classification, tolling operations, traffic surveys, parking facility management, and roadway hazard identification.

Chapter 4 describes sensor requirements, types of data, accuracies, and sampling intervals that support several traffic management strategies. Surveys of operations personnel and literature searches can often assist in defining data requirements. Advanced signalized intersection control, freeway incident detection and management, and freeway metering are discussed to show how a particular application can influence present and future input data requirements and hence sensor specifications. Compilation of real-time data accuracy requirements for a variety of traffic operations, planning, and traveler information services may guide transportation management personnel in the planning, procurement, and design of detection stations.

The operating principles of traffic flow sensor technologies are explained in Chapter 5. Traffic flow sensors are often divided into two broad categories: those mounted on or under the roadway surface and those mounted above the roadway on sign bridges or to the side of the roadway on poles and other structures. The first category is also referred to as intrusive sensors because they infringe on the roadway pavement, while the second category is referred to as nonintrusive. The traffic flow sensors described include inductive loops, magnetometers, magnetic sensors, video detection systems, presence-detecting microwave radars, microwave Doppler sensors, acoustic sensors, lidar (also referred to as laser radar or an active infrared sensor), passive infrared sensors, ultrasonic sensors, and sensors that utilize combinations of several technologies. It is critical that sensors selected for a first-time application be field tested under actual operating conditions that include variations in traffic flow rates, day and night lighting, and inclement weather before they are purchased in large quantities for operational use.

Chapter 6 explores the installation and sensitivity of ILDs as they are one of the most prevalent types of traffic flow sensors. The first portion of the chapter summarizes loop installation guidelines, while the second describes the methods for calculating the threshold loop system sensitivity. This calculation is necessary to ensure that the sensitivity of the electronics unit (also called the detector) is greater than the threshold sensitivity of the loop and lead-in wires and cable so that the loop can detect vehicles with high undercarriages or small metal content. Properly installed ILDs perform well for vehicle passage and presence detection if two important items are attended to, namely, following proper loop installation guidelines and maintaining the integrity of the roadbed.

The overhead sensor installation and initialization procedures described in Chapter 7 illustrate the variety of methods employed by manufacturers to instruct personnel in the use of their products. Before installing any sensor, the manufacturer or authorized representative should be contacted to ensure that the latest installation and user manuals and software have been obtained. The procedures in this chapter are not prescriptive and should not be relied on to contain all of the information required to successfully install and operate a sensor. Not all overhead sensor types and models are discussed as this would be a daunting task. However, many details for installing and initializing video detection systems and microwave radar sensors are given as these are two of the most popular types of overhead sensors in use. Other sensors whose installation is discussed are the multilane acoustic sensor and passive infrared sensor.

Chapter 8 contains examples of roadside sensor evaluations under actual operating conditions. The chapter introduces the reader to sensor testing, required test site documentation, and the types of data usually sought and obtained. Roadside assessments are recommended before large-scale purchases occur in order to validate sensor performance under operational traffic flow conditions that often vary with season, vehicle mix, unique road configurations, lighting, and weather.

In Chapter 9, we discuss four topics related to testing and evaluating sensor performance. First, presence detection, the most ubiquitous application of traffic detection systems on freeways and surface street arterials, is explored in terms of the consequences that can occur when a sensor is used to detect the presence of a vehicle. These are the result of the sensor either correctly detecting the vehicle or failing to detect the vehicle. The three possible outcomes are correct detection of an actual vehicle, false detection when no vehicle is present, and failure to detect an actual vehicle. The second topic is a review of the information available in testing standards such as those developed through ASTM International. The third examines the concepts of confidence intervals and confidence levels that should be included in any standard or specification that is prepared for sensor accuracy. The fourth topic concerns interoperability as it relates to institutions, policies and procedures, and technical concerns such as interfacing with other components and data transfer among devices.

Chapter 10 begins with a description of global navigation satellite systems deployed by several countries around the world. Then, with the U.S. GPS as an example, it examines the operation of global navigation satellite systems and INSs. The subjects treated are GPS architecture, GPS augmentation approaches, the GPS modernization program, differential GPS, performance of combined GPS–INS positioning systems in areas where GPS signal reception is weak, and use of Bluetooth media access control (MAC) address readers for obtaining travel time information. GPS applications are diverse, consisting of tracking of transit vehicles and taxis; hazmat, police, fire, and paramedic service vehicles; street and highway work zone vehicles and personnel; tree harvesters in forests; snow plows; package delivery vehicles to ensure timely delivery of merchandise and efficient operations; and search and rescue aircraft. Future applications include air traffic control systems and real-time tracking of vehicles and pedestrians in support of connected and cooperative vehicle programs.

Chapter 11 delves into several methods of classifying the automation levels and combinations of automated features making their appearance in vehicles. These are the ten-level human–computer decision automation, the Society of Automotive Engineers (SAE) six-level automation, the German Federal Highway Research Institute (BASt) five-level automation, and the U.S. National Highway Traffic Safety Administration's (NHTSA) five-level automation taxonomies. The latter scheme was set aside by NHTSA in 2016 in favor of the SAE categories. The chapter also discusses the driving environment in which these vehicles operate; legal issues associated with autonomous vehicle operation; the 2016 NHTSA automated vehicles policy that proposes an approach to hasten the safe development, testing, and operation of highly automated vehicles; currently available driver assist and automation options being offered by a variety of vehicle manufacturers; and the Mobility as a Service concept.

Chapter 12 describes U.S. and European connected and cooperative vehicle programs and the major results of tests that studied their feasibility. The overriding purpose of connected vehicle and similar programs is to increase driver and pedestrian safety. This is achieved by two-way communication of data and information from vehicle-to-vehicle, vehicle-to-pedestrian, and vehicle- and pedestrian-to-infrastructure using wireless devices. The chapter explores technology, security, policy (technical, legal, and implementation), and institutional issues being addressed by vehicle manufacturers, governmental agencies, and professional transportation organizations as they develop driver assist and automated vehicle functions and systems along with the infrastructure that lead to a fully autonomous or self-driving vehicle. The 2016 NHTSA Notice of Proposed Rulemaking for Vehicle-to-Vehicle Communications and the 2016 FHWA Vehicle-to-Infrastructure Deployment Guidance for state and local transportation agencies are also examined. The final sections discuss connected vehicle pilot test and evaluation programs in the United States; the European Union approach for cultivating a shared vision for the interoperable deployment of cooperative vehicles, namely, the Platform for the Deployment of Cooperative Intelligent Transport Systems in the European Union (C-ITS Platform); and cooperative vehicle pilot and operational projects in Europe.

Chapter 13 illustrates the importance of applying systems engineering principles throughout the life cycle of a transportation or traffic management system. It explains the need to understand and acknowledge stakeholder wishes and system functionality early in the development stage by documenting requirements and system design options, obtaining stakeholder buy-in for the proposed system design, and only then proceeding with the subsystem and component-level designs, and system validation. Fundamental to stakeholder buy-in is the creation of a concept of operations. It provides the initial definition of the system, documents the way the envisioned system is to operate, and shows how the envisioned system will meet the demands and expectations of the stakeholders. The concept of operations describes system operation from multiple viewpoints that take into account the needs of the owner, operators, users (drivers, riders, pedestrians, cyclists, transit and commercial freight operators and drivers, emergency response providers), maintenance personnel, and managers. A concept of operations for a multimodal intelligent traffic signal system is presented as an example of how this narrative is created. System design and operation are evaluated through a series of performance measures that compare system performance with design goals. Another important area of system design is interface and standards specification and definition to ensure that components and subsystems integrate properly and data and information get accurately transmitted over all the required communications channels.

Chapter 14 examines National Intelligent Transportation System Architectures in the United States, European Union, Japan, and Canada. National Architectures provide a framework for planning, programming, and implementing ITSs. In the United States, it facilitates the ability of local, regional, state, and interstate jurisdictions to operate collaboratively and to harness the benefits of a regional approach to transportation challenges. In other countries, the National Architectures serve local and other types of governing entities and service providers (e.g., countries, prefectures, provinces, and districts). The architecture structure allows it to evolve and incorporate technological improvements and changing user needs as brought about, for example, by connected and cooperative vehicles. The National Architectures define the functions that are required for ITS, the physical entities or subsystems where these functions reside, the information flows and data flows that connect the functions and physical subsystems together into an integrated system, and the communications approaches employed for the accurate and timely exchange of information between systems. They do not prescribe specific technologies to implement functions or physical entities in order to leave the technology decision open to the latest innovations.

Chapter 15 describes a derivative architecture and several applications that benefit from information communicated by connected vehicles. To this end, we examine the Connected Vehicle Reference Implementation Architecture (CVRIA) first used in Southeast Michigan, a simulation of a traffic signal control system for connected vehicles that was evaluated with a calibrated model of a test network of four intersections, and a lane management system for connected and conventional vehicles that alerts drivers to when it is productive and safe to change lanes on a controlled-access highway. The CVRIA is developed from four viewpoints, namely, enterprise, functional, physical, and communications. The enterprise viewpoint addresses the relationships between organizations and the roles of those organizations in the delivery of services in the connected vehicle environment. The functional viewpoint focuses on the behavior, structure, and interaction of the functions performed within the connected vehicle environment. The physical view consists of a set of integrated physical objects that interact and exchange information to support a particular connected vehicle application. The communications view concerns itself with the design and implementation of protocols and communications standards, including implementation choices, and specification and allocation of communications functionality to physical objects in the physical view. The traffic signal control system relies on connected vehicles to wirelessly transmit their positions, headings, and speeds. Finally, the lane management system functions as a decision support system that provides lane changing advice to drivers of conventional and connected vehicles.

Multisensor data fusion, the subject of Chapter 16, can bring many benefits to traffic management. It aids in the interpretation of information gathered from a complex environment characterized by the presence of different types of vehicles, unexpected objects such as debris or a pedestrian darting across a roadway, inclement weather, vehicles changing lanes, and roadside structures or weather effects that interfere with the normal observation of traffic patterns and the gathering of needed data. The chapter begins with a review of the definitions of sensor and data fusion and their role in improving the effectiveness of traffic management strategies. Factors that influence the selection of a sensor and data fusion architecture are described. The U.S. Department of Defense Joint Directors of Laboratories (JDL) six-level data fusion model is discussed, followed by taxonomies for object detection, classification, and identification algorithms and for state estimation algorithms utilized to track objects.

Two of the widely applied detection, classification, and identification algorithms are examined in detail in Chapter 17, namely, Bayesian inference and Dempster–Shafer evidential reasoning, and several examples of their applications to traffic management are provided.

REFERENCES

1. W. McCall and W.C. Vodrazka Jr., *States' Successful Practices Weigh-In-Motion Handbook*, *Center for Transportation Research and Education (CTRE)*, Iowa State University, Ames, IA, December 15, 1997, http://www.ctre.iastate.edu/research/wim_pdf/. Accessed December 26, 2015.

2. L.A. Klein, *Sensor Technologies and Data Requirements for ITS*, Appendix F, Artech House, Norwood, MA, June 2001.

3. R. Murphy, R. Swick, and G. Guevara, *Best Practices for Road Weather Management*, Version 3.0, FHWA-HOP-12-046, Federal Highway Administration, U.S. Department of Transportation, Washington, DC, June 2012. http://www.ops.fhwa.dot.gov/publications/fhwa-hop12046/fhwahop12046.pdf. Accessed December 21, 2016.

4. L.A. Klein, D. Gibson, and M.K. Mills, *Traffic Detector Handbook: Third Edition*, FHWA-HRT-06-108 (Vol. I) and FHWA-HRT-06-139 (Vol. II), Federal Highway Administration, U.S. Department of Transportation, Washington, DC, October 2006. www.fhwa.dot.gov/publications/research/operations/its/06108/06108.pdf and www.fhwa.dot.gov/publications/research/operations/its/06139/06139.pdf. Accessed December 13, 2013.

5. *Manual on Uniform Traffic Control Devices*, 2009 Edition with Revisions 1 and 2, U.S. Department of Transportation, Federal Highway Administration, Washington, DC, May 2012. http://mutcd.fhwa.dot.gov/pdfs/2009r1r2/mutcd2009r1r2edition.pdf. Accessed April 4, 2016.

6. K. Fitzpatrick, M. Brewer, G. Hawkins, P. Carlson, V. Iragavarapu, J. Pline, and P. Koonce, Potential MUTCD Criteria for Selecting the Type of Control for Unsignalized Intersections, National Cooperative Highway Research Program (NCHRP) Web-Only Document 213, Contractor's Final Report for NCHRP Project 03-109, Transportation Research Board, Washington, DC, March 2015. http://onlinepubs.trb.org/onlinepubs/nchrp/nchrp_w213.pdf. Accessed April 4, 2016.

Chapter 2

Freeway traffic management centers

Traffic or transportation management centers (TMCs), also referred to as traffic operations centers or transportation management and operations centers (TMOCs), are a key component of transportation systems management and operations (TSM&O). They are the heart of most freeway or limited-access highway and integrated corridor management systems, and arterial traffic signal systems that support traffic-responsive and traffic-adaptive signal control. Major cities and metropolitan areas in the United States usually contain a TMC according to a 2010 survey by the Research and Innovative Technology Administration of the U.S. Department of Transportation. It found that there were 266 TMCs in the United States, with one in almost every large city as illustrated in Figure 2.1 [1].

TMCs facilitate day-to-day traffic management and operations, network monitoring, strategy implementation, traffic incident management and response coordination, and responses to other events and disturbances (e.g., weather-related episodes such as hurricanes and snow storms) to the transportation system. Here, real-time data and situational awareness information are monitored, collected, processed, fused, shared, and acted upon to improve the operating efficiency of arterial and freeway networks. TMCs operate the field-located devices and communications media that transmit data and information to and from the devices, and implement the policies and procedures that address transportation- and travel-related events impacting the system.

2.1 TMC BENEFITS

Benefits produced by TMC operations vary depending on the purpose and functionality of the TMC. Benefits are created across multiple areas that include safety, mobility, and the environment. The range of benefits depicted in Figure 2.2 is collected from data compiled by the U.S. Federal Highway Administration (FHWA). They show decreases that vary from 11% to 69% for incident clearance time, from 54% to 88% for delay, 50% for queue length, 27.5% for crashes, and 10% for travel time [2].

2.2 FREEWAY TMC OPERATIONAL MODELS, JURISDICTIONS, AND ROLES

The category of TMC focused on in the rest of this chapter is the freeway or limited-access highway TMC. Many of its attributes are similar to those found in centers that manage arterials and other types of transportation modes.

Freeway management is performed through a combination of human and physical elements. The human elements are the managers and operators in the agencies who plan

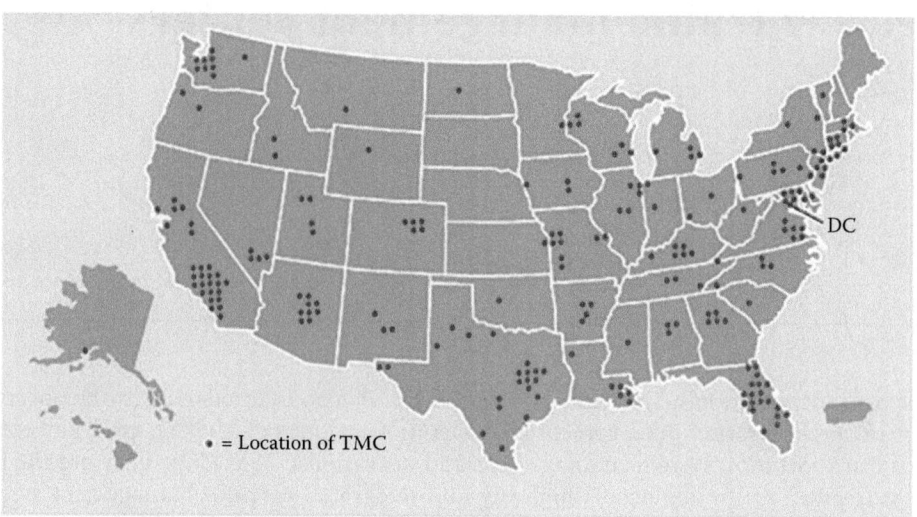

Figure 2.1 Locations of TMCs in the United States as of 2010. (From J. Chu and L. Radow, "Behind the Scenes at TMCs," FHWA-HRT-12-005, *Public Roads Magazine*, 76:1, July/August 2012.)

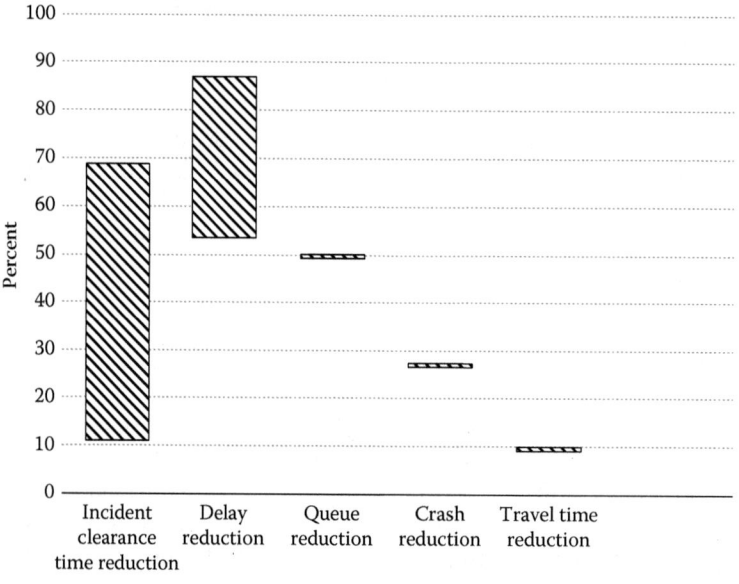

Figure 2.2 TMC benefits. (From G. Hatcher et al. *Intelligent Transportation Systems Benefits, Costs, and Lessons Learned: 2014 Update Report*, FHWA-JPO-14-159, ITS Joint Program Office, Washington, DC, June 2014.)

and perform control functions, while the physical elements (e.g., cameras; changeable or dynamic message signs [CMS or DMS]; ramp meters; sensors; electronic toll tag, license plate, and media access control [MAC] address readers; highway advisory radio; and communications systems) are the individual components and systems that assist the operators in performing their functions. The human and physical elements are brought together at the TMC.

2.2.1 Operational and business model

In addition to the field elements and communication systems, transportation management systems utilize business processes and associated tools that help maximize the effectiveness of the transportation system. The California Department of Transportation (Caltrans) business model in Figure 2.3 incorporates selective system expansion, operational improvements, traveler services, traffic control, incident management, demand management, maintenance and operations, and system monitoring and evaluation to help meet its productivity goals [3].

The difference between this approach and other models is an emphasis on the middle sections of the figure (traveler information, traffic control, and incident management), which are all operational processes. The success of these processes depends on the availability of real-time performance information (provided by the system monitoring illustrated at the base of the figure), reflecting a focus on maximizing the system's productivity measured in terms of vehicle flow in units of vehicles/hour.

Success also depends on Caltrans and regional and local agencies working closely together and coordinating technology initiatives and funding priorities. Therefore, addressing system issues such as congestion requires a multipronged approach that includes adding new capacity, maintaining present infrastructure, investing in and encouraging the use of alternate travel modes such as bus and rail transit, promoting bicycling, and utilizing effective and forward-looking transportation management strategies such as active transportation and demand management, integrated corridor management, and application of crowd sourcing and connected vehicle data and information.

Freeway TMC roles vary depending on the roadway type, region or area monitored, and collocation strategy, if any, that involves other agencies and services. In TMCs where public safety and law enforcement dispatch, service patrols, transit dispatch, public information and communications personnel, and other such functions are located, the capabilities and sphere of influence of a TMC are increased. Information about real-time conditions and response status is shared

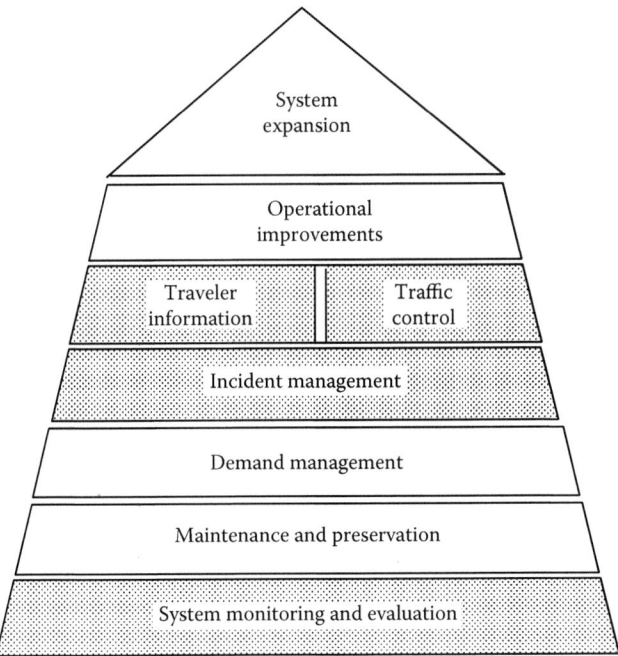

Figure 2.3 Caltrans transportation management system operational and business model.

with partner agencies and with the public through various traveler information dissemination strategies, for example, CMS, variable speed limit and lane usage signs, radio, Internet Web sites, kiosks, highway advisory radio, and in the future, connected (cooperative) vehicles. TMCs can serve as the technical and institutional hub for bringing these interests together within a metropolitan area, region, state, or multistate levels in the United States [4]. Other countries may have different organizational levels for their TMCs such as city, region or province, and national. Because it is critical to the efficient operation of a limited-access highway system (and the broader surface transportation network), it is essential that the TMC be planned, designed, commissioned, staffed, and maintained in a manner that promotes this goal.

2.2.2 Urban area focus

A TMC in the United States that administers freeway or limited-access highway facilities in an urban area typically involves a state-run department of transportation (DOT) or toll authority responsible for monitoring, operating, managing, and responding to conditions on its roadways. The entity is concerned with the devices and infrastructure that are part of its network, and does not normally have responsibility for monitoring or managing roads outside of the network. Urban freeway management systems are often characterized by the geographic boundaries of agency-owned and -operated infrastructure (such as detection devices, surveillance cameras, and CMS), although new sources of more ubiquitous data are providing TMCs with more situational awareness information about speeds and potential incidents well outside of traditional infrastructure boundaries (e.g., cellular-provided information from travelers and connected vehicle data). The urban area TMC can include one or more collocated partners, such as law enforcement dispatch, local agency traffic operations and management, or traffic reporting. Examples of this type of operation include the North Carolina DOT, Caltrans urban districts (San Francisco, Los Angeles and Ventura Counties, Orange County, and San Diego), and Houston TranStar shown in Figure 2.4, which contains representatives from the City of Houston, Harris County, Metropolitan Transit Authority, and the Texas DOT.

2.2.3 Urban area and statewide focus

TMCs with combined urban area and statewide jurisdiction are responsible for operations and management strategies of urban area freeways and networks outside of the urban area up to and including statewide. Often, the monitoring and controlling of devices in the extended area are limited to closed-circuit television (CCTV) cameras, CMS, road-weather

(b)

(a)

Figure 2.4 (a) Houston TranStar operations personnel at a coordination meeting and (b) TranStar video wall and work stations. (From (a) Houston TranStar, http://www.houstontranstar.org/ and (b) LAK.)

information systems, and environmental sensors. Operation of these resources outside the urban area might be in partnership with district or regional staff in closer proximity to the devices. Similar to the urban freeway focus, this category of TMC may include one or more collocated partners. Examples include the Wisconsin DOT Statewide Traffic Operations Center in Milwaukee with statewide operations, management, and maintenance responsibility; Arizona Department of Transportation (ADOT) Traffic Operations Center in Phoenix responsible for the Phoenix Freeway Management System and statewide devices, not including the Tucson metropolitan area; and the Utah Department of Transportation (UDOT) Traffic Operations Center in Salt Lake City.

2.2.4 Multiregion or multistate focus

TMCs of this type include operational responsibility or shared operations within a region, multiple regions, or a region that includes multiple states. Implementation requires a cooperative governance and organizational model that provides for operational responsibility outside of the traditional single entity-owned infrastructure. An example of this category is the Kansas City Scout System and TMC illustrated in Figure 2.5, which oversees the Kansas City metropolitan area and freeway system management and operations in both Kansas and Missouri.

2.3 COLLOCATION CONSIDERATIONS

Freeway management TMCs may function in two general ways. The first utilizes a single entity operations facility such as a toll authority or state DOT with collocated operations partners such as law enforcement, other emergency response providers, and freeway service patrol dispatch. The second operates with multiple jurisdictional agencies that may not be collocated, such as those responsible for freeway response other than law enforcement, signalized arterial operations, or other regional transportation management. These TMC operating models each have their advantages and disadvantages, and policy, budget, and other constraints.

Agencies that have opted to collocate do so for a variety of reasons such as those below:

- Improved coordination among partners for traffic management, traffic incident management, planned special event management, and response to emergencies or road and weather hazards.

Figure 2.5 Kansas City Scout TMC (http://www.kcscout.net).

- Improved relationships among partner agencies as a result of collocation.
- Economies of scale for equipment such as the video wall and telecommunications.
- Central operations and information technology management.
- Economies of scale for facility management and operations, including capital and operating costs.

While cited as beneficial, collocation requires understanding and often accommodation of the unique policies, human resource needs, information technology systems, security requirements, business rules, and operating procedures of each agency. The risks and challenges associated with collocating agency services require consideration of the following factors:

- Governance structure in a combined facility for decision-making that influences all collocated agencies.
- Staffing ownership for operations.
- Individual agency policies within a multiagency environment.
- Creation of new agency policies, processes, and protocols that are only used in that multiagency environment. This involves establishing agreements for processes, shared funding strategies (if required), and communication and understanding of business processes employed by the agencies.

2.3.1 Concept of operations for collocation

A tool that aids planning and implementation of TMCs with collocated departments and agencies is the concept of operations (ConOps). The ConOps defines the roles and responsibilities of each collocating or sharing agency and provides a structure for managing operations and processes once the collocation is completed. Summaries of risks, challenges, and necessary procedural changes by the ConOps identify the agency's and department's functional relationships with each other and with external entities, such as the media and traveling public. For some agencies, collocation is a natural method that leads to improved operations and coordination between services that would already need data and information or status sharing on a daily or event basis. For others, collocation may require some justification in order to quantify the benefits of a move to a collocated facility, such as cost and functional analyses that show why collocation of operations may be necessary. ConOps development and use is described further in Chapter 13.

2.3.2 Collocated center examples from the United States

The combination of agency operations and services provided to the public by collocated facilities is unique based on the needs of the locale and the capabilities of the individual agencies. While some regions have moved toward full agency collocation of services or entire departments, others have opted for liaison relationships and added staffing in order to create the link to external agency services that relate to transportation operations. Advances in technology are enabling improved coordination and communication opportunities for those liaisons and partner agencies located in separate facilities. Secure virtual connections can promote interoperability, although this does not necessarily replace the personal connections that a collocated environment can support. Table 2.1 contains a sampling of collocated centers in the United States, corresponding partner agencies, and the services they offer [4]. Additional details concerning these centers are found in the paragraphs below.

Table 2.1 Collocated U.S. TMC examples

Collocated center	Managing agency	Partner agencies	Services
Combined Transportation, Emergency and Communications Center (CTECC)	City of Austin	City of Austin	Police, Fire, Emergency Medical Services, Transportation Department, Office of Homeland Security and Emergency Management (HSEM)
		Travis County	Sheriff, Constables, Office of Emergency Management
		Texas Department of Transportation	Courtesy Patrol, ITS Freeway Management
		Capital Metropolitan Transportation Authority	Transit Dispatch
Virginia Public Safety and Transportation Operations Center (PSTOC)	Fairfax County	Department of Public Safety Communications	Fairfax County 9-1-1 Communication Center
		Office of Emergency Management	Fairfax County Emergency Operations Center
		Fairfax County Fire and Rescue and Police Departments	Fire and rescue dispatch Police forensics facility housed in a separate building connected to the PSTOC
		VDOT Northern Region Transportation Operations Center and Signal System	Traffic monitoring and operations center
		Virginia Department of State Police Division 7	Dispatchers and communications center
Utah Department of Transportation (UDOT) Joint Center	UDOT	UDOT	Field responders
		Utah Highway Patrol (UHP)	CAD listing of incidents Dispatchers
Freeway and Arterial System of Transportation (FAST) TMC	Regional Transportation Commission (RTC) of Southern Nevada	Nevada Highway Patrol (NHP) Southern Command and Dispatch	NHP Dispatch and Command and Control
		Nevada Department of Transportation and the RTC	Arterial and freeway traffic operations, traffic management
		Nevada DOT	Maintenance and construction personnel
Regional Transportation Management Center (RTMC)	Minnesota DOT (MnDOT)	MnDOT	Office of Traffic, Security, and Operations MnDOT FIRST (Freeway Incident Response Safety Team)

(Continued)

Table 2.1 (Continued) Collocated U.S. TMC examples

Collocated center	Managing agency	Partner agencies	Services
		MnDOT Metro District Highway Maintenance Dispatch	Snow and ice control, pothole repair, mowing, guardrail repair, highway debris removal, paint striping
		Minnesota Department of Public Safety	State Patrol Dispatch for incident management
Traffic Operations Center (TOC)	Arizona DOT (ADOT)	ADOT	Information dissemination to public via a Highway Condition Reporting System
		Arizona Department of Public Safety (AZ DPS)	CAD, information dissemination to public

2.3.2.1 Austin Combined Transportation, Emergency and Communications Center

Opened in 2003, the Austin Combined Transportation, Emergency and Communications Center (CTECC) appearing in Figure 2.6 improves regional emergency response coordination and cooperation by providing a centralized public safety facility sustaining the operations of shared, critical emergency communications and transportation management. The primary goal is to receive and process 911 calls for service and emergencies. The center consists of a 75,000 ft^2 emergency operations building and 5600 square feet in utility and support buildings. The Emergency Operations Center (EOC) is 4200 ft^2. Typical staffing consists of 150 emergency provider staff and 100 additional staff during EOC deployment.

2.3.2.2 Virginia Public Safety and Transportation Operations Center

The Virginia Public Safety and Transportation Operations Center (PSTOC), opened in 2008, gives public safety dispatch and response access to Virginia Department of Transportation

Figure 2.6 Austin CTECC. (From https://drive.google.com/file/d/0B0SQ9WFkWPyVMmdBMEJTNFNkRWM/view.)

(VDOT) cameras and images, and assists in coordinating responses to incidents on arterials and freeways in the Fairfax County, Northern Virginia, and greater Washington DC areas.

2.3.2.3 Utah Department of Transportation and Utah Highway Patrol Joint Center

Opened in 1999, the UDOT and Utah Highway Patrol (UHP) Joint Center in Salt Lake City serve the regions immediately north and south of Salt Lake City as well as Salt Lake City itself. As this UDOT TMC is the only 24/7 facility in the state, it is staffed with three shifts of three operators each, plus an overnight operator. The TMC monitors both the UHP computer-aided dispatch (CAD) log and the radio frequencies used by UHP troopers and manages the Incident Management Team (IMT) specialists. The TMC has a statewide event tracking system that receives filtered data from five different CAD systems, including the Utah State Police, 911, and Salt Lake City Police and Fire. This provides UDOT operators with current incident information affecting statewide highways and key arterial routes in the metropolitan area.

2.3.2.4 Freeway and Arterial System of Transportation Traffic Management Center

The Freeway and Arterial System of Transportation (FAST) in Las Vegas houses the Nevada Highway Patrol Southern Command and Dispatch, and arterial and freeway traffic management and operations. Funding for operations and management of FAST is provided jointly by Nevada DOT and RTC. The TMC, opened in Spring 2005, is approximately 66,670 ft^2, of which FAST occupies approximately 20,000 ft^2. The main control room is 3200 ft^2, and includes 10 FAST operator consoles and 10 NHP dispatch consoles. The FAST center features public accessibility to the NHP Command room where crash records, reports, tickets, and so on are made available. As such, there is no perimeter security or limited access through the front door. A media room, public information office, and a meeting and training room are in the publicly accessible lobby. NHP does not have control of the cameras, but can request a specific camera image to be viewable on the monitors. FAST and NHP operate independently of each other. Each maintains its own information technology support, although there is a shared computer and equipment room.

2.3.2.5 Minneapolis–St. Paul Regional Transportation Management Center

Opened in Spring 2003, the Minneapolis–St. Paul Regional Transportation Management Center (RTMC) is located in Roseville, Minnesota next to the Metropolitan District headquarters of the Minnesota Department of Transportation (MnDOT). State Patrol Dispatch is able to view CCTV and provide coordination with MnDOT of freeway incident management. The Minnesota State Patrol operates a 24/7 CAD system from the RTMC. State Patrol automatic vehicle location (AVL) and message display terminal (MDT) systems are monitored from the RTMC.

2.3.2.6 Arizona Department of Transportation Traffic Operations Center

The ADOT TOC contains Arizona Department of Public Safety (AZ DPS) trooper and public information officer (PIO) personnel who interface with incident responders. ADOT has 15 operators working various shifts to ensure 24-h coverage in the control room. There are three AZ DPS PIOs and one AZ DPS supervisor collocated at the ADOT TOC 24 h a day

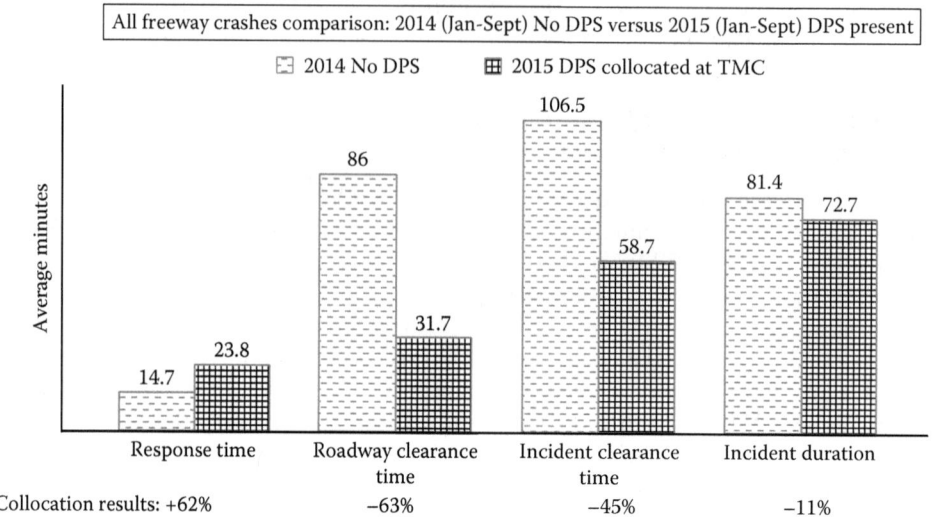

Figure 2.7 Roadway clearance, incident clearance, and duration time benefits produced by collocation of Arizona DPS trooper and dispatch with TOC personnel.

Monday through Friday. ADOT also has PIOs on-site to distribute notifications and updates via Twitter and Facebook social media. ADOT and AZ DPS PIOs coordinate dissemination of information to the public in the event of a major freeway incident. All employees at the ADOT TOC are state employees.

A 3-year pilot begun in January 2015 collocates one AZ DPS trooper alongside four ADOT staff within the TOC. Figure 2.7 shows the roadway clearance time, incident clearance time, and incident reduction benefits just 9 months into the program [5]. The most striking result is the 63% reduction in time to clear the roadway compared with a year earlier in spite of a 23% growth in crashes, which increased the average response time from 14.7 to 23.8 min. Roadway clearance time is defined as the average time taken to clear the travel lanes. Incident clearance time is defined as the time between the arrival of response vehicles and the time at which all vehicles and debris are removed from the crash scene.

2.4 WORK STATIONS AND VIDEO DISPLAY WALLS

The control room, workstations, video wall, and operators observing real-time traffic flow allow a typical freeway TMC to perform the tasks listed in Table 2.2 [4]. The work stations monitor the operation of the roadside sensors; change messages on signs; transmit highway advisory radio broadcasts; control dynamic lane accessibility, speeds in specific lanes, and other active traffic management devices; and inform information service providers of incidents and delays. A video wall with a large number of screens displays traffic flow conditions on the monitored road network to the staff. Often, there are more roadside cameras than there are monitors. Therefore, the staff has the ability to present the imagery from any camera in the system on the wall monitors. Displays of information other than traffic may be available. For example, in California, there are monitors that show earthquake data and reports and forest fire status during fire season. Other work stations may be available for agencies that share responsibility for safety on the network, such as police and other emergency response providers, and supply information to the public, such as information service providers.

Table 2.2 Freeway management tasks performed by TMCs

Task	Implementation
Traffic incident management	The TMC manages and coordinates the DOT's response to incidents on the freeway network by detecting incidents using visual identification, automated system alerts, public safety CAD, or other notification, e.g., cell phone reports and connected vehicle messages.
	Many TMCs have formed relationships with state and local law enforcement to access their automated CAD incident data. Established response protocols initiate response strategies, coordination with partners, information dissemination to the public, request for response resources, and ongoing situation updates as response, clearance, and restoration progress.
	Some TMCs have a role in either requesting safety service patrol resources or dispatching those resources to an incident. In some multiagency centers that include law enforcement or emergency management dispatch, the TMC acts as the situation or war room for multiagency response and coordination.
	Responses to large-scale weather events can be similar to an incident response from a TMC perspective. Many TMCs actively monitor resources such as the National Weather Service or contracted weather forecast companies for weather forecast information, and obtain real-time pavement and atmospheric condition information from their road-weather and environmental sensor stations.
Emergency traffic management	Similar to traffic incident management, the TMC's role for emergency traffic management and operations leverages the real-time monitoring and control capabilities, coordination with partners, and network status update functions.
	Some TMCs may also serve as an emergency operations coordination point with connectivity to state or regional EOCs.
Planned special events	The TMC acts as the nerve center for management of planned special events, particularly during event execution. This includes monitoring traffic flow, coordinating with staff in the field, updating traveler information tools, and supporting event egress.
	In some multiagency centers, the TMC acts as the situation or war room for multiagency response and coordination.
Active traffic management (ATM)	The TMC serves as the primary operations point for ATM strategies. The TMC monitors traffic and travel conditions, initiates response to traffic conditions, and in some instances monitors and verifies automated ATM strategies (such as variable speed limits, lane accessibility, or dynamic ramp meter operations).
	Some ATM strategies might also warrant coordinating with other partners, such as law enforcement or emergency management dispatch. Where lane controls are part of ATM strategies (such as shoulder operations during peak travel periods or incidents), notifications or coordination is essential to inform partners such as law enforcement, emergency response, freeway service patrols, transit, and others. ATM strategies will often leverage existing capabilities within the TMC, including freeway traffic management, incident management, and traveler information dissemination.
Integrated corridor management (ICM)	ICM strategies can involve multiple TMCs, including freeway-focused TMCs, arterial-focused TMCs, and transit operations centers.
	ICM strategies leverage existing freeway operations and management functions at the TMC, and may also include additional decision-support tools to coordinate traffic operations across jurisdictional corridors within an ICM network. With ICM, there is an increased expectation for improved coordination among operators (and systems) at TMCs, with or without decision-support capabilities.
Managed lanes	Lane management systems, e.g., overhead lane control signals or CMS, are typically administered, monitored, controlled, and maintained by a combination of TMC personnel and computer systems. Lane management operations utilize the data collected from detection and surveillance components in the field that are processed and made available at the TMC.

(Continued)

Table 2.2 (Continued) Freeway management tasks performed by TMCs

Task	Implementation
Information dissemination	The TMC is the location where real-time traffic and pre-planned event information (such as work zone or special event) are fused and then distributed to stakeholders via the conventional roadside infrastructure (e.g., CMS and Highway Advisory Radio) or connected vehicle technology. Many TMCs actively share information with partner agencies to coordinate public notifications or responses to events that impact the transportation network, such as weather hazards or incidents. TMCs also distribute traveler information to independent service providers and the media for distribution to the public. Many DOT systems provide data and information to 511 systems operated by the DOT or contractors. Information dissemination includes extensive use of social media for alerts and notifications (most common is Twitter).
Performance monitoring	As the data and information hub for DOT real-time systems, TMCs are often the central point of data acquisition, data fusion, and data storage to support performance monitoring and management strategies for freeway operations. Performance monitoring data are typically analyzed in non-real-time, and are compiled into weekly, monthly, or quarterly performance reports. Trends can be identified, as can areas requiring more focus based on the outcomes. As more active freeway operations and management strategies are implemented, there is an opportunity to integrate real-time freeway system performance outcomes into operations strategies.

An issue that TMC staff may have to address is how to view all of the available images in a timely manner as the number of cameras on the monitored road network proliferates. One answer may lie in a new generation of cameras that incorporates machine vision capabilities. These cameras digitize and then detect and track vehicles in their field of view, automatically alerting TMC personnel to anomalies that require their attention and simultaneous viewing of a particular camera's imagery on a monitor.

2.5 PHYSICAL AND VIRTUAL ATTRIBUTES OF A TMC

Today's TMCs encompass both *physical attributes* and *virtual attributes* [4]. Typical physical and virtual TMC elements are described below.

2.5.1 Physical attributes

Most TMCs contain the following *physical attributes*:

- *Central operating and reporting systems*: Consist of the software programs that monitor and enable control of devices, data processing software and systems, and associated hardware and peripherals.
- *Staff resources*: Consist of the engineers, technicians, and other support personnel that are integral to the TMC.
- *Control room*: Accommodates the operator workstations and workstation equipment (computers and monitors), video wall, and perhaps office, visitor, and additional workspace.
- *Communications room*: Contains the servers, networking hubs, and acts as the distribution point for transmitting data to the system users and stakeholders.
- *Common areas*: Includes reception and lobby area, meeting rooms, break rooms, locker rooms, and storage.

- *Office space*: Can be outside of the TMC or designed as office workspace within the control room.
- *Maintenance equipment area*: Space for storing additional parts, offering minor on-site maintenance and repair, housing maintenance staff, and providing associated offices and work areas.
- *Multiagency coordination room*: If the TMC also serves as an EOC, it might include a multiagency coordination room with additional monitors, phones, communications equipment, and meeting space. The coordination room might also be available for the print and broadcast media.
- *Garage or other vehicle parking and storage*: Some TMCs provide space for parking maintenance vehicles, incident response and freeway service patrol vehicles, and staff vehicles.

2.5.2 Virtual attributes

Virtual attributes of a TMC include those that may not be under the direct control of the TMC. Modern communications networks allow for geographic expansion of a TMC's operational influence well beyond the boundaries of traditional communications and infrastructure footprints. New data sources (including cellular, Wi-Fi, Bluetooth®, and probe and connected vehicle data) are enabling greater amounts of situational awareness information concerning freeway and highway networks to be obtained than was possible in prior generations of traditional infrastructure-based data collection. In addition, the ability to gather and share information with partners, in real time using Web-based systems and networks, provides a significant advantage over earlier systems that required hard-wired dedicated servers to distribute data and information needed for effective system operations, network reliability, and the depth and accuracy of information requested by the public.

Remote capabilities also allow for a potential sharing of data access or control that was once confined to a TMC network, making it easier for DOT regions or districts to access information from personnel offices or computers as long as proper network security protocols and standard operating procedures are in place to sustain those functions. The transition to Web-based applications for condition and incident reporting, 511 traveler information system updates, and alert notifications (among others) helps to support proactive operation of TMCs and DOTs. These benefits have been incorporated into the U.S. National ITS Architecture through its ability to plan and incorporate the virtual aspects of a TMC.

To summarize, virtual attributes allow TMCs to

- Benefit from potential compatibility among programs and systems with standard data formats.
- Expand their geographic footprint with wireless communications.
- Establish backup operations to handoff operations to a different center or additional DOT staff.
- Create a remote operations capability where the staff does not need to be physically in the TMC to monitor, view, or control many key systems.
- Implement thresholds for automating functions, such as updates to travel-time messages in near real time, or distributing alerts to designated partners if conditions meet preestablished criteria.

2.6 FREEWAY MANAGEMENT CENTER SYSTEMS AND SOFTWARE

Freeway management and operations centers (FMOCs) require operating systems and software to support the implementation of real-time monitoring and control strategies [4].

A variety of software systems are found at FMOCs throughout the United States. Some represent first- or second-generation operating environments; others support new active management strategies and various combinations in between. Most centers rely on multiple software packages, presenting operators with a significant challenge as they navigate the many systems. An integrated system that provides the needed control, information management, and strategy implementation is an ideal solution, but not an easily achievable one. The integrated approach is limited by the significant investments agencies have made in legacy freeway operations and management software and systems, and the operations processes designed around specific software.

2.6.1 Operating systems

Typical operating systems found in FMOCs include:

- Closure and restriction reporting systems and event management and tracking systems.
- Real-time freeway management systems that contain modules to control ramp meters, CMSs, and sensor data processing.
- Road-weather management and information systems (RWISs).
- Information-processing packages and algorithms (such as to compute travel times from sensor data, vehicle probes, cellular phones, Bluetooth and license-plate readers, and connected vehicles).
- Traffic signal management software.
- Specialized software for 511 traveler information and social media monitoring.
- Enterprise operating systems for email, agency network servers, consumer-oriented cloud storage options such as SharePoint and ShareFile, and Web browsers.

In addition to the systems found in a particular FMOC, there are external systems in other agencies that could potentially interact with the TMC such as

- Law enforcement, public safety, 911 CAD.
- National Weather Service and National Oceanic and Atmospheric Administration (NOAA) forecast data.
- Neighboring agency traffic management software.
- Emergency management systems.
- Third-party data acquisition and analysis systems.

2.6.2 Software applications and interfaces

Software applications that reside on servers, workstations, and field device processors provide the functionality (including the user interface) for a freeway management system. Several recommendations for enhancing software and interface selection are presented below.

- Web browsers and environments supporting Web browsers are multi-platform, allowing for application hardware and software to be chosen from among a variety of providers.
- Using a generic user-interface device with a stable local configuration reduces downtime from device failure. When such a device fails, it is easily replaced because there is little or no customization in the environment.
- Server downtime that affects all clients can be addressed with redundant equipment supporting the server.

- Some ITS functions and applications may require performance that cannot be achieved using only HyperText Markup Language (HTML). For example, a full-featured GIS-based display may be required. Indicating equipment status change updates on a map by reloading the entire map image is slow (and possibly unusable by an operator) and wastes network bandwidth. Solutions such as plug-ins and small programs written to be activated only when needed can address this issue.
- Server software that resides at a TMC can be configured to accommodate almost any security scheme (to prevent unauthorized users from gaining access to the system) depending on what portion of the system is exposed to clients other than core users. A center-to-center user interface is straightforward to implement and support and can be realized by allowing personnel to reference a user interface at the other facility through their Web browser. Strategies at many TMCs provide data to external entities via systems that reside outside of the agency network configurations. This limits risk and exposure since there is not a direct interface into agency systems. Network security precautions also allow agencies to access data from external systems, such as third-party data providers or other agencies. Anyone with a browser who is connected to the network potentially has access to the functions of the traffic management system through a Virtual Private Network (VPN). Thus, it is relatively easy to distribute any portion of the information contained in the system to users outside of the network, either within or outside of the organization.

2.6.3 Connected vehicle impacts

Introduction of connected vehicles on roadway facilities will require improved network awareness for FMOCs with respect to congestion, end-of-queue location, weather and pavement conditions, and traveler information. Furthermore, personal and mobile communication devices will be generating and receiving data from the TMC. Therefore, TMCs will need to upgrade existing capabilities, processes, and staff resources or incorporate new ones to analyze and deliver user-generated content and data. These resources will include the following [6]:

- *Data management*: Information management and analytics will be a high priority requiring personnel capable of managing the requirements of a Big Data environment, and effectively supporting the integration of that data into TMC functions and processes.
- *Broadening of responsibilities*: A more ubiquitous data environment, such as from connected vehicles and travelers, could broaden the current boundaries of the region served by a TMC, which may necessitate additional staff resources.
- *Skill set focus to support operations rather than numbers of employees*: Exponentially increasing the amount of data available to sustain transportation operations decision-making will create a need for new skill sets beyond what exist in many TMCs today. TMCs should consider shifting to a staffing approach based on needed skill sets versus number of full-time equivalent (FTE) employees, as this could support redefining positions and potentially acquiring the needed skills to enhance system operations in a connected vehicle environment.
- *Creativity in providing staffing*: In addition to contracting for specific skill sets to supplement agency FTE staff within the TMC, it may be necessary to gain access to regional resources and create a multiagency environment. For example, smaller cities may not be able to individually acquire specific resources, but there may be economies of scale for a regional agency (or one agency in a region) to take that responsibility.
- *Additional maintenance support*: There may be additional responsibility for communications network maintenance to sustain additional layers of infrastructure.

2.7 ROLE OF TMCs IN SUPPORTING EMERGENCY MANAGEMENT

Because of the heightened focus on transportation infrastructure security and the evolution of TMC capability to monitor and manage emergency events on transportation networks, the role of the TMC in emergency operations has increased. The specific TMC attributes that make this responsibility possible are its ability to provide situational awareness of the road network through mechanisms such as CCTV, detection devices, road-weather sensors, and external data sources; the capacity to assess impacts across a broad region; and the management of a rather robust communications system that provides connectivity to devices and potentially to other agencies [4].

Accordingly, the day-to-day resources for traffic monitoring and management, traffic incident management, traveler information, data sharing, and road-weather management can be leveraged to support emergency operations. During emergency conditions, directives are typically issued from state or regional EOCs. Recent years have seen more collaboration and coordination between TMCs and EOCs, largely due to the capabilities of the TMC to provide real-time situational awareness information and implement control strategies. This is true for TMCs that include emergency response partners such as the ones described in Table 2.1, and for transportation-only TMCs such as the Ohio DOT Statewide TMC and the Oregon DOT Traffic Operations Centers.

Two guidebooks from the FHWA highlight the role and importance of the TMC in supporting emergency management and operations, include information to support multiagency planning, coordination, and preparation, and mitigate potential challenges.

2.7.1 Role of Transportation Management Centers in Emergency Operations Guidebook

The first of the guidebooks discusses various planning and preparedness activities for TMCs to better support emergency operations that involve other partners [7]. These situations range from responding to a localized traffic incident to major regional events such as hurricane evacuations. The key is to remove the technical and institutional barriers that prevent TMCs from fully supporting emergency operations. TMCs should undertake advance planning and preparation consistent with the National Incident Management System (NIMS) and the Incident Command System (ICS). Table 2.3 lists the components of the strategic, tactical, and support level functions that are suggested by the NIMS as part of incident planning at a TMC [8].

Table 2.3 Strategic, tactical, and support level components of the NIMS process

Level	Function	Approach
Strategic	Preparedness	Planning Training and exercises Personnel and equipment certification Mutual-aid agreements
	Resource management	Identify and type resources Mobilize resources Reimbursement
Tactical	Incident Command System	On-scene command and control procedures
Support	Communications and information management	Information policies Interoperability standards Common technology utilization Communications systems development

Checklists are provided in the *Emergency Operations Guidebook* to support emergency preparedness needs assessment, continuity of operations planning, response and recovery, large planned event and national special security events, post-event debriefings, and maintenance of emergency operations equipment. Case studies highlight a TMC's role in collaborative emergency response and offer guidance for establishing an emergency preparedness working group. The guidebook outlines the role of the TMC in support of the Department of Homeland Security and emergency operations and management centers in providing critical monitoring and status assessments for damage to transportation infrastructure, coordinating and implementing strategies for emergency traffic management, identifying alternative routes, and informing decision-making by emergency management officials. Moreover, the guidebook references agreements and legal authorities that govern emergency management operations and decision-making, and provides recommended agreement language and examples. Because an emergency incident requires cooperation from a variety of agencies, some TMCs and EOCs collocate to provide greater communication and coordination, and allow for the leveraging of resources as discussed earlier. TMCs should explore the feasibility of collocating with the state or region EOC, which may depend on the laws, operational configurations, and available facilities.

2.7.2 Information-Sharing Guidebook for Transportation Management Centers, Emergency Operations Centers, and Fusion Centers

The second guidebook describes the critical role of information sharing among key operations and management centers during emergency preparedness, operations, and restoration [9]. It provides definitions and distinguishes among the roles of TMCs, EOCs, and fusion centers (FCs), and the types of information generated and shared by each. The role of TMCs has already been described. EOCs manage and respond to emergencies of all kinds that threaten or result in significant impact on public health and safety, infrastructure, commerce, and national security. They typically are physical communications centers where responsible government, law enforcement, fire, hazardous materials, emergency medical services (EMS), and infrastructure management authorities gather to coordinate emergency response. EOCs usually define and tier coordination and leadership roles along jurisdictional lines. Operations at these centers are conducted according to defined criteria for declaring emergency conditions.

An FC is a collaborative effort of two or more agencies that provide resources, expertise, and information to the center with the goal of maximizing their ability to detect, prevent, investigate, and respond to criminal and terrorist activity. Some FCs address specific laws such as driver licensing, banking crime, or specific critical infrastructure elements. These centers may also synthesize information and focus on a much wider set of public safety and national security challenges such as terrorism, major criminal activities, public health risks, major economic risks, critical infrastructure protection, and major natural hazards.

TMCs, EOCs, and FCs behave similarly in gathering, processing, and synthesizing at least three basic kinds of information that assist in making operational decisions or reaching conclusions concerning needed actions. These information classes, shown with examples and their applications to the centers in Table 2.4, are as follows:

- *Operational information* (*situational*): Critical for making fast and informed operational decisions and for communicating accurate alerts and notifications on incidents, threats, and emergencies.

Table 2.4 Information shared across TMC, EOC, and FC

Information class	Transportation information examples	Application		
		TMC	EOC	FC
Operational	Traffic flows, video feeds, localized surface weather	Traffic control, snow and ice tactics	Assessment of emergency situation and risks	Real-time threat, risk assessment
Recorded	Incident logs, video records, traffic records	Traffic safety assessment and planning	After-action assessment	Law enforcement, investigation
Physical infrastructure	Maps, physical feature data	Work zone management, resource deployment	Data and framework for decisions, communications	Framework for threat and risk assessment

- *Recorded information*: Basis for operational assessments, investigation, planning, and after-action reporting.
- *Physical infrastructure information*: Framework for setting and communicating priorities, determining risks, and deploying field resources.

TMCs, EOCs, and FCs interconnect with partner agencies and deployed assets (e.g., cameras, sensors, and control systems) via landline, wireless, and Internet links. Key external communications links for TMCs and EOCs also include weather services, 911 centers, law enforcement dispatch systems (e.g., CAD and similar systems), and the traffic reporting media as illustrated in Figure 2.8 [9].

Furthermore, the *Information-Sharing Guidebook* identifies potential types of data exchanges and communications media that may be of value to the three categories of centers and the challenges that exist in sharing multiagency information. These include privacy laws as related to agency information technology, data, security policies, and classified or

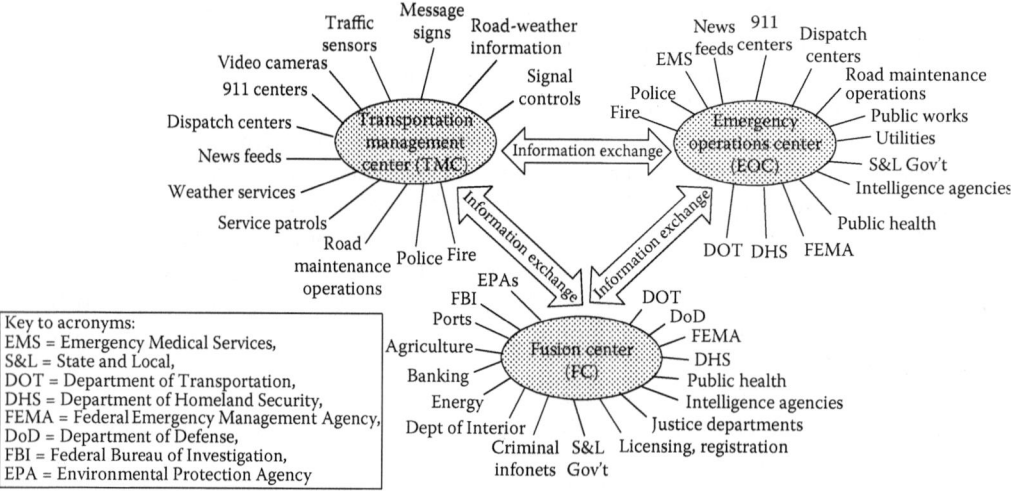

Figure 2.8 Information sharing among TMC, EOC, and FC. (From N. Houston et al., *Information-Sharing Guidebook for Transportation Management Centers, Emergency Operations Centers and Fusion Centers*, FHWA-HOP-09-003, Federal Highway Administration, U.S. Department of Transportation, Washington, DC, June 2010.)

Table 2.5 Technology impacts on TMC operations

Trend	Description	Strategies
1. A nimble service-oriented program mindset and organizational structure	Being positioned to successfully select and rapidly adopt changing technologies and processes to address growing and changing expectations from travelers for efficiency and communication	• Foster an agency culture of embracing technological change • Create a TMC operator training program • Enhance operational communication, which will promote a culture of open communications among staff • Develop memoranda of understanding and interagency agreements facilitating multiagency (sometimes multistate) cooperation and operations • Create new technology piloting and testing programs, including those for connected vehicles • Develop skill sets for TMC managers in areas of contracting, privacy, security, and intellectual property • Adopt standards on TMC-related equipment and processes • Use open-source or nonproprietary software when possible • Require application programming interfaces and document for future development • Require documentation on all systems and software—include search capabilities and provide remote accessibility • Follow the systems engineering process
2. Active transportation and demand management (ATDM) concept and toolkit	Engaging a variety of tools at one's disposal to proactively make operations more efficient through staff and technology utilization	• Implement a suite of emerging transportation concepts, coordinating as necessary • Integrated corridor management • Active traffic management that may include lane use control, variable speed limits, and hard shoulder running • High occupancy toll (HOT) lanes • Portable work zone ITS systems • Regional or multistate coordination of detours and traveler information • Provide real-time travel-time estimates on full range of devices and systems available • Display transit information on parallel route DMS (possibly with comparative travel time and parking availability) • Parking management including dissemination of real-time garage space on DMS and through mobile device apps • Arterial management with ITS devices such as CCTV cameras, DMS, and remote access to traffic signal controllers • Integrate ramp metering schemes with adjacent arterial signal timing to minimize conflicts with ramp queues • Adaptive signal control technologies • Transit signal priority • Road-weather integration • Weather-responsive signal and ramp meter timing plans • Develop protocols and maintenance program to address increased number and complexity of ITS field devices • Collocate freeway and arterial transportation management • Promote coordination with arterial management agencies • Seek out opportunities to share resources with other agencies (e.g., communication networks, cameras, DMS)

(Continued)

Table 2.5 (Continued) Technology impacts on TMC operations

Trend	Description	Strategies
3. Accommodate toll and other pricing operations in TMCs	Integrating pricing into operations to encourage revenue capture through tolling and financing infrastructure expansion	• Develop protocols for operations (such as pricing and operations for diversions to HOT lanes during major main-lane incidents) during early feasibility planning • Develop protocols for joint operation of freeways and toll roads during early feasibility planning • Develop protocols for operations and implementation of HOT lanes with variable pricing based on congestion during early feasibility planning • Develop protocols for operations for cordon pricing for congested areas during early feasibility planning • Consider increased network reliability and data security needs
4. Performance monitoring and management	Increasing data collection and analysis to improve operations, enhance customer service, and document effectiveness of TMC actions	• Apply results of performance monitoring related to agency goals to support funding requests • Proactively develop performance metrics based on staff priorities and agency goals • Use multiple data sources to monitor system congestion, including support of travel-time estimation • Consolidate efforts to develop data management tools across agencies • Frequently process and distribute measures of effectiveness to operators to improve operational effectiveness • Utilize features in software to track and report performance • Utilize onboard device data from agency vehicles to monitor pavement condition • Train TMC operators how to use performance monitoring and how to populate the data needed for performance monitoring
5. Automation and related tools to increase efficiency	Incorporating new technologies to increase productivity through improvements in system management and cost effectiveness	• Use advanced graphical user interfaces to increase operator efficiency • Develop decision-support systems • Install remote power cycling of field devices • Install automatic power cycling of field devices • Specify automation features in software contracts • Consolidate interfaces to or consolidate alert systems across agencies • Develop default sets of traveler information messages across devices (such as DMS) and media for quick implementation during recurrent special events or incident types and sites • Utilize low-cost, low-infrastructure-impact devices such as solar-powered pole-mounted traffic sensors with wireless communications • Utilize predictive analysis and forecasting for anticipating congestion • Because the private sector often develops the automation tools, support strong participation with industry to provide better tailored tools • Include options for manual verification and override to be applied as operators fine-tune and gain confidence in new applications • Develop a data fusion engine to merge data from multiple sources, such as travel-time information from toll-tag and license-plate readers, Bluetooth sensors, and third-party providers

(Continued)

Table 2.5 (Continued) Technology impacts on TMC operations

Trend	Description	Strategies
6. Involvement of third parties in data and traveler information	Utilizing data services that third-party vendors provide to manage roadway traffic and deliver traveler information to the public	• Develop pre-qualifications or standards regarding data accuracy and validation (potentially both for data received and for data provided) • Provide real-time data to third-party app developers • Share data among agencies • Develop protocols for data privacy and confidentiality to keep the media and other agencies collocated in the TMC from observing restricted material • Utilize private sector meteorological services or in-house meteorological resources • Research solutions that others have used to solve similar problems • Use multiagency procurement for economies of scale • Train TMC operators how to interpret alternate data sources to support operations decision-making • Consider use of applicable standards to simplify data exchange, such as xml
7. Mobile communications and wireless networks	Incorporating advances in wireless technology to provide options to modernize field equipment and increase data coverage	• Coordinate with information technology staff to develop firewalls and other security protocols that are effective without limiting functionality • Efficiently expand field device coverage and operations cost using wireless networks • Allow appropriate remote access into TMC software or devices (primarily for maintenance staff and appropriate coordinating staff from partner agencies) • Utilize commercial mobile devices and apps to support collaboration between freeway service patrol and other emergency responders, TMC operations staff, and field maintenance staff for improved communication and enhanced field collaboration • Operate mobile command centers or satellite centers with TMC software access
8. Social media for traveler information and crowd sourcing	Using social networking tools to receive and distribute information among agencies, travelers, and third parties	• Develop procedures and protocols for use of social media • Foster relationships among agency public relations groups • Collocate traveler information provider staff with TMC staff and agency public relations staff • Support two-way information exchange via social media • Designate a larger or statewide TMC to take responsibility for social media alerts on behalf of multiple agencies in a region • Provide information through social media and mobile apps focused on pre-trip planning to minimize driver distraction (near term) • Utilize en-route social media (including crowd sourcing) as voice activation becomes more common • Utilize crowd sourcing for traffic information, incident information, feedback on department performance, and pavement roughness • Provide incentive for drivers to participate in crowd sourcing • Partner with the private sector to facilitate social media outlets and realize cost efficiencies • As more traveler information content is available to travelers through third-party apps, TMCs can focus on providing content on core mission (such as upcoming construction and estimated time to reopen lanes)

restricted data and information. It also acknowledges the need for training TMC personnel working in EOCs or FCs.

2.8 FUTURE TRENDS FOR TMCs

The most transforming implications for TMC operations are based on the proliferation of wireless communication, increased awareness of social media, involvement of third parties, and the emergence of connected or cooperative vehicles. Together they create massive two-way data and communication streams throughout the transportation network. New classes of real-time holistic data become available to TMC operations, often through third parties. This enables unprecedented real-time understanding of the transportation network that can be leveraged into increasingly sophisticated control strategies. As travelers access personalized and user-friendly commercial information through their mobile device apps, their expectations for transportation system information increase.

An FHWA study identified and analyzed potential impacts of technology advancements on TMC operations through approximately the mid-2020s [10]. Table 2.5 describes the eight key trends that were highlighted. The first four represent trends and technologies emerging from within the transportation community, while the next four represent those from outside the transportation community that can be adapted to fulfill TMC needs.

2.9 SUMMARY

Freeway management systems in urbanized areas of the United States represent an established and mature concept, although freeway management and operations strategies and systems continue to evolve. Many TMCs that were implemented as part of first-generation freeway management systems have gone (or will undergo) some significant changes as technologies, systems, business models, and freeway operations approaches evolve. In some cases, regions are collocating various TSM&O functions within a TMC. In others, components of freeway management and operations are being supplemented or outsourced to non-agency staff. Modern networking and Web-based capabilities allow functions that were once restricted to the TMC operating floor to be distributed to other entities or shared as part of joint operating strategies. Future TMC operations will be greatly affected by the proliferation of wireless communication, increased incorporation of social media data, involvement of third parties, and of course the Connected Vehicle Program and others similar to it worldwide. TMCs will have to develop strategies to effectively utilize the new sources and quantities (e.g., Big Data) of data to fulfill the traveler's expectations of increased roadway safety and mobility.

REFERENCES

1. J. Chu and L. Radow, "Behind the Scenes at TMCs," FHWA-HRT-12-005, *Public Roads Magazine*, 76:1, July/August 2012. http://www.fhwa.dot.gov/publications/publicroads/12julaug/02.cfm. Accessed March 5, 2016.
2. G. Hatcher, C. Burnier, E. Greer, D. Hardesty, D. Hicks, A. Jacobi, C. Lowrance, and M. Mercer, *Intelligent Transportation Systems Benefits, Costs, and Lessons Learned: 2014 Update Report*, FHWA-JPO-14-159, ITS Joint Program Office, Washington, DC, June 2014.

3. *Caltrans Transportation Management System Master Plan*, System Metric Group, Inc., Prepared for California Department of Transportation, Sacramento, CA, February 2004.
4. *Freeway Management and Operations Handbook*, Final version, Chapter 15, "Transportation Management Centers," Federal Highway Administration, U.S. Department of Transportation, Washington, DC, January 2016. https://drive.google.com/file/d/0B0SQ9WFkWPyVMmdBMEJTNFNkRWM/view. Accessed March 4, 2016.
5. "Cooperation Cuts Delays Caused by Freeway Crashes," *ITS International*, NAFTA11–NAFTA12, March/April 2016.
6. *Traffic Management Centers in a Connected Vehicle Environment: Task 3. Future of TMCs in a Connected Vehicle Environment Final Report*. Prepared for CTS Pooled Fund Study, University of Virginia by Kimley-Horn and Associates and Noblis, Dec. 2013. http://www.cts.virginia.edu/wp-content/uploads/2014/05/Task3._Future_TMC_12232013_-_FINAL.pdf. Accessed March 7, 2016.
7. D. Krechmer, A. Samano III, P. Beer, N. Boyd, and B. Boyce, *Role of Transportation Management Centers in Emergency Operations Guidebook*, FHWA-HOP-12-050, Federal Highway Administration, U.S. Department of Transportation, Washington, DC, October 2012. http://ops.fhwa.dot.gov/publications/fhwahop12050/index.htm. Accessed March 4, 2016.
8. J. Ang-Olson and S. Latoski, *Simplified Guide to the Incident Command System for Transportation Professionals*, FHWA-HOP-06-004, Federal Highway Administration, U.S. Department of Transportation, Washington, DC, February 2006. http://www.ops.fhwa.dot.gov/publications/ics_guide/ics_guide.pdf. Accessed March 5, 2016.
9. N. Houston, J. Wiegmann, R. Marshall, R. Kandarpa, J. Korsak, C. Baldwin, J. Sangillo, S. Knisely, K. Graham, and A.V. Easton, *Information-Sharing Guidebook for Transportation Management Centers, Emergency Operations Centers and Fusion Centers*, FHWA-HOP-09-003, Federal Highway Administration, U.S. Department of Transportation, Washington, DC, June 2010. http://ops.fhwa.dot.gov/publications/fhwahop09003/tmc_eoc_guidebook.pdf. Accessed March 4, 2016.
10. A. Mizuta, K. Swindler, L. Jacobson, and S. Kuciemba, *Impacts of Technology Advancements on Transportation Management Center Operations*, FHWA-HOP-13-008, U.S. Department of Transportation, Federal Highway Administration, January 2013.

Chapter 3

Sensor applications to ITS

Many traffic management applications require the collection of traffic flow data. The types of data and their corresponding accuracies are dependent on the application and the data processing algorithm. The following sections describe several common traffic management applications and their input traffic data requirements. They include local isolated intersection signal control, interconnected intersection signal control based on either selecting from among either prestored signal timing plans or plans developed in real time, ramp and freeway metering, travel time estimation, wrong-way vehicle detection, freeway incident detection and congestion monitoring, active transportation and demand management (ATDM), and traffic data collection for planning and archival or historical purposes. Sensor data similarly support vehicle classification, tolling operations, traffic surveys, parking facility management, and roadway hazard identification. Other technologies such as the Global Positioning System (GPS), Bluetooth® and toll-tag media access control (MAC) address readers, and license plate readers also provide sources of traffic data.

3.1 LOCAL ISOLATED INTERSECTION CONTROL

Local isolated intersection control manages arterial traffic flow independently of adjacent traffic signals. Two types of local isolated control exist: pretimed and actuated [1–4]. The type of control selected is frequently subject to local policy and practice. Offset is not a controlled parameter when isolated intersection control is implemented since each signal operates independently.

3.1.1 Pretimed control

Sensors are not required for pretimed control as right-of-way is assigned based on a predetermined fixed-time duration for all signal display intervals. Therefore, pretimed control is generally inefficient for controlling intersections that experience changes in demand. In fact, pretimed control is normally suitable only in areas with closely spaced interconnected intersections, such as central business districts that have consistent volumes on the low-demand approaches.

3.1.2 Actuated control

Actuated control requires sensors to provide data to a local traffic signal controller. Sensors are typically located at stoplines (A), upstream of the stopline (B), left-turn lanes (C), and at positions to detect emergency (D) and transit vehicles (E) as illustrated in Figure 3.1a [3].

Two types of actuated control are used in practice: semi-actuated and fully actuated. In semi-actuated control, the major street operates in a non-actuated mode such that green is

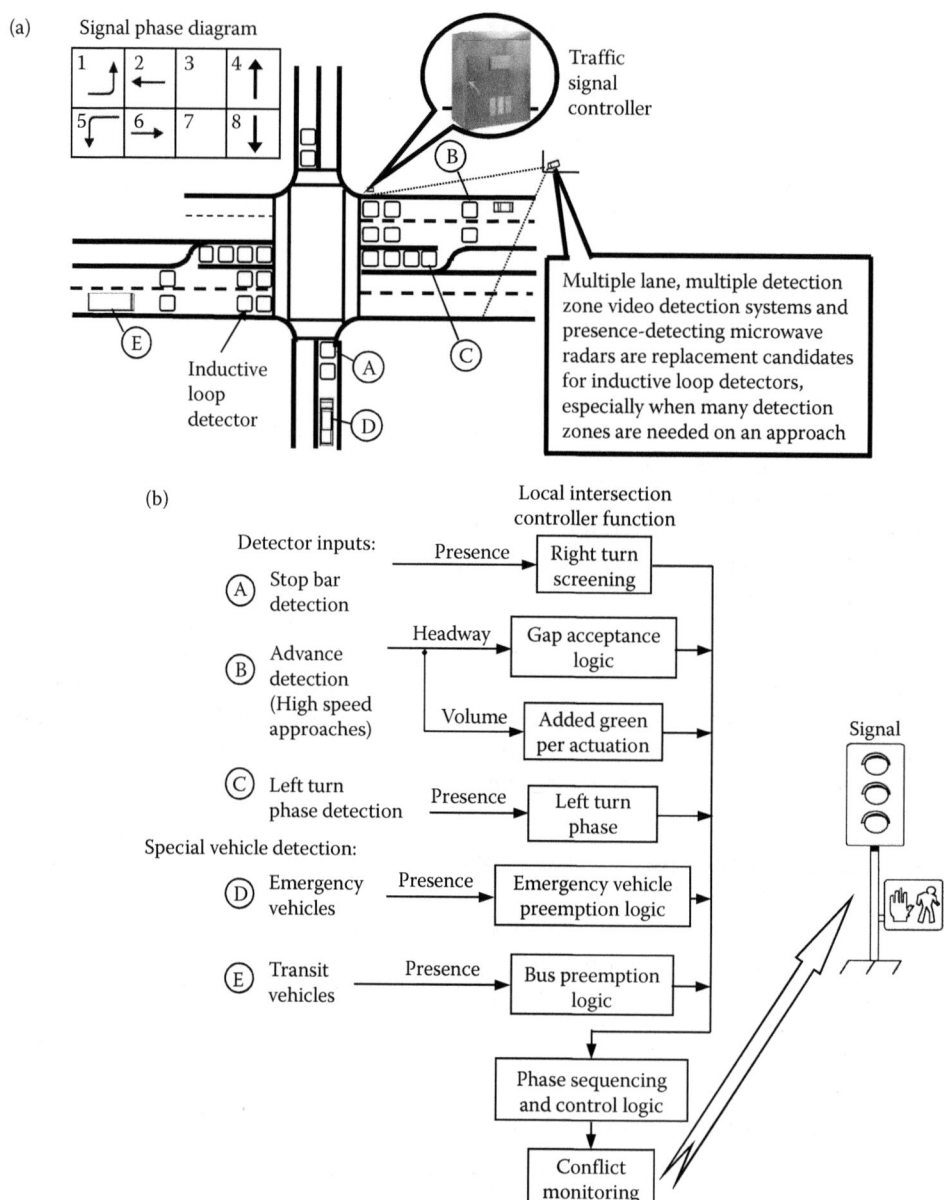

Figure 3.1 Isolated intersection signal control. (a) Inductive loop detector locations provide specific types of data for signal control. (b) Signal control functions supported by the data.

always present unless a minor street actuation is received. Therefore, sensors are needed only for the minor cross-street phases. In the absence of cross-street demand, semi-actuated signals are recalled to the major street phase. Semi-actuated operation is appropriate when vehicles on the minor streets approach the intersections in a random manner, that is, where platoons (groups of closely spaced vehicles traveling at the same speed) cannot be sustained. Such a condition is likely where there are long distances between signalized intersections, unpredictable or relatively low minor-street volumes (e.g., less than 20% of volumes on the major street), and a large proportion of turning movements.

Figure 3.2 Separate traffic signals indicate bicycle right-of-way. (From LAK.)

Fully actuated control operates with traffic detection on all approaches to the intersection for all signal phases. It is the most widely applied control strategy for isolated intersections. Because the cycle length varies from cycle to cycle, it can be utilized at street intersections with sporadic and varying traffic distribution. The information gathered by the sensors can be processed as indicated in Figure 3.1b or in another manner depending on the particular traffic management requirements and strategies. Fully actuated control is also used to grant passage to bicycles when they are regular conveyances on a roadway. Specialized bike signal heads depicted in Figure 3.2 may be used for this purpose.

3.1.2.1 Presence-detecting microwave radar sensor application to actuated control

Figure 3.3 shows the application of a multi-zone or multilane presence-detecting microwave radar sensor to actuated signal control. In this particular configuration, the side-looking sensor is mounted on a pole.

Several cautions should be observed when using this type of sensor. The first is that tall vehicles could occlude other vehicles in the lanes further from a side-mounted sensor including the left-turn lane. Forward-looking multilane models, such as the one shown in Figure 5.18, may not be subject to occlusion from vehicles in other lanes. Second, the nature of radar detection as a statistical detection process may result in the sensor missing a vehicle detection at some detection opportunities. Many presence-detecting radar sensors have "sensitivity" settings to minimize the probability of a missed detection. The suitability of sensors installed above the roadway, also referred to as nonintrusive sensors, for a specific application should be evaluated through field testing by the responsible agency to ensure

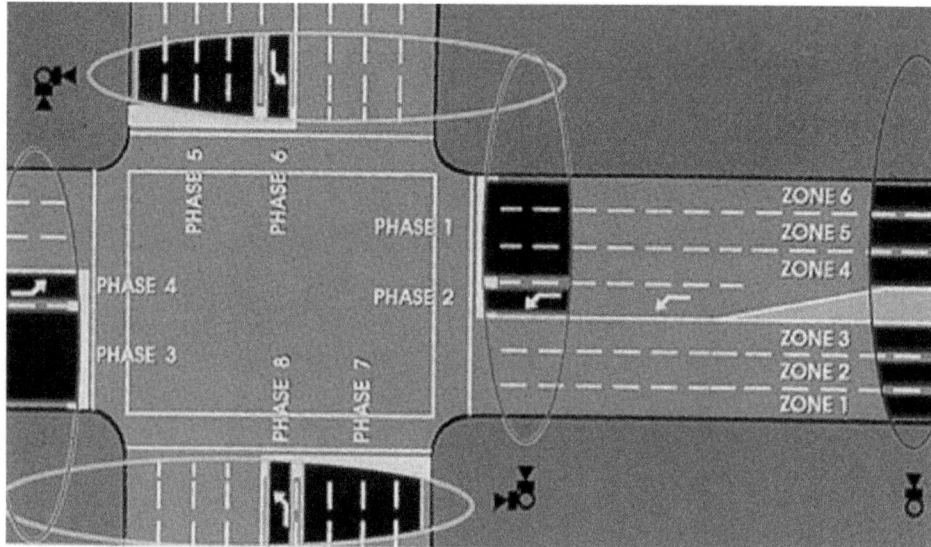

Figure 3.3 Intersection control using multi-zone presence-detecting microwave sensors.

that the required calls are provided reliably. Additional information about the merits and limitations of below-ground and above-ground mounted sensors is found in Chapter 5, which describes the various sensor technology options.

3.1.2.2 Video detection system application to actuated control

Video detection systems (VDSs) are becoming more ubiquitous as a means of detecting vehicles at intersections for traffic signal actuation. The cameras in such a system should be mounted as high as possible, preferably between 25 and 40 ft (8–12 m), over the center of the monitored lanes as depicted in Figure 3.4a. Figure 3.4b shows an alternate camera mounting location on the higher luminaire mast arm rather than on the signal mast, but offset to the side. Cautions should be observed to minimize sun glint by adjusting the sun shield and pointing the camera downward as much as possible while still keeping the required detection area in the field of view. Sun glint is especially problematic when the cameras are aligned in an east–west direction. Another consideration is to have adequate lighting for reliable night-time operation of the detection system. Some VDSs may be better suited than others to operate in areas with high winds that may degrade the detection zone calibrations in the system. Additional information concerning the installation of VDS is found in Chapter 7.

3.2 INTERCONNECTED INTERSECTION CONTROL

Interconnected intersection control provides signal progression that allows platoons of vehicles to proceed along arterial routes without stopping. It also offers area-wide control to minimize total delay and number of stops over an entire network. This type of control is effective when traffic moves in platoons and their arrival time can be predicted at downstream intersections.

Interconnected intersection control can function in two general ways. The first method, traffic responsive control, selects from among a library of prestored signal timing plans that

(a)

(b)

Figure 3.4 Camera mounting options for VDSs deployed at signalized intersections. (a) Camera mounted on signal mast arm over center of monitored lanes. (b) Camera mounted higher on luminaire off to side of monitored lanes. (From LAK.)

best match current traffic flow conditions. The prestored plans are generated offline from average or historical data as in the Urban Traffic Control System (UTCS). In the second method, signal timing plans are generated online in real time based on current traffic flow conditions. These traffic adaptive plans are updated incrementally at each signal cycle.

Sensors are utilized in interconnected intersection control to gather traffic flow data for signal timing plan selection, critical intersection control (CIC—a type of traffic adaptive control), and other traffic adaptive control algorithms. The signal timing selection process is similar for arterial and network systems. Signal timing plan operation is determined by the roadway configuration and the goals of the corresponding plans.

3.2.1 Urban Traffic Control System

The U.S. FHWA-developed UTCS selects from among prestored timing plans that best match the volume and occupancy conditions on the roadway [3]. Traffic system control is achieved by choosing a plan based on (1) time of day, (2) direct operator selection, or (3) best matching recently measured traffic volumes and occupancies acquired from roadway traffic flow sensors. UTCS provides the capability to implement area-wide, locally actuated, and critical intersection traffic control.

3.2.1.1 Volume plus weighted occupancy

When recently measured volumes and occupancies are used for plan selection, UTCS first calculates smoothed volume V plus weighted smoothed occupancy KO, where K is the weighting factor and O is the smoothed occupancy. Experience supports the assumption that volume alone is insufficient to select appropriate timing plans because volume will drop when a link is oversaturated. The inclusion of weighted occupancy assures that the plan selection parameter will continue to recognize high traffic demand during the oversaturated condition. The raw volume and occupancy data are smoothed to suppress the random component in the cycle-to-cycle data caused, for example, by lane changes, debris in the roadway, or some other annoyance to the driver that produces erratic behavior.

3.2.1.2 Comparison and selection of UTCS plans

Figure 3.5 describes UTCS timing plan selection when recent volume and occupancy measurements are utilized to pick a prestored plan. This process is typical of a network-based traffic responsive signal control strategy. UTCS chooses a timing plan by evaluating a function that compares smoothed values of the actual volume and weighted occupancy reported by the system sensors against the stored volume and weighted occupancy that correspond to a particular timing plan. The comparison operation is repeated at user-selected intervals, for example, 4–15 min. If a potential timing plan is found to be more favorable than the current plan (i.e., the new plan comparison function is smaller than the existing one), then the new plan is subjected to an anti-hunting test. The purpose of the anti-hunting test is to verify that the new plan is sufficiently better (by a predefined amount) to warrant implementation. This prevents needless transitions between timing plans that have similar benefits [3].

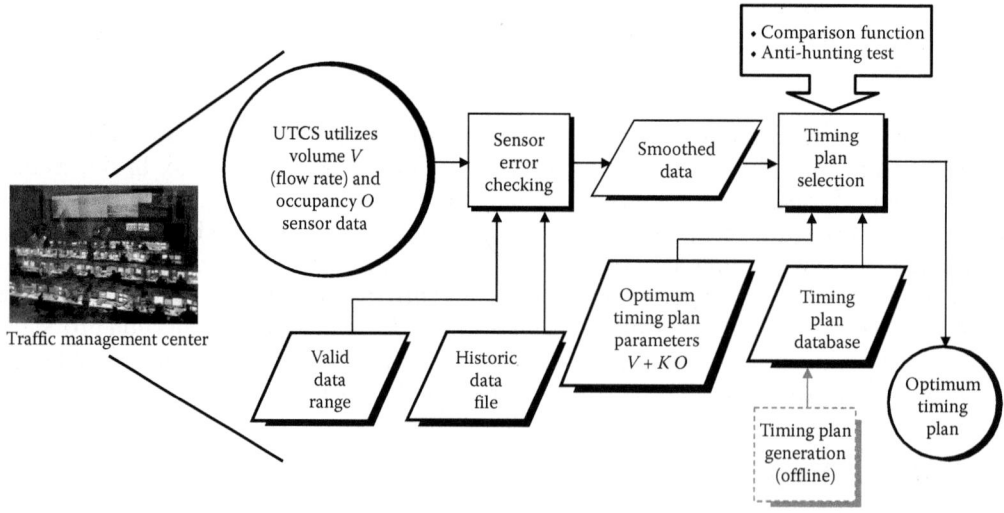

Figure 3.5 UTCS timing plan selection procedure. (From LAK.)

3.2.2 Critical intersection control

The enhanced version of UTCS supports CIC. CIC is a traffic adaptive algorithm that alters splits in a coordinated system to accommodate varying traffic demand on conflicting inter-section approaches. Splits are varied every cycle, while cycle length and offset remain fixed to maintain coordination within the system. Green times for each phase are adjusted based on the traffic demands of the approaches, which are calculated from smoothed volume and occupancy data. Safe intersection operation is mandated through designated minimum greens for each phase.

When CIC is active, green demand is calculated for each traffic signal phase that has detection. For phases with multiple sensors, the green demand value from the sensor with the greatest demand is used. Total green demand for the intersection is the sum of individual phase demands. The demand calculation for a phase may be bypassed and a default phase time substituted when a phase has no demand. The latter situation may not occur frequently since smoothed values of volume and occupancy, which incorporate data from several past cycles, are used as the demand input.

Current phase times are calculated based on the previously determined demand for each phase. The calculations take into account the following:

- Reductions in the total computed green demand and cycle length to satisfy minimum phase times specified in the timing plans.
- Allocations of the remaining cycle time based on the percentage of green demand per phase.
- Allocations of any remaining time that results from rounding errors in a round robin manner by phase length after phases have been arranged in descending order.

Volume change in the 5%–10% range impacts the signal timing assuming that the goal of CIC is to set the timing to approximately the nearest second. The CIC algorithm typically requires occupancy measurement within $\pm 2.5\%$ accuracy at 25% occupancy levels during each cycle.

3.2.3 Other traffic adaptive control algorithms

Traffic adaptive systems are designed to overcome several limitations of signal control sys-tems that rely on prestored timing plans. For example, prestored timing plans developed offline are best suited for traffic flow on a normal day or for events that produce predictable traffic patterns. Their major disadvantage is that they are developed from specific traffic flow scenarios and, therefore, cannot respond to situations that are significantly different from those used to generate them. Furthermore, data collection and manpower costs limit the ability of many traffic management organizations to maintain timing plans that are representative of current traffic volumes and patterns. Traffic adaptive systems attempt to overcome these limitations by providing signal timing that is more responsive to real-time traffic flow sensor data [3].

Traffic adaptive systems generally require a greater number of sensors than conventional first-generation traffic signal control systems and may also need extensive initial calibration and validation. Therefore, it is prudent to calculate the total system life-cycle costs, includ-ing software licensing, purchase of local controllers and central computers, and ongoing operating and field maintenance costs, and compare these to expected benefits when evalu-ating traffic signal operating strategies. Nevertheless, continued advancements in sensor and computer system technology plus improving traffic adaptive control algorithms are making

Table 3.1 Traffic adaptive signal system capital costs

System	Initial capital cost per intersection ($)	System developer	System distributor
SCOOT	30,000–60,000	Transport Research Laboratory, UK	Siemens UK
SCATS	25,000–30,000	Road Transit Authority, Sydney, NSW	TransCore
OPAC	20,000–50,000	University of Massachusetts, Lowell, MA	PB Farradyne
RHODES	30,000–50,000	University of Arizona, Tucson, AZ	Siemens ITS
ACS-Lite	8000–12,000	FHWA, USA	Siemens ITS
LA-ATCS	30,000–60,000[a]	Los Angeles Department of Transportation	McTrans Center
InSync	25,000–35,000	Rhythm Engineering, Lenexa, KS	Rhythm Engineering

[a] Bill Shao (personal communication).

traffic adaptive systems increasingly attractive as compared to conventional systems when traffic volumes and roadway network design warrant their use. The ability to adapt to changes in traffic flow patterns over long-term intervals (i.e., respond to aging of prestored timing plans) frequently make traffic adaptive systems cost effective. Operators of these systems can view maintenance costs as facilitating proactive maintenance of detection and communications systems, and operations costs as shifting operations from reactive complaints to proactive performance management.

Table 3.1 gives the initial capital cost for several traffic adaptive systems [5] and the developer and distributor of the system [6], while Table 3.2 displays the cost per intersection for a selection of traffic adaptive projects in the United States [7]. On average, the costs of installing a traffic adaptive signal system in 2014 were approximately $28,725 per intersection [7]. In addition to software and computer costs, installation costs depend on the linear foot of conduit for inductive loop detectors (ILDs), costs of other types of sensors that may be used, need to replace existing controllers and cabinets, and local labor costs.

Table 3.3 lists the sensor technologies, sensor locations, data collected, data processing characteristics, and backup provisions for a number of traffic adaptive algorithms whose characteristics are summarized below. Often when technologies other than ILDs are used as sensors, the algorithm must be recalibrated to account for the difference in detection area of the alternative sensor as compared to the loop, especially if occupancy is required. Additional details concerning the operation of these algorithms are found in Klein [3] and

Table 3.2 Cost per intersection for traffic adaptive projects in the United States

Project date	Total project cost ($)	Number of intersections	Cost per intersection ($)	Region
January 2013	28,725	1	28,725 (average based on responses from 8 agencies)	Nationwide
July 2012	176,300	8	22,037	Colorado
July 2012	905,500	11	82,318 (includes infrastructure upgrades)	Colorado
2010	65,000	1	65,000	Nationwide
2010	1,708,029	18	94,890 (includes infrastructure upgrades)	Georgia

Source: Adapted from A. Stevavovic, *Adaptive Traffic Control Systems: Domestic and Foreign State of Practice*, NCHRP Synthesis 403, National Cooperative Highway Research Program, Transportation Research Board, Washington, DC, 2010.

Table 3.3 Sensor technologies, locations, data collected, data processing characteristics, and backup provisions for several traffic adaptive algorithms

Algorithm	Sensor technologies	Sensor locations	Data collected	Data processing interval and location	Backup if real-time sensor data not available
SCOOT	ILD (2 m [6.6 ft] in direction of travel), VDS, possibly presence-detecting microwave radar	Approximately 15 m (49 ft) downstream of the previous intersection	Volume, occupancy	Second by second Central	Default parameters based on time of day (TOD) and day of week
SCATS	ILD 6-ft (1.8-m) wide by 16-ft (5-m) long VDS can also be used	Immediately in advance of the stopline Minor intersections require side-street sensors only	Volume, occupancy in most lanes of the subsystem's critical intersection	Second by second Central for overall network control Distributed for local control of green phase	Use data from an upstream critical intersection to control cycle length and sometimes splits throughout the subsystem TOD plan
OPAC	ILD, VDS, presence-detecting microwave radar, magnetometers with RF data links	Upstream, about 8–12 s from stopline or upstream of worst queue of all through phases	Volume, occupancy, speed	Second by second Distributed except for central control of cycle length	Activated in event of central system or communications failure Use average of data from remaining sensors, TOD plan, or historical data
RHODES	ILD, VDS, presence-detecting microwave radar	Stop bar presence with queue estimate Upstream passage	Volume	Second by second Intersection decision frequency: 7–15 s Distributed and central	Use a plan previously identified by the traffic management center
ACS-Lite	ILD, VDS, presence-detecting microwave radar	At least one sensor per phase at stopline and at least one mid-block sensor	Volume, occupancy	5–15 min Central	TOD or use of historical data collected by the algorithm
InSync	Video, although ILDs, presence-detecting microwave radar, and magnetometers may be used	Detection from stopline to approximately 250 ft (76 m) upstream of stopline	Volume (queue), delay	Second by second	Historical optimization data collected from the previous 4 weeks of operation
LA-ATCS	ILD (6-ft diameter) is primary sensor for system, advance, queue, and limit line detectors Other loop configurations used for bike lane, train, and transit priority detection Some VDS and magnetometers used	Stopline presence Passage detection sensor ≤250 ft upstream from stopline	Volume, occupancy, and derived approximate spot speed	Second by second Central	Local- and system-level backup Local controller firmware, Traffic Signal Control Program, has CIC capability comparable to UTCS, along with transit signal priority (TPS) for bus and light rail, in addition to TOD System-level TOD also provided

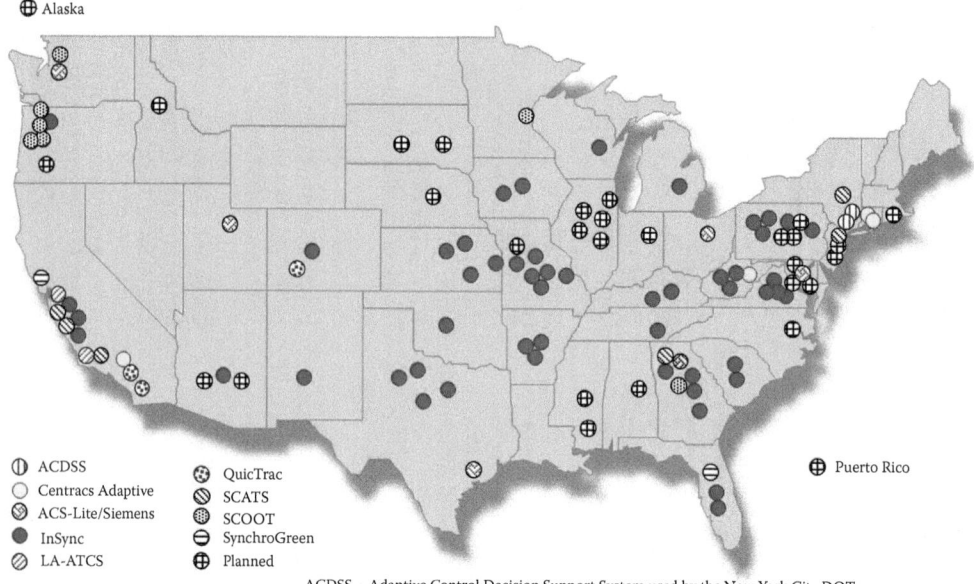

Figure 3.6 Traffic adaptive signal control systems in the United States for the 2010–2012 period. (Adapted from J.A. Lindley, *Applying Systems Engineering to Implementation of Adaptive Signal Control Technology*, ITS World Congress, October 25, 2012.)

Stevavovic [6]. Figure 3.6 shows the locations of traffic adaptive signal control systems in the United States for the 2010–2012 period.

- SCOOT continuously measures traffic demand on all approaches to intersections in the network and optimizes the signal timings (cycle lengths, splits, and offsets) to minimize delay and stops. If SCOOT detects significant changes in the flow profiles during a signal cycle, the signal optimizer makes frequent but small alterations to the timing.
- Like SCOOT, SCATS adjusts cycle lengths, splits, and offsets in response to real-time traffic demand and system capacity. The principal goal of SCATS is to minimize overall stops and delay when traffic demand is less than system capacity. When demand approaches system capacity, SCATS maximizes throughput and controls queue formation.
- OPAC utilizes dynamic programming to minimize the total intersection delay and stops over a user-specified rolling horizon interval. It finds local solutions that produce near-optimal operation along an arterial.
- RHODES utilizes the natural stochastic variations in traffic flow to improve the performance of a signalized arterial network. It proactively responds to the stochastic variations by predicting the flow for the next signal cycle and adjusting the start and end times of the signal phases.
- ACS-Lite adapts the splits and offsets of signal control patterns in a closed-loop system. Changes to cycle time occur on a time-of-day schedule as in traditional traffic control systems. At each optimization step, which occurs about every 10 min, the system changes the splits and offsets a small amount (e.g., 2–5 s) to accommodate changes in traffic flow.
- InSync employs local and global optimization of delay and stops. At the local level, InSync minimizes summed delay at all approaches to an intersection by constantly

measuring volume (number of vehicles) and delay (time vehicles spend waiting) at each intersection, and then makes instant decisions about how to best reduce those numbers. Locally, the signal is adapted to the demand in three different ways: phasing, green time allocation, and sequencing. Globally, the algorithm minimizes vehicle stops by synchronizing all traffic signals in a network of intersections, such as on a corridor, to move platoons of vehicles at a desired speed through a progression of green signals. Gaps in time between passing platoons of vehicles are used by the local optimization algorithm to move traffic on the minor movements [9].

- LA–Adaptive Traffic Control System (ATCS) adjusts signal timing on a cycle-by-cycle basis through changes in cycle length, splits (for CIC), and offset (for critical link control). At least one sensor per phase collects volume and occupancy data every second that are used every cycle.

3.3 BENEFITS OF TRAFFIC ADAPTIVE SIGNAL CONTROL

Figure 3.7 shows the ranges of reported benefits from traffic adaptive signal control systems. Benefits extend across several measures including safety, mobility, and environmental improvements [7]. These encompass travel time reductions of 9%–19%; increase in average speed by 7%–22%; and reduction in fuel consumption by 2%–7%, emissions by 0%–7%, stops by 23%–34%, delay by 10%–40%, and crashes by 28%. In general, benefits are produced by the continuous distribution of green time equitably for all traffic movements, improvement of travel time reliability by progressively moving vehicles through green signals, reduction of congestion by creating smoother flow, and prolonging the effectiveness of traffic signal timing.

In July 2012, the Colorado Department of Transportation (CDOT) released its evaluation of two different adaptive signal systems on two different corridors. The mobility benefits for

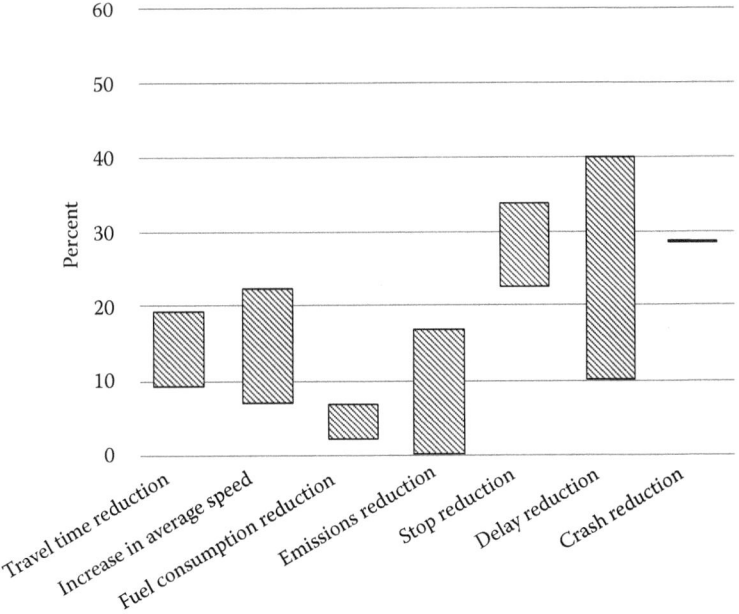

Figure 3.7 Benefits of adaptive signal control over conventional signal systems. (From G. Hatcher et al. *Intelligent Transportation Systems Benefits, Costs, and Lessons Learned: 2014 Update Report*, FHWA-JPO-14-159, ITS Joint Program Office, Washington, DC, June 2014.)

both corridors combined included 9%–19% improvement in travel times and an increase in average speed by 7%–22%. The environmental benefits found by CDOT included a 2%–7% reduction in fuel consumption and a reduction of pollution emissions by up to 17% [7].

3.4 SENSOR PLACEMENT FOR INTERSECTION CONTROL

Sensor placement depends on local and system requirements for the intersection and of course the type of control algorithm used. Local intersection requirements include phases to be actuated, approach speed measurement, and length of queues expected. System requirements depend on the type of traffic responsive or traffic adaptive coordination strategy employed. A review of existing detector placement is required whenever any of these factors change [4,10,11].

3.5 MEASURES OF EFFECTIVENESS AND DATA SOURCES FOR VALIDATING COMMON OPERATIONAL OBJECTIVES FOR ADAPTIVE SIGNAL CONTROL AND TRADITIONAL SIGNAL TIMING

Smooth-flow operation of a roadway can be evaluated with vehicle re-identification systems, GPS probe runs, and occupancy data from advance detectors connected to the signal controller [12,13]. Each data source has benefits and limitations for computing performance. Vehicle re-identification systems can provide a wealth of data 24/7, but only for point-to-point travel time. GPS probes offer more detailed information concerning link-by-link performance and can more easily pinpoint trouble areas. However, this method of collecting information is expensive when generating a large data set. Measures of effectiveness (MOEs) from the signal controller produce link-by-link performance 24/7 and also efficiently identify trouble spots as do probes, but different techniques are needed to aggregate these data into information concerning the performance of a route. Furthermore, many signal controllers are not equipped with such capabilities.

Table 3.4 contains the MOEs frequently utilized to evaluate the benefits of adaptive signal control and traditional signal timing systems. These include smooth flow, access equity, throughput, and travel time reliability. A balance between access equity and smooth-flow operation of the roadway network is common in most suburban settings with some variation in agency and locality preferences. Access equity is found by measuring sensor occupancy and green time data from stop-bar sensors connected to the signal controller. The main challenge in many systems is that some agencies do not utilize stop-bar detection for phases that are coordinated 24/7. If advance detection zones or loops are reasonably close to the stop bar, some anecdotal research indicates that green occupancy ratio (GOR) measures can be computed and compared with stop-bar zones from side-street sensors at the stop bar. Additional research is needed to bolster validation of this approach. Reliability of the GOR and served volume-to-capacity ratio (V/C) are also important metrics for determining the range of performance between operational strategies.

Throughput is the number of vehicles passing through a section of roadway in a given time. It provides a better measure of traffic system efficiency than total traffic volume. This MOE is obtained using tube counters or other traffic counting equipment (video, microwave radar, etc.) deployed at a specific location. In addition, counting detectors connected to the signal controller can also be utilized if they are located far enough from the stop bar so that queues do not habitually form on top of those zones. Exit detection is particularly suited

Table 3.4 MOEs and data sources for adaptive signal control and traditional signal timing

MOE	Data sources	Operational measure
Smooth flow	Travel time data from vehicle re-identification scanners Trajectory data from GPS probes High-resolution signal timing and sensor data	Route travel time Route travel delay Route average speed Link travel time and delay Number of stops per mile on route Percent arrivals on green by link Platoon ratio by link
Access equity	High-resolution signal timing and sensor data	Green occupancy ratio (GOR) Minimum, maximum, and standard deviation of GOR Served volume-to-capacity ratio by movement
Throughput	Count data from sensor file	Total traffic volume on route Time to process equivalent volume
Travel time reliability	Travel time data from Bluetooth scanner Trajectory data from GPS probe High-resolution signal timing and sensor data	Buffer time Planning time Minimum, maximum, and standard deviation of platoon ratio Minimum, maximum, and standard deviation of percent arrivals on green

Source: D. Gettman et al. *Measures of Effectiveness and Validation Guidance for Adaptive Signal Control Technologies*, FHWA-HOP-13-031, U.S. Department of Transportation, Federal Highway Administration, Office of Operations, Washington, DC, July 2013.

for counting vehicles when the distance to the next intersection is significant. Since data are taken at a specific point, this measure addresses both through traffic and turning traffic and does not directly reflect throughput on a specific route.

Travel time reliability can be measured from vehicle re-identification data, GPS probes, and sensor occupancy data supplied by advance sensors connected to the signal controller. For GPS probes and vehicle re-identification systems, buffer time is the primary measure of route reliability. Percent arrivals on green and platoon ratio can be computed from signal controller data to estimate reliability. Additional methods are needed to synthesize link-by-link statistics into reliability of performance along a route. Vehicle re-identification systems have a significant advantage for reliability estimation since they collect data 24/7.

3.6 RED-LIGHT RUNNING VEHICLE DETECTION

Although not strictly a traffic flow sensor application, red-light running cameras are found at many intersections where signal control exists. Table 3.5 lists the pros and cons of using red-light running cameras. Many traffic engineers believe that alternatives to these devices should be explored before considering their installation. Others believe that the vast majority of serious red-light running accidents are caused by driver error due to impaired vision of the intersection, poorly engineered signal timing, tiredness and distraction, and medical emergency.

Alternatives to the cameras include improved signal timing, such as increasing yellow by as little as 0.3 s to reduce violations by 70%–80% (CA), increasing yellow by 1 s over the Institute of Transportation Engineers (ITE) standard to reduce violations by 53% and crashes by 40% (TX), and adding an all-red clearance interval. The downside is that efficiency is lost in moving traffic through intersection. Alternatives to signal timing changes are the deployment of roundabouts and upgrades to the intersection geometry and improving

Table 3.5 Pros and cons of red-light running camera installation

Case for red-light running cameras	Case against red-light running cameras
IIHS review of international deployments showed that [14] • Cameras lower violations by 40%–50% • Cameras reduce injury crashes by 25%–30% • Cameras give an estimated 13%–29% reduction in all types of injury • Cameras reduce by 24% more serious right-angle injury crashes IIHS study of 57 U.S. cities with and without red-light cameras from 1992 and 2014 observed that cameras result in [15] • 21% fewer fatal red-light running crashes per capita • 14% fewer fatal crashes of all types per capita at signalized intersections IIHS study of 14 U.S. cities that ended their camera programs between 2010 and 2014 found they experienced [15] • 30% more fatal red-light running crashes per capita • 16% more fatal crashes of all types per capita at signalized intersections Pennsylvania 2011 study claims 48% reduction in red-light running within 12 months of enforcement and 24% reduction in total number of crashes at 10 intersections where 3 years of data were available	California state legislature report documents a 325% increase in rear-end collisions after cameras were installed [14] At one Los Angeles intersection, rear-end collisions increased by 80% [14] Other studies show a 15% increase in rear-end collisions, but a 25% decrease in right-angle crashes [14] Virginia Department of Transportation study of more than 3,500 crashes over a 7-year period (1998–2004 inclusive) at 28 intersections with cameras and 44 intersections without cameras in Northern Virginia found [16,17] • Rear-end crash rates increased by an average of 27% for the entire study area • Red-light running crash rates decreased by 42% for the entire study area Declining support by the courts because violations were automatically sent to registered drivers, even though they may not have been driving Camera operators in some states are incentivized to give more citations because their revenue depends on number of violations

visibility through restriping and clearing of vegetation. Insurance Institute for Highway Safety (IIHS) studies show roundabouts typically reduce overall collisions by 35%–61%, injury collisions by 25%–87%, and fatality collisions by 90% at intersections where stop signs or signals were previously used [18,19].

Figure 3.8 displays additional benefits of red-light running cameras as reported by the ITS Joint Program Office [7]. They are reduction in crash injuries by 20%–98%, reduction in left-turn/right-angle crashes by 3%–85%, reduction in serious and fatal injuries by 8%–83%, reduction in red-light violations by 20%–86%, and reduction in per capita rate of fatal red-light running crashes by 35%. However, as stated above, before installing these cameras, the intersection geometry and sight distances should be reviewed to ensure proper design and visibility for the driver as these measures alone may ameliorate the unsafe condition.

3.7 TRAVEL TIME NOTIFICATION

Traffic flow sensors improve the effectiveness of information dissemination programs by providing more timely and accurate travel time data for the impacted highways and by identifying alternate routes. Figure 3.9 illustrates an example of travel time notification to freeway motorists using a changeable message sign. Current information about the extent of stop-and-go traffic resulting from incidents is important to motorists because of its impact on alternate route selection. Thus, incident management applications may require travel time accuracies to the nearest minute and end-of-queue tracking to the nearest half mile or to some other critical ramp spacing that applies to a particular segment of highway.

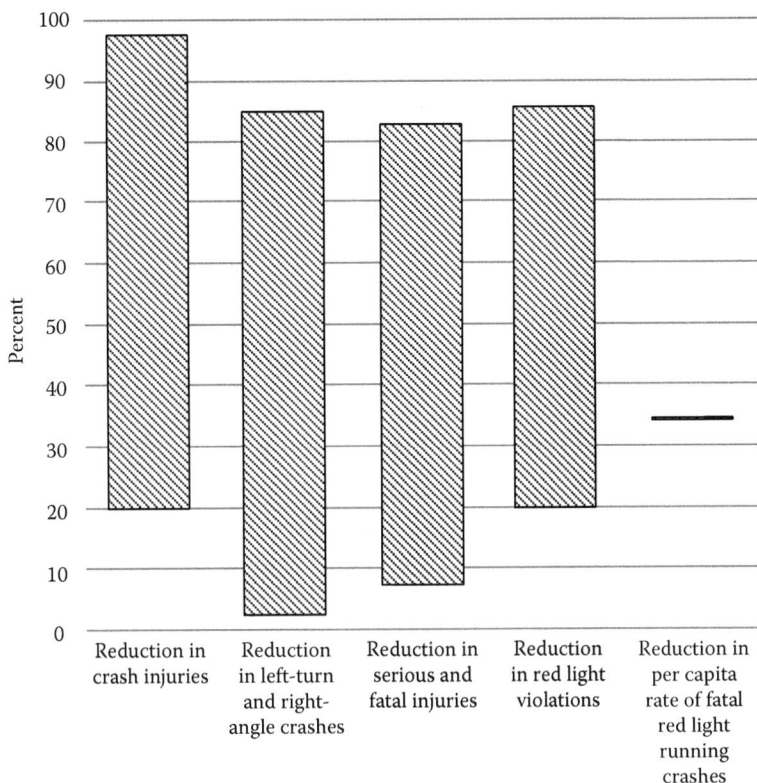

Figure 3.8 Benefits of red-light running cameras. (From G. Hatcher et al. *Intelligent Transportation Systems Benefits, Costs, and Lessons Learned: 2014 Update Report*, FHWA-JPO-14-159, ITS Joint Program Office, Washington, DC, June 2014.)

3.8 RAMP METERING

The most common technique for addressing recurring congestion on freeways is ramp metering as illustrated in Figure 3.9. It limits the rate at which vehicles enter the freeway mainline so that downstream mainline capacity is not exceeded. Ramp metering redistributes the freeway demand over space and time. Excess demand is either stored on the ramp or diverted. The diverted vehicles may choose less traveled alternate routes or their occupants may select another mode of transportation.

Metering regulates ramp traffic by dispersing platoons of vehicles released from nearby signalized intersections. By allowing a limited number of vehicles into the mainline traffic stream, turbulence is reduced in the merge zone. This leads to a reduction in sideswipe and rear-end accidents associated with stop-and-go traffic flow. Maximum mainline flow rates can be achieved by controlling ramp flow rates such that freeway traffic moves at or near optimum speed throughout the network. An algorithm determines the vehicle entrance rate based on mainline volume, speed, and queue length.

A secondary benefit associated with ramp metering is management of nonrecurring congestion created by freeway incidents. Once an incident is detected, ramp metering can potentially reduce the number of vehicles impacted by the incident. For example, meters upstream of a detected incident can be adjusted to allow fewer vehicles to enter the affected

Figure 3.9 Sensors (not shown) determine travel times displayed on changeable message signs and ramp meter signal cycle times. (From LAK.)

facility, potentially diverting some trips to on-ramps downstream of the incident location. Conversely, the downstream ramps can operate with relaxed metering rates in order to accommodate the increased demand.

Figure 3.10 depicts the location of vehicle detection sensors used for ramp metering [20]. Sensors on the mainline upstream of the ramp determine the parameters used by the metering algorithm, for example, volume, speed, and occupancy. Ramp metering algorithms such as ALINEA require occupancy data downstream of the ramp. The sensors also collect historical volume and occupancy data. An advance queue sensor informs the meter if the queue is spilling over onto the arterial that feeds the ramp. If this is the case, the meter can flush the ramp allowing all vehicles on the ramp to enter the freeway. A sensor, typically an ILD, near the ramp entrance also acts a queue sensor and informs the controller of a building queue. A demand sensor on the ramp indicates the arrival of a vehicle at the stop bar and the commensurate start of the metering cycle. A passage sensor detects when the vehicle clears the stop bar and returns the ramp signal to red for the next vehicle. The passage sensor can also be used to monitor meter violations (i.e., drivers who ignore the red stop signal) and provide historical data about the violation rate at each ramp [2]. Ramps that contain two metered lanes or one metered and one unmetered high-occupancy vehicle (HOV) lane add a count sensor after the passage sensor to obtain the total count of vehicles entering the mainline.

Another technique for managing recurring congestion is to meter freeway-to-freeway connector ramps as shown in Figure 3.11. Experiences in Minneapolis, MN and San Jose, San Diego, Los Angeles, and Orange County, CA indicate that significant benefits can be achieved with connector metering under conditions similar to those associated with ramp metering. Freeway connectors often have per lane flow rates greater than 900 veh/h (the maximum possible with single entry metering). Metering rates exceeding this number can be obtained with two-lane metering or possibly platoon metering. Such configurations work best when there is an added lane downstream from the on-ramp.

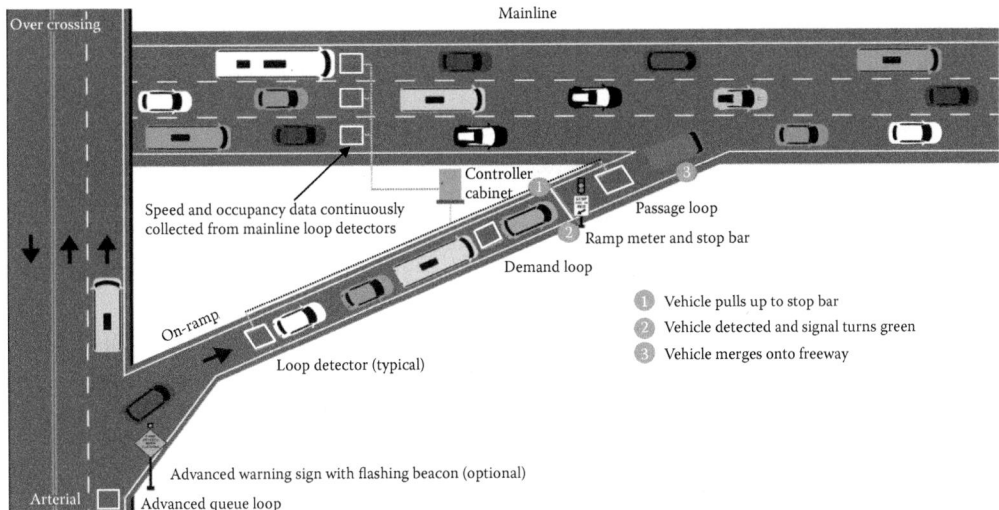

Figure 3.10 Ramp metering operation (notional). This illustration is meant only to illustrate the placement and function of sensors used to implement ramp metering. It should not be used as a design drawing to construct a freeway ramp as the ramp entrance lane does not have sufficient vehicle storage capacity, while the freeway merge lane is too short to allow vehicles to properly accelerate to prevailing freeway speeds. (From *Active Transportation and Demand Management Webinar Series*, Webinar #3: Ramp Metering Benefits, Opportunities, and Keys for Overcoming Common Challenges, U.S. Department of Transportation, Federal Highway Administration, Washington, DC, December 10, 2014.)

Figure 3.11 Freeway-to-freeway metering on the connector from the eastbound I-105 Freeway onto the southbound I-605 Freeway in Norwalk, CA. Meters on the two I-105 lanes allow three vehicles per lane during each green cycle to enter the four lanes of the I-605 mainline. One of the two I-105 ramp lanes continues as a fifth mainline lane. (From LAK.)

Figure 3.12 Mainline metering of vehicles heading west onto the San Francisco–Oakland Bay Bridge. (From Caltrans District 4, Oakland CA.)

Figure 3.12 demonstrates still another strategy to control flow rates by metering traffic on the freeway mainline. Mainline metering manages traffic demand at a mainline control point to maintain a desired level of service (LOS) on the freeway downstream of the control location. The desired LOS is selected to achieve one or several of the following objectives:

- Flow maximization through a downstream bottleneck.
- High LOS downstream of the control point.
- Distribution of total delay on the freeway system more equitably.
- Diversion of traffic to other routes or modes.
- Increase in the overall safety of the facility.

3.8.1 Ramp metering benefits

Ramp metering benefits as measured by increase in vehicle speed, travel time reduction, collision reduction, emission reduction, and other advantages vary from city to city as Figure 3.13 and Table 3.6 illustrate [20,21]. Reasons for the variation can include differences in ramp metering algorithms and driver behavior in each city. However, the overall benefits do support the implementation of ramp metering as a tool for congestion and accident reduction.

3.8.2 Minnesota Department of Transportation evaluation of ramp metering benefits

The State of Minnesota Department of Transportation (MnDOT) conducted an extensive evaluation of ramp metering benefits in the Minneapolis–St. Paul metropolitan area in 2000 [21]. First, the ramp meters were turned off for 6 weeks to establish a baseline condition. An extensive planning and policy review effort followed to modify the region's

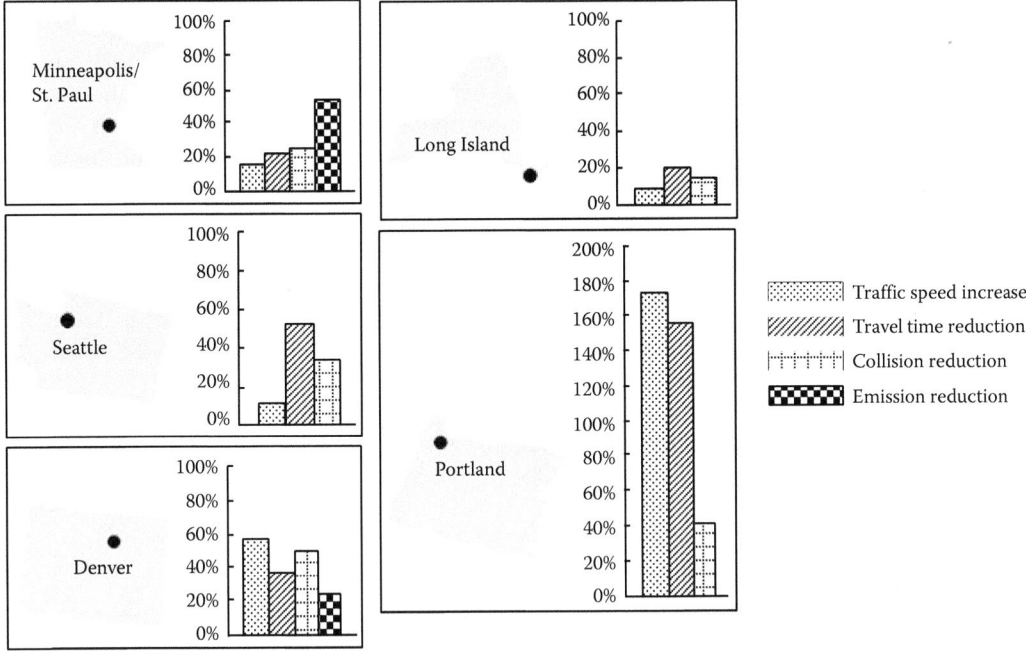

Figure 3.13 Ramp metering benefits. (From *Active Transportation and Demand Management Webinar Series*, Webinar #3: Ramp Metering Benefits, Opportunities, and Keys for Overcoming Common Challenges, U.S. Department of Transportation, Federal Highway Administration, Washington, DC, December 10, 2014.)

Table 3.6 Ramp metering benefits as reported by additional U.S. cities

Performance measure	Location and improvement
Travel time	Atlanta—10% decrease in peak period Houston—22% decrease in peak period Arlington—10% decrease in peak period
Travel speed	Milwaukee—35% increase in peak period Portland—155% increase in peak period Detroit—8% increase in peak period Los Angeles—15 mi/h increase
Crash rate	Phoenix—16% decrease during metered hours Milwaukee—15% decrease in peak period
Crash frequency	Portland—43% decrease Sacramento—50% decrease Los Angeles—20% decrease
Driver hours saved	Sacramento—50% decrease Los Angeles—8470 h per day
Vehicle volume	Milwaukee—22% increase in peak period Sacramento—5% increase in peak period Detroit—14% increase in volume Los Angeles—increase of 900 vehicles per day
Gallons of fuel saved	Portland—700 gallons per weekday
Emissions reduction	Minneapolis—1160 tons annual reduction
Benefit-to-cost ratio	Atlanta—about 4:1 in Year 1; about 20:1 after 5 years

metering system to better balance the needs of system operators and regional travelers. Finally, the meters were turned back on in accordance with the new policies.

The impacts observed during the experiment supported MnDOT's assertions that the ramp metering system provided substantial benefits. However, market research revealed that many residents were dissatisfied with certain operational aspects of the system and did not necessarily understand the trade-off between more restrictive metering and improved freeway performance. Through these findings, MnDOT became more aware of the importance of public information and education campaigns in promoting the operation of ramp meters. The result of the evaluation was the implementation of modifications that achieved a better balance of operational efficiency of the system with travelers' perceptions and expectations. This effort was combined with an increased focus on public outreach to promote the benefits of the system, which in the end were

- Improved throughput by 25%.
- Improved freeway travel times by 20%.
- Improved travel time reliability by 90%.
- Reduced crashes by 25%.
- Reduced congestion resulting in reduced emissions and fuel consumption.

3.8.3 Conditions under which to install ramp metering

A warrant is a set of criteria used to justify the implementation of a traffic control or traffic management device or strategy. The three warrants in Table 3.7 were developed to assist agencies in determining when ramp meters should be installed to mitigate freeway traffic congestion [26]. Conditions in all three warrants should be satisfied before installing ramp meters.

3.8.4 Deployment challenges for ramp meters

Figure 3.14 contains several ramp metering deployment challenges that may arise because of existing road geometry and heavy ramp volume. Geometric considerations that deter ramp metering include lack of adequate acceleration length, mainline weaving distance, and sight distances. These issues were experienced by 58% of agencies surveyed. Heavy ramp volume challenges arise from long queue length, arterial backup, and lack of ramp storage. These issues were experienced by 25% of agencies surveyed [21].

3.9 WRONG-WAY VEHICLE DETECTION

Sensors such as microwave radars, VDSs, inductive loops, and magnetometers may be used to detect vehicles traveling in the wrong direction. The upper left photograph in Figure 3.15 shows applications to wrong-way vehicle detection, driver warning, and prevention of vehicle entry onto limited-access highways from exit ramps. Wrong-way vehicle detection is also applicable to monitoring the entrances to one-lane bridges and tunnels. The lower left photograph illustrates the application of video detection to monitoring traffic flow in a tunnel. In this instance, a vehicle has careened off the tunnel wall and come to a stop facing in the wrong direction. This event was automatically detected by the VDS. The photograph on the right applies to monitoring reversible traffic flow lanes. In this case, the sensors are used to ensure that vehicles do not enter these lanes traveling in the wrong direction.

Some sensors (microwave radars and VDSs) may come equipped with built-in algorithms for detecting wrong-way vehicle flow. Others such as the inductive loop and magnetometer

Table 3.7 Warrants that justify installation of ramp meters to mitigate freeway traffic congestion

Warrant 1	Warrant 2	Warrant 3
1a. The freeway operates at speeds less than 50 mi/h for a duration of at least 30 min for 200 or more calendar days per year [22]. OR 1b. There is a high frequency of crashes (collision rate along the freeway exceeds mean collision rate in the subject metropolitan area) near the freeway entrances because of inadequate merge area or congestion [22]. OR 1c. The ramp meter will contribute to maintaining a specific LOS identified in local transportation plans and policies [22]. OR 1d. The ramp meter will contribute to maintaining a higher-level vehicle occupancy through the use of HOV preferential treatments as identified in the region's transportation system management (TSM) plan [22]. OR 1e. The ramp meter will contribute to balancing demand and capacity at a system of adjacent ramps entering the same freeway facility [22]. OR 1f. The ramp meter will mitigate predictable sporadic congestion on isolated sections of freeway because of short peak period loads from special events or from severe peak loads of recreational traffic [22].	2a. The total mainline-ramp design hour volume (mainline volume plus ramp volume) exceeds the following [23]: Two mainline lanes in one direction—2650 veh/h Three mainline lanes in one direction—4250 veh/h Four mainline lanes in one direction—5850 veh/h Five mainline lanes in one direction—7450 veh/h Six mainline lanes in one direction—9050 veh/h OR 2b. The total volume of the sum of traffic in the right most lane and the ramp exceed 2100 veh/h during the design hour [23]. OR 2c. Platoons from signalized intersections are recognized to adversely impact the ramp in consideration. If hourly volume, based on maximum 30-s volume readings projected to hourly values, exceed 1100 veh/h regardless of overall hourly volume [24]. *Note:* Overall hourly volume entering from arterials may be relatively low, for example, 700 veh/h. However, Warrant 2c is considered met during periods when platoons arrive if 30-s readings of volumes are 1100 veh/h or greater.	*Functionality factors.* Volumes at ramps being considered for meters fall within the range of 240–900 veh/h/ln during peak periods [25]. *Note:* The length and geometry of the ramp is a factor in the final decision of whether to deploy a ramp meter. The current guideline for ramp meters does not address this factor, as it is believed the analysis of the ramp will be a part of the preliminary and final design. The focus of the guideline is on whether or not a ramp meter is needed, not on whether a ramp meter can be designed at the location as that is determined during the design process.

require external logic to be implemented in the controller to indicate a vehicle traveling in the wrong direction. For example, if the trailing loop or magnetometer is activated before the leading sensor, this is an indication that a vehicle is traveling in the wrong direction.

3.10 INCIDENT DETECTION

The importance of incident detection in traffic management led to the development of several categories of automatic incident detection (AID) algorithms that rely on roadside sensors to gather required input data. Sensors are valuable assets not only for incident detection,

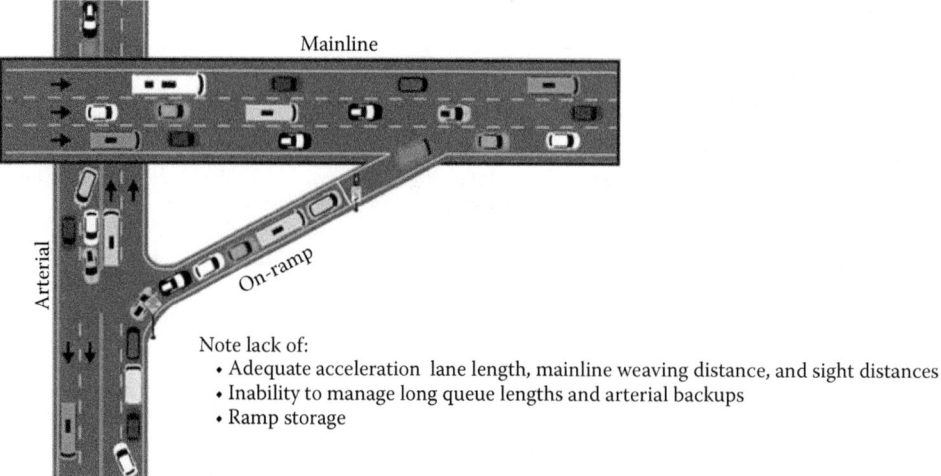

Figure 3.14 Example of challenging ramp geometry. (From Active Transportation and Demand Management Webinar Series, Webinar #3: Ramp Metering Benefits, Opportunities, and Keys for Overcoming Common Challenges, U.S. Department of Transportation, Federal Highway Administration, Washington, DC, December 10, 2014.)

Traficon-Stopped vehicle in Leutenbach Tunnel, Germany (https://www.youtube.com/watch?v=AvJAoJ1EeN4)

Monitoring reversible lanes

Figure 3.15 Applications of wrong-way vehicle detection.

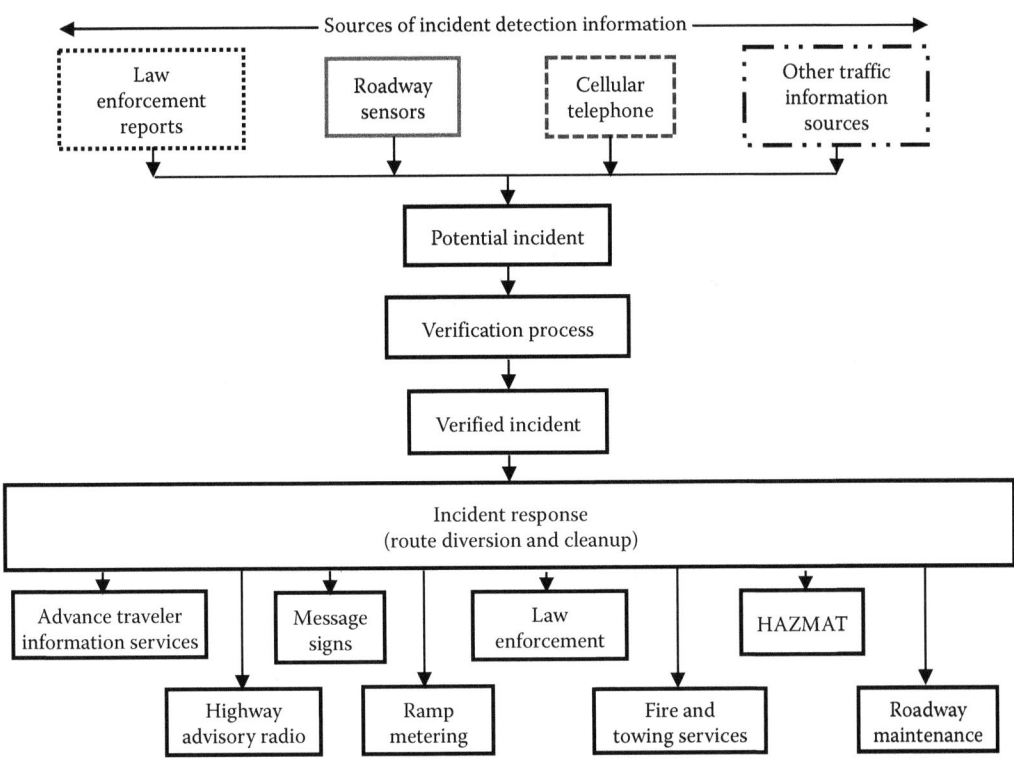

Figure 3.16 Incident detection, verification and identification, and response processes. Sensors and other sources of information are relied on to detect and verify incidents.

but also for the verification of the type of incident and the kind of assistance needed, for example, by providing video imagery of the scene. The key steps of the incident detection, verification and identification, response, removal, and recovery process are illustrated in Figure 3.16. Other sources of incident detection information are indicated in the top row of the figure. Verification is frequently provided by emergency responders as well as from surveillance cameras that may be located near the incident site. Response and removal can involve several types of emergency service responders depending on the nature of the incident, for example, highway patrol, fire department services, ambulance, HAZMAT personnel, and tow trucks. As indicated toward the bottom of the figure, ramp metering can also be a resource that contributes to incident recovery by controlling the number of vehicles entering the mainline. Other assets that assist in expediting the recovery process are traveler information services such as 511 advisories, kiosk posts, cell phone incident alert applications, highway advisory radio, highway message signs that advise motorists of the incident location and delay period, and roadway maintenance and cleanup crews.

Traffic flow characteristics during a freeway incident can be defined in terms of four flow regions as shown in Figure 3.17 [27]. Flow region A is far enough upstream of the incident so that traffic moves at normal speeds with normal density. Flow region B, depicted in Figure 3.18 along with Region C, is the area located directly behind the incident where vehicles are queuing if traffic demand exceeds the restricted capacity caused by the incident. Region B is characterized by the upstream propagation of a shock wave where speeds are generally lower and a greater vehicle density may exist. Flow region C is the region directly downstream from the incident where traffic is flowing at a metered rate, or incident flow

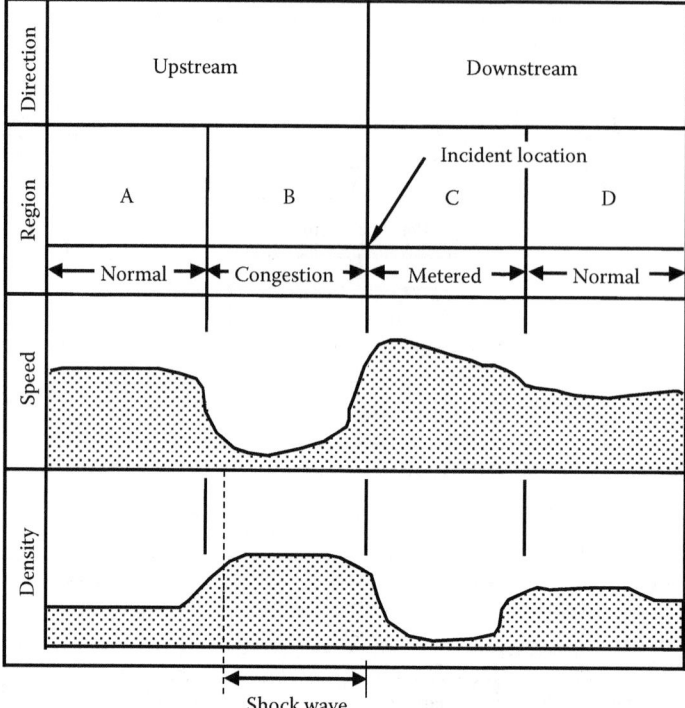

Figure 3.17 Regions of different traffic flow characteristics during an incident. (From JHK and Associates, *Intelligent Vehicle Highway Systems: The State of the Art*, Prepared for Massachusetts Department of Highways, March 1993.)

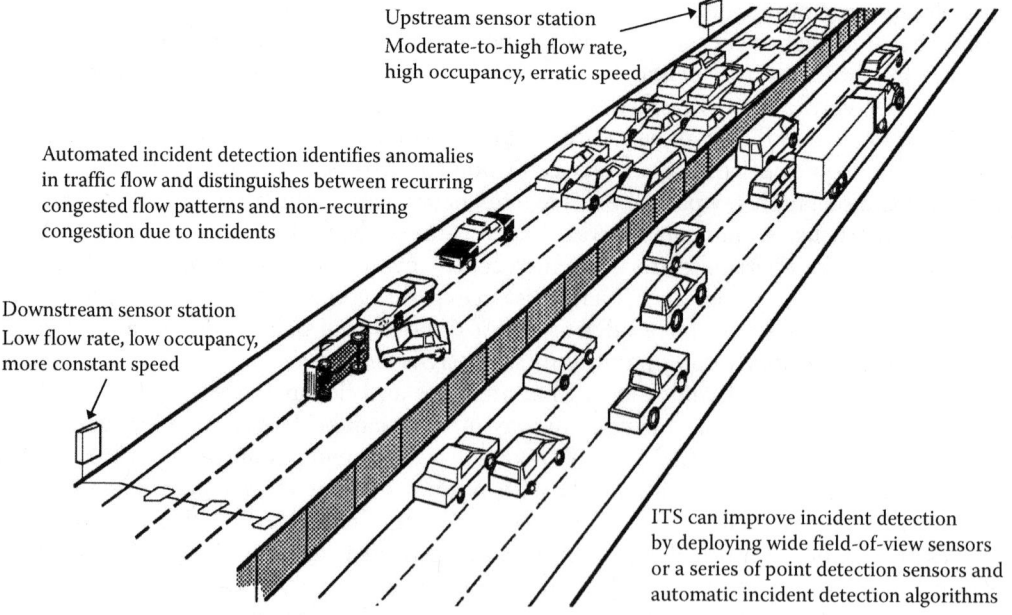

Figure 3.18 Freeway incident detection using roadway sensors.

rate, due to the restricted capacity caused by the incident [3]. Depending on the extent of the capacity reduction, traffic density in Region C can be lower than normal, while the corresponding traffic speed can be higher than normal. Flow region D is far enough downstream from the incident such that traffic in D flows at normal density and speed as in Region A.

The California algorithms detect an incident by comparing the increase in the occupancy of an upstream sensor to the decrease in the occupancy of a sensor downstream of a potential incident. Multiple thresholds are utilized to enhance the probability that the occupancy measurements correspond to a true incident. The McMaster algorithm, derived from catastrophe theory, separates traffic flow parameters such as volume, occupancy, and speed into different states that are typical of congested and uncongested traffic flow. If significant changes in the traffic state are observed, an alarm is triggered. Time series algorithms use statistical indicators derived from volume and occupancy to provide short-term forecasts of traffic parameters. An alarm is triggered when a significant deviation occurs between observed and forecast traffic parameter values. The high occupancy (HIOCC) algorithm triggers an alarm when the occupancy increases abnormally during consecutive 1-s observation intervals. Artificial neural networks and fuzzy logic are additional techniques used for AID. Chapter 4 discusses these algorithms further.

Advancements in communications and detection technologies have created new approaches to incident detection that promise faster detection times, reduction of false alarms, and higher detection rates. These technologies incorporate GPS, roadside MAC address readers, cellular telephony, VDSs, and lidar, microwave, infrared, acoustic, and ultrasonic sensors. New sensor technologies and the introduction of connected vehicles can potentially provide a richer data set from which more effective incident detection algorithms can be developed.

3.11 ACTIVE TRANSPORTATION AND DEMAND MANAGEMENT

Active transportation and demand management (ATDM) is the dynamic management, control, and influence of travel demand, traffic demand, and facility demand on the entire transportation system and over a traveler's entire trip. Under an ATDM approach, the transportation system is continuously monitored as shown in Figure 3.19 [20,28,29]. Using real-time and archived data along with predictive methods, ATDM supports real-time actions that achieve or maintain traffic flow and influence traveler behavior to prevent or delay breakdown conditions, improve safety, promote sustainable travel modes, reduce emissions, or maximize system efficiency. Rush-hour congestion reduction benefits of ATDM are displayed in Table 3.8. Noteworthy are the 35% and 25% reductions in congestion attributed to flexible working hours and working at home, respectively.

3.11.1 The active management cycle

Key features of the performance-based feedback loop in Figure 3.19 are as follows:

- Continuously assessing system performance by monitoring the entire transportation system, that is, highways, arterials, public transit, parking availability.
- Evaluating and recommending real-time responses based on the assessed state of the system.
- Implementing these dynamic actions to improve system performance, which is fed back as part of the monitoring function.
- Basing all of these decisions on performance objectives and indicators such as person throughput, reliability, and safety.

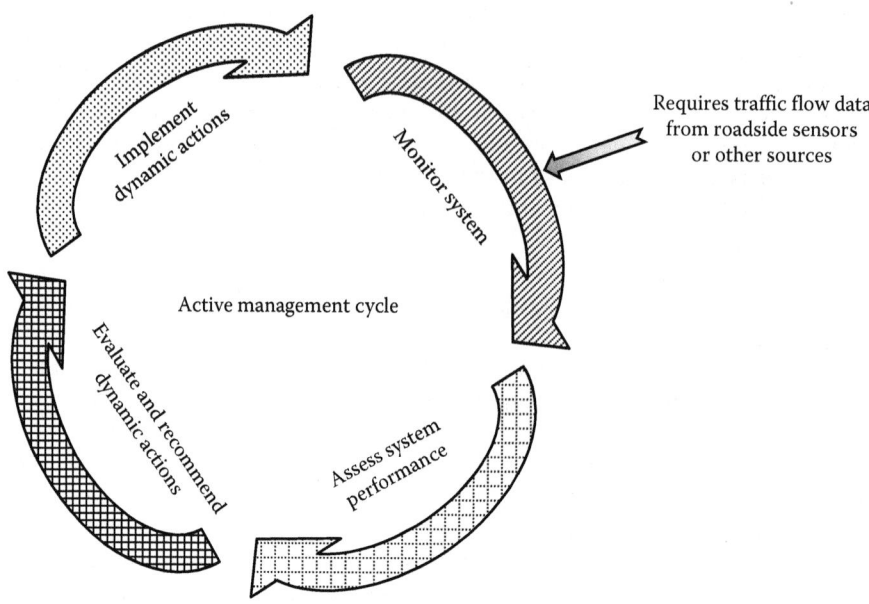

Figure 3.19 ATDM as a dynamic process requiring traffic flow data from sensors and other sources to support the system monitoring process. (From *Active Transportation and Demand Management Webinar Series*, Webinar #5: National ATDM Program Research, U.S. Department of Transportation, Federal Highway Administration, Washington, DC, March 26, 2015.)

Since ATDM is dynamic, changing conditions can require dropping some transportation management strategies and adding others, even the ones that involve travel mode changes for the traveler as illustrated in Figure 3.20 [28,29]. The process begins by considering overall travel demand (i.e., the destination and time-of-day decisions) and moves through traffic demand (i.e., the mode choice decision), and finally facility demand (i.e., the route and lane and facility use decisions). By using the various ATDM approaches at each stage of the trip, agencies can influence travel behavior and the resulting demand on the system to optimize performance.

Figure 3.21 displays the stair-step approach that explains how transportation operations departments can progress from static to responsive to proactive management of the transportation system. The simplest transportation management method is a static one that uses time-of-day operations involving minimal real-time adjustments. Moving to responsive management involves minimal risk. This approach only works well when there is little variability in the traffic flow patterns. Responsive management involves taking actions to respond to current traffic conditions and to reduce the time of degraded operations. This is

Table 3.8 Potential of ATDM strategies to mitigate rush-hour congestion

ATDM strategy	Amount of congestion reduction (%)
Car pooling	5
Alternate route	5
Flexible working hours	35
Working at home	25
Working at another location	5
Traveling with public transportation	10
Cycling	15

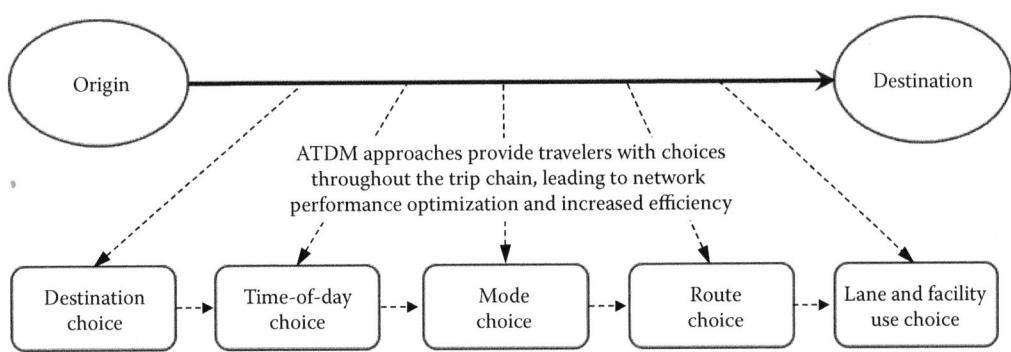

Figure 3.20 Dynamic nature of ATDM may require introduction of new strategies within a trip. (From *Active Transportation and Demand Management Webinar Series*, Webinar #5: National ATDM Program Research, U.S. Department of Transportation, Federal Highway Administration, Washington, DC, March 26, 2015.)

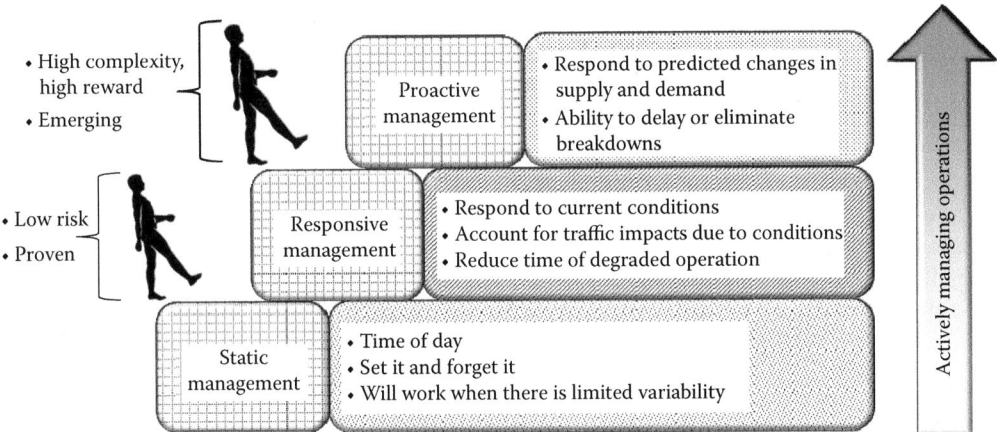

Figure 3.21 Moving toward active management of the transportation system. (From *Active Transportation and Demand Management Webinar Series*, Webinar #5: National ATDM Program Research, U.S. Department of Transportation, Federal Highway Administration, Washington, DC, March 26, 2015.)

a widely accepted tactic and there are many successful examples of responsive management. Proactive management has a higher barrier of entry as it involves a higher risk and is more experimental. However, there is the potential for big rewards and success in reducing congestion. Proactive management involves the highest level of active management and includes active responses in anticipation of changing supply and demand.

3.11.2 ATDM strategies

ATDM consists of a series of strategies that can be grouped into three main categories: active demand management (ADM), active traffic management (ATM), and active parking management (APM) [30]. ADM focuses on travelers creating more fluid decisions, especially at the mode, destination, and route levels. Moving beyond traditional transportation demand management (TDM) approaches of carpooling and ride matching, new strategies such as congestion pricing, financial incentives, advanced traveler information, dynamic ridesharing, and on-demand transit are used to increase the day-to-day choices available to the travelers.

ATM includes strategies intended to mitigate the effects of recurring and nonrecurring congestion on the roadway with technology to dynamically control speeds, lane usage, and junctions. Strategies include implementations of dynamic speed limits, queue warning systems, dynamic shoulder use, junction control through ramp metering and lane control, and transit signal priority.

APM focuses on improved parking management and information systems to more efficiently utilize parking resources. Strategies include pricing parking at a market rate, displaying parking space availability in real time using smart phone applications and on-street signs, and implementing electronic parking reservation systems.

Variable speed limit (VSL) or dynamic speed limit systems are used in a number of countries, first in Europe but now also in the United States, to improve flow and increase safety. VSL systems use sensors to collect data on current traffic and weather conditions. Posted speed limits are then dynamically updated to reflect the conditions that motorists are actually experiencing. The speed limits may even vary by travel lane. Presenting drivers with speed limits that are appropriate for current conditions may reduce speed variance, a concept sometimes called speed harmonization. If properly designed, VSL systems reduce crash occurrence and can also reduce system travel time and vehicle emissions through increased uniformity in traffic speeds. Getting motorists to comply with posted speed limits is sometimes an issue that has to be resolved by education and enforcement.

3.11.3 Safety benefits of ATM

A Strategic Highway Research Program (SHRP) study sponsored by the U.S. Transportation Research Board produced a combined safety versus congestion relationship for the Seattle, Minneapolis–St. Paul, and Sacramento metropolitan areas [31]. The curves in Figure 3.22 were developed by rendering the data to reflect the average freeway crash rate for the three metropolitan areas and then averaging the individual data points. The results are

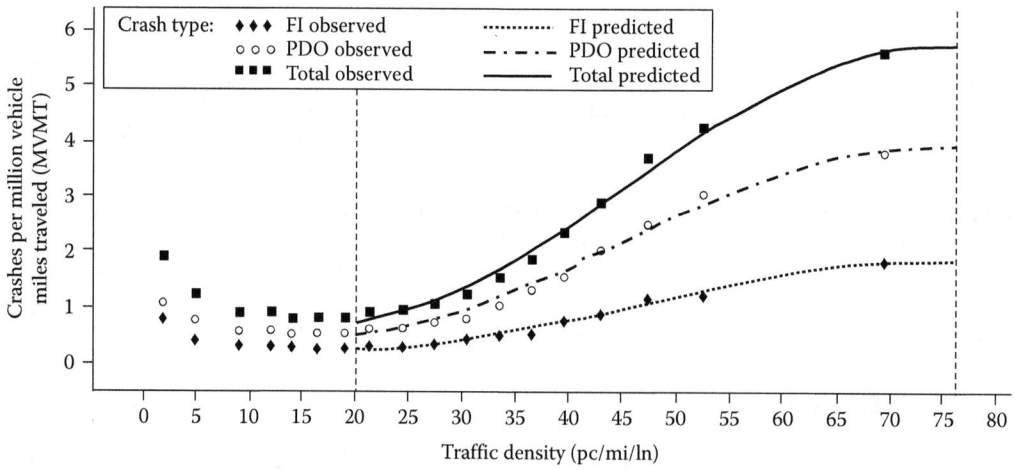

Figure 3.22 Observed and predicted fatal-and-injury (FI), property-damage-only (PDO), and total-crash (all crash severity levels combined) rates versus traffic density using combined Seattle, Minneapolis–St. Paul, and Sacramento data. (From I.B. Potts et al. *Further Development of the Safety and Congestion Relationship for Urban Freeways, SHRP 2 Report S2-L07-RR-3:* Figure 1.1, p. 4. Transportation Research Board, Washington, DC, 2015. Reproduced with permission of the Transportation Research Board.)

representative of a freeway system with a total crash rate of 1.86 crashes per million vehicle miles traveled (MVMT), a fatal-and-injury (FI) crash rate of 0.42 crashes per MVMT, and a property-damage-only (PDO) crash rate of 0.82 crashes per MVMT, which represents the average freeway crash rate for the three metropolitan areas giving equal weight to each area. The portion of the safety versus congestion relationship that is most relevant to the SHRP project objectives is the range from LOS C to LOS F (20 to approximately 75 passenger cars/ mile/lane [pc/mi/ln]) indicated by the vertical dashed lines in Figure 3.22. In this region, the data show that freeway crash rates can be reduced by decreasing congestion.

As in the original Phase II research, the best fit to the safety versus congestion relationship in this range is given by a family of cubic equations that allow the system manager and others to predict the total number of crashes, number of fatal-and-injury crashes, and number of PDO crashes per MVMT as a function of vehicle density in pc/mi/ln. Table 3.9 contains the equations for the number of crashes in units of MVMT. When the vehicle density is less than 20 pc/mi/ln or more than 76 pc/mi/ln, the crash types have a constant value. In between those limits, they are described by the cubic equations.

Besides accident reduction, ATM has the potential to smooth the flow of traffic on limited-access highways and thus increase throughput or flow rate during rush-hour periods. This conclusion is based on the well-known Greenshield relationship for vehicle speed versus vehicle density that predicts a reduction in speed and flow rate (vehicles/hour) when the density reaches some critical value.

Washington State is one of several in the United States that uses a multifaceted traffic operations center to manage traffic flow and congestion. Among the tools employed are inductive loop detectors, closed circuit cameras, variable message signs, highway advisory radio, freeway patrols, weather stations, traveler information systems, and ATM. About 25% of traffic congestion in the Seattle area is due to events such as collisions or disabled vehicles. The ATM system uses overhead lane signs similar to the ones shown in Figure 3.23 to provide advance notice of conditions that will modify traffic flow patterns. The sign categories include the following:

- VSL signs to direct drivers to incrementally reduce their speeds in response to congestion, incidents, unfavorable weather, or other conditions that warrant this approach.
- Symbols to direct drivers to change lanes when a lane is blocked.
- Overhead message signs to warn drivers of slowdowns, backups, and collisions ahead.

Such measures decrease last-second avoidance maneuvers and panic braking, both of which are important contributors to collisions. Emergency responders report high compliance with the "lane blocked/lane closed" symbols. These control symbols, including red Xs and yellow merge arrows, provide extra time for vehicles to move over and provide a gap between emergency responders and moving traffic. Washington State Police patrolling the corridors report feeling safer when working on the roadway.

Table 3.9 Regression equations for total crashes, FI, and PDO in MVMT as a function of vehicle density

Crash type	Density region		
	$D < 20$ pc/mi/ln	20 pc/mi/ln $\leq D \leq 76$ pc/mi/ln	$D > 76$ pc/mi/ln
Total crashes	0.72	$2.190 - 0.1979 \times D + 0.00728 \times D^2 - 5.34 \times 10^{-5} \times D^3$	5.77
FI	0.24	$0.831 - 0.0718 \times D + 0.00246 \times D^2 - 1.76 \times 10^{-5} \times D^3$	1.86
PDO	0.48	$1.359 - 0.1261 \times D + 0.00482 \times D^2 - 3.58 \times 10^{-5} \times D^3$	3.91

Figure 3.23 Seattle ATM VSL example. (From http://www.wsdot.wa.gov./smarterhighways/.)

3.11.4 ATM tools

Express lanes for HOVs, toll-paying vehicles, and buses are another ATM tool. Examples are shown in Figure 3.24. To counter gridlock in the Washington, DC metropolitan area, new express lanes were opened in December of 2014 along 46.6 km (29 mi) of I-95 between Garrisonville Road in Stafford County and the Edsall Road area of Fairfax County in Northern Virginia. The almost $1 billion (U.S.) project to convert HOV lanes to express lanes started in August 2012 and was funded through a public–private partnership (PPP) between the Virginia Department of Transportation, the Virginia Department of Rail and Public Transportation, the Federal Highway Administration, and Transurban. The project also expanded the previous two-lane layout to three, extended the system by a further 14.5 km (9 mi), and created additional entry and exit points.

Vehicles with three or more occupants can travel on the express lanes free of charge (once their driver has switched the tag to HOV mode) as can motorcycles, and although buses are permitted, trucks are not. All-electronic tolling removes the need for booths on the express lanes. Dynamic pricing adjusts toll charges from $0.124/km ($0.20/mi) during the quieter periods, to approximately $0.5/km ($0.80/mi) in some sections during rush hour.

High-occupancy vehicle lanes   Express lanes in Minneapolis, MN   Express lanes on CA SR-91

Figure 3.24 HOV and express lane ATM strategies.

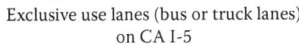

Exclusive use lanes (bus or truck lanes) on CA I-5   Reversible lanes   Reversible lanes (middle section) on I-95 in Washington DC metro area

Figure 3.25 Exclusive-use lanes and reversible lanes ATM strategies.

An example of the express lanes' effectiveness is the travel time savings experienced by southbound vehicles traveling the full length of the I-95 Express Lane during the peak period. These vehicles save an average of between 33 min on Mondays and Fridays and 13 min on Wednesdays, while maximum time savings are between 2 and 2.5 h on Mondays and Fridays and 30 min on Wednesdays.

Figure 3.25 displays other ATM tools. On the left are exclusive-use lanes for buses or trucks. These allow slower moving vehicles not to hamper traffic flow on uphill grades and provide runaway ramps on downhill grades. Reversible lanes that support traffic flow in either of two directions are illustrated in the middle and right photos. The ability to reverse lanes by moving a barricade or using signs to indicate the current traffic direction provides additional capacity in the dominant travel direction during heavily traveled rush-hour periods.

When stationary, the movable reinforced concrete barriers are linked to form a continuous barrier wall. The road zipper machine in the middle of Figure 3.25 lifts 1-m sections of the barriers and passes them through a conveyor system. In one pass, a barrier is transferred up to 7.3 m (24 ft) and set down in its new position. The zipper machine is capable of traveling at speeds up to 16 km/h (10 mi/h).

Utilization of highway shoulder and breakdown lanes during rush hours and other times of day is another ATM strategy that increases highway capacity [32]. Figure 3.26 illustrates three implementations of this technique, first by authorized transit vehicles at all times, second by general-purpose traffic during fixed times of a day, and third by dynamic shoulder use whereby general-purpose traffic is allowed to travel on the shoulder as dictated by real-time traffic conditions.

Authorized transit vehicles use shoulder at all times   General-purpose traffic use shoulder as a lane during fixed times of a day   General-purpose traffic use shoulder dynamically as a lane as needed based on real-time traffic conditions

Figure 3.26 Shoulder use as an ATM strategy. (Extreme left image from USDOT.)

3.11.5 Open questions concerning ATDM

In the United States, traffic management agencies still have questions about the effectiveness of ATDM for their day-to-day operations. Typical of these are the following:

1. What will be the impact of increased prediction accuracy of congestion and its onset, additional implementation of active management strategies, and robust behavioral predictions on mobility, safety, and environmental benefits?
2. What ATDM strategy or combinations will have the most impact on short-term versus long-term behaviors and under what operational conditions?
3. Are ATDM strategies more beneficial when implemented in isolation or in combination, for example, combinations of ADM, ATM, and APM strategies?
4. Which ATDM strategy or combination yields the most benefits for specific operational conditions?
5. Which ATDM strategy or combination will benefit most through reduced latency and under what operational conditions?
6. Which ATDM strategy or combination will have the most benefits for individual facility deployment versus system-wide deployment versus region-wide deployment and under what operational conditions?
7. Which ATDM strategy or combination will yield the most benefits through changes in short-term behaviors versus long-term behaviors and under what operational conditions?

Answers are being obtained by using focused testbeds in California, Arizona, Texas, and Illinois as shown in Figure 3.27 and Table 3.10 [28]. Although not indicated in Table 3.10, an additional testbed in San Diego, CA is analyzing potential ATDM applications that include queue warning, speed harmonization, intelligent signal control, dynamic lane use

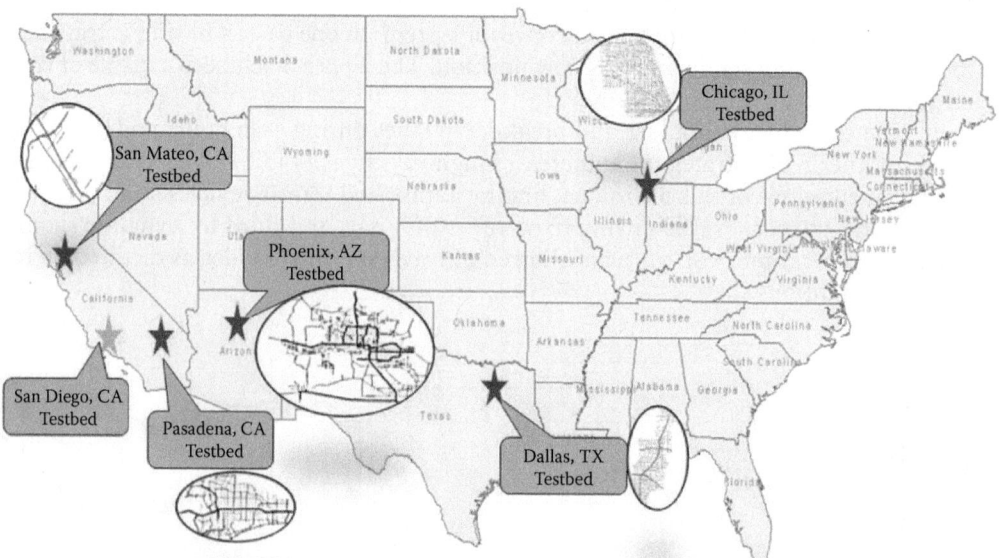

Figure 3.27 ATDM testbeds in the United States. (From Active Transportation and Demand Management Webinar Series, Webinar #5: National ATDM Program Research, U.S. Department of Transportation, Federal Highway Administration, Washington, DC, March 26, 2015.)

Table 3.10 ATDM testbed applications

ATDM strategy	Application	San mateo	Phoenix	Dallas (ICMᵃ)	Pasadena	Chicago
Active traffic management	Dynamic shoulder lanes	–	–	√	√	√
	Dynamic lane use control	–	–	–	√	√
	Dynamic speed limits	–	–	–	√	√
	Queue warning	–	–	–	√	–
	Adaptive ramp metering	–	√	√	√	–
	Dynamic junction control	–	–	–	√	–
	Adaptive traffic signal control	–	√	√	√	√
Active demand management	Predictive traveler information	–	√	√	–	√
	Dynamic routing	–	√	√	√	√
Active parking management	Dynamic priced parking	–	–	√	–	–
Dynamic mobility	DMAᵇ program evaluation	√	–	–	–	–

[a] Integrated corridor management.
[b] Dynamic mobility Intelligent Network Flow Optimization (INFLO) application consisting of queue warning, speed harmonization, and cooperative adaptive cruise control, and the Multi-Modal Intelligent Traffic Signal Systems application.

control, dynamic speed limits, dynamic merge control, predictive traveler information, managed lanes, and dynamic routing. Additional ATDM references available from the FHWA are found in [33–41].

3.12 TRAFFIC DATA COLLECTION

Prior to the advent of ITS with its emphasis on sensors and communications, traffic management agencies relied on floating cars, manual observations, and manual recording of traffic flow data to manage traffic. ITS allows the automatic monitoring of roadways using modern sensors and communications media to transmit even more information to a traffic management and operations center as summarized in Table 3.11. Information is also received from travelers by reading the MAC addresses on their personal cellular phones, toll-tag transponders, and GPS devices. For example, travel times and speeds can be measured by re-identifying the MAC address on a Bluetooth-enabled device after a traveler has moved some distance or by using license plate readers. Traffic volumes and origin–destination pairs can also be obtained from these devices. Connected vehicles using vehicle-to-vehicle (V2V) and vehicle-to-infrastructure (V2I) communications will provide even more data and information to road network operators.

Much of the information collected to execute the real-time functions of a traffic management system is also available as a valuable data resource for operations analyses and reviews, new construction and safety planning, highway research, and other administrative and planning services as indicated in Table 3.11. These offline applications support a broad spectrum of stakeholders, each requiring specific data, accuracy, precision, and spatial and temporal resolution.

Table 3.12 lists several overhead sensor applications to traffic management, the assumptions or conditions that lead to an appropriate selection of a sensor, and the types of technologies that meet these criteria. Purchase, installation, and maintenance costs have not been included in determining the sensor options presented in this table. This discussion continues in Chapter 5 where sensor technologies are described.

Table 3.11 Traffic data collection methods pre- and post-ITS

Application	Data required	Pre-ITS approach	ITS approach	Benefits
Planning and operations analysis	Flow rate	Manual counts	Wide area detection sensors such as VDSs and microwave radar	Increased accuracy Reduced cost Increased count accuracy by lane
	Approach flow rate	Pneumatic tubes	Advanced portable sensors	
Vehicle classification	Axle counts	Manual surveys	Inductive loop-piezo systems Lidar sensors	Reduced costs Long-term data collection
	Weight	Staffed weight scales	Weigh-in-motion sensors	Reduced delay for truck drivers
	Vehicle height Vehicle length		Lidar sensors Ultrasonic sensors Automatic vehicle identification systems, for example, electronic tags	Increased compliance Automated overweight fee collection
Speed data Traffic surveys Speed limit compliance	Minimum speed Maximum speed Mean speed Median speed 85th percentile speed	One- or two-loop (or rode tube) configurations Staffed roadside radar	Automated radar and camera systems	Improved highway speed monitoring Increased safety
Parking facility management	Parking occupancy via entrance and exit counts Space availability	Parking attendant	Automatic vehicle identification Sensor counting technologies	Automated fee collection Automated parking availability information Efficient management of parking resources
Roadway hazard identification	Ice, snow, fog, surface water, end of queue, over height and weight vehicle, vehicle identification	Visual observation	Automated environmental sensing (RWIS)[a] Wide area detection sensors including VDSs Fog sensors Weigh in motion	Increased safety Improved traffic management capability

[a]   Road-weather information systems.

3.13 CONCLUSIONS

Traditional and evolving ITS applications and strategies are dependent on the availability of accurate and timely sensor data. While early ITS implementations relied on roadside sensor data, newer sources of data, such as from MAC address, toll-tag, and license plate readers, are available and even more will be forthcoming as connected and cooperative vehicle programs are deployed and expanded. Traffic management system owners and operators should be aware of all data resources and design their systems such that they can avail themselves of the data no matter what their source. Furthermore, provisions should be made to archive traffic data for later use in planning and forecasting future needs.

Table 3.12 Overhead sensor technology applications to traffic management

Application	Assumptions	Overhead sensor technologies[a]
Signalized intersection control	Detect stopped vehicles Weather not a major factor	VDSs Presence-detecting microwave radar Passive infrared Lidar
	Detect stopped vehicles Inclement weather	Presence-detecting microwave radar
Traffic adaptive signal control (real time)	6-ft × 6-ft inductive loops are most frequently used Sensor locations are dependent on specific control algorithm	VDSs[b] Presence-detecting microwave radar[b] Passive infrared[b]
Direction of vehicle travel	Detection of stopped vehicles not required	VDSs Doppler microwave sensor Presence-detecting microwave radar Lidar with two detection zones
Vehicle counting (surface street or freeway)	Detect and count vehicles traveling at speeds >2.5–5 mi/h (4–8 km/h)	VDSs Presence-detecting microwave radar Doppler microwave sensor Passive infrared
Vehicle speed measurement	Detect and count vehicles traveling at speeds >2.5–5 mi/h (4–8 km/h)	VDSs Presence-detecting microwave radar Doppler microwave sensor Lidar
Vehicle classification	By length	VDSs[c] Presence-detecting microwave radar[c] Lidar
	By length, weight, and axle count By profile	Inductive loop-piezo sensor system Lidar Inductive loop with high-frequency excitation, special detector card, and signal processing software

[a] Purchase, installation, and maintenance costs have not been included in determining the sensor options presented in this table.
[b] Algorithm must be recalibrated to account for difference in the size of the detection zones of the inductive loops and the overhead sensors.
[c] Limited number of length bins.

REFERENCES

1. *Highway Capacity Manual, Special Report 209*, 3rd Ed., Transportation Research Board, National Research Council, Washington, DC, 1994.
2. R.L. Gordon, R.A. Reiss, H. Haenel, E.R. Case, R.L. French, A. Mohaddes, and R. Wolcottet, *Traffic Control Systems Handbook*, FHWA-SA-95-032, Federal Highway Administration, U.S. Department of Transportation, Washington, DC, February 1996.
3. L.A. Klein, *Sensor Technologies and Data Requirements for ITS*, Artech House, Norwood, MA, June 2001.
4. L.A. Klein, D. Gibson, and M.K. Mills, *Traffic Detector Handbook: Third Edition*, FHWA-HRT-06-108 (Vol. I) and FHWA-HRT-06-139 (Vol. II), Federal Highway Administration, U.S. Department of Transportation, Washington, DC, October 2006. www.fhwa.dot.gov/publications/research/operations/its/06108/06108.pdf and www.fhwa.dot.gov/publications/research/operations/its/06139/06139.pdf. Accessed December 13, 2013.

5. E. Curtis, Adaptive Signal Control Technology Overview. http://www.fhwa.dot.gov/everyday-counts/events/trb/docs/asct/asct_breakout_overview_11232010.pdf.
6. A. Stevavovic, *Adaptive Traffic Control Systems: Domestic and Foreign State of Practice*, NCHRP Synthesis 403, National Cooperative Highway Research Program, Transportation Research Board, Washington, DC, 2010.
7. G. Hatcher, C. Burnier, E. Greer, D. Hardesty, D. Hicks, A. Jacobi, C. Lowrance, and M. Mercer, *Intelligent Transportation Systems Benefits, Costs, and Lessons Learned: 2014 Update Report*, FHWA-JPO-14-159, ITS Joint Program Office, Washington, DC, June 2014.
8. J.A. Lindley, Applying systems engineering to implementation of adaptive signal control technology, *Proceedings of 19th ITS World Congress*, Vienna, Austria, October 25, 2012.
9. R. Chandra and C. Gregory, *InSync Adaptive Traffic Signal Technology: Real-Time Artificial Intelligence Delivering Real-World Results*, Rhythm Engineering, Lenexa, KS, March 2012.
10. R.L. Gordon and W. Tighe, *Traffic Control Systems Handbook*, FHWA Report FHWA-HOP-06-006, Dunn Engineering Associates, Federal Highway Administration, Washington, DC, October 2005.
11. P. Koonce, L. Rodegerdts, S. Quayle, S. Beaird, C. Braud, J. Bonneson, P. Tarnoff, and T. Urbanik, *Traffic Signal Timing Manual*, Kittelson & Associates, Inc., Report FHWA-HOP-08-024, Federal Highway Administration, Washington, DC, June 2008.
12. D. Gettman, E. Fok, E. Curtis, K. Kacir, D. Ormand, M. Mayer, and E. Flanigan, *Measures of Effectiveness and Validation Guidance for Adaptive Signal Control Technologies*, FHWA-HOP-13-031, U.S. Department of Transportation, Federal Highway Administration, Office of Operations, Washington, DC, July 2013.
13. K. Fehon, M. Krueger, J. Peters, R. Denney, P. Olson, and E. Curtis, *Model Systems Engineering Documents for Adaptive Signal Control Technology Systems—Guidance Document*, FHWA-HOP-11-27, U.S. Department of Transportation, Federal Highway Administration—HOTM, Washington, DC, 94, August 2012.
14. Running the reds, *Traffic Technology International*, 35–39, October/November 2014.
15. Turning Off Red Light Cameras Costs Lives, New Research Shows, Insurance Institute for Highway Safety, *IIHS News*, July 28, 2016. http://www.iihs.org/iihs/news/desktopnews/turning-off-red-light-cameras-costs-lives-new-research-shows. Accessed November 30, 2016.
16. N.J. Garber, J.S. Miller, R.E. Abel, S. Eslambolchi, and S.K. Korukonda, *The Impact of Red Light Cameras (Photo-Red Enforcement) on Crashes in Virginia*, FHWA/VTRC 07-R2, Virginia Transportation Research Council, Charlottesville, VA, June 2007.
17. J. Masters, Green light for safety cameras, *ITS International*, NAFTA5–NAFTA7, September/October 2016.
18. Q&A: Roundabouts, Insurance Institute for Highway Safety, Arlington, VA, July 2012.
19. W. Hu, Public opinion, traffic performance, the environment, and safety after construction of double-lane roundabouts, *Proceedings of the Transportation Research Board 93rd Annual Meeting*, Transportation Research Board, Washington, DC, January 2014.
20. *Active Transportation and Demand Management Webinar Series*, Webinar #3: Ramp Metering Benefits, Opportunities, and Keys for Overcoming Common Challenges, U.S. Department of Transportation, Federal Highway Administration, Washington, DC, December 10, 2014.
21. L. Jacobson, J. Stribiak, L. Nelson, and D. Sallman, *Ramp Management and Control Handbook*, FHWA-HOP-06-001, Office of Transportation Management, Federal Highway Administration, Washington, DC, January 2006.
22. *Manual on Uniform Traffic Control Devices*, Federal Highway Administration, Washington, DC, 2003 Edition.
23. *Freeway Management System—Design Guidelines*, Kimley-Horn and Associates, Arizona Department of Transportation, 2002.
24. Discussions re: Ramp Meter Warrants Development Process, Minnesota Department of Transportation, St. Paul, MN, 2009.
25. Ramp Meter Design Manual, California Department of Transportation, Traffic Operations, Sacramento, CA, 2000.

26. *Ramp Meter Guideline 2—Localized Freeway Traffic Issues, Planning Guidance for the Installation and Use of Technology Devices for Transportation Operations and Maintenance*, ENTERPRISE Transportation Pooled Fund Study. http://enterprise.prog.org/itswarrants/ramp-meter2.html. Accessed June 7, 2015.

27. JHK and Associates, *Intelligent Vehicle Highway Systems: The State of the Art*, Prepared for Massachusetts Department of Highways, March 1993.

28. *Active Transportation and Demand Management Webinar Series*, Webinar #5: National ATDM Program Research, U.S. Department of Transportation, Federal Highway Administration, Washington, DC, March 26, 2015.

29. B. Kuhn, D. Gopalakrishna, and E. Schreffler, *The ATDM Program: Lessons Learned*, FHWA-HOP-13-018, U.S. Department of Transportation, Federal Highway Administration, Washington, DC, March 2013.

30. R. Dowling, R. Margiotta, H. Cohen, and A. Skabardonis, *Guide for Highway Capacity and Operations Analysis of ATDM Strategies*, FHWA-HOP-13-042, U.S. Department of Transportation, Federal Highway Administration, Washington, DC, June 2013.

31. I.B. Potts, D.W. Harwood, C.A. Fees, K.M. Bauer, and C.S. Kinzel, *Further Development of the Safety and Congestion Relationship for Urban Freeways*, SHRP 2 Report S2-L-07-RRB-3, Transportation Research Board, Washington, DC, 2015. http://onlinepubs.trb.org/onlinepubs/shrp2/SHRP2_S2-L07-RR-3.pdf.

32. P. Jenior, R. Dowling, B. Nevers, and L. Neudorff, *Use of Freeway Shoulders for Travel—Guide for Planning, Evaluating, and Designing Part-Time Shoulder Use as a Traffic Management Strategy*, FHWA-HOP-15-023, U.S. Department of Transportation, Federal Highway Administration Office of Operations, Washington, DC, January 2016.

33. W. Berman, D. Differt, K. Aufschneider, P. DeCorla-Souza, A. Flemer, L. Hoang, R. Hull, E. Schreffler, and G. Zammit, *Managing Travel Demand: Applying European Perspectives to US Practice*, FHWA-PL-06-015, U.S. Department of Transportation, Federal Highway Administration, Washington, DC, May 2006.

34. M. Mirshahi, J. Obenberger, C.A. Fuhs et al., *Active Traffic Management: The Next Step in Congestion Management*, FHWA-PL-07-012, U.S. Department of Transportation, Federal Highway Administration, Washington, DC, March 2007.

35. C. Fuhs, *Synthesis of Active Traffic Management Experiences in Europe and the United States*, FHWA-HOP-10-031, U.S. Department of Transportation, Federal Highway Administration, Washington, DC, May 2010.

36. D. Sallman, E. Flanigan, K. Jeannotte, C. Hedden, and D. Morallos, *Operations Benefit/Cost Analysis Desk Reference*, FHWA-HOP-12-028, U.S. Department of Transportation, Federal Highway Administration, Washington, DC, May 2012.

37. M. Kaufman, M. Formanack, J. Gray, and R. Weinberger, *Dynamic Parking Pricing Primer*, FHWA-HOP-12-026, U.S. Department of Transportation, Federal Highway Administration, Washington, DC, May 2012.

38. D. Gopalakrishna, E. Schreffle, D. Vary, D. Friedenfeld, B. Kuhn, C. Dusza, R. Klein, and A. Rosas, *Integrating Demand Management Into the Transportation Planning Process: A Desk Reference*, FHWA-HOP-12-035, U.S. Department of Transportation, Federal Highway Administration, Washington, DC, August 2012.

39. J. Atkinson, J. Bauer, K. Hunt, K. Mullins, M. Myers, E. Rensel, M. Swisher, and R. Taylor, *Designing for Transportation Management and Operations: A Primer*, FHWA-HOP-13-013, U.S. Department of Transportation, Federal Highway Administration, Washington, DC, February 2013.

40. Texas A&M Transportation Institute, Battelle, Kimley-Horn and Associates, Inc., and Constance Sorrell, *Planning and Evaluating Active Traffic Management Strategies*, NCHRP 03-114, September 2014.

41. A. Mizuta, K. Roberts, L. Jacobsen, and N. Thompson, *Ramp Metering Primer*, FHWA-HOP-14-020, U.S. Department of Transportation, Federal Highway Administration, Washington, DC, October 2014.

Chapter 4

Sensor data requirements

Several methods may be utilized to develop sensor requirements and sensor specifications. These include a formal systems engineering analysis of the traffic management or transportation system; referring to studies and reports issued by federal, state, county, and city agencies; and relying on results of tests performed by universities and other entities that meet or approximate the operating conditions of your project. The techniques and the information they provide are explored in this chapter and also in Chapter 8 that describes the results of several sensor field tests. Chapter 13 provides additional details concerning the systems engineering process.

4.1 SYSTEMS ANALYSIS APPROACH TO DETERMINING SENSOR SPECIFICATIONS

Figure 4.1 displays a formal systems analysis process for developing traffic sensor requirements, which begins with the identification of the overall transportation system requirements, shown as inputs to the systems analysis phase at the top left side of the figure [1,2]. Sensor requirements are normally a derivative of the systems engineering process that determines the design of an ITS architecture and its related components. During systems analysis, a detailed investigation is performed to identify and specify all the subsystems, including the detection subsystem, that are part of the traffic management system. Potential alternative solutions are examined to find the one that best meets stakeholder needs and preestablished performance measures [3,4].

Once the systems analysis phase is complete, the detection subsystem design phase can begin. Its major tasks, illustrated on the right side of the figure, are location of the sensor stations, selection of the sensor technologies (there may be more than one), and definition of sensor station configurations. The selection of the sensor technologies may include consideration of the specific signal processing algorithms that will be used, types of data required for the application or algorithm (count, speed, occupancy, queue length, etc.), data accuracies, data collection interval, and spatial resolution and coverage area.

Figure 4.2 enumerates further steps that are often involved in choosing a sensor technology, namely, consideration of stakeholder desires and restrictions, selection of sensor screening criteria, evaluation of candidate sensor technologies and models, installing and verifying sensor operation, maintaining the sensor system, and monitoring its performance [5]. If no evaluation results are available for the selected sensor models, sensor field tests are needed to ensure that the sensors will function properly under all anticipated vehicle mix, weather, lighting, and traffic flow conditions that characterize the particular facility for which they were selected, for example, congested limited-access highways and arterials, tunnels, bridges, hills, and curved sections. A testbed also can be used to compare the performance of different sensor technologies under the same operating conditions. The limitation of a testbed is that it may not be able to replicate the full range of operational conditions, which

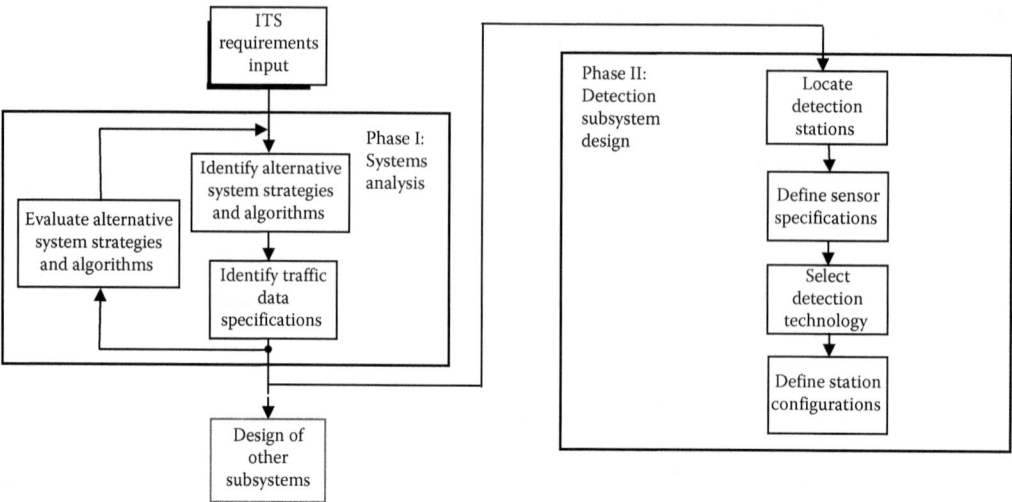

Figure 4.1 Systems analysis process for developing traffic sensor requirements.

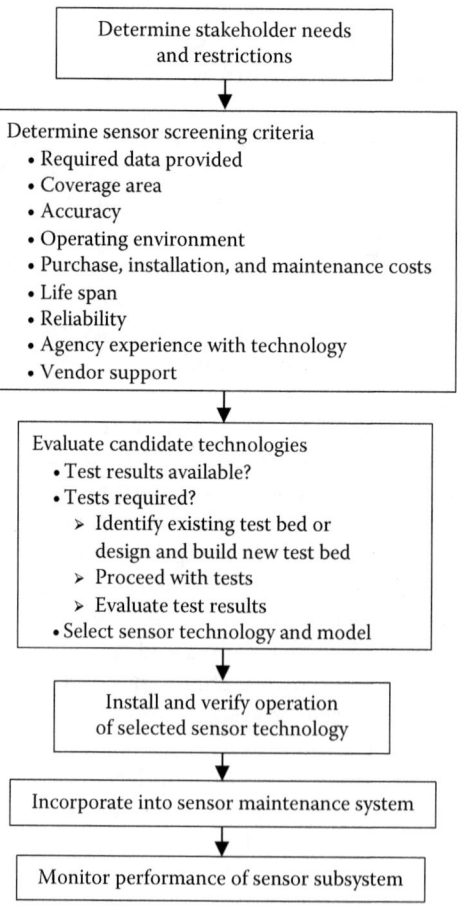

Figure 4.2 Sensor selection, installation, and maintenance process.

can change significantly from site to site. However, with practice, departments of transportation or their contractors can learn to quickly set up test facilities on different roads and highways that encompass a wide range of operational conditions. Chapter 9 describes tools that assist with the preparation of sensor specifications and testing protocols.

Improved testing and installation procedures may be needed to assure that new or upgraded detection stations are installed and operating properly. There are a number of factors that lead to faulty installation of detection equipment. These include limited available time in the right of way, lack of expertise with the particular detection technology among contractors, and sequencing of construction activity that leaves sensor installation and testing until the end. The last factor often limits the time available to properly inspect and test equipment before projects are accepted.

Life-cycle costs that include maintenance and expected years of service should also be considered when making a sensor selection. Several sensor technologies promise reduced costs of installation and maintenance. Since many transportation agencies have limited experience with these technologies, they frequently rely on vendor-supplied information instead of conducting their own rigorous testing program.

4.2 CALTRANS SENSOR SPECIFICATIONS FOR REAL-TIME TRAFFIC MANAGEMENT OPERATIONS, PLANNING, AND TRAVELER INFORMATION

Tables 4.1 through 4.3 show the California Department of Transportation (Caltrans) real-time data accuracy requirements for traffic operations, planning, and traveler information, respectively [6]. There is significant overlap in all three categories. For example, the weather-related data listed under the traveler information category is also useful to operations personnel. Required data elements are listed along with the definition of the units. System accuracy estimates originally presented in a single column were split into three groups: accuracy as currently practiced (based primarily on user needs interviews), state-of-the-art sensor accuracy or tolerance, and system accuracy desired by Caltrans, which is based on previous detection plans and user needs interviews.

These tables can be used as a quick reference guide by transportation management personnel involved in the planning, procurement, and design of detection stations. As a formalized, rigorous independent testing and evaluation process is implemented and results are accumulated, the state-of-the-art accuracies and tolerances can be estimated and expectations adjusted accordingly. The accuracy desired by the agency to implement a particular function can also be tailored to meet its particular needs. Overstating this accuracy should be avoided to the extent possible as this may increase the cost of the sensor and costs associated with analyzing the data.

4.3 TRAFFIC PARAMETER ACCURACIES FROM FHWA DETECTION TECHNOLOGY FOR IVHS PROGRAM

Table 4.4 summarizes typical accuracies for traffic parameters that were studied in FHWA's Detection Technology for IVHS Program in the early to mid-1990s [1]. They apply to signalized intersection control, freeway incident management, and freeway ramp metering. The accuracies are meant to be rules of thumb and may have to be modified for specific applications, depending on the particular details of the data processing algorithm being used.

Table 4.1 Data accuracy for real-time operations

Symbol	Primary use of data	Required data	Units	System accuracy or tolerance as used in current practice	System accuracy or tolerance—state of the art[a]	System accuracy or tolerance desired by Caltrans[b]
O-1	Traffic signal actuation	Presence	Yes/no	100% (goal)		99.8%
O-2	Ramp metering actuation	Presence	Yes/no	100% (goal)		99.8%
O-3	Local responsive ramp metering	Occupancy	Percent of time vehicle present	±2% at 25% occupancy		±1% per 30-s interval
O-4	System-wide ramp metering and incident detection	Speed	mi/h	±5%		±5 mi/h per 30-s interval
O-5	Mainline metering	Classification	Percent of trucks in vehicle volume	90%–95%		±1% of volume for trucks
O-6	System-wide incident detection, system-wide and local traffic responsive ramp metering, mainline speed estimation	Occupancy	Percent of time vehicle present	±1% at 25% occupancy per 20- or 30-s interval		±1% per 30-s interval
O-7	Interconnected intersection control	Occupancy	Percent of time vehicle present	±2.5% at 25% occupancy per 20- or 30-s interval		
O-8	Ramp metering and local isolated intersection control	Headway	Seconds	±0.05 s (based on sampling interval of 0.1 s)		±0.1 s
O-9	System-wide incident detection and ramp metering	Volume	Number of vehicles per hour	Ramp metering: ±2 veh/min at 2000 veh/h for upstream volume; equivalent to 94% at 2000 veh/h. Incident detection: ±1 veh/min at 2000 veh/h; equivalent to 97% at 2000 veh/h		95% per 30-s interval. 95% per 30-s interval
O-10	Local traffic responsive ramp metering	Volume	Number of vehicles per hour	±2 veh/min at 2000 veh/h for upstream volume; equivalent to 94% at 2000 veh/h		

(Continued)

Table 4.1 (Continued) Data accuracy for real-time operations

Symbol	Primary use of data	Required data	Units	System accuracy or tolerance as used in current practice	System accuracy or tolerance— state of the art	System accuracy or tolerance desired by Caltrans
O-11	Interconnected intersection control—timing plan selection	Volume	Number of vehicles per hour	±2.5% at 600 vehicles per hour per lane (vphpl)		
O-12	Interconnected intersection control—critical intersection control (CIC) where timing is set to nearest second	Volume	Number of vehicles per hour	±1 veh per lane at 600 vphpl		
O-13	Tolls	Classification	Percent of trucks in vehicle volume	>99% (driven by revenue considerations)		±1% of volume for trucks
O-14	Vehicle probes, emergency vehicle response	Vehicle location	GPS coordinates		1–3 m at 95% confidence level, horizontal	4 m at 95% confidence level, horizontal[c] (3 m)[d]
O-15	Timely incident clearance	Incident detection	Distance from nearest road-side mileage marker or other known object			Identify location and severity within 2 min of occurrence

Source: Cambridge Systematics, Inc. and L.A. Klein, Data quality evaluation memorandum, In *Statewide Detection Plan*, California Department of Transportation, Sacramento, CA, July 2, 2008.

[a] State-of-the-art sensor accuracy or tolerance is found from manufacturer- and vendor-independent evaluations of sensor products. These evaluations may be conducted by independent facilities run by universities, government or private laboratories, or transportation agencies. The state-of-the-art sensor accuracy or tolerance is a function of the hardware and software versions that are in the device when it is tested. The state-of-the-art sensor accuracy or tolerance needs to be revaluated at intervals representative of the product improvement cycle for that device.

[b] TMS Detection Plan, December 2002. Prepared by IBM for Caltrans.

[c] Teleconference, March 28, 2008 with Martha Styer (Caltrans), Joe Palen (Caltrans), Dan Krechmer (Cambridge Systematics), and Lawrence Klein (Consultant to Cambridge Systematics).

[d] Data Detection Quality—An Informational Report for the Caltrans Traffic Management System Business Plan, Cambridge Systematics, 2003.

Table 4.2 Data accuracy for planning

Symbol	Primary use of data	Required data	Units	System accuracy or tolerance as used in current practice	System accuracy or tolerance— state of the art[a]	System accuracy or tolerance desired by Caltrans[b]
P-1	Traffic studies (for hi-resolution modeling)	Volume	Number of vehicles per hour	>95%		95% per 30-s interval
P-2	Census	Volume	Number of vehicles per hour	To be coordinated with other studies		95% per 30-s interval
P-3	Performance measurement	Volume	Number of vehicles per hour	90%		95% per 30-s interval
P-4	Performance measurement	Speed	mi/h	±5%		±5 mi/h per 30-s interval
P-5	Pavement and truck studies, permits	Weight	lbs/axle			95%
P-6	Pavement and truck studies, permits	Height	in.			±4 in.
P-7	Truck counts, pavement studies, permits, vehicle class	Classification	Percent of trucks in vehicle volume	90–95%		±1% of volume for trucks
P-8	Determination of trip or route origin and destination (O/D), fleet management	O/D pairs	Location of entry and exit points for freeways and major arterials	≈90%		95% between adjacent detection points
P-9	Predictive traffic models	O/D pairs	Location of entry and exit points for freeways and major arterials			Based on model needs for adjacent detection point accuracy

Source: Cambridge Systematics, Inc. and L.A. Klein, Data quality evaluation memorandum, In *Statewide Detection Plan*, California Department of Transportation, Sacramento, CA, July 2, 2008.

[a] State-of-the-art sensor accuracy or tolerance is found from manufacturer- and vendor-independent evaluations of sensor products. These evaluations may be conducted by independent facilities run by universities, government or private laboratories, or transportation agencies. The state-of-the-art sensor accuracy or tolerance is a function of the hardware and software versions that are in the device when it is tested. The state-of-the-art sensor accuracy or tolerance needs to be revaluated at intervals representative of the product improvement cycle for that device.

[b] TMS Detection Plan, December 2002. Prepared by IBM for Caltrans.

Table 4.3 Data accuracy for traveler information

Symbol	Primary use of data	Required data	Units	System accuracy or tolerance as used in current practice	System accuracy or tolerance— state of the art[a]	System accuracy or tolerance desired by Caltrans[b]
T-1	Incident management, traveler information	Road surface temperature	°F or °C		±0.2°C	95%
T-2	Incident management, traveler information	Road surface condition identification	Yes/no (qualitative)			95%
T-3	Incident management, traveler information	Water depth, ice or snow depth, amount of "melting" chemical present	cm or percent (quantitative)		Greater of ±1 cm or ±0.4% of distance to vehicle	95%
		Rainfall	in.		±1% at ≤2 in./h or less	
		Visibility	Meteorological optical range in km[c]		±2% to ±20%	95%
T-4	Traveler information	Travel time	Minutes or seconds	≈90%		±20% 90% of the time
T-5	Traveler information	Speed	mi/h			Accuracy consistent with T-4 travel time calculation requirement
T-6	Incident management, traveler information	Air temperature	°C		±0.3°K	95%
T-7	Incident management, traveler information	Relative humidity	Percent		±1.5%	95%

Source: Cambridge Systematics, Inc. and L.A. Klein, Data quality evaluation memorandum, In *Statewide Detection Plan*, California Department of Transportation, Sacramento, CA, July 2, 2008.

[a] State-of-the-art sensor accuracy or tolerance is found from manufacturer- and vendor-independent evaluations of sensor products. These evaluations may be conducted by independent facilities run by universities, government or private laboratories, or transportation agencies. The state-of-the-art sensor accuracy or tolerance is a function of the hardware and software versions that are in the device when it is tested. The state-of-the-art sensor accuracy or tolerance needs to be revaluated at intervals representative of the product improvement cycle for that device.

[b] TMS Detection Plan, December 2002. Prepared by IBM for Caltrans.

[c] Meteorological optical range (MOR) is calculated by the user by converting the received signal strength (which is given by the instrument in terms of extinction coefficient γ) using Koschmeider's formula: MOR (km) = $3.91/\gamma$, where γ is in units of Np/km.

Table 4.4 Sensor requirements for applications studied during the Detection Technology for IVHS Program

Application	Use of data	Data collection interval	Parameter and accuracy
Local isolated intersection control	Inter-vehicle gap detection on intersection approach	Sampled every 0.1 s	Detect inter-vehicle gaps of ≈3–4 s duration to an accuracy of ±0.05 s
	Stopline presence and passage detection	Sampled every 0.1 s	100% vehicle detection
Interconnected intersection control	Timing plan selection	5 min or signal cycle	Flow rate within ±2.5% at 600 vplph[a] Occupancy within ±2.5% at 25% occupancy
	System performance measure of effectiveness	5 or 15 min	Flow rate, average vehicle length, occupancy within ±10% Average vehicle speed within ±5%
	Critical intersection control where goal is to set signal timing to nearest second	Signal cycle	Flow rate to ±1 vehicle per lane at 600 vplph[a] Occupancy within ±2.5% at 25% occupancy
Traffic adaptive intersection control	SCOOT split optimization that requires generation of traffic flow cyclic profiles	Signal cycle	Vehicle detection within ±2 veh/cycle for 90% of the signal cycles
Freeway incident management	Incident management and decision support	5 min	Vehicle detection within ±1 vehicle for 90% of the of the 5-min intervals
	Incident detection algorithms	20 or 30 s	Occupancy within ±1% at 25% occupancy Flow rate within ±1 veh/min at 2000 veh/h
Freeway ramp metering	Ramp metering based on mainline traffic flow	1 min	Downstream occupancy within ±2% at 25% occupancy Upstream flow rate within ±2 veh/min at 2000 veh/h Upstream occupancy within ±2% at 25% occupancy

[a] vplph, vehicles per lane per hour.

4.3.1 Data requirements for incident detection algorithms

Four incident detection algorithms were analyzed for their input data requirements, namely, the California comparative-type algorithms, the McMaster algorithm, time series algorithms, and the high occupancy (HIOCC) algorithm. Knowing how these incident detection algorithms function allows their input data needs to be specified, as shown in Table 4.5 [7].

The incident detection logic in California comparative-type algorithms is based on spatial variations in lane-specific values of occupancy or speed between successive upstream and downstream detection stations for a given direction of travel [8,9]. Multiple thresholds are utilized to enhance the probability that the occupancy measurements correspond to a true incident. Occupancy values derived from inductive loop detector data normally range between 10% and 30% when traffic conditions vary from level of service B to level of service E. The required accuracy of the occupancy measurements is to the nearest 1%.

Modifications to the California algorithms resulted in more reliable incident detection. The modifications included a decision tree that determines whether the shock compression wave is caused by recurring congestion or an incident, a persistence check to require a traffic discontinuity to continue for a specified time before the incident alarm is given, and a procedure that

Table 4.5 California, McMaster, time series, and high-occupancy incident detection algorithm input data and characteristics

	Traffic variables			Data collection interval		Number of stations	
Algorithm	Flow rate	Occ.	Speed	I s	30–60 s	Single	Adjacent
Comparative		•			•		•
McMaster	•	•	•		•	•	•
Time series	•	•			•	•	•
HIOCC		•		•		•	

Source: Y.J. Stephanedes, A.P. Chassiakos, and P.G. Michalopoulos, *Transportation Research Record* 1360, Transportation Research Board, 50–57, 1993.

detects incidents in light-to-moderate traffic. The original algorithms were further augmented by filtering the input data, optimizing the thresholds with artificial neural networks and Bayesian inference to maximize the probability of a correct decision, and other improvements [2].

The McMaster algorithms are based on catastrophe theory. They utilize differences in parameter values at the same station at successive time intervals or at adjacent stations to detect an incident. The algorithm separates traffic flow parameters such as volume, occupancy, and speed into different states that are typical of congested and uncongested traffic flow [10,11]. If significant changes in the traffic state are observed, an alarm is triggered. For instance, an incident is declared if the average speed at the sensor station is below a specified threshold and the combination of volume and occupancy indicates congestion. Such a congested condition occurs if a high-occupancy threshold and a low-volume threshold are exceeded for some number of consecutive sampling periods. If only one criterion is satisfied for the given number of periods, then another sample is required before the incident is declared. Less efficient operation occurs without the speed information. Volume is needed to within ±1 veh/min at flows of 2000 veh/h. The required lane occupancy accuracy is ±1%.

The advantages of using data from a single detection station are twofold:

1. Maintenance of the same sensitivity and tuning at adjacent sensor stations is not critical. In fact, sensor failures at adjacent stations do not have any effect beyond their own location.
2. Sensors based on different technologies may be used at different sensor stations. Calibration procedures for the McMaster algorithm are site specific in any case.

Time series algorithms use statistical indicators derived from volume and occupancy to provide short-term forecasts of traffic parameters. An alarm is triggered when a significant deviation occurs between observed and forecast traffic parameter values [12–16].

The HIOCC algorithm triggers an alarm when the occupancy increases abnormally during consecutive 1-s observation intervals [17].

4.3.2 Sensor specifications for future ITS applications

Sensor specifications for three future ITS applications were also examined as part of the FHWA Detection Technology for IVHS Program. These were signalized intersection control, freeway incident detection and management, and freeway metering control. Each application had three categories of specifications, tactical, strategic, and historic, depending on the time frame over which the data are used.

Tactical parameters are generally collected over short time intervals (usually of the order of a few seconds) since tactical decisions are made in quick response to changing real-time traffic variables. Because of the shorter intervals, fewer vehicles are included in each sample and variation from sample to sample occurs due to the random nature of vehicle arrivals. The limited sample size usually imposes increased accuracy and precision on the measurement of tactical parameters.

Strategic traffic data parameters support strategic-level planning. Strategic-level decisions generally operate at a higher level in the system hierarchy and are often broader in geographic scope than tactical decisions. For example, strategic decisions activate preplanned traffic management strategies that respond to broad indicators of traffic flow conditions and thus change the mode of an entire system or a large subsystem. Strategic traffic parameters are collected over a period of minutes rather than seconds; therefore, the sample size is larger.

Historic traffic data parameters maintain or update online historic traffic databases. These databases typically include information collected over periods of 5 min or greater and are archived by time-of-day and day-of-week or by time-of-day and date. The primary purpose of historic databases is to provide information for offline planning and design operations. Historic data also find application as inputs to online tactical and strategic decision processes as, for example, to predict future near-term traffic flow demands.

Available literature contains limited information regarding the required accuracy of traffic parameters that support these ITS applications. Consequently, the traffic parameter ranges, collection intervals, and accuracies for signalized intersection control, freeway incident detection and management, and freeway metering presented in Tables 4.6 through 4.8 are based on (1) values derived or inferred from those needed to support an existing algorithm (when one was known), (2) experience with current operating systems, and (3) sensitivity analyses developed during the Detection Technology for IVHS Program or found in the literature [1,2]. However, a detailed analysis is recommended to develop the traffic flow data specifications for a specific system design or for other applications not discussed.

Table 4.6 contains the tactical, strategic, and historic data and corresponding accuracies that support innovative approaches to signalized intersection control. The tactical parameters include those relating to flow rate, speed, occupancy, delay, and stop measurements. Typical flow-related parameters may include cyclically collected intersection approach flow rates, flow profile data, and turning flow rates that are normalized into hourly rates. This minimizes the short-term parameter fluctuations caused by inconsistencies between data collection intervals and whole number multiples of cycle length. This issue can also be resolved by maintaining weighted running averages and by other smoothing techniques.

Strategic-level parameters most often used by intersection signal control logic include smoothed flow rate, occupancy, and average speed. Some systems also tabulate parameters such as average approach delay and percent of vehicles stopping or total stops by approach. Strategic data are normally stored as smoothed values (weighted running averages) with time constants ranging from 1 to 5 min. They are computed from data samples generally collected over 10 s to 1 min or 1 cycle. In most instances, strategic flow rate data are collected to tabulate current demands for network links. Similarly, occupancy parameters are regularly used to monitor the extent of current congestion on the roadway network. Strategic traffic parameters can be useful for implementing incident management strategies designed for surface street applications.

Historic parameters for intersection signal control include link-based flow rate, occupancy, and speed. Turning movement and origin–destination pair patterns are also important as inputs to demand prediction algorithms.

Table 4.7 identifies selected traffic flow parameter specifications for freeway incident detection and management. Tactical parameters support automated incident detection

Table 4.6 Signalized intersection control traffic parameter specifications

Tactical parameters	Units	Range	Collection interval	Allowable error
Approach flow profiles	veh	0–3	1 s	±2 veh/signal cycle
Turning movement flow rate	veh	0–200	1 cycle	±2 veh/signal cycle
Average link travel time	s	0–240	1 cycle	±2 s
Average approach speed	mi/h	0–100	1 cycle	±2 mi/h (0–55 mi/h)
Queue length	veh/lane	0–100	1 s	±2 veh
Demand presence	Yes/no	–	10 Hz (minimum)	No missed vehicles
Average approach delay	s/veh	0–240	1 cycle	±2 s
Approach stops	Stops	0–200	1 cycle	±5% of stops

Strategic parameters	Units	Range	Smoothing or filtering interval (min)	Allowable error
Flow rate	veh/h for each approach	0–2500	5	±2.5% at 500 veh/h
Occupancy	%/lane	0–100	5	±5% occupancy
Average speed	mi/h	0–100	5	±2 mi/h (0–55 mi/h)
Average delay	s/veh	0–240	5	±2.5 s
Percent stops	%	0–100	5 (approx.)	±5%

Historic parameters	Units	Range	Collection interval (min)	Allowable error
Turning movement flow rate	veh/movement	0–2000	15	±2.5% at 500 veh/h
Flow rate	veh/h for each approach	0–2500	15	±2.5% at 500 veh/h
Occupancy	%	0–100	15	±5% occupancy
Average speed	mi/h	0–100	15	±2 mi/h (0–55 mi/h)

Source: L.A. Klein and M.R. Kelley, *Detection Technology for IVHS, Final Report*, FHWA-RD-95-100, U.S. Federal Highway Administration, McLean, VA, December 1996. http://ntl.bts.gov/lib/jpodocs/repts_te/6184.pdf.

algorithms. Basic tactical inputs consist of lane-specific mainline flow rate, occupancy, and average speed. Tactical parameters derived from these basic parameters include spatial occupancy differential and spatial average speed differential.

Strategic-level parameters function as traffic monitoring inputs to the overall incident management process, including congestion management, changeable message sign displays, and status information transfer to information service providers. Strategic-level parameters include mainline lane-specific flow rate, occupancy, average speeds, and freeway on-ramp and off-ramp flows. Alternative route data are also collected when applicable. As a minimum, flow rates and link speed or travel times should be maintained for significant alternate routes in the system. Strategic parameters are generally maintained online as 5-min running averages.

Table 4.7 Freeway incident detection and incident management traffic parameter specifications

Tactical parameters (detection)	Units	Range	Collection interval (s)	Allowable error
Mainline flow rate	veh/h for each lane	0–2500	20	±2.5% at 500 veh/h
Mainline occupancy	% (by lane)	0–100	20	±1% occupancy
Mainline speed	mi/h (by lane)	0–80	20	±1 mi/h
Mainline travel time	min	–	20	±5%

Strategic parameters (incident management)	Units	Range	Smoothing or filtering interval (min)	Allowable error
Mainline flow rate	veh/h for each lane	0–2500	5	±2.5% at 500 veh/h
Mainline occupancy	%	0–100	5	±2% occupancy
Mainline speed	mi/h	0–80	5	±1 mi/h
On-ramp flow rate	veh/h for each lane	0–1800	5	±2.5% at 500 veh/h
Off-ramp flow rate	veh/h for each lane	0–1800	5	±2.5% at 500 veh/h
Link travel time	s	–	5	±5%
Current O-D patterns	veh/h	–	5	±5%

Historic parameters (planning)	Units	Range	Collection interval	Allowable error
Mainline flow rate	veh/h for each lane	0–2500	15 min or 1 h	±2.5% at 500 veh/h
Mainline occupancy	%	0–100	15 min or 1 h	±2% occupancy
Mainline speed	mi/h	0–80	15 min or 1 h	±1 mi/h
On-ramp flow rate	veh/h for each lane	0–1800	15 min or 1 h	±2.5% at 500 veh/h
Off-ramp flow rate	veh/h for each lane	0–1800	15 min or 1 h	±2.5% at 500 veh/h
Link travel times	s	–	15 min or 1 h	±5%
Current O-D patterns	veh/h	–	15 min or 1 h	±5%

Source: L.A. Klein and M.R. Kelley, *Detection Technology for IVHS, Final Report*, FHWA-RD-95-100, U.S. Federal Highway Administration, McLean, VA, December 1996. http://ntl.bts.gov/lib/jpodocs/repts_te/6184.pdf.

Historic parameters that support freeway incident detection and management are similar to the strategic-level parameters. Only here they are collected over longer time intervals that range from 15 min to 1 h.

Table 4.8 contains selected traffic flow parameter specifications for freeway metering control. Tactical parameters for this application include queue length estimates, demand presence, passage count, queue overflow presence, mainline flow rate, occupancy, and speed. Queue length as an input parameter to meter rate control algorithms is typically estimated from approach and passage flow rates or derived from data produced by one or more occupancy sensors on the approach to the metering signal. Other tactical inputs to metering control algorithms are mainline occupancy, speed, and flow rate, which are also utilized for freeway incident management.

Strategic parameters for metering include mainline and metered traffic flow rates. Mainline values are typically lane specific and include flow rate, occupancy, and average speeds. Derived average freeway speeds based on flow rate and occupancy data from a single inductive loop detector give reasonable results for strategic decisions because data

Table 4.8 Freeway metering control traffic parameter specifications

Tactical parameters (local responsive control)	Units	Range	Collection interval (s)	Allowable error
Ramp demand	Yes/no	–	0.1	0% (no missed vehicles)
Ramp passage	Yes/no	–	0.1	0% (no missed vehicles)
Ramp queue length	veh	0–40	20	±1 veh
Mainline occupancy	%	0–100	20	±2% occupancy
Mainline flow rate	veh/h for each lane	0–2500	20	±2.5% at 500 veh/h
Mainline speed	mi/h	0–80	20	±5 mi/h

Strategic parameters (central control)	Units	Range	Smoothing or filtering interval (min)	Allowable error
Mainline occupancy	%	0–100	5	±2% occupancy
Mainline flow rate	veh/h for each lane	0–2500	5	±2.5% at 500 veh/h
Mainline speed	mi/h	0–80	5	±5 mi/h

Historic parameters (pretimed options)	Units	Range	Collection interval	Allowable error
Mainline occupancy	%	0–100	15 min or 1 h	±2% occupancy
Mainline flow rate	veh/h for each lane	0–2500	15 min or 1 h	±2.5% at 500 veh/h
Mainline speed	mi/h	0–80	15 min or 1 h	±5 mi/h
On-ramp flow rate	veh/h for each lane	0–1800	15 min or 1 h	±2.5% at 500 veh/h
Off-ramp flow rate	veh/h for each lane	0–1800	15 min or 1 h	±2.5% at 500 veh/h

Source: L.A. Klein and M.R. Kelley, Detection Technology for IVHS, Final Report, FHWA-RD-95-100, U.S. Federal Highway Administration, McLean, VA, December 1996. http://ntl.bts.gov/lib/jpodocs/repts_te/6184.pdf.

smoothing procedures are normally used and collection intervals are typically 5 min or longer. The relatively long collection intervals generally ensure that the data samples contain a representative mix of vehicle lengths and, hence, a satisfactory estimate of average vehicle speed.

Historic parameters of value in freeway metering include on-ramp and off-ramp flow rates and those already identified in the strategic category. The collection intervals for historic data are lengthened to 15 min to 1 h, which are the same as the intervals for freeway incident detection and management.

4.4 SURVEYS FOR DETERMINING SENSOR DATA REQUIREMENTS

Sometimes, the sensor data requirements are not known or need to be verified. Under those circumstances, a survey such as the one outlined in Table 4.9 may be used to solicit input from the stakeholders who will use the data. This activity can produce a complete set of requirements that will assist in sensor selection. The accuracy of the acquired data not only depends on the sensor accuracy, but also depends on deterministic and random errors. The latter are introduced when an estimator for a traffic control parameter is computed

Table 4.9 Survey to determine sensor data requirements

Which types of data are collected at least once a year? Please check as appropriate.

	Urban freeway	Rural freeway	Major arterial	Other arteries	Rural highway
Volume					
Speed					
Classification					
Occupancy					
Weight					

What methods are used to collect these data? 1 = manual, 2 = tube, 3 = loop, 4 = camera imagery only, 5 = VDS, 6 = other (please explain in the cell or comment below)

	Urban freeway	Rural freeway	Major arterial	Other arteries	Rural highway
Volume					
Speed					
Classification					
Occupancy					
Weight					

Comments: _____

Please use a rating scheme from 1 to 5, where 1 is top priority and 5 is lowest priority to rank data needs based on your personal experience and requirements.

	Urban freeway	Rural freeway	Major arterial	Other arteries	Rural highway
Volume					
Speed					
Classification					
Occupancy					
Weight					

If you are in charge of data gathering or prioritization of aggregate needs for your section or department, please perform the ranking below (again on a 1–5 scale); otherwise, go to the next question.

	Urban freeway	Rural freeway	Major arterial	Other arteries	Rural highway
Volume					
Speed					
Classification					
Occupancy					
Weight					

(Continued)

Table 4.9 (Continued) **Survey to determine sensor data requirements**

What coverage should the sensors provide?
1 = each individual lane, 2 = all lanes per direction, 3 = specific lanes only

	Urban freeway	Rural freeway	Major arterial	Other arteries	Rural highway
Volume					
Speed					
Classification					
Occupancy					
Weight					

Some combinations of facilities and data may require a higher accuracy.
1A = highest accuracy all the time, 1P = highest accuracy during peak times, 2A = high accuracy all the time, 2P = high accuracy during peak times, 3A = less accuracy all the time, 3P = less accuracy during peak times, 4A = least accuracy all the time, 4P = least accuracy during peak times.
(For example, highest accuracy may lie at 99% and above, high accuracy in the 95%–99% range, less accuracy in the 90%–95% range, and least accuracy in the 80%–90% range.)
Please indicate your requirements in the cells below

	Urban freeway	Rural freeway	Major arterial	Other arteries	Rural highway
Volume					
Speed					
Classification					
Occupancy					
Weight					

What should the maximum acceptable error be? Specify only for those cells that you have knowledge about.
1 = less than 1%, 2 = 1%–5%, 3 = 5%–10%, 4 = 10%–20%, 5 = other: please specify in the cell.

	Urban freeway	Rural freeway	Major arterial	Other arteries	Rural highway
Volume					
Speed					
Classification					
Occupancy					
Weight					

Specify the need for data by interval.
1 = nearly instantaneous (e.g., at 5-30 s intervals for incident detection or real-time adaptive signal control)
2 = short intervals for storage and analysis (e.g., 1-min interval data)
3 = long intervals for storage and analysis (e.g., 15-min interval data)

	Urban freeway	Rural freeway	Major arterial	Other arteries	Rural highway
Volume					
Speed					
Classification					
Occupancy					
Weight					

(Continued)

Table 4.9 (Continued) Survey to determine sensor data requirements

Specify the need for data by collection frequency.
1 = continuous, year-round, 2 = about once a week, 3 = about once a month, 4 = about once a year

	Urban freeway	Rural freeway	Major arterial	Other arteries	Rural highway
Volume					
Speed					
Classification					
Occupancy					
Weight					

Specify the desired density of coverage.
1 = sensors every 0.5 mi or less, 2 = about every mile, 3 = at key locations only (sparse)

	Urban freeway	Rural freeway	Major arterial	Other arteries	Rural highway
Volume					
Speed					
Classification					
Occupancy					
Weight					

How many of the sensors should be portable? 1 = none, 2 = some, 3 = all

	Urban freeway	Rural freeway	Major arterial	Other arteries	Rural highway
Volume					
Speed					
Classification					
Occupancy					
Weight					

What type of vehicle classification is used?
1 = FHWA 13 classes, 2 = based on number of axles, 3 = three classes based on vehicle length, 4 = other: please explain

	Urban freeway	Rural freeway	Major arterial	Other arteries	Rural highway
Classification					

Explanation: _____

Is presence detection required at stoplines or on freeways, for example, for signal control or incident detection?
1 = No, 2 = Yes at intersections, 3 = Yes at freeway ramp meters, 4 = Yes on freeway mainline

	Urban freeway	Rural freeway	Major arterial	Other arteries	Rural highway
Presence					

Please use the space below to describe some of the uses for traffic data. Also please describe any other requirements.

_____
_____
_____

Please provide a desirable figure of sensor cost per lane per type of data
(e.g., a sensor that costs $2100 and measures volume and speed over a maximum of three lanes has a cost per lane per type of data of $350.)

$ _____

(Continued)

Table 4.9 (Continued) **Survey to determine sensor data requirements**

Rank the following techniques for travel time data collection from 1 to 5, with 1 being best in terms of lowest cost and best accuracy:

Instrumented vehicle with 1-person crew	
Vehicle with 2-crew: driver and stop-watch timer/recorder	
Speed detection every ~0.5 mi (0.8 km) and integration over space	
Analytical technique based on volumes, geometrics, etc. (Potential references include the *Highway Capacity Manual* and National Cooperative Highway Research Program publications)	
Computer simulation	
Vehicle probes with transponders and receptors	

Please rate the importance of the characteristics or statements listed below using a 1 to 5 scale, with 1 being the top priority or the most important feature and 5 being the least important. Several items may have the same priority. For example, 1 = all the time, 2 = most of time, 3 = average priority, 4 = sometimes, 5 = infrequently

Sensors should rely on supervising computers as little as possible	
Sensors should be remotely controlled (by computer)	
Sensors should not require a computer for setup	
Sensors should have local data storage for several days	
Sensors should have wireless communication option for data retrieval	
Sensors should have solar power option for remote locations	
Sensors should have remote status check and trouble shooting	
Mounting height in excess of 30 ft (9.1 m) is likely to be a problem	
Overhead mounting at some locations would be a problem	
Immune to damage from re-surfacing and other roadway construction	
Installation and data retrieval crew safety	
Manpower requirement for data retrieval	
Manpower requirement for inspection, cleaning, maintenance	
On-site assistance by vendor for installation and calibration	
Sensor warranty period	
Guarantees for operation in humid and corrosive environment	
Mean time between failures (MTBF)	
Cost per lane per type of data gathered	

from the available data, as required by some traffic management applications. Thus, the maximum acceptable error specified in survey instruments such as these tables should be the composite requirement that includes the three error components, namely, deterministic and random errors incurred in estimating the required control parameter and the sensor measurement error itself.

A detection system does not act alone. It relies on other entities to maintain the sensors and transmit data and information to the TMC, where they are analyzed and the results sent on to other departments, agencies, and service providers for their particular uses.

4.5 SUMMARY

A detailed systems analysis is a valuable if not a necessary procedure to properly specify the requirements for the vehicle detection subsystem found in many traffic management systems. The types of data, corresponding accuracies, and sampling intervals are often prescribed by the data processing algorithm and certainly by the traffic management strategy. Surveys of operations personnel can often assist in defining data requirements. Algorithms are an integral part of an ITS solution and are dependent upon state-of-the-art sensor technology to provide data with the requisite spatial and temporal resolutions, accuracy, precision, repeatability, and sampling frequency. Once the required traffic flow parameters, accuracies, and temporal and spatial collection intervals have been defined, then the type of sensor and sensor technology can be selected. The Caltrans real-time data accuracy requirements for traffic operations, planning, and traveler information were presented as an example of the application of the systems engineering process to determining sensor requirements. Several traffic management applications (advanced signalized intersection control, freeway incident management, and freeway metering) and data categories (tactical, strategic, and historical) were discussed to show how they influence present and future input data requirements and hence sensor specifications. The latter results were developed as part of the Detection Technology for IVHS Program. Future data and sensor requirements that will support as yet undefined algorithms and paradigms may likely be different from the ones reviewed here.

REFERENCES

1. L.A. Klein and M.R. Kelley, *Detection Technology for IVHS, Final Report*, FHWA-RD-95-100, U.S. Federal Highway Administration, McLean, VA, December 1996. http://ntl.bts.gov/lib/jpodocs/repts_te/6184.pdf. Accessed December 10, 2015.
2. L.A. Klein, *Sensor Technologies and Data Requirements for ITS*, Artech House, Norwood, MA, June 2001.
3. L.A. Klein, N.A. Rantowich, C.C. Jacoby, and J. Mingrone, IVHS Architecture Development and Evaluation Process, *IVHS Journal*, 1(1):13–34, 1993.
4. *Systems Engineering Guidebook for Intelligent Transportation Systems*, Ver. 3.0, U.S. Department of Transportation, Federal Highway Administration, California Division and California Department of Transportation, Sacramento, CA, November 2009. http://www.fhwa.dot.gov/cadiv/segb/. Accessed November 9, 2015.
5. Cambridge Systematics, Inc. and C.A. MacCarley, Detection system testing. In *Statewide Detection Plan*, California Department of Transportation, Sacramento, CA, September 23, 2008.
6. Cambridge Systematics, Inc. and L.A. Klein, Data quality evaluation memorandum, In *Statewide Detection Plan*, California Department of Transportation, Sacramento, CA, July 2, 2008.
7. Y.J. Stephanedes, A.P. Chassiakos, and P.G. Michalopoulos, Comparative Performance Evaluation of Incident Detection Algorithms, *Transportation Research Record* 1360, Transportation Research Board, Washington, DC, 50–57, 1993.
8. H.J. Payne, E.D. Helfenbein, and H.C. Knobel, *Development and Testing of Incident Detection Algorithms, Vol. 2: Research Methodology and Detailed Results*, FHWA-RD-76-20, PB-259-237/6/XPS, U.S. Federal Highway Administration, Washington, DC, Apr. 1976.
9. H.J. Payne and S.C. Tignor, Freeway Incident Detection Algorithms Based on Decision Trees with States, *Transportation Research Record* 682, Transportation Research Board, National Research Council, Washington, DC, 30–37, 1978.
10. B.N. Persaud and F.L. Hall, Catastrophe theory and patterns in 30-Second traffic data—Implications for incident detection, *Transportation Research*, 23A(2):103–113, 1989.

11. B.N. Persaud, F.L. Hall, and L.M. Hall, *Congestion Identification Aspects of the McMaster Incident Detection Algorithm*, Transportation Research Record 1287, Transportation Research Board, Washington, DC, 1990.

12. C.L. Dudek and C.J. Messer, Incident Detection on Urban Freeways, *Transportation Research Record* 495, Transportation Research Board, Washington, DC, 12–24, 1974.

13. A.R. Cook and D.E. Cleveland, Detection of Freeway Capacity-Reducing Incidents by Traffic-Stream Measurements, *Transportatikon Research Record* 495, Transportation Research Board, National Research Council, Washington, DC, 1–11, 1974.

14. S.A. Ahmed and A.R. Cook, Application of Time-Series Analysis Techniques to Freeway Incident Detection, *Transportation Research Record* 841, Transportation Research Board, National Research Council, Washington, DC, 19–21, 1982.

15. S.A. Ahmed and A.R. Cook, Discrete dynamic models for freeway incident detection systems, *Transportation Planning and Technology*, 7(4):231–242, 1982.

16. B. Park, C.J. Messer, and T. Urbanik II, Short-Term Freeway Traffic Volume Forecasting Using Radial Basis Function Neural Network, *Transportation Research Record* 1651, Transportation Research Board, National Research Council, Washington, DC, 39–47, 1998.

17. J.F. Collins, C.M. Hopkins, and J.A. Martin, *Automatic Incident Detection—TRRL Algorithms HIOCCV and PATREG*, TRRL Supplementary Report 526, Transport and Road Research Laboratory, Crowthorne, UK, 1979.

Chapter 5

Modern traffic flow sensor technologies

Traffic flow sensors are often divided into two broad categories: those mounted on or under the roadway surface and those mounted above the roadway on sign bridges or to the side of the roadway on poles and other structures. The first category is also referred to as intrusive sensors because they infringe on the roadway pavement, while the second category is referred to as nonintrusive. Many traffic flow sensors function by detecting electromagnetic energy in some form, for example, radio frequency (RF) spectrum, visible spectrum, infrared spectrum, microwave spectrum, and millimeter-wave spectrum. Several of the sensors though detect acoustic and ultrasonic energy. The sensors may be either passive or active. Passive sensors only receive energy, that is, they do not transmit energy of their own. The energy they receive is a combination of energy emitted and reflected into their aperture by motorized vehicles, bicycles, pedestrians, other objects of interest, the road surface, and road divider structures, as well as from extraneous sources such as trees, buildings, billboard signs, and bridges. Active sensors both transmit and receive energy. The received energy is a portion of the transmitted energy that is scattered back into the aperture of the sensor by vehicles, the road, or other objects of interest and from extraneous objects. Some active sensors, such as inductive loops, detect the change in a property of the surroundings in which they are located, for example, the inductance of the electric circuit of which the loop is a part. Table 5.1 lists the sensor technologies included in each category.

Sensor selection depends on many factors such as the following:

- Types of data required.
- Life-cycle cost inclusive of sensor hardware and software, installation, and maintenance.
- Agency culture.
- Vendor support.
- Availability of overhead mounting.
- Condition of roadbed.
- Weather and climate conditions.
- Restrictions on pavement destruction or overhead mounting.

These and other selection criteria are discussed further as each sensor technology is described in the sections that follow.

5.1 INDUCTIVE LOOP DETECTOR

The inductive loop detector (ILD) is the most common sensor used in traffic management applications. Its size and shape vary with the detection objective, for example, automobiles; scooters, motorcycles, and bicycles; long vehicles and large high-bed trucks; queue detection

Table 5.1 In-roadway and over-roadway mounted sensors

In-roadway technologies (intrusive)		Above-roadway technologies (nonintrusive)	
Inductive loop	Active	Video detection systems	Passive
Magnetometer	Passive	Microwave:	
Magnetic sensor, also referred to as a magnetic detector	Passive	• Presence-detecting radar	Active
		• Doppler sensor	Active
		Acoustic	Passive
		Lidar (laser radar)	Active
		Passive infrared	Passive
		Ultrasound	Active
		Technology combinations	Active and passive

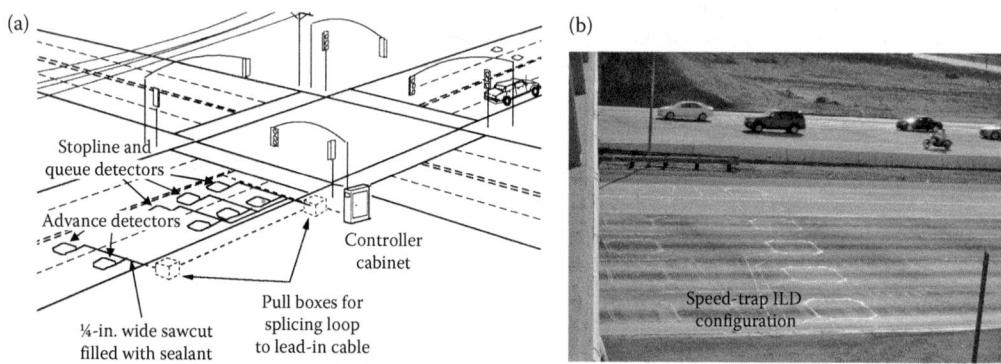

Figure 5.1 Inductive loop configurations at a signalized intersection and a limited-access highway (typical). (a) Inductive loops installed for signal control, (b) Inductive loops in speed-trap configuration on a California freeway. (From LAK.)

at freeway off ramps; vehicle counting; and safety and congestion applications that require speed measurements. On the left of Figure 5.1 is a typical configuration of loops as might be found at a signalized intersection, while the right shows a speed-trap configuration of loop pairs (i.e., two loops in a lane spaced a known distance apart) used to measure vehicle speed in each lane of a limited-access highway.

ILD configurations include the 5-ft by 5-ft (1.5-m by 1.5-m) or 6-ft by 6-ft (1.8-m by 1.8-m) square loops, 6-ft (1.8-m) diameter round loops, and rectangular configurations having a 6-ft (1.8-m) width and variable length. Quadrupole loop configurations, which divide the cross-lane loop dimension in half, are utilized to enhance motorcycle and bicycle detection (by increasing the electromagnetic field strength in the center of the lane) and to eliminate adjacent lane detection in high-sensitivity inductive loop systems. The increased field of the quadrupole loop is due to the doubling of the number of windings at the lane center. Diamond-shaped loops also enhance motorcycle detection by extending the field to the lane edges where motorcycles sometimes drive to avoid oil spots that are more prevalent at the lane center. Other loop configurations that extend the detection area to full lane widths or increase the sensitivity for motorcycle and bicycle detection are discussed by Klein et al. [1].

The popularity of the ILD is due, in part, to its mature technology and low unit cost. Reliability of the wire loop has improved through better packaging and installation techniques. These include delivery of loops already encased by the manufacturer in protective materials, more thorough cleaning of debris from the sawcut, and the use of improved sealant in the installation process. The loop detector system, however, may still suffer from poor reliability. Contributing factors are poor connections made in the pull boxes, failure to twist wire pairs properly leading to crosstalk, and faulty sawcut cleaning and sealant application procedures. These problems are accentuated when loops are installed in poor pavement or in areas where utilities frequently dig up the roadbed. Recommended procedures for installing ILDs in sawed slots in roadway pavement are found in Chapter 6 and also in Klein et al. [1] and the ASTM Standard Practice for the Installation of Inductive Loop Detectors E 2561 [2].

5.1.1 Operation of inductive loops

An inductive loop requires an AC excitation voltage whose nominal frequency varies from 15 to 100 kHz. Special purpose detector electronics units (often referred to as simply the "detector") operate at frequencies up to 150 kHz or beyond. The inductive loop system behaves as a tuned electrical circuit in which the wire loops, lead-in wire, and lead-in cable are the inductive elements [1]. When a vehicle passes over the loop wires or is stopped within the loop, the vehicle induces eddy currents in the wire loops, which decrease their inductance. The decreased inductance increases the actual loop-system frequency. The change in frequency is sensed by the detector electronics unit, which then generates a pulse output from a solid-state optically isolated device. The pulse indicates to the controller the passage or presence of the vehicle. In addition to the optically isolated output device, the electronics unit contains a tuning network, oscillator, and a mechanism to adjust its sensitivity. Loops produce accurate vehicle counts and presence indication when properly installed and maintained in good pavement. Table 5.2 lists the salient features of inductive loops as used for traffic management.

5.1.2 Speed measurement using inductive loops

Inductive loops can be utilized to measure speed in two ways: in a speed-trap configuration and as a single loop. The speed-trap method depicted on the right of Figure 5.1 is the more accurate. With this method, vehicle speed S is calculated as

$$S = \frac{d}{\Delta T},$$

(5.1)

Table 5.2   Inductive loop typical output data, installation location, advantages, and limitations

Typical output data	Installation	Advantages	Limitations
Count Presence Lane occupancy Average vehicle speed: • With one loop and an assumed vehicle length • With two loops in a speed-trap configuration Queue length with multiple loops Vehicle class when activated with a high-frequency (typically >100 kHz) electronics unit	Embedded in roadway	Low per unit cost Mature, well-understood technology Standardization of loop and detector electronics configurations	Traffic interrupted for installation and repair Susceptible to damage by heavy vehicles and road or utility repair Reliability is a strong function of the skill of the installers

where d is the distance between the leading edges of the loop pair and ΔT is the time difference between the pulse produced by the electronics unit when the vehicle is first detected by the leading loop and the time the vehicle is detected by the trailing loop.

Single loops can also be utilized to give a less accurate measure of vehicle speed. This technique is based on an estimate of the average length of a vehicle using the facility. The accuracy of the estimate of the vehicle length can vary by lane and vehicle mix (trucks are more frequently found in the rightmost two lanes of a roadway), location of the road (urban or rural), type of road (arterial or freeway), season, time of day, weather, and occurrence of special events. The other information required to implement the single-loop speed measurement approach is measured values of vehicle count and occupancy, and the effective loop length. Effective loop length may be different from the physical dimension of the loop as detection may occur as soon as the vehicle enters the electromagnetic field of the loop, which in effect lengthens the effective detection area. Average vehicle speed $\overline{S}$ from the single-loop method is given by

$$\overline{S} = \frac{0.6818 V_C \, (L_L + V_L)}{O}, \tag{5.2}$$

where $\overline{S}$ is speed in mi/h, 0.6818 is constant that converts ft/s into mi/h, V_C is vehicle count during the measurement period, L_L is effective loop length in ft, V_L is vehicle length in ft, and O is seconds of lane occupancy during the measurement period.

Current ILD technology provides single-loop speed estimates that differ from the true value by as much as 30%. To obtain even these relatively crude measurements, vehicle count, vehicle length, and occupancy must be known to within values having an error no greater than $\pm 10\%$. Of these, vehicle length is the most difficult to estimate accurately for the reasons mentioned previously.

The allowable error in vehicle length can be determined for any desired speed measurement accuracy by forming the differential of vehicle speed $\Delta \overline{S}$ with respect to vehicle length from Equation 5.2 and then calculating the value of the factors that multiply the vehicle length differential error ΔV_L. Accordingly, the differential speed with respect to vehicle length is found as

$$\Delta \overline{S} = \frac{0.6818 V_C}{O} \Delta V_L \tag{5.3}$$

Therefore, in order to estimate speed to within $\pm 10\%$ when vehicle count is 12.5 over a 30-s measurement interval (equivalent to a per lane flow rate of 1500 veh/h) and occupancy is 6 s (equivalent to 20%) requires vehicle length to be known to within 7% of its true value. This calculation may be repeated for other values of vehicle count, occupancy, and desired speed accuracy, as illustrated in Figure 5.2. A similar analysis may be performed for the vehicle count, effective loop length, and occupancy parameters to determine the influence of errors in knowledge of their values on the speed estimate provided by a single ILD.

Occupancy values greater than 25% reliably indicate the onset of congestion, while values greater than 35% represent very heavy congestion or queuing. Therefore, occupancy measurements utilized in ITS applications are generally most important below 35% as indicated in the shaded portion of Figure 5.2.

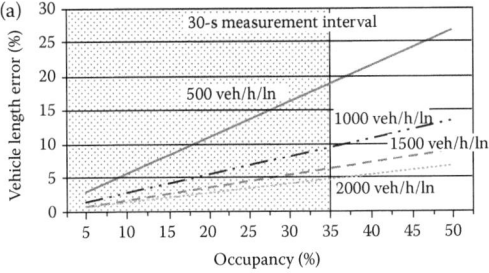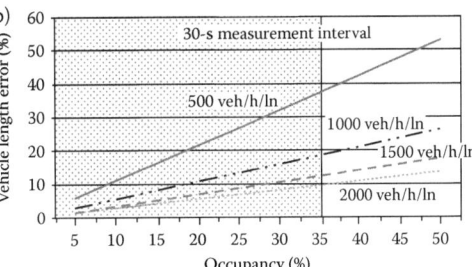

Figure 5.2 Allowable error in vehicle length estimate as a function of lane occupancy, flow rate, and desired speed measurement accuracies of 5% and 10%. (a) Tolerable vehicle length error for estimating speed within 5% over a 30-s measurement interval, (b) tolerable vehicle length error for estimating speed within 10% over a 30-s measurement interval.

5.1.3 Vehicle classification using inductive loops

ILD electronic modules vary in the number of loops they support and their functionality. The module in Figure 5.3 is capable of classifying the traffic stream into the initial 13 classes depicted in Figure 5.4. It can also re-identify vehicles for measurement of travel time and vehicle speed. Classes 14 through 23 represent vehicles with unique characteristics. The FHWA 13 classes are based on whether the vehicle carries passengers or commodities. Non-passenger vehicles are further subdivided by the number of axles and number of units, including both power and trailer units. The addition of a trailer to vehicle classes 1–5 does not change the classification of the vehicle.

Special configurations of inductive loops are available to detect axles and their relative position in a vehicle. The arrangement in Figure 5.5 can be used at toll plazas to elicit the correct payment for the vehicle class [3]. The first 15 classes in Figure 5.4 are standard outputs of the system. The configuration in Figure 5.5 situates an axle loop array between main loops 1 and 2 to detect axle presence. The relative position of the axles in the vehicle is determined

I-Loop duo card

I-Loop duo card specification

Signal bandwidth	15–150 kHz (loop signals)
Controller data output rate	300–921.6 kbps (Serial and Bluetooth®), 100 Mbps (Ethernet), 12 Mbps (SPI/USB), 100–2500 Hz (digital control lines)
Adjustable sample rate	100–2500 Hz
Controller compatibility	170/2070 or NEMA TS-1/TS-2 (jumper selectable) Also supports stand-alone operation (outside of card cage)
Frequency selections	16 (software controllable)
Frequency selection method interfaces	Front panel by user or through one of five communication connections: USB, Bluetooth®, SPI, UART, or Ethernet
Sensitivity selection method	Front panel by user or through five communication interfaces: USB, Bluetooth®, SPI, UART, or Ethernet
Communication speeds	USB 2.0, Ethernet 10/100, Bluetooth® up to 921.6 kbps, SPI up to 12 Mbit, UART to 921.6 Kbps
Concurrent communications connections	Yes
Data storage	2 Gbit(128 Mb) to 16 Gbit(2 Gb) on board flash storage
Firmware upgradable	Via USB, Bluetooth®, or Ethernet
Performance monitoring system compatible	Yes
Customization	Yes. On-board real-time data processing for vehicle signature and parameter calculations via dual-core ARM Cortex 32-bit safety-core processor operating at 220 MHz

Figure 5.3 I-Loop Duo Card for classifying or re-identifying vehicles. (Photograph courtesy of CLR Analytics and Diamond Traffic Products.)

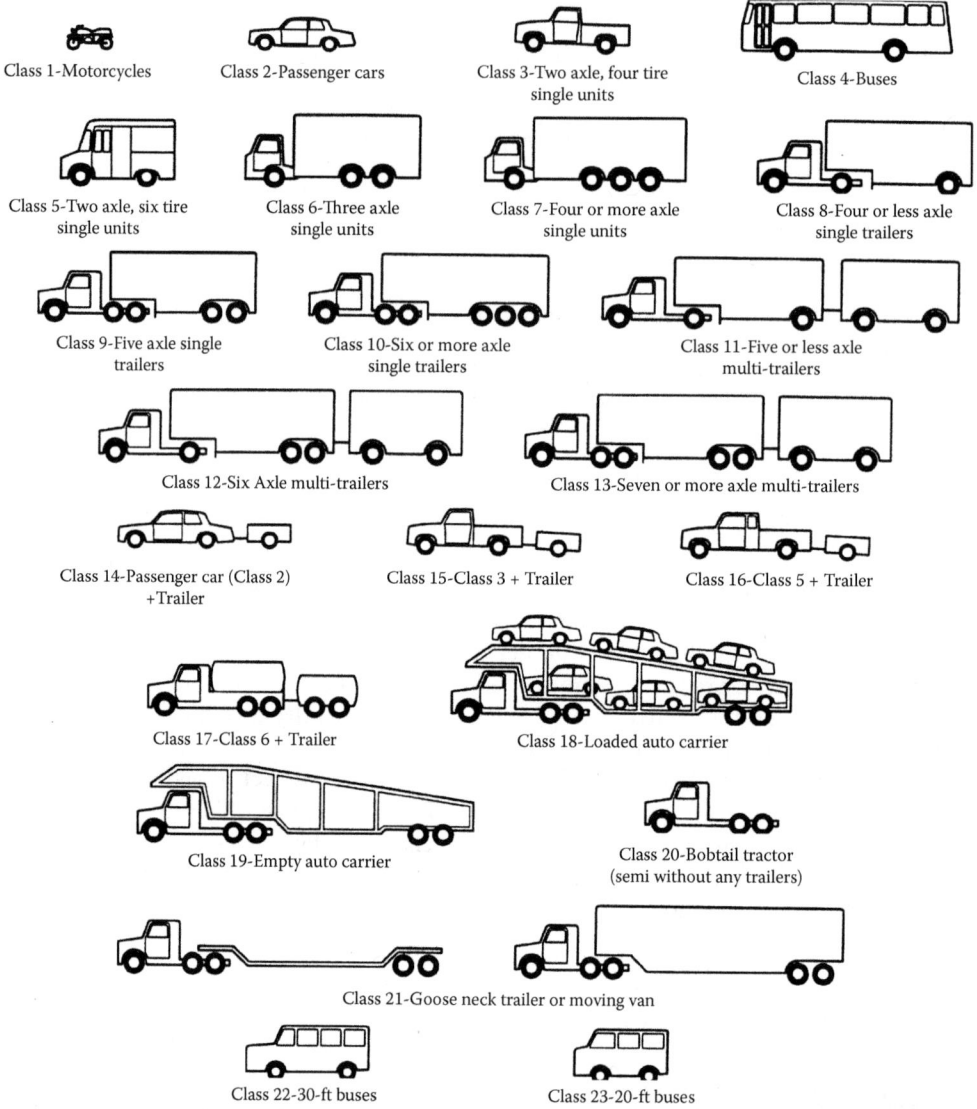

Figure 5.4 Vehicle classification categories. (Figure courtesy of Intersection Development Corporation, Downey, CA.)

by the main loop signatures. The data provided are vehicle length, speed, acceleration, vehicle type, number of axles, and axle separation. Profile information can also be obtained to refine and validate classification in ambiguous cases. This system, as well as the one in Figure 5.3 can be used to identify transit buses and provide priority treatment at traffic signals.

5.2 MAGNETOMETER SENSORS

Magnetic sensors are passive devices that indicate the presence of a metallic object by detecting the perturbation (known as a magnetic anomaly) in the Earth's magnetic field created by the object. Two types of magnetic field sensors are used for traffic flow parameter

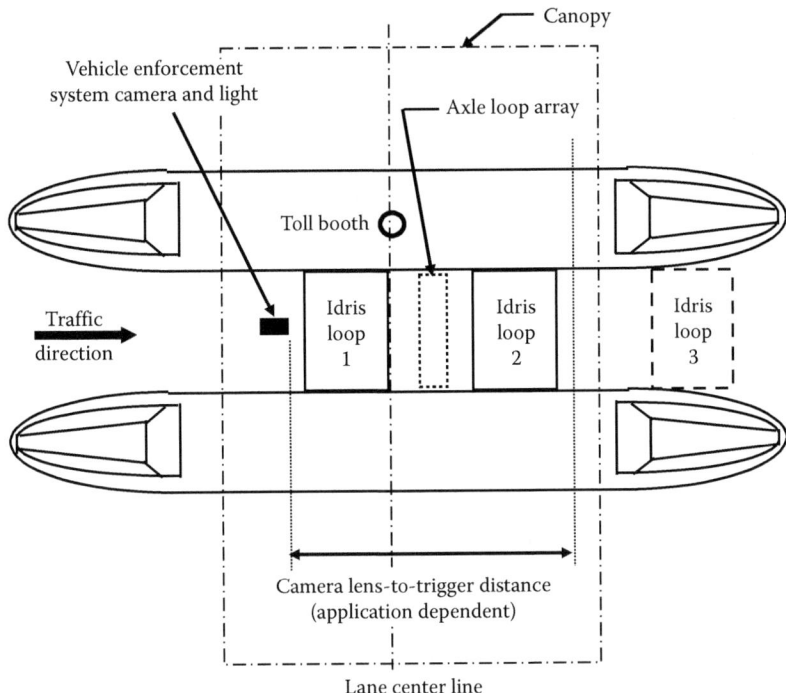

Figure 5.5 Idris® Traffic Axle Location and Vehicle Classification System for classifying or re-identifying vehicles configured for a toll plaza application.

measurement. The first type, the two-axis and three-axis fluxgate magnetometer, detects changes in the vertical and horizontal components of the Earth's magnetic field produced by a ferrous metal vehicle. The two-axis fluxgate magnetometer contains a primary winding and two secondary "sense" windings on a bobbin surrounding a high permeability soft magnetic material core. In response to the magnetic field anomaly created by the magnetic signature of a vehicle, the magnetometer's electronics circuitry measures the output voltage generated by the secondary windings. The sensor declares a vehicle present when the voltage exceeds a predetermined threshold. In the presence mode of operation, the detection output is maintained until the vehicle leaves the detection zone.

Magnetometers can be used on bridge decks where ILDs may be affected by the steel support structure or simply cannot be installed. Arrays of three-axis fluxgate magnetometers can gather vehicle signatures in support of vehicle classification. Magnetometer sensors, such as those in Figure 5.6, supply vehicle flow data such as presence, passage, count, and lane occupancy. Vehicle speed can be measured when they are installed in a speed-trap configuration as illustrated in Figure 5.7. Since magnetometers have a relatively limited sensing area, several may be required across a lane to guarantee 100% vehicle detection at a signal stopline or to determine queue length, especially on a curved section of road where vehicles may be partially in one lane and partially in another.

5.3 MAGNETIC DETECTORS

The second type of magnetic field sensor is the magnetic detector, more properly referred to as an induction or search coil magnetometer. It detects the vehicle signature by measuring

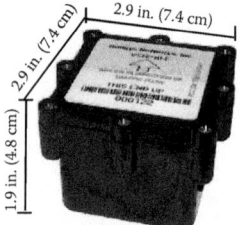

2.9 in. (7.4 cm)

2.9 in. (7.4 cm)

1.9 in. (4.8 cm)

Sensys networks flush-mounted magnetometer

- Sensing area is approximately that of a 6-ft diameter round loop.
- Flush-mounted sensor: Installed by coring 4-in. (10-cm) diameter × approximately 2¼ in. (6.5 cm) deep hole, inserting the sensor into the hole to align it with the direction of traffic flow, and sealing the hole with fast drying epoxy.
- 10-year battery life based on 300 million detections.

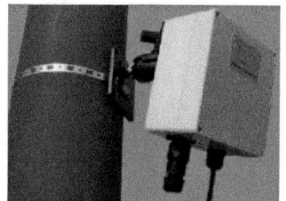

Sensys networks access point:
Range up to 150 ft (46 m).

(Photographs courtesy of
Sensys Networks, Berkeley, CA)

Trafficware pod magnetometer sensor

- 900 MHz wireless frequency.
- Installed by coring a 4.5-in. (11-cm) diameter × 2.75-in. deep hole sealed with fast-drying epoxy.
- 10-year battery life with an average of 700 activations per hour, 24/7.

(Photograph courtesy of
Trafficware, Sugar land, TX)

Figure 5.6 Magnetometer sensors provide traffic flow data such as presence, passage, count, and lane occupancy.

the distortion in the magnetic flux lines induced by the change in the Earth's magnetic field produced by a moving ferrous metal vehicle. These devices contain a single coil winding on a permeable magnetic material rod core. Like the fluxgate magnetometer, magnetic detectors generate a voltage when a ferromagnetic object perturbs the Earth's magnetic field. Most magnetic detectors do not detect stopped vehicles since they require a vehicle to be moving or otherwise changing its signature characteristics with respect to time. However, multiple units of some magnetic detectors can be installed and utilized with specialized signal processing software to generate vehicle presence data.

The Model 231 magnetic detector in Figure 5.8 is installed by inserting it into a trench under the roadway. It requires an amplifier, either a Model 201 or 232, to be installed in a

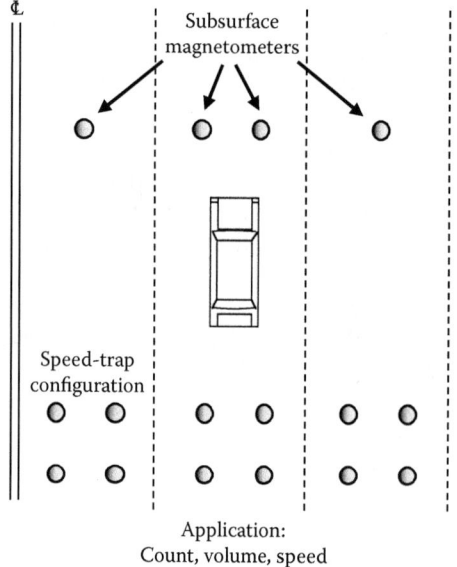

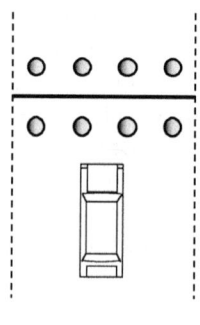

Figure 5.7 Magnetometer applications.

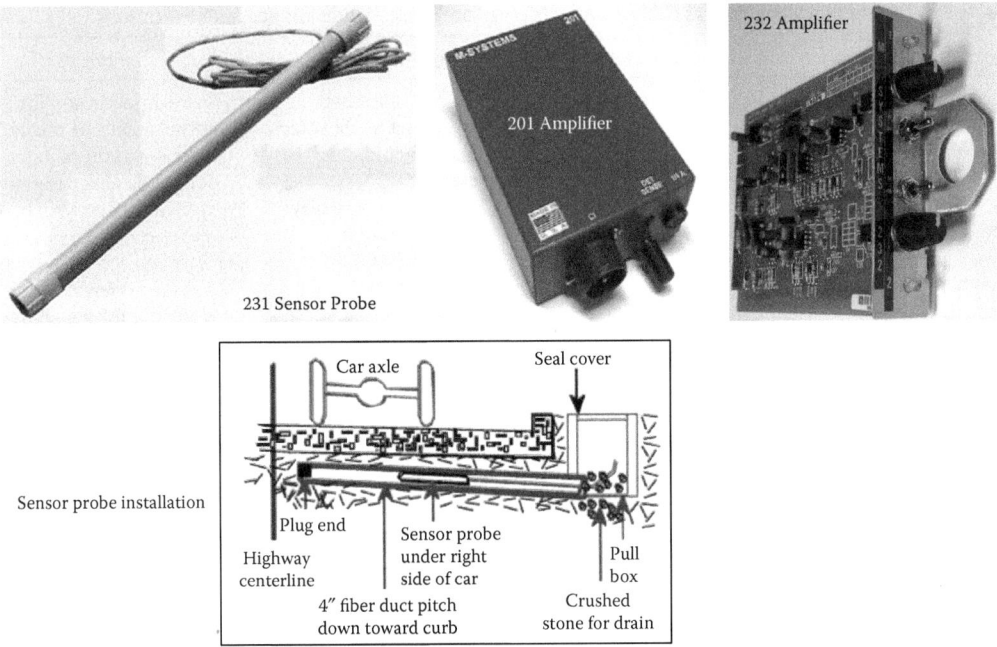

Figure 5.8 Model 231 magnetic detector. (Photographs courtesy of M-Systems, Inc., Newtown, CT.)

controller cabinet. Model 701 microloop sensors, shown in Figure 5.9, are inserted vertically in 1-in. (2.5-cm) holes and placed 18–24 in. (46–61 cm) below the roadway surface. Up to four 701 sensors can be connected in series. The Model 702 is an example of a magnetic detector that can be utilized with specialized signal processing software to generate vehicle presence data. It is installed by inserting it into a 3-in. (7.6-cm) nonferrous Schedule 80 conduit. The conduit is placed 21 ± 3 in. (53.3 ± 7.6 cm) below the road surface using horizontal directional drilling or open trenching techniques.

Model 701 microloop sensor

Model 702 microloop sensor can detect stopped vehicles using application-specific software and an array of sensors

Figure 5.9 Model 701 and 702 magnetic detectors. (Photographs courtesy of Global Traffic Technologies, LLC, St. Paul, MN.)

Table 5.3 Magnetic sensor typical output data, installation location, advantages, and limitations

Typical output data	Installation	Advantages	Limitations
Count Presence: • Magnetometers—yes • Magnetic detectors—no Lane occupancy Average vehicle speed with two sensors in a speed-trap configuration Queue length with multiple sensors Vehicle class with sensor arrays and special signal processing	Embedded in roadway	Low per unit cost Can detect small vehicles including bicycles Arrays of magnetometers may provide vehicle classification	Traffic interrupted for installation and repair Discrimination of longitudinal separation between closely spaced vehicles May need more than one sensor across lane to detect lane straddlers and motorcycles, especially when 100% presence detection is required (e.g., at actuated traffic control signals)

Table 5.3 summarizes the prominent attributes of magnetic sensors as used for traffic management. Now that we have discussed sensors that are mounted on or in the roadway bed, let us explore sensors that are mounted above or to the side of the roadway.

5.4 VIDEO DETECTION SYSTEMS

Video detection systems (VDSs), such as those in Figure 5.10, are an example of a passive sensor that transmits no energy of its own. Video cameras were introduced to traffic

RackVision pro 2 Encore Solo terra RackVision terra RackVision terra system 16

Autoscope® (Photographs courtesy of Econolite Control Products, Anaheim, CA)

Peek VideoTrak-IQ® Four channel card (Photograph courtesy of Peek Traffic, Houston, TX)

MediaCity (Photograph courtesy of Citilog, Philadelphia, PA)

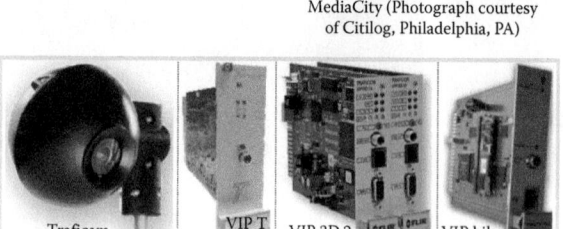

Traficam VIP T VIP 3D.2 VIP bike

Traficon(Photographs courtesy of FLIR, Marke, Belgium; FLIR USA, Wilsonville, OR)

Iteris Vantage™ Family (Photograph courtesy of Iteris, Santa Ana, CA)

Figure 5.10 VDSs offered by a variety of manufacturers.

management for roadway surveillance based on their ability to transmit closed-circuit television imagery to a human operator for interpretation.

Present-day traffic management applications utilize video image processing to automatically analyze the scene of interest and extract information for traffic surveillance, traffic management, and signal control. A VDS typically consists of one or more cameras, a microprocessor-based computer for digitizing and processing the imagery, and software for interpreting the images and converting them into traffic flow data. Such a system can replace several in-ground inductive loops, provide detection of vehicles across several lanes, and perhaps lower maintenance costs. Some process data from more than one camera and further expand the area over which data are collected. A VDS classifies vehicles by their length and reports vehicle presence, volume, lane occupancy, and speed for each class and lane. Vehicle density, link travel time, and origin–destination pairs are potential traffic parameters that can be obtained by analyzing data from a series of image processors installed along a section of roadway.

5.4.1 VDS image processing

Figure 5.11 demonstrates the use of vehicle detection zones to collect the required data. VDS may utilize different shaped detection zones to differentiate between vehicles and bicycles and collect different types of data, for example, down lane, stopline, and speed detection zones. Other systems track vehicles through the entire field-of-view of camera using Kalman filtering techniques to update vehicle position and velocity estimates. The time trace of the position estimates yields a vehicle trajectory. By processing the trajectory data, local traffic parameters (e.g., flow and lane change frequency) can be computed. These parameters, together with vehicle signature information (e.g., time stamp, vehicle type, color, shape, position, and speed), can then be communicated to the traffic management center. Tracking vehicle sub-features such as edges, corners, and two-dimensional patterns, rather than entire vehicles, has been proposed to make the VDS robust to partial occlusion of vehicles in congested traffic. VDSs that track vehicles can also register turning movements.

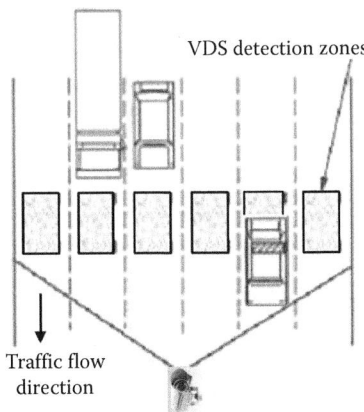

Video detection system field-of-view is determined by focal length of lens, camera mounting height, and viewing angle

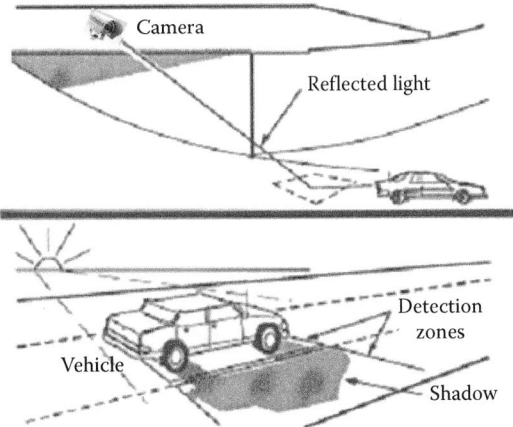

Reflected light from wet pavement and shadows may cause false detections

Figure 5.11 VDSs are a popular option for obtaining vehicle counts, presence, and speed. Camera setup should avoid or minimize susceptibility to reflected light and shadows that can lead to false or missed detection events.

VDSs identify vehicles and compute traffic flow parameters by analyzing the imagery from a traffic scene to determine changes between successive frames, as illustrated in the conceptual algorithm sequence of Figure 5.12. The image processing algorithms that analyze black and white imagery examine the variation of gray levels in groups of pixels (picture elements) contained in the video frames. The algorithms are designed to remove gray-level variations in the image background caused by weather conditions, shadows, and daytime or nighttime artifacts and retain objects identified as automobiles, trucks or buses, motorcycles, and bicycles. While these algorithms are not perfect, they do allow VDSs to function adequately under most circumstances and have shown continuous performance improvement over time. Some VDSs contain software that limits the effects of wind-induced camera movement artifacts.

Once the camera imagery is digitized and stored, a detection process is utilized to identify data that exceed one or more thresholds. This process limits and segregates the data passed on to the rest of the algorithms indicated in Figure 5.12 that classify and identify the vehicles and calculate their traffic flow information. It is undesirable to severely limit the number of potential vehicles during detection, for once data are removed they cannot be recovered. Therefore, false vehicle detections are permitted at the detection stage since the declaration of actual vehicles is not made at the conclusion of the detection process. Rather algorithms that are part of the classification, identification, and tracking processes still to come are relied on to eliminate false vehicles and other objects and retain the real ones.

Image segmentation is used to divide the image area into smaller regions (often composed of individual vehicles) where features can be better recognized. The feature extraction process examines the pixels in the regions over some time period for pre-identified characteristics that are indicative of vehicles. When a sufficient number of these characteristics are present and recognized by the processing, a vehicle is declared present and its flow parameters are calculated.

The term classification as used here implies the set to which the vehicle belongs (e.g., automobile, pick-up truck, 18-wheeler, or bus). Identification refers to the vehicle's description up to the limit of the sensor's ability to differentiate one vehicle from another and should include the vehicle's manufacturer and model and perhaps color (e.g., Toyota Corolla, Ford Fusion, or Mercedes SUV GLK-class). Generally, higher resolution and signal-to-noise ratio are needed to perform identification as compared to classification [4].

Color imagery can also be exploited to obtain traffic flow data. Chromatic information can enhance vehicle discrimination in inclement weather or when camera mounting conditions are not ideal, differentiate vehicles from shadows, or identify features on individual and groups of vehicles.

Figure 5.13 illustrates the operation of one video detection algorithm that tracks vehicles through multilane detection areas during daytime and nighttime operation. Daytime

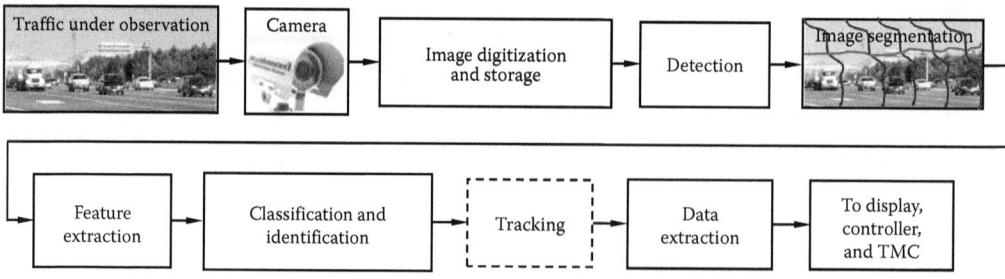

Figure 5.12 Conceptual image detection, classification, and tracking algorithm.

(a)  (b)

Figure 5.13 Vehicle detections during (a) daytime and (b) nighttime operation of one VDS. (Photographs courtesy of TrafficVision, Anderson, SC.)

detection identifies vehicle features that assist in classifying vehicles into four groups, motorcycle, passenger vehicle, truck, and large truck. Nighttime detection relies on identifying a vehicle's headlights. In both cases, the square symbols in the photographs outline the detected portions of the vehicles.

5.4.2 Infrared VDS

As infrared cameras become more cost competitive, they are being utilized by traffic management agencies to detect the heat signature of vehicles, bicycles, and pedestrians without having to combat the glare than can accompany visible spectrum imagery. The longer wavelength infrared cameras see in darkness or poor lighting conditions, and through smoke and light fog. They can also detect vehicles that may be difficult to distinguish in shadows.

Infrared image features, whether for automobile, bicycle, or pedestrian detection, are generated by reflected and emitted energy that is captured by a camera operating in an infrared wavelength band. There are three commonly cited infrared bands: near-infrared band from 0.87 to 1.5 μm, mid infrared band from 3 to 5 μm, and long-wavelength thermal energy band from 8 to ≥12 μm. By contrast, features in visible spectrum (0.4–0.7 μm) images are formed by reflected sunlight or light captured from headlights and taillights that enters the camera lens. Images in the near-infrared band are generated predominantly by reflected energy and appear similar to visible wavelength images to the human eye. Mid infrared images begin to take on emissive characteristics, where features are proportional to the emissivity of the radiating surface and its absolute temperature. Long-wavelength infrared images are predominantly formed by energy emitted from the objects in field-of-view of the camera. These images have little reflected energy component and appear different than visible wavelength images to the human eye.

The long-wavelength or thermal infrared cameras are not subject to sun glint and the effects of inadequate lighting as are video systems that operate in the visible spectrum. The difference in detection behavior of a visible spectrum and thermal infrared VDS is illustrated in Figure 5.14. The white areas near the bottom of the vehicle or the tires in the infrared images are produced by thermal energy emitted by the engines or hot tires and represent the hottest regions in these images.

Figure 5.15 shows an infrared camera detection system on the left designed for intersection signal control, inverse direction detection, and bicycle counting. Contact closures are initiated by the sensor to open and close a pair of isolated contacts in the controller

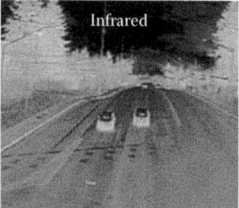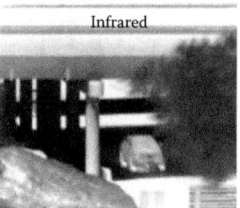

Thermal imaging cameras see in total darkness and show more scene detail

Thermal imaging cameras can see into shadows

Thermal imaging cameras see through glare and backlighting, improving signal control

Figure 5.14 Visible and thermal image comparison. (Photographs courtesy of FLIR Systems, Marke, Belgium.)

- ThermiCam uses long wavelength (7–14 µm) thermal energy emitted from vehicles and bicyclists to distinguish between them. The sensor provides the traffic signal controller with vehicle and bike presence, which allows green times to adapt to bikes and other vehicles
- Intersection control provides vehicle and bicycle detection at and nearby the stop bar. Typical intersection applications are green on demand and green time extension
- Inverse direction detection senses wrong-way drive on highways and their entrances and exits
- Vehicle and bicycle counting occurs simultaneously with presence detection
- Detection range: 0–90 m, depending on focal length of lens
- Frame rate: 30 fps (http://www.flir.co.uk/cs/display/?id=61843)

Vehicle detection

Bicyclist detection

Figure 5.15 Long-wavelength (thermal) infrared camera detection system. (Photographs courtesy of FLIR Systems, Marke, Belgium.)

cabinet, in response to a vehicle or bicycle detection as indicated in the right portion of the figure. The contact closure thus provides information concerning the number of vehicles passing the sensor per hour or the presence of a vehicle or bicycle in the detection area of the sensor.

5.4.3 General guidelines for installing VDS cameras

Four general guidelines for installing cameras used with VDSs are the following:

1. Maximize camera height to minimize vehicle occlusion and headlight reflection artifacts, and thus maximize the VDS measurement accuracy of vehicle count, speed, and

Figure 5.16 VDS camera mounting options include luminaire or signal mast arm mounting. (From LAK.)

presence detection. Cameras are often mounted on luminaire masts (preferred because they are higher) or on the traffic signal masts (often on an extension arm) over the centers of the lanes being monitored as displayed in Figure 5.16. In this particular four-approach intersection, four cameras are used, one for each approach direction. Sometimes, additional cameras are added to detect vehicles in left-turn pockets or other areas that are difficult for a single camera to monitor.

2. Select the camera location to avoid or minimize occlusion.
3. Center the camera over the lanes to be monitored.
4. Adjust the sun shield to minimize glint during sunrise and sunset with east–west facing cameras.

When operating VDS at night, adequate street lighting should be provided to ensure reliable vehicle detection by the VDS [5]. Most VDSs issue a recall to the controller if a vehicle is not detected within some specified time period to prevent the vehicle from being trapped at the intersection. Additional information concerning the installation of VDS is found in Chapter 7. Table 5.4 reviews the prominent characteristics of VDS as used for traffic management. Several of the sensor's limitations were noted in tests conducted several years ago and may not be present or as prominent in newer systems [6]. However, the weaknesses are worth noting so that potential users of these devices can make informed purchase and field-test design decisions.

5.5 MICROWAVE RADAR SENSORS

There are two types of microwave sensors used in traffic management applications: presence detecting and Doppler. The presence-detecting models detect stopped vehicles, while the Doppler models usually require the vehicle speed to be greater than some minimum value for detection. Presence-detecting radars find application in signalized intersection control, especially as an advance sensor, wrong-way vehicle detection, and freeway incident detection. Doppler microwave sensors are used to determine vehicle speeds on many limited-access highways.

Table 5.4   VDS typical output data, camera installation location, advantages, and limitations

Typical output data	Installation	Advantages	Limitations
Count Presence Lane occupancy Speed Vehicle and queue length Vehicle classification by length (up to 3) Alarms Expanded traffic parameter data set	Overhead (forward looking) or to side of roadway	Installation and repair need not interrupt traffic (for side-mounted cameras) Single camera and processor can service multiple lanes Rich array of data available Detection areas are easily reconfigurable	Large vehicles project their image into adjacent lanes, sometimes leading to false detection Large vehicles can also mask trailing vehicles and vehicles in lanes further from a side-mounted camera Performance may be affected by shadows (false calls or masking if a vehicle is in an area of shadow), reflections from wet pavement (false calls), day/night transitions, headlight beams protruding past stop bar (dropped calls) or into adjacent lanes (false calls), relative color of vehicles and background (failure to detect), camera vibration, sun glint for east–west facing cameras, weather (effects ameliorated by recall modes) Reliable nighttime signal actuation (based on video imagery) requires street lighting High camera mounting needed for more accurate data

5.5.1 Presence-detecting microwave radar sensors

Figure 5.17 contains examples of microwave presence-detecting radar sensors that detect stopped vehicles and therefore have presence detection outputs. These sensors are active devices that transmit energy and receive a portion of it that is scattered back into their aperture or sensing area. They operate in the 24-GHz band and can differentiate vehicles in multiple lanes from a side-mounted configuration. The presence-detecting microwave sensors in Figure 5.18 are capable of differentiating vehicles in up to six lanes from a forward-looking mounting position when they are mounted on a signal mast arm. The models in these figures are meant to be representative of products and capabilities that are currently offered. These and other manufacturers should be contacted by interested purchasers to obtain information about the latest devices and features.

5.5.2 Doppler microwave sensors

These sensors do not generally detect stopped vehicles because they rely on the Doppler principle to sense a vehicle, that is, the vehicle must be moving at a speed greater than some minimum established by the manufacturer. However, there are some Doppler microwave sensors that do report the positions of stopped vehicles due to their electronics designs or the manner in which their data processing algorithms are constructed. Figure 5.19 shows examples of Doppler microwave sensors that measure the speed and other characteristics of vehicles in support of various applications.

RTMS™ Sx-300 multizone presence-detecting microwave radar. Detection range up to 76 m (250 ft) in up to 12 lanes. Optional HD camera is available for visual confirmation of setup. Provides presence indication and measurements of volume, occupancy, gap, headway, average speed, 85th percentile speed, and classification (up to 6 classes). (Photograph courtesy of ISS, St. Paul, MN)

SmartSensor HD™ multizone presence-detecting microwave radar. Detection range 1.8–76.2 m (6–250 ft) in up to 22 lanes. Provides per lane data of volume, average speed, occupancy, classification counts (up to 8 classes), 85th percentile speed, average headway, average gap, speed bin counts (up to 15 bins), direction counts. Provides per vehicle data of speed, length, class, lane assignment, range. (Photograph courtesy of Wavetronix, Lindon, UT)

UMRR-OC Traffic sensor family. Models offer a variety of detection ranges capable of 1000 ft and greater for cars, with field of views up to 100-deg azimuth. Provides speed, range gate, and angle gate data corresponding to moving and stationary (option) vehicles. Multiple vehicle tracking gives x and y components of position and speed. Applications include stop bar and advance zone detection; vehicle count, headway, occupancy, queue length, and classification (up to 4 classes); red light enforcement. (Photograph courtesy of Smart Microwave Sensors GmbH, Braunschweig, DE)

AGD model 316 multizone presence-detecting microwave radar. Optimized for detection of stationary vehicles at the stopline, Individual vehicle tracking, User adjustable zone position. Typical applications include: dual-lane stopline vehicle detection at intersection and single-lane stop-line detection. (Photograph courtesy of AGD, Gloucestershire, UK)

Figure 5.17 Presence-detecting microwave radar sensors.

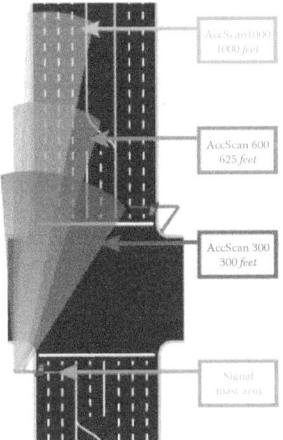 

❑ AccuScan 300 microwave sensor
  • Stop bar vehicle detection for up to 6 lanes
  • Counting and classification
  • Wrong way detection
  • Speed measurement
❑ Installation Parameters
  • Traffic direction: Approaching and receding
  • Typical mounting height: 18–24 ft (6–8 m)
  • Typical stop bar detection distance: 60–150 ft (20–50 m)
❑ Sensor performance data
  • Maximum range typ. 344 ft (105 m)
  • Range accuracy: typ. <±2.5% or <±0.25 m (bigger of)
  • Speed accuracy: typ. <±0.28 m/s or ±1% (bigger of)
  • Update time: ≤50 ms
  • Track initialization time: 6–10 cycles
  • Simultaneously tracked objects: up to 64

❑ AccuScan 600 microwave sensor
  • Stop bar and advance detection for up to 4 lanes
  • Counting and classification
  • Wrong way detection
  • Speed measurement
❑ Installation parameters
  • Traffic direction: Approaching and receding
  • Typical mounting height: 18–24 ft (6–8 m)
  • Typical stop bar detection distance: 60–150 ft (20–50 m)
  • Typical advance detection distance: 150–480 ft (50–160 m)
❑ Sensor performance data
  • Maximum range typ. 525 ft (160 m)
  • Range accuracy: typ. <±2.5% or < ±0.25 m (bigger of)
  • Speed accuracy: typ. <±0.28 m/s or ±1% (bigger of)
  • Update time: ≤50 ms
  • Track initialization time: 6–10 cycles
  • Simultaneously tracked objects: up to 64

AccuScan 24 GHz Multilane, forward-looking microwave radar sensors

Figure 5.18 Forward-looking, multilane, presence-detecting microwave radar sensors. (Photographs courtesy of Econolite, Anaheim, CA.)

SpeedInfo DVSS-100 solar-powered Doppler sensor. Measures the speed of vehicles on both sides of the highway from a single device. (Photograph courtesy of SpeedInfo, San Jose, CA)

ASIM 334 24 GHz Doppler sensor. Detects vehicles moving into or through its field of view out to 45 m (≈150 ft). Minimum detectable speed: 4 or 8 km/h (2.5 or 5 mi/h). Applications include direction-selective detection and green phase request and extension. (Photograph courtesy of ASIM Xtralis Technologies, Kiel, DE)

ADEC TDD1 24 GHz Doppler series sensor. Detects vehicles moving through its field of view up to 75 m (≈250 ft). Minimum detectable speed: 4 or 8 km/h (2.5 or 5 mi/h). Applications include direction-selective detection and green phase request and extension. (Photograph courtesy of ADEC Technologies, Eschenbach, CH)

AGD Model 318 FMCW, 24 GHz radar
Speed measurement from 4 km/h to 300 km/h across multiple lanes. Vehicle range measurement from 6 – 150 m. Typical applications include: multi-lane highway vehicle detection, single lane detection for traffic control, dual line detection for rail, wrong-direction detection for highways, sign activation. (Photograph courtesy of AGD, Gloucestershire, UK)

Figure 5.19 Doppler microwave sensors.

5.5.3 Microwave radar operation

Radar detection is a stochastic process, that is, the radar sensor may or may not produce a detection event when a vehicle is present in the sensor's field-of-view. A radar sensor may also produce a detection event when an object other than a vehicle is in the field-of-view of the sensor, for example, clutter or some other object not of interest such as a metallic feature on a bridge or fence. The latter detections are called false alarms and may be reduced by requiring the signal corresponding to valid vehicle detections to exceed a threshold voltage and contain a number of characteristics common to those of the vehicles of interest. The probability of a valid detection is specified through a detection probability and the probability of a false detection event through a false alarm probability. These parameters are determined by the manufacturer of the radar.

Figure 5.20 describes how an overhead-mounted microwave radar transmits energy toward an area of roadway and detects a vehicle. The beamwidth or area in which the radar energy is concentrated is controlled by the size and the distribution of energy across the aperture of the antenna. These design constraints are usually established by the sensor

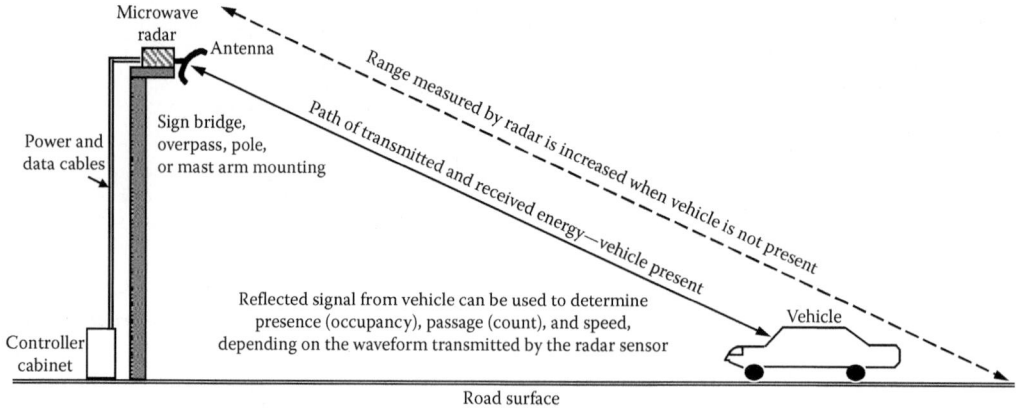

Figure 5.20 Vehicle detection by a presence-detecting microwave radar sensor.

manufacturer. When a vehicle passes through the antenna beam, a portion of the transmitted energy is reflected back toward the antenna. The presence of a vehicle is detected by the change in distance to the energy reflecting surface measured by the radar when the vehicle appears. The signal that enters the sensor's receiver is used to calculate vehicle and traffic flow data such as volume, lane occupancy, speed, and vehicle length.

Forward-looking radars with large antenna beamwidths acquire data representative of the composite traffic flow in one direction over multiple lanes. Forward-looking radars with narrow antenna beamwidths can monitor a single lane or several individual lanes of traffic flowing in one direction, depending on the model. Side-mounted, multiple-detection zone radars project their detection area (i.e., footprint) perpendicular to the traffic flow direction and provide traffic data from several lanes.

5.5.4 Types of transmitted waveforms

The top of Figure 5.21 illustrates the constant frequency waveform transmitted by a continuous wave (CW) Doppler microwave sensor. The constant frequency signal (with respect to time) allows vehicle speed to be measured using the Doppler principle. Vehicle speed is proportional to the frequency change f_D between the transmitted and received signals given by

$$f_D = \frac{2Sf\cos\theta}{c}, \tag{5.4}$$

where S is vehicle speed, f is transmitted frequency, f_D is Doppler frequency, c is the speed of light, and θ is the angle between the direction of propagation of the sensor energy and the direction of travel of the vehicle.

The frequency of the received signal is given by $f \pm f_D$, where $\pm$ corresponds to whether the vehicle is moving toward or away from the sensor. Accordingly, the frequency of the

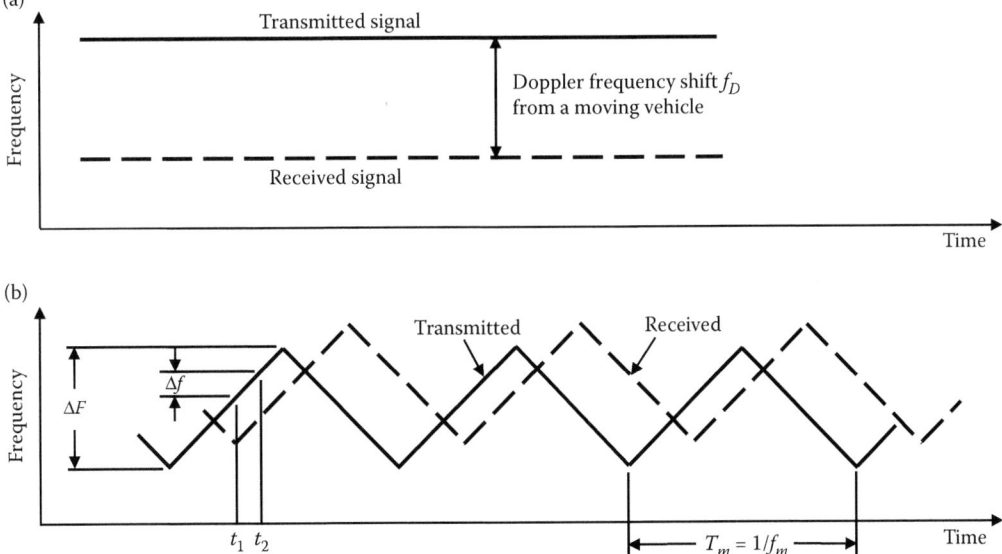

Figure 5.21 Waveforms transmitted by a Doppler microwave sensor and a presence-detecting microwave radar. (a) Doppler microwave sensor transmits a constant-frequency signal, (b) Presence-detecting microwave radar transmits an FMCW signal.

received signal is increased by a vehicle moving toward the sensor and decreased by a vehicle moving away from the sensor. Vehicle passage or count is denoted by the presence of the frequency shift. Vehicle presence cannot be measured with the constant frequency waveform as only moving vehicles are detected by most sensors of this type.

The bottom of Figure 5.21 shows the waveform transmitted by many presence-detecting microwave radars. This frequency-modulated continuous wave (FMCW) signal is characterized by a transmitted frequency that is constantly changing with respect to time. It is this feature that enables the radar to measure range and detect vehicle presence, that is, stopped vehicles. The range R to the vehicle is proportional to the difference in the frequency Δf of the transmitter at the time t_1 when the signal is transmitted and the time t_2 when a portion of the transmitted energy is received back at the sensor or, equivalently, the time difference $t_2 - t_1$. This relation is written as

$$R = \frac{c(t_2 - t_1)}{2}.$$

(5.5)

The equation for range may also be written in terms of the other parameters that appear when designing an FMCW radar as

$$R = \frac{c\Delta f}{4\Delta F f_m},$$

(5.6)

where Δf is instantaneous difference in frequency of the transmitter at the times the signal is transmitted (t_1) and received (t_2), ΔF is RF modulation bandwidth (a design parameter), and f_m is RF modulation frequency (another design parameter).

5.5.5 Range resolution

Range resolution ΔR, the minimum distance resolved by an FMCW radar, is given by

$$\Delta R = \frac{c}{2\Delta F}.$$

(5.7)

Therefore, if the radar sensor operates in the 10.500- to 10.550-GHz band and the bandwidth ΔF is limited to 45 MHz to limit signal spillover outside the band, the range resolution is 10.8 ft (3.3 m). A radar sensor operating in the 24-GHz band with 75 MHz of bandwidth has a range resolution of 6.6 ft (2 m).

Speed or Doppler resolution Δf_D is

$$\Delta f_D = 2f_m = \frac{2}{T_m},$$

(5.8)

where f_m is defined above and T_m is the reciprocal of f_m.

5.5.6 Range bins

Forward-looking and side-mounted radars employ range bins to assist in measuring speed and in differentiating vehicles traveling in different lanes as illustrated in the top portion

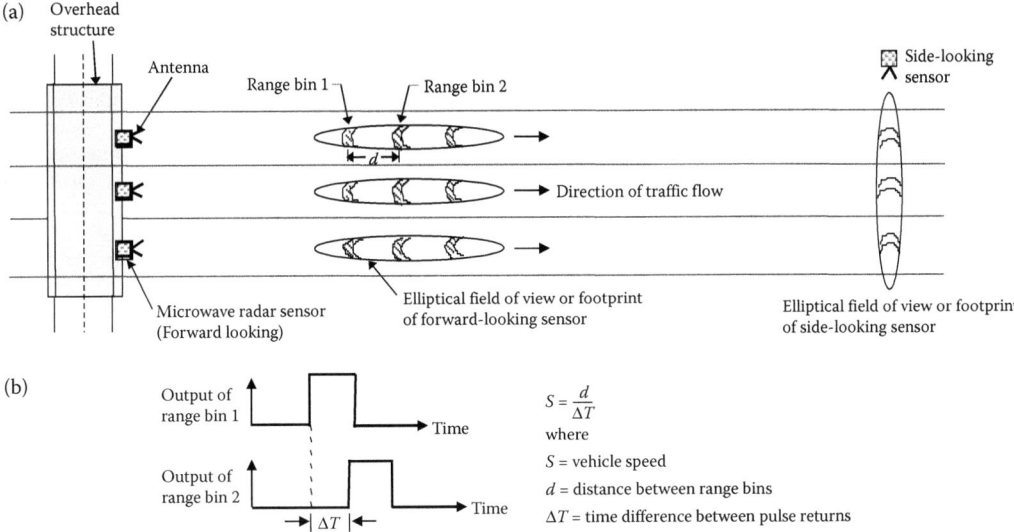

Figure 5.22 Forward- and side-looking presence-detecting radars use range bins for improved spatial resolution and speed measurement accuracy. (a) Range bins in forward- and side-mounted presence-detecting radars, (b) Vehicle speed measurement in a forward-looking radar.

of Figure 5.22. The range bins sort the received signals by time of arrival relative to the transmitted signal. This technique allows vehicle flow data to be collected over one lane or multiple lanes in the case of forward-looking devices or over multiple lanes with side-mounted devices, depending on the manufacturer and model of the radar. The range bins in a forward-looking sensor operate similarly to the speed-trap configuration of inductive loops for measuring vehicle speed. When a vehicle enters the leading range bin, a pulse is created by the signal processing electronics. A similar pulse is formed when the vehicle enters the trailing range bin. The vehicle speed is found by dividing the distance between the range bins (a known design quantity) by the time difference between the start of the leading and trailing pulses as indicated in the bottom of Figure 5.22 [7]. The range bins in a side-looking sensor are used to differentiate vehicles traveling in different lanes and hence measure their traffic flow parameters.

Table 5.5 summarizes the characteristics of presence-detecting microwave radar sensors as used for traffic management. Table 5.6 contains similar information for microwave Doppler sensors.

5.6 PASSIVE INFRARED SENSORS

Figure 5.23 displays examples of passive infrared (PIR) sensors. This type of sensor transmits no energy of its own. Rather it detects energy from two sources: (1) energy emitted from vehicles, road surfaces, and other objects in their field-of-view and (2) energy emitted by the atmosphere and reflected by vehicles, road surfaces, or other objects into the sensor aperture. The energy detected by infrared sensors is focused by an optical system onto an infrared-sensitive material mounted at the focal plane of the optics. With infrared sensors, the word detector takes on another meaning, namely, the infrared-sensitive element that converts the reflected and emitted energy into electrical signals. Real-time signal processing

Table 5.5 Microwave radar typical output data, installation location, advantages, and limitations

Typical output data	Installation	Advantages	Limitations
Count Presence with FMCW waveform Lane occupancy with FMCW waveform for stopped and moving vehicles Speed Range with FMCW waveform "Pseudo" traffic density calculated from point data with FMCW models Vehicle class by length (up to some limit), where length is found as the time a vehicle is in the radar beam multiplied by vehicle speed	Overhead (forward looking) or to side of roadway	Installation and repair need not interrupt traffic (when side-mounted) Direct measurement of speed Multilane data collection Day/night operation Detection areas are reconfigurable in many models Not affected by inclement weather	With side mounting, possibility of missed detections if tall vehicles occlude more distant lanes when congestion is heavy—therefore some models not recommended for stopline detection (may not be a significant effect in other applications) Vehicle undercounting may increase in heavy congestion Offset mounting distance must be accommodated Stochastic property of radar detection may cause device to miss detecting a vehicle (using multiple-detection zones, when available, and device design features such as a high FMCW repetition frequency may ameliorate this issue)

Table 5.6 Microwave Doppler sensor typical output data, installation location, advantages, and limitations

Typical output data	Installation	Advantages	Limitations
Count Lane occupancy for moving vehicles Speed	Side of roadway	Installation and repair need not interrupt traffic Direct measurement of speed Day/night operation Not affected by inclement weather	Generally, cannot detect stopped or very slow-moving vehicles

ASIM IR 254/5. Multi-zone sensor counts vehicles, measures speed, classifies vehicles by length, and detects vehicle presence.

ASIM IR 300 series. Detects moving vehicles to request and extend the green phase of traffic signals. Also detects moving people, animals, and other objects.

(Photographs courtesy of ASIM Xtralis, Kiel, DE)

ADEC TDC1. Multi-zone sensor counts vehicles, measures speed, classifies vehicles by length (3–5 classes), and detects vehicle presence.

(Photograph courtesy of ADEC Technologies, Eschenbach, CH)

ITMS-100. Solar-powered multilane sensor provides vehicle count, speed, and classification by lane. Also road surface temperature for determining snow and ice conditions. Range up to 250 ft (76 m).

(Photograph courtesy of SpeedInfo, San Jose, CA)

Figure 5.23 Passive infrared sensors.

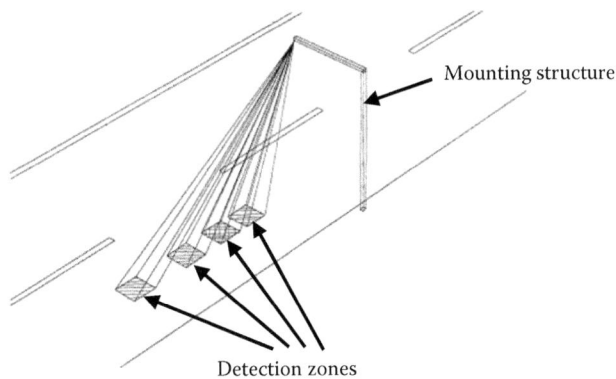

Figure 5.24 Multiple-detection zones on road surface as created by a passive infrared sensor.

is applied to analyze the received signals for the presence of a vehicle. Infrared sensors are utilized for signal control; volume, speed, and vehicle class measurement; detection of pedestrians in crosswalks; and transmission of traffic information to motorists.

Multichannel (i.e., more than one type of sensor technology) and multizone (i.e., more than one detection region) passive infrared sensors measure speed and vehicle length as well as the more conventional volume and lane occupancy. Figure 5.24 illustrates the multiple-detection zone concept. These models are designed with dynamic and static thermal energy detection zones that provide the functionality of two inductive loops. The time delays between the signals from three dynamic zones are utilized to measure speed. The vehicle presence time from the fourth zone is used to calculate the lane occupancy of stationary and moving vehicles.

The Planck radiation law governs the emission of energy detected by a passive infrared sensor. The emitted energy is produced by the nonzero surface temperature of emissive objects in the sensor's field-of-view. Emission occurs at all frequencies by objects not at absolute zero (−273.15°C). If the emissivity of the object is perfect, that is, the emissivity is 1, the object is called a blackbody. Most objects have emissivities less than 1 and, hence, are termed gray bodies. Passive sensors can be designed to receive energy at any frequency. Cost considerations make the infrared band a good choice for vehicle sensors that incorporate a limited number of pixels. Some models operate in the long-wavelength infrared band from 8 to 14 μm and thus minimize the effects of sun glint and changing light intensity from cloud movement. Electronics in the sensor is designed to detect the difference between energy emitted from the road surface when no vehicle is present and energy emitted when a vehicle is present.

5.6.1 Planck radiation law

Planck's radiation law describes the emission of energy from blackbody objects not at a temperature of absolute zero. These objects emit energy E per unit volume and per unit frequency at all wavelengths according to

$$E = \frac{8\pi h f^3}{c^3} \frac{1}{\exp(hf/k_B T) - 1} \quad \text{J/m}^3\text{Hz,} \tag{5.9}$$

where J represents units of energy in Joules, h is Planck's constant (6.6256×10^{-34} J-s), k_B is Boltzmann's constant (1.380662×10^{-23} J/K), c is the speed of light (3×10^8 m/s), T is the physical temperature of the emitting object in degrees K, and f is frequency at which the energy is measured in Hz.

Upon expanding the exponential term in the denominator, Planck's radiation law becomes

$$E = \frac{8\pi h f^3}{c^3} \frac{1}{(hf / k_B T) + (hf / k_B T)^2 + \cdots} \quad \text{J/m}^3\text{Hz}. \tag{5.10}$$

For frequencies f less than $k_B T / h$ ($\approx 6 \times 10^{12}$ Hz at 300 K), only the linear term in temperature is retained and Planck's radiation law reduces to the Rayleigh–Jeans law given by

$$E = \frac{8\pi f^2 k_B T}{c^3} \quad \text{J/m}^3\text{Hz}. \tag{5.11}$$

In the Rayleigh–Jeans approximation, temperature is directly proportional to the energy of the radiating object, making passive sensor calibration simpler. This form of Plank's radiation law is valid for passive microwave sensors, but not for passive infrared sensors.

With perfect emitters or blackbodies, the physical temperature of the object T is equal to the brightness temperature T_B that is detected by a radiometer such as a passive infrared vehicle sensor. However, the surfaces of real objects do not normally radiate as blackbodies (i.e., they are not 100% efficient in emitting the energy predicted by the Planck radiation law). To account for this nonideal emission, a multiplicative emissivity factor is added to represent the amount of energy radiated by the object, now referred to as a gray body. Emissivity ε is equal to the ratio of T_B to T where $0 \leq \varepsilon \leq 1$.

Because of emission from molecules not at absolute zero, the atmosphere emits energy that passive sensors detect either directly or indirectly (such as by reflection of energy from surfaces whose emissivity is not unity) as they view objects through the atmosphere. Application of radiative transfer theory allows the determination of the brightness temperature change caused by a vehicle passing through a sensor's field-of-view.

5.6.2 Radiative transfer theory

When a vehicle enters a passive infrared sensor's field-of-view, the detected energy changes due to the presence of the vehicle [8]. Radiative transfer theory describes the contributions of cosmic, galactic, atmospheric, and ground-based emission sources to the passive signature of the detected objects. As shown in Figure 5.25, the cosmic, galactic, and atmospheric emission sources can be lumped into a term denoted as T_{sky}, while the ground-based radiation sources derive from emission from the road and vehicle surfaces and the reflection of the cosmic, galactic, and atmospheric sources into the sensor's aperture.

In Figure 5.25, the emissivities of the vehicle and road surface in the wavelength region of interest are denoted by ε_V and ε_R and their surface temperatures in degrees kelvin by T_V and T_R, respectively. Emission from the vehicle contributes a brightness temperature T_{BV} given by

$$T_{BV}(\theta, \phi) = \varepsilon_V T_V + (1 - \varepsilon_V) T_{sky}, \tag{5.12}$$

where θ and ϕ are the nadir and azimuth angles, and T_{sky} is a function of atmospheric, galactic, and cosmic emission.

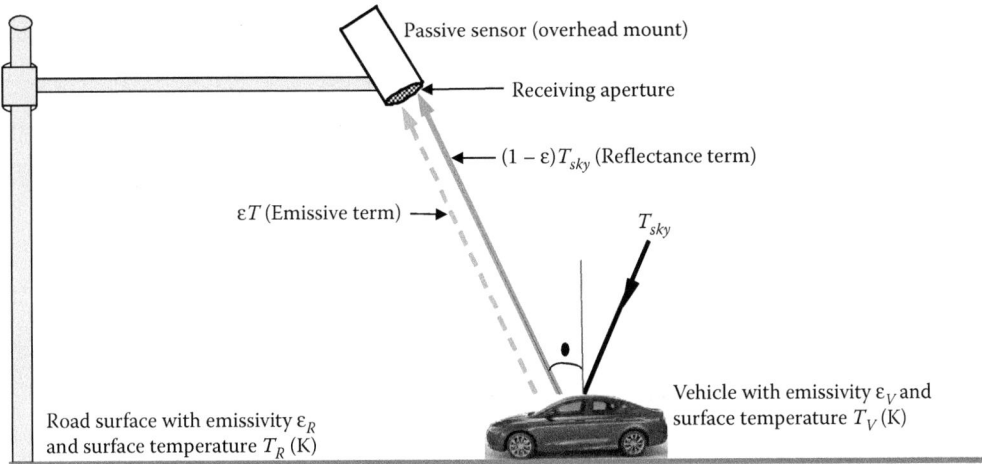

Figure 5.25 Radiative transfer theory is used to calculate the change in apparent temperature sensed by a passive infrared sensor when a vehicle enters its field-of-view.

Emission from the road surface contributes a brightness temperature T_{BR} as

$$T_{BR}(\theta,\phi) = \varepsilon_R T_R + (1 - \varepsilon_R)T_{sky}. \tag{5.13}$$

Finally, the difference in brightness temperature $\Delta T_B (\theta, \phi)$ with and without a vehicle in the field-of-view is

$$\begin{aligned} \Delta T_B(\theta,\phi) &= T_{BV} - T_{BR} = \varepsilon_V T_V + T_{sky} - \varepsilon_V T_{sky} - \varepsilon_R T_R - T_{sky} + \varepsilon_R T_{sky} \\ &= \varepsilon_V T_V - \varepsilon_R T_R + (\varepsilon_R - \varepsilon_V)T_{sky} = \varepsilon_V(T_V - T_{sky}) - \varepsilon_R(T_R - T_{sky}). \end{aligned} \tag{5.14}$$

When $T_R = T_V$,

$$\Delta T_B(\theta,\phi) = (\varepsilon_R - \varepsilon_V)(T_R - T_{sky}). \tag{5.15}$$

5.6.3 Passive infrared sensor summary

Table 5.7 lists the output data, installation location, advantages, and performance limitations of passive infrared sensors. Several disadvantages of passive infrared sensors are sometimes cited. Glint from sunlight may cause unwanted and confusing signals in some infrared sensor wavelength bands. Atmospheric particulates and inclement weather can scatter or absorb energy that would otherwise reach the focal plane of the sensor. The scattering and absorption effects are sensitive to water concentrations in fog, haze, rain, and snow as well as to other obscurants such as smoke and dust.

At the relatively short operating ranges encountered by infrared sensors in traffic management applications, these concerns may not be significant. However, some performance degradation (e.g., undercounting) in heavy rain and snow has been reported. A rule of thumb for determining when a sensor operating in the near-infrared wavelength band may experience difficulty detecting a vehicle is to note if a human observer can see the vehicle under

Table 5.7 Passive infrared sensor typical output data, installation location, advantages, and limitations

Typical output data	Installation	Advantages	Limitations
Count Presence Lane occupancy Speed with multiple- detection zone models Queue with multiple sensors or detection zones	To side of roadway	Installation and repair need not interrupt traffic Compact size Day/night operation Potentially better performance with long-wavelength infrared than visible wavelength sensors in some fog conditions	Performance possibly degraded by heavy rain, fog, snow One per lane required

the same circumstances. If the observer can see the vehicle, there is a high probability the infrared sensor will detect the vehicle.

5.7 LIDAR SENSORS

Lidar sensors illuminate detection zones with energy transmitted by laser diodes operating in the near-infrared region of the electromagnetic spectrum at a wavelength of 0.85 μm. A portion of the transmitted energy is reflected or scattered by vehicles back toward the sensor. The sensors are mounted overhead to view approaching or departing traffic as exemplified by the left and middle lidar models in Figure 5.26. They can also be mounted in a side-looking configuration, which may be preferred when classifying vehicles for tolling applications or axle counting as shown on the right of the figure. Lidars provide vehicle presence at traffic signals, volume, speed, length assessment, queue measurement, and classification.

Lidar sensors can project their detection zones across a lane in two ways. The first technique scans beams across a lane as depicted in Figure 5.27. The transmitting optics split the pulsed laser diode output into two beams, separated by several degrees, which are scanned across the lane with a rotating mirror. The LaserScan sensor on the left of Figure 5.26 is typical of this design. The receiving optics has a wider field-of-view so that it can more effectively capture the energy scattered from the vehicles, which is focused by an optical system onto a detector array mounted at the focal plane of the optics.

The two-beam approach allows lidars to measure vehicle speed by recording the times at which the vehicle enters the detection area of each beam. Since the beams are a known distance apart, the speed is given by the ratio of the distance to the time difference corresponding to the vehicle's arrival at each beam or, in other words, the leading edge of each range bin. Other configurations of lidar sensors employ several laser diodes to transmit multiple

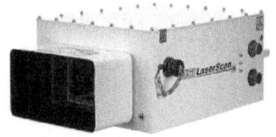

LaserScan 615: 7.65 m range,
4.1 m lane-width coverage
(Photograph courtesy of OSI Laserscan,
Hawthorne, CA)

SICK LMS211:
10–30 m range
(Photograph courtesy of SICK AG,
Waldkirch, Germany)

AxleLightLaser Axle Sensor: axle detection in 1 to 4 lanes
(Photographs courtesy of Peek Traffic, Palmetto, FL)

Figure 5.26 Lidar sensors.

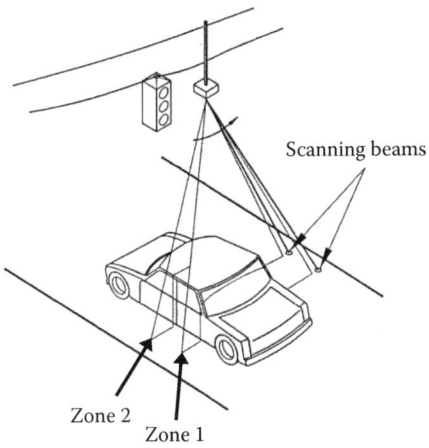

Figure 5.27 Lidar sensor using mirrors to scan beams across a lane.

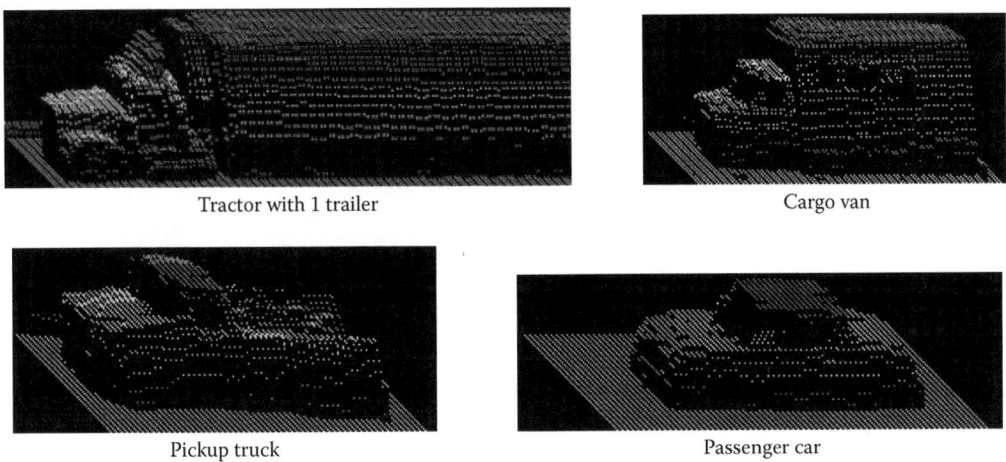

Figure 5.28 3D vehicle imagery produced by lidar sensors. (Photographs courtesy of OSI Laserscan [formerly Schwartz Electro-Optics], Hawthorne, CA.)

side-by-side beams over the monitored roadway lane and, thus, do not require a mirror to scan the beams across the lane.

With suitable signal processing software, modern lidar sensors produce two- and three-dimensional imagery of vehicles suitable for vehicle classification as illustrated in Figure 5.28. Table 5.8 summarizes the output data, installation location, advantages, and performance limitations of lidar sensors.

5.8 PASSIVE ACOUSTIC ARRAY SENSORS

Acoustic sensors are passive devices that transmit no energy of their own. Passive acoustic sensors measure vehicle passage, presence, and speed by detecting acoustic energy or audible sounds produced by vehicular traffic from a variety of sources within each vehicle, such as the engine and the interaction of the vehicle's tires with the road. When a vehicle

Table 5.8 Lidar sensor typical output data, installation location, advantages, and limitations

Typical output data	Installation	Advantages	Limitations

Count
Presence
Lane occupancy
Speed
Vehicle length
Vehicle classification with
  2D and 3D imaging
  models
Range
	Overhead
	Side of roadway for toll
	  road applications
		Installation and repair
		  need not interrupt
		  traffic when installed
		  at side of road
		Day/night operation
		Multilane operation
		  depending on model
			Performance degraded
			  by heavy fog and
			  blowing snow where
			  visibility $\leq$20 ft (6 m)
			Poor foliage penetration

Figure 5.29 SAS-1 passive acoustic sensor. (Photograph courtesy of SmarTek Systems, Inc., Woodbridge, VA.)

passes through the detection zone, an increase in sound energy is recognized by the signal processing algorithm and a vehicle presence signal is generated. When the vehicle leaves the detection zone, the sound energy level drops below the detection threshold and the vehicle presence signal is terminated. An array of microphones and signal processing algorithms impart spatial directivity to the detected sounds and have the ability to recognize them as originating from vehicles traveling in different lanes as indicated in Figure 5.29. Signals emanating from locations outside the detection zone are attenuated and ignored. Table 5.9 itemizes the output data, installation location, advantages, and performance limitations of passive acoustic array sensors.

5.9 ULTRASONIC SENSORS

Ultrasonic sensors are active sensors that transmit pressure waves of sound energy at a frequency between 25 and 50 kHz, which are above the human audible range. The most accurate data are obtained when they are mounted over the center of the monitored lane. An alternate mounting location at the lane edge (especially if the monitored lane is the rightmost lane) is sometimes used. They can also be mounted in a horizontal position when

Table 5.9 Passive acoustic array sensor typical output data, installation location, advantages, and limitations

Typical output data	Installation	Advantages	Limitations
Count Presence Lane occupancy Average vehicle speed for a selectable update period (1–220 s) by using multiple-detection zones or data processing algorithm that assumes an average vehicle length	To side of monitored lanes (up to 5 lanes)	Installation and repair need not interrupt traffic Day/night operation Multilane data collection	May undercount in congested flow Performance may degrade in heavy rain or cold weather

used as a vehicle detection trigger, for example, to prevent a gate in a parking structure from closing on top of a vehicle.

Most ultrasonic sensors operate with pulse waveforms and provide vehicle count, presence, and occupancy information. Pulse-shape waveforms measure distances to the road surface and vehicle surface by detecting the portion of the transmitted energy that is reflected toward the sensor from an area defined by the transmitter's field-of-view. When a distance other than that to the background road surface is measured, the sensor interprets that measurement as the presence of a vehicle. Pulsed ultrasonic sensors use detection gates (similar to those in microwave radar sensors) to assist in range measurement. The received ultrasonic energy is converted into electrical energy that is analyzed by signal processing electronics that is either collocated with the transducer or placed in a roadside controller.

Constant frequency ultrasonic sensors that measure speed using the Doppler principle are also manufactured. However, these are more expensive than pulsed models. The speed-measuring Doppler ultrasonic sensor is designed to interface with the highway infrastructure in Japan. It is mounted overhead facing approaching traffic at a 45° incidence angle. It has two transducers, one for transmitting and one for receiving a signal. The constant frequency ultrasonic signal is analogous to the constant frequency electromagnetic signal transmitted by a Doppler microwave sensor as illustrated in Figure 5.21a. The Doppler ultrasonic sensor detects the passage of a vehicle by a shift in the frequency of the received signal. Vehicle speed can be calculated from the pulse width of an internal signal generated by the sensor's electronics that is proportional to the speed of the detected vehicle.

Pulse-waveform ultrasonic sensors transmit a series of pulses of width T_p (typical values are between 0.02 and 2.5 ms) and repetition period (time between bursts of pulses) T_0, typically 33–170 ms, as described in Figure 5.30. The sensor measures the time it takes for the pulse to arrive at the vehicle and return to the transmitter. The receiver is gated on and off with a user-adjustable interval that differentiates between pulses reflected from the road surface and those reflected from vehicles. The detection gates of various models are adjusted to detect objects at distances greater than approximately 0.5–0.9 m above the road surface. This is achieved by closing the detection gate several milliseconds before the reflected signal from the road surface arrives at the sensor. A hold time T_h (composite values from manufacturers range from 115 ms to 10 s) is built into the sensors to enhance presence detection.

Figure 5.31 shows the reflection of ultrasonic pulses from a vehicle in its field-of-view. The heights at which the reflected pulses are detected can produce a height profile image of a vehicle if the pulse repetition period and vehicle speed are appropriate.

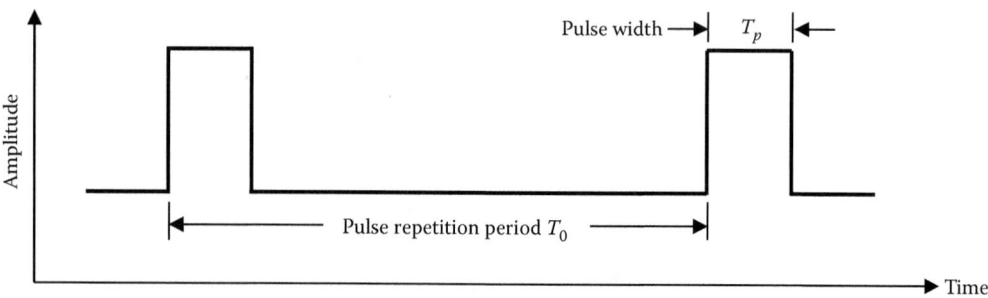

Figure 5.30 Pulse waveform for presence-measuring ultrasonic sensor.

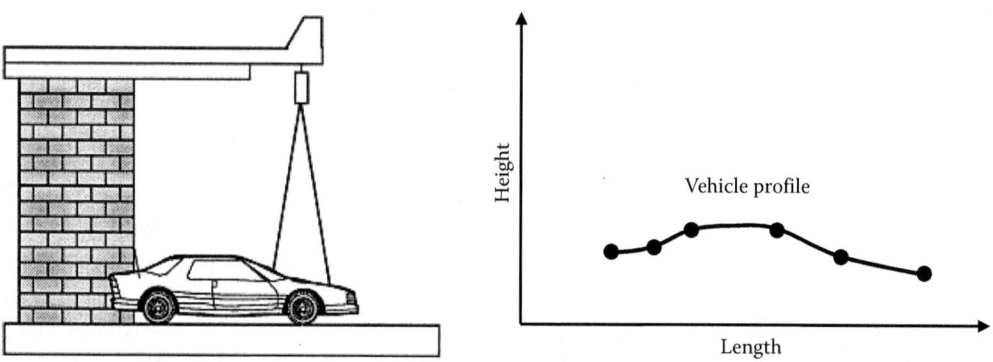

Figure 5.31 Ultrasonic sensor pulse returns are representative of a vehicle height profile.

Table 5.10 lists the output data, installation location, advantages, and performance limitations of ultrasonic sensors. Temperature change and extreme air turbulence may affect the performance of ultrasonic sensors. Temperature compensation is built into some models. Large pulse repetition periods may degrade occupancy measurement on freeways when vehicles are traveling at moderate to high speeds as an insufficient number of pulses are transmitted and reflected from the vehicle while in the sensor's detection zone.

Table 5.10 Ultrasonic sensor typical output data, installation location, advantages, and limitations

Typical output data	Installation	Advantages	Limitations
Count Presence Lane occupancy Speed with Doppler model or multiple-detection zone model Range with pulse model Queue with multiple single zone models	Most accurate when mounted overhead	Day/night operation Compact size	Traffic interrupted with overhead installation and maintenance May undercount in congested flow Performance may be degraded by variations in temperature and extreme air turbulence Low pulse repetition frequency (PRF) may degrade occupancy measurement on freeways with moderate to high speed vehicles One per lane required (generally)

5.10 SENSOR TECHNOLOGY COMBINATIONS

Several manufacturers combine two or more technologies in a single sensor unit for specialized applications or enhanced operational characteristics. Figures 5.32 through 5.34 display a number of these devices and their principal characteristics.

5.11 SENSOR OUTPUT DATA, BANDWIDTH, AND COST

Table 5.11 summarizes the types of data typically available from each sensor technology, lane coverage area, communication bandwidth requirements, and purchase costs. Several technologies are capable of supporting multiple-lane, multiple-detection zone applications with one or a limited number of units. These devices may be cost effective when larger numbers of detection zones are needed for the traffic management application. Other traffic flow parameters are provided by wider field-of-view sensors such as VDSs and multilane microwave radar sensors.

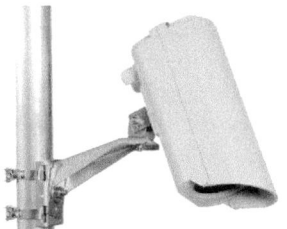

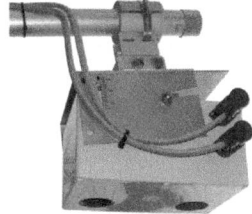

DT 372 Infrared-ultrasound sensor. Provides presence detection and vehicle counts.
Ultrasound frequency: 40 kHz
Pulse frequency: 3.5–30 Hz
Pyroelectric infrared detector: 8–14 μm side mounted.

DT 351 Infrared-Doppler sensor. Provides vehicle counts, speed, length classification
(2 classes), true presence and queue detection.
K-band Doppler: 24.05–24.25 GHz
Pyroelectric infrared detector: 8–14 μm side mounted.

TT 290 Series infrared-Doppler-ultrasound sensor. Provides per vehicle classification, speed, length; vehicle counts; true presence and queue detection; occupancy; headway.
K-band Doppler: 24.05–24.25 GHz
Ultrasonic frequency: 50 kHz
Pulse frequency: 10–30 Hz
Pyroelectric infrared detector: 8–14 μm mounted above lane and aimed along direction of travel.

Figure 5.32 ASIM sensor technology combinations. (Photographs courtesy of ASIM by Xtralis, Kiel, Germany.)

TDC3 series infrared, Doppler, ultrasound sensor provides vehicle counts, individual vehicle speed (Doppler), vehicle class (ultrasound and PIR), presence, queue and wrong-way driver detection, occupancy, headway, and time gap.
K-band Doppler: 24.05–24.25 GHz
Ultrasonic frequency: 40 kHz
Ultrasonic pulse rate: 10–30 pulses/s
PIR spectral response: 6.5–14 μm
Recommended mounting: Gantries or other overhead structures above the lane center.

TDC3 series sensor

TDC4 series sensor

TDC4 series infrared, Doppler, ultrasound, video sensor provides vehicle counts, individual vehicle speed, vehicle class, presence, queue and wrong-way driver detection, occupancy, headway, time gap, and visual verification of traffic flow anomalies.
VGA color video: 640 × 480 max.
(Provides snapshot pictures transmitted over 9K6 bps RS 485 for visual verification of wrong-way drivers and queues, and photos for outstation command.)
K-band Doppler: 24.05–24.25 GHz
Ultrasonic frequency: 40 kHz
Ultrasonic pulse rate: 10–30 pulses/s
PIR spectral response: 6.5–14 μm
Recommended mounting: Same as for TDC3.

Figure 5.33 ADEC sensor technology combinations. (Photographs courtesy of ADEC Technologies, Eschenbach, Switzerland.)

- Video and microphone array (acoustic) sensor combination
- Solar operation
- Outputs include:
  - Vehicle count
  - Vehicle speed
  - Lane occupancy
  - Lane by lane vehicle class
- Remote programming of flow and speed thresholds

(Photograph courtesy of Neavia Technologies, Créteil, FR)

Presence-detecting 24 GHz radar and video detection system for signal actuation. It provides simultaneous traffic data collection, IP-based communications, and MPEG-4 streaming video.
(Photograph courtesy of Econolite, Anaheim, CA)

Figure 5.34 Neavia Technologies EagleVia sensor (left) and Autosccope® Duo (right) technology combinations.

Communication bandwidth is low to moderate if only data and control commands are transmitted between the sensor, controller, and traffic management center. The bandwidth is larger if real-time video imagery is transmitted at 30 frames/s. The transmission rate is also affected by the numbers of sensors, roadside information devices such as changeable message signs and highway advisory radio, and frequency of signal timing plan updates needed to implement traffic management strategies.

The range of purchase costs for a particular sensor technology reflects cost differences among specific sensor models and capabilities. If multiple lanes are to be monitored on a lane-by-lane basis and a sensor is capable of only single detection zone operation, then the sensor cost must be multiplied by the number of monitored lanes. Installation and life-cycle maintenance costs also contribute to the true cost of any sensor selection.

5.11.1 Life-cycle cost considerations

Direct hardware and software purchase costs are only one portion of the expense associated with a sensor. Installation, maintenance, and repair should also be factored into the sensor selection decision. Installation costs include fully burdened costs for technicians to prepare the road surface or subsurface (for inductive loops or other surface or subsurface sensors), install the sensor and mounting structure (if one is required), provide power if none is available at the site, close traffic lanes, divert traffic, provide safety measures where required, and verify proper functioning of the device after installation is complete. Environmental concerns may warrant providing for the removal of cutting water and debris from the site.

Maintenance and repair estimates may be available from manufacturers and from other agencies and localities that have deployed similar sensors. Some of the above-roadway mounted sensors are designed with a mean time between failures of 64,000 and 90,000 hours. Thus, maintenance and replacement costs for these devices may be significantly less than for inductive loops over a 10-year period, especially if commercial vehicle loads, poor subsoil, weather, and utility improvements frequently require road resurfacing and loop replacement. The technologies listed in Table 5.11 are mature with respect to current traffic management applications, although some may not provide the data required for a specific application or may not perform as needed under the local weather and other environmental conditions. Some technologies, such as VDSs and presence-detecting radars, continue to evolve by adding capabilities that measure additional traffic parameters, track vehicles, link data from one sensor to another, improve resolution, operate from solar energy, or remove susceptibility to factors that once affected their operation.

Table 5.11 Sensor output data, lane coverage options, communication bandwidth, and cost

| Technology | Output data | | | | | | Communication bandwidth | Sensor purchase cost[a] (each in 2010 or later $US) |
	Count	Presence	Speed	Occupancy	Classification	Multiple-lane, multiple-detection zone data		
Inductive loop	X	X	X[b]	X	X[c]		Low to moderate	Low[d] ($600 to $900)
Magnetometer (2- or 3-axis)	X	X	X[b]	X			Low	Moderate to high[d] ($2100 to $31,000)
Magnetic (induction coil)	X	X[e]	X[b]	X			Low	Low[d] ($180 to $400)
Microwave radar	X	X[f]	X	X[f]	X[f]	X[f]	Moderate	Moderate ($3900 to $5800)
Lidar	X	X	X[g]	X	X	X	Low to moderate	Moderate to high ($4000 to $10,500)
Passive infrared	X	X	X[g]	X			Low to moderate	Low ($480 to $650)
Ultrasonic	X	X		X			Low	Moderate[h] ($1300 to $2900)
Acoustic array	X	X	X	X		X	Low to moderate	Moderate ($3850 to $4500)
Video detection system	X	X	X	X	X	X	Low to high[i]	Moderate to high ($2375 to $6000 per approach)

a   Budgetary prices. Installation, maintenance, and repair costs must also be included to arrive at the true cost of a sensor solution. Quantity discounts usually apply.
b   Speed measured by using two sensors a known distance apart or by knowing or assuming the effective detection zone and vehicle lengths.
c   With specialized high-frequency sampling electronics unit (detector).
d   Includes wire loops, lead-in wire and cable, and detector electronics purchase and installation. Detector electronics options are available for multiple-sensor, multiple-lane coverage.
e   With special sensor layouts and signal processing software.
f   From presence-detecting microwave radar sensors that transmit FMCW waveform and have appropriate signal processing.
g   With multi-detection zone passive or active infrared sensors.
h   Cost is for ultrasonic in combination with other senor technologies.
i   Depends on whether higher-bandwidth raw data, lower-bandwidth processed data, or video imagery is transmitted to the traffic management center.

5.11.2 Relative cost of a sensor solution

A satisfactory cost comparison between various sensor technologies can only be made when the specific application is known. A relatively inexpensive ultrasonic, microwave, or passive infrared sensor (assuming that the data types, accuracy, and other requirements are fulfilled by all the candidates) may seem to be the low-cost choice for instrumenting a surface street intersection if ILDs are not desired. But when the number of sensors needed is taken into account along with the limited amount of directly measured data that may be available (e.g., speed is not measured directly by a single zone infrared sensor), a more expensive sensor type such as a VDS may be the better choice. Consequently, if it requires 12–16 conventional ILDs (or ultrasonic, microwave, or infrared, etc. sensors) to fully instrument an intersection, the cost becomes comparable to that of a VDS. Furthermore, the additional traffic data and visual information made available by the VDS may more than offset any remaining cost difference. In this example, the VDS is assumed to meet the other requirements of the application, such as the desired 100% detection of vehicles at the intersection. Similar arguments can be made for freeway applications using multiple sensors and requiring information not always available from the less-expensive sensors.

Still other applications, such as simple monitoring of multilane freeway traffic flow or surface street vehicle presence and speed, may be performed by two to four multi-detection zone presence-detecting radars mounted in a forward- or side-looking configuration. In this instance, the microwave sensors replace a greater number of loops or other subsurface sensors that otherwise need be installed in the travel lanes. Furthermore, the microwave sensor potentially provides direct measurement of speed at a greater accuracy than provided by the loops.

5.12 SUMMARY

Knowledge of the theory of operation of modern traffic flow sensors gives traffic management personnel the capability to understand the attributes of each of the technologies and thus make an informed decision as to which is appropriate for a particular application. A sample of sensor models representing inductive loop, magnetic, and above-roadway mounted technologies was described to show that the data and information for supporting current traffic management applications are available from a variety of sources. Since new sensors are constantly reaching the market, the capabilities that have been described may be superseded by those of newer models or by the introduction of new products by other manufacturers. Additional operating and installation information should be obtained from the manufacturer or their representatives before making the final sensor selection.

The higher cost of the above-roadway mounted sensors is often offset by the costs associated with installing and maintaining multiple lower-cost sensors such as inductive loops. Mounting location is critical to the selection and proper operation of a traffic sensor. Experience by state transportation agencies indicates that suitable mounting locations must be available with the proper elevation and proximity to the roadway in order for the above-roadway sensors to function properly. Sensors selected for a first-time application should be field tested under actual operating conditions that include variations in traffic flow rates, day and night lighting, and inclement weather before large-scale purchases of the device is made.

REFERENCES

1. L.A. Klein, D. Gibson, and M.K. Mills, *Traffic Detector Handbook: Third Edition*, FHWA-HRT-06-108 (Vol. I) and FHWA-HRT-06-139 (Vol. II), Federal Highway Administration, U.S.

Department of Transportation, Washington, DC, Oct. 2006. www.fhwa.dot.gov/publications/research/operations/its/06108/06108.pdf and www.fhwa.dot.gov/publications/research/operations/its/06139/06139.pdf. Accessed December 13, 2013.

2. *Standard Practice for the Installation of Inductive Loop Detectors*, E 2561-07a, ASTM International, 100 Barr Harbor Drive, PO Box C700, West Conshohocken, PA, July 2007.

3. ADR 6000 *Traffic Counter/Classifier Brochure*, Peek Traffic Corporation, Houston, TX, 2012.

4. L.A. Klein, *Millimeter-Wave and Infrared Multisensor Design and Signal Processing*, Artech House, Norwood, MA, August 1997.

5. A. Rhodes, D. Bullock, J. Sturdevant, and Z. Clark, Evaluation of Stop Bar Video Detection Accuracy at Signalized Intersections, Presented at 84th Transportation Research Board Annual Meeting, Washington, DC, January 9–13, 2005.

6. A. Rhodes, E. Smaglik, and D. Bullock, *Vendor Comparison of Video Detection Systems Final Report*, FHWA/IN/JTRP-2005/30, Joint Transportation Research Program, Purdue University, West Lafayette, IN, May 2006.

7. L.A. Klein, *Sensor Technologies and Data Requirements for ITS*, Artech House, Norwood, MA, June 2001.

8. L.A. Klein, *Sensor and Data Fusion: A Tool for Information Assessment and Decision Making*, Second Edition, Press Monograph 222, SPIE, Bellingham, WA, Appendix A, 2012.

Chapter 6

Inductive loop installation and loop system sensitivity

Proper inductive loop installation practice ensures that the detection system functions as desired and maintenance costs are minimized. This chapter summarizes wire loop installation guidelines and describes the methods for computing the threshold loop system sensitivity. This calculation is necessary to adjust the sensitivity of the electronics unit (also called the detector) to accommodate the threshold sensitivity of the installed wire loop, lead-in wire, and lead-in cable combination. For example, if the sensitivity of the electronics unit was not set properly and a large change in inductance was required to ensure vehicle detection, then vehicles with high undercarriages or small metal content would be difficult to detect.

6.1 GENERAL LOOP INSTALLATION GUIDELINES

The procedures that appear in this chapter for installing inductive loop detectors in slots sawed into roadway pavement were compiled from inductive loop specifications of several states and the *Traffic Detector Handbook* [1]. Although the procedures are not intended for installing preformed loops, they are of value for this type of loop as they discuss the number of turns of loop wire, number and direction of twists in the lead-in wires and lead-in cable, splice location (if needed), and grounding options. Additional details concerning inductive loop design and installation may be found in the *Traffic Detector Handbook* and the reader should consult this valuable resource before planning and fabricating an inductive loop detection system.

Figure 6.1 illustrates an inductive loop detector system composed of one or more wire loops embedded in the pavement, a splice between the lead-in wire and the lead-in cable in the pull box, lead-in cable (usually in a conduit) connecting to the terminal strip in the controller cabinet, cable from the terminal strip to the inductive loop electronics unit, and, finally, the electronics unit itself.

The major steps for installing an inductive loop detector system are

1. Preparing plans and specifications.
2. Securing the work zone.
3. Installing underground conduit and pull box.
4. Cutting a slot for the loop wire and lead-in wires.
5. Installing the wires.
6. Twisting the lead-in wires.
7. Testing for proper operation of the wire loop and lead-in wires.
8. Sealing the sawcuts.
9. Splicing the lead-in wires to the lead-in cable in a pull box.

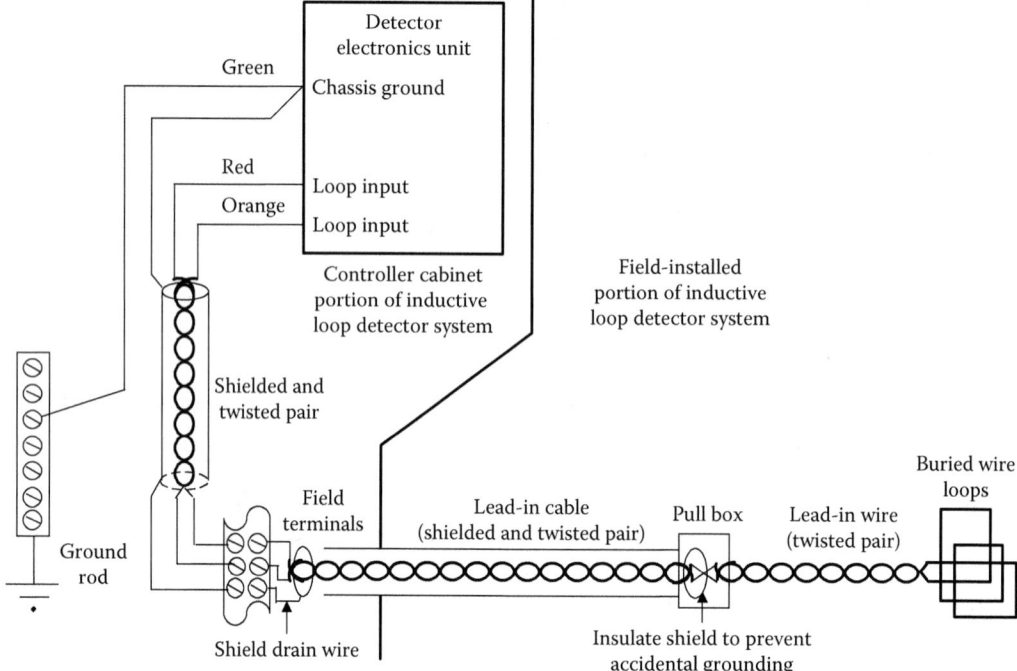

Figure 6.1 Inductive loop detector system (typical). (From L.A. Klein, D. Gibson, and M.K. Mills, *Traffic Detector Handbook: Third Edition*, FHWA-HRT-06-108 (Vol. I) and FHWA-HRT-06-139 (Vol. II), Federal Highway Administration, U.S. Department of Transportation, Washington, DC, October 2006. www.fhwa.dot.gov/publications/research/operations/its/06108/06108.pdf and www.fhwa. dot.gov/publications/research/operations/its/06139/06139.pdf.)

10. Connecting the lead-in cable to the terminal strip in the cabinet.
11. Testing for proper operation of the wire loop, lead-in wires, and lead-in cable assembly.
12. Connecting the terminal strip to the electronics unit.

The general installation guidelines given in Table 6.1 apply under many circumstances; however, specific locations may require other designs, installation procedures, testing, lightning protection, and other modifications or additions to the guidelines. For example, agencies may require that loop detectors on overlay or new pavement locations be cut before the final pavement is applied so that it covers and seals the sawcuts [2].

6.1.1 Loop dimensions and number of turns

Loop dimensions and number of turns are selected according to the types of vehicles to be detected, vehicle undercarriage height, lane width, length of lead-in cable, and, for some applications, the data desired. Inductive loops should not be wider than 6 ft (183 cm) in a 12-ft (366-cm) lane. Loops should not be less than 5 ft (152 cm) wide because the detection distance between the road surface and the vehicle undercarriage becomes limited as the detection distance is approximately equal to one-half to two-thirds of the loop width (i.e., the minimum loop dimension). Since the inductance of the loop must be greater than the inductance of the lead-in cable (e.g., 21 μH per 100 ft [69 μH per 100 m] of #14 AWG lead-in cable) for the loop system to have sufficient sensitivity, the *Traffic Detector Handbook*

Table 6.1 General installation guidelines for installing inductive loop detectors

Design
- The width of the loop should be tailored to the width of the lane.
- Loops should not be over 6 ft (1.8 m) wide in a 12-ft (3.7-m) lane.
- Loops should not be less than 5 ft (1.5 m) wide (detection height is approximately 1/2 to 2/3 of the loop width).
- All loops should have a minimum of two turns of wire in any sawcut except in a quadrupole.
- One additional turn of wire may be specified for loops installed in reinforced concrete or over 2 in. (5 cm) deep.

Installation of loop wire and lead-in cables
- The corner of loop sawcuts should be cored, chiseled, beveled, or diagonally cut to eliminate sharp turns.
- Sawcut should be deep enough to provide for a minimum of 1 in. (2.5 cm) of sealant over uppermost wire.
- Sawcut should be cleaned out with high-pressure water after cutting and then dried with compressed air.
- If a 1/4-in. sawcut is used, select the wire size to allow encapsulation of the wires (AWG #14 or #16).
- Loop wires should have high-quality insulation such as cross-linked polyethylene or polypropylene.
- Wire should be laid in sawcuts using the same rotation (clockwise or counterclockwise) in each loop.
- Loop wires should be tagged to indicate start (S) and finish (F) and should indicate the loop number in the pull box to facilitate series splicing with alternate polarity connections.
- Sawcuts for the loop lead-in wire should be at least 12 in. (30 cm) from adjacent loop edges.
- The loop lead-in wire from the loop to the pull box should be twisted a minimum of 3–5 turns per foot.
- Splices of loop lead-in wire to lead-in (home-run) cable must be soldered, insulated, and waterproofed to ensure environmental protection and proper operation.
- The lead-in cable should be twisted, shielded, and waterproofed.
- The cable selected should have a polyethylene jacket.
- For most installations, the lead-in cable should not be connected to earth ground at the pull box, but left insulated and floating. Manufacturer's recommendations should be followed concerning whether the cabinet end of the cable is grounded (per National Electrical Manufacturers Association [NEMA] recommendations).

Testing
- Prior to filling sawcuts with sealant, loops should be tested with an ohmmeter for continuity and loop and lead-in wires in pull boxes should be tested with a 500 V DC Megger to confirm insulation resistance >100 MΩ.
- Loops should be tested with a direct reading inductance meter at the pull box to confirm the number of turns of wire in any loop. The following formula provides a simple method to calculate the approximate inductance of any loop configuration and confirm the number of turns in the loop:

  Inductance $(L) = K \times$ feet of sawcut, where K is a function of the number of turns of wire as indicated below:

No. of turns	K (μH/ft)
1	0.5
2	1.5
3	3.0
4	5.0
5	7.5

- The electrical splice configuration of multiple loops should be confirmed with the inductance meter to assist in the selection of the correct sensitivity setting on the electronics unit.

Continued

Table 6.1 (Continued) General installation guidelines for installing inductive loop detectors

Connections
- All spade lug connections in the loop circuit should be soldered.
- Multiple loops connected to the same channel of an electronics unit should be connected in series or series-parallel. Some manufacturers dispute the use of series-parallel. An example of a series-parallel loop connection is two loops connected in series that are then connected in parallel with another series connection of two loops.
- Series splices should be verified with inductance measurement prior to connecting to the lead-in cable.
- Multiple loops connected to the same channel of an electronics unit should be connected with alternate polarity (clockwise–counterclockwise) to improve noise immunity and stability.
- Loops in adjacent lanes should be connected to the same multiple-channel electronics unit.

Source: L.A. Klein, D. Gibson, and M.K. Mills, *Traffic Detector Handbook: Third Edition*, FHWA-HRT-06-108 (Vol. I) and FHWA-HRT-06-139 (Vol. II), Federal Highway Administration, U.S. Department of Transportation, Washington, DC, October 2006. www.fhwa.dot.gov/publications/research/operations/its/06108/06108.pdf and www.fhwa.dot.gov/publications/research/operations/its/06139/06139.pdf.

recommends that the inductance of single loops and series, parallel, or series-parallel combinations of loops be greater than 50 μH to ensure stable operation of the inductive loop detector system [1]. Guidance for the number of turns needed to produce the required inductance value is given as "If the loop perimeter is less than 30 ft (9 m), use three turns of wire; if the loop perimeter is greater than 30 ft (9 m), use two turns of wire." Appendix C of the *Traffic Detector Handbook* contains tables showing the inductance values for various size loops and shapes (i.e., rectangular, quadrupole, and circular) [1].

6.1.2 Loop quality factor

The resonant efficiency of a circuit containing inductance, resistive losses, and capacitive coupling, such as that in Figure 6.2a, is expressed through the dimensionless quality factor Q. The resistor R_S in series with the equivalent lossless inductor L_S represents the total energy loss in the inductive portion of the wire loop. The quality factor Q represents the ratio of the inductive reactance to the resistive losses of the inductor. If the losses of the inductor are large, the Q is low. A perfect inductor has no losses; therefore, there is no dissipation of energy within the inductor and the quality factor is infinite. The excitation frequency of the loop must be specified when measuring quality factor since inductive reactance (i.e., the electrical resistance of an inductor to current flow when used in an AC circuit) is a frequency-dependent quantity. The capacitive coupling C_P is created by two factors: the loop wires themselves and the interaction of the loop wires with the sidewalls of the sawcut slot.

The electronics unit adds an additional resistive element R_P to the circuit, which acts in parallel with C_P. Thus, the equivalent circuit for an inductive loop with capacitive coupling and resistive losses attributed to the electronics unit becomes that shown in Figure 6.2b. The loaded quality factor Q_L for this circuit is given by Equation 6.1 as

$$Q_L = \frac{Q_P Q_0}{Q_P + Q_0},$$

(6.1)

where

$$Q_0 = \sqrt{\frac{1}{(R_S)^2}\frac{L_S}{C_P} - 1},$$

(6.2)

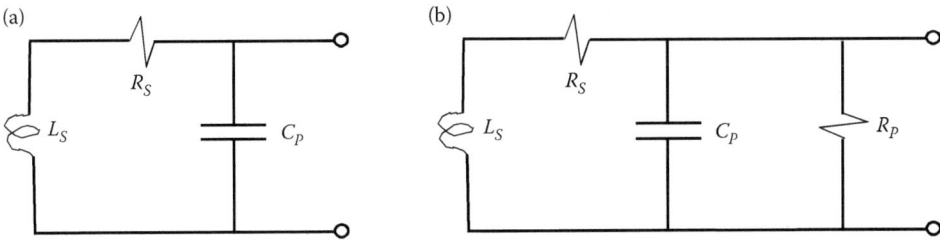

Figure 6.2 Equivalent electrical circuit for an inductive loop with capacitive coupling from the loop wires and the sidewalls of a sawcut slot: (a) not including resistive losses from the detector electronics unit and (b) including such losses.

$$Q_P = \omega_0 C_P R_P, \tag{6.3}$$

$$\omega_0 = \frac{Q_0 R_S}{L_S}, \tag{6.4}$$

which is the resonant frequency of the inductive loop equivalent electrical circuit in Figure 6.2a, and L_S, R_S, and C_P are the self-inductance, resistance, and coupling capacitance of the loop, respectively.

Equation 6.1 is intended for applications in which losses are low, quality factor is high, and frequency f, loop inductance L_S, and loop resistive loss R_S can be readily measured.

Inductive loop detectors installed in roadways, on the other hand, are not as adaptable to the above analysis because the inductance is distributed over the loop and lead-in cable and is difficult to measure. Calculation of the quality factor is further complicated by the larger actual resistances of the loop wire and lead-in cable as compared to the series value measured with an ohmmeter. The extra losses are due to the high-frequency excitation and ground currents in the pavement associated with the loop configuration and the roadway environment near the wire. As a result, the Q of an identical inductive loop configuration will vary from location to location.

Table 6.2 illustrates the method for calculating the inductive loop system quality factor using Q_0 and Q_P. Calculated quality factors for rectangular, quadrupole, and circular inductive loops are found in Tables 2-2 through 2-4, respectively, of the *Traffic Detector Handbook* [1]. Loops are excited at 20 kHz in these tables, with conductor and quadrupole lateral spacing of 200 mils. All inductance and quality factors are apparent values (i.e., loop capacitance and resistance are included).

Quality factors of 5 and above are recommended when installing inductive loop detectors since oscillators in most electronics units will not operate with low Q. Moisture in the pavement and subgrade can increase the loop ground resistance such that the Q of the inductive loop system decreases below 5, thereby reducing the sensitivity of most electronics units. Loop capacitance will also reduce Q.

6.1.3 Loop location and laying of the loop wire in the sawcut

To protect the integrity of the pavement and loop installation, cracks and joints in the roadway pavement should not be located closer than 18 in. (45 cm) upstream or downstream of the inductive loop detector being installed. Some agencies relax this constraint to 1 ft (0.3 m) [3]. Sawcuts for other wire loops or other in-roadway sensors must not be

Table 6.2 Loop system quality factor Q calculation

Assumptions

Loop type: 3-turn, 6 ft × 6 ft (1.8 m × 1.8 m) of #14 AWG wire

Loop inductance: 74 µH at 20 kHz from Appendix C of *Traffic Detector Handbook* [1]

Loop resistance (in air): 0.0025 Ω/ft (0.0083 Ω/m) from Appendix D of *Traffic Detector Handbook* [1]

Lead-in cable type: 100 ft (30 m) of Belden 8718 #12 AWG

Lead-in cable inductance: 0.20 µH/ft (0.67 µH/m) from Appendix D of *Traffic Detector Handbook* [1]

Lead-in cable resistance: 0.0031 Ω/ft (0.0103 Ω/m) from Appendix D of *Traffic Detector Handbook* [1]

Operating frequency: 20 kHz

Total loop system series inductance: 74 µH + 20 µH = 94 µH

Total loop system series resistance: 0.25 Ω + 0.62 Ω = 0.87 Ω

Note: Wire length for resistance calculation is per wire, i.e., twice the cable length.

Total inductive loop system capacitance

$$C_P = \frac{1}{\omega_0^2 L_S} = \frac{1}{(2\pi \times 20 \times 10^3)^2 (94 \times 10^{-6})} = 6.74 \times 10^{-7} \, F$$

Quality factor of inductive loop system

$$Q_0 = \sqrt{\frac{1}{R_S^2} \frac{L_S}{C_P} - 1} = \sqrt{\frac{94 \times 10^{-6}}{(0.87)^2 \times (6.74 \times 10^{-7})} - 1} = 13.54$$

This value is the unloaded inductive loop system quality factor with 100 ft (30 m) of Belden 8718 #12 AWG lead-in cable.

Assume that the detector electronics unit adds a parallel resistance of 1000 Ω. Then

$$Q_P = \omega_0 C_P R_P = (2\pi \times 20 \times 10^3) \times (6.74 \times 10^{-7}) \times 1000 = 84.70.$$

Therefore, the total loaded loop system quality factor is

$$Q_L = \frac{Q_P Q_0}{Q_P + Q_0} = \frac{84.70 \times 13.54}{84.70 + 13.54} = 11.67.$$

located closer than 2 ft (0.6 m) upstream or downstream of the inductive loop detector being installed [4]. A 6 in. (15 cm) minimum distance between lead-in sawcuts is recommended until the sawcuts are within 1 ft (0.3 m) of the edge of the pavement or curb, at which point they may be placed closer together [5]. Lead-in sawcuts should not be closer than 12 in. (30 cm) from adjacent loop edges [4].

Figure 6.3 shows the winding details for a typical square loop, round loop, and rectangular quadrupole loop [6]. Placing of the loop wire in the sawcut proceeds as follows. Starting at the pull box location (or first entered pole or pedestal), allow 3–5 ft (91–152 cm) for slack and then lay the wire alongside the lead-in wire sawcut and run it to the point where the lead-in wire sawcut meets the loop sawcut. Place the wire into the loop sawcut and wrap it the prescribed number of turns and direction around the sawcut to form the loop. The first turn of loop wire is placed in the bottom of the sawcut, with each subsequent turn placed on top of the preceding one [5]. If a bead of sealant is placed at the bottom of the cut to aid encapsulation, apply the sealant before inserting the turns of wire. Each turn of a given loop

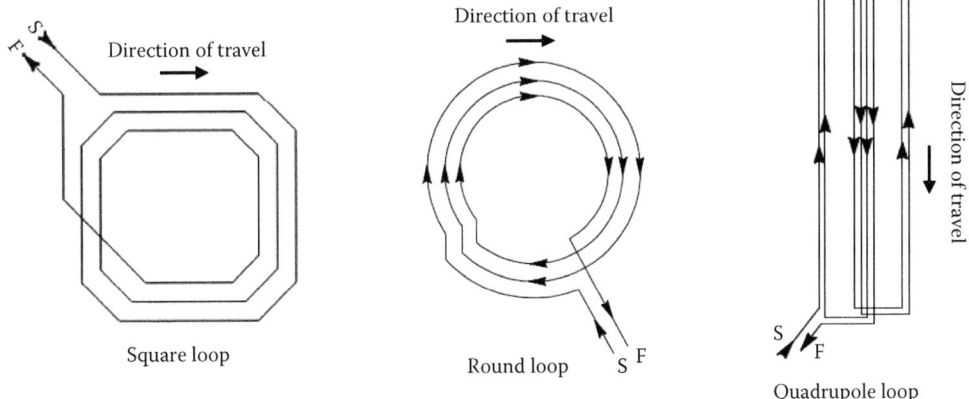

Figure 6.3 Inductive loop winding detail for square, round, and rectangular quadrupole loops (typical). S and F indicate the start and finish of the loop, respectively. (From *Electrical Systems (Detectors) Standard Plan ES-5B*, California Department of Transportation, Sacramento, CA, May 1, 2006. http://www. dot.ca.gov/hq/esc/oe/project_plans/HTM/stdplns-US-customary-units-new06.htm#electrical.)

must be wound in the same direction. Be sure to count the number of turns in the slot as it is a common error to miscount. Adjacent loops using the same electronics unit are wound in the opposite direction to minimize interference. Run the remaining length of wire alongside the lead-in wire sawcut to the pull box location. Cut the wire remembering to keep 3–5 ft (91–152 cm) of slack at each end.

6.1.4 Crosstalk

When two loops constructed of the same wire diameter have the same loop dimensions, number of turns, and lead-in length, they have the same resonant frequency. When these loops are near each other or when the lead-in wires from the loops are in close proximity (perhaps running in the same conduit), a phenomenon known as crosstalk can occur. This effect is caused by an electrical coupling between the two loop channels and will often manifest itself as brief, false, or erratic actuations when no vehicles are present. The most common technique utilized to prevent crosstalk is a frequency selection switch that varies the operating frequency of the adjacent loop channels.

6.1.5 Sealant application techniques

Common practices for sealing the loop wire are depicted in Figure 6.4. The procedure on the left of the figure consists of applying a layer of sealant to the floor of the sawcut after thoroughly cleaning and drying the slot. The loop wires are then laid in the slot and covered with a second, final layer of sealant. This method tends to fix the position of the loop wires in the middle of the sawcut, protecting them on the top and bottom. Some agencies believe that this procedure, although costlier, protects the loop wires from water intrusion.

In the technique illustrated in the middle procedure in Figure 6.4, the wire is simply laid in the slot and covered with sealant. There is no way to control the positioning of the wire in the slot. In a three-wire installation, the three layers of wire may form a triangle on the bottom of the slot or may stack over each other.

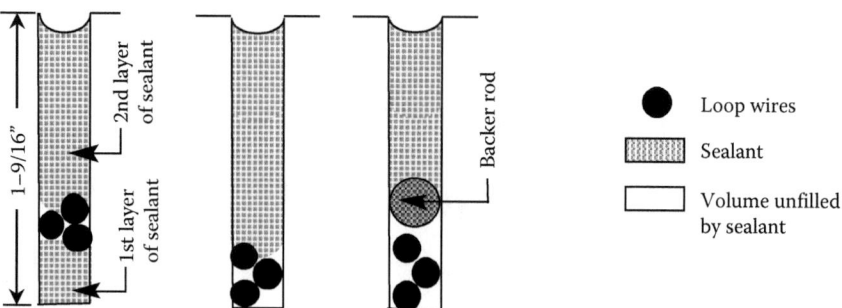

Figure 6.4 Methods of applying sealant.

The backer rod–sealant combination shown in the procedure on the right is based on the theory that stresses on the sealant caused by elongation are reduced if the sealant has less depth. With this method, the wires are placed in the slot and then a backer rod (generally a closed-cell polyethylene rope) is forced into the slot over the wires. The remainder of the slot is then filled with sealant. The backer rod assures a shallow layer of sealant, reducing tensile stresses and leaving the wires free to adapt to shifting of the pavement. An alternative method is to insert short pieces of backer rod of approximately 1 in. (2.5 cm) in length every foot (30 cm) or two (61 cm) to anchor the wire in the slot before applying the sealant.

No published evidence of the superiority of one sealant application method over another has appeared. Most inductive loop detector installers agree that the neat arrangement of wires that appear in many published illustrations is simply not indicative of actual installations; rather, the lay of the wires is random in the slot. They agree that complete encapsulation by the sealant is seldom achievable. Some installers also argue that placing sealant in the bottom of the sawcut (as depicted on the left in Figure 6.4) before laying the wire is time consuming and requires more road-closure time. *Installers indicated that even when this method of installation is specified, it is unlikely to be followed unless an agency inspector was actually overseeing the installation.* Therefore, inspection of the installation by the responsible agency is mandatory [1].

On the other hand, proponents of placing sealant on the bottom of the sawcut report that the extra protection afforded by the sealant bed prevents the intrusion of water through small pavement cracks. It also avoids the possibility of sharp edges or rocks becoming dislodged and piercing the installation. Others feel that this is a remote possibility, particularly if the sawcut is well cleaned of debris, which is not always the case. Alternatively, some agencies specify the placement of a layer of sand rather than sealant at the bottom of the sawcut. This provides a smooth bed but does not prevent the intrusion of water through pavement cracks.

The amount of sealant applied should be sufficient to completely fill the sawcut, but not overfill. A trowel or another tool should be used to ensure that the sealant is slightly below the pavement surface and to remove any excess sealant. Poor installation procedures result in overfilling, underfilling, and air bubbles in the sealant. All three conditions can lead to inductive loop detector failure and should be corrected during the installation process.

Sealant may be applied with a special applicator or by hand directly from a container. A paint stirrer can be inserted into the slot to hold the wire down while the sealant is being applied. Other techniques can also be used (e.g., backer rod strips and nylon rope) to hold the wire securely in place as sealant is added. The sealant application procedure is completed by removing any excess material from the pavement and dusting talc or sand on the fresh sealant before opening the lane to traffic. This prevents tracking of the sealant during its curing process and allows earlier opening of the traffic lane.

Some agencies in hotter regions of the United States use sand as the sealant by tamping it into the slot after the wire is placed in the sawcut. However, the sand is easily tracked out of the slot and the wires may become dislodged. Therefore, this practice is not recommended [1].

6.1.6 Splicing the lead-in wire to the lead-in cable

Another critical step in the loop installation process is splicing the loop lead-in wire to the lead-in cable that connects to the electronics unit in the controller cabinet. This splice, located in the pull box, should be the only splice in the loop system. The splice is frequently the cause of inductive loop detector system failure. However, if proper splicing procedures are used, the splice should not pose a problem. There are two steps to creating a splice: the physical connection of the wires and the environmental sealing of the connection.

Methods for physically connecting the lead-in wires with the lead-in cable vary among agencies. The two preferred methods are twisting and soldering or crimping and soldering. Most electronics unit manufacturers specify a solder connection in their installation procedures. The argument for soldering is that it provides a connection with lower resistance and has less susceptibility to corrosive degradation. The soldered connection will, therefore, require less maintenance in the long run.

While pressure connectors (crimping) without soldering may have been generally acceptable in the past, the use of solid-state electronic assemblies now makes soldered connections preferable. These assemblies operate at low voltage levels and minimum current loads. Because of this, they are susceptible to even slight voltage drops, which occur where poor electrical connections cause high resistance in a circuit.

Once the wires are spliced, it is essential that the splice be environmentally sealed against weather, moisture, abrasion, and other harmful effects. A variety of methods are used, including heat-shrinkable tubing, special sealant kits, special forms that are filled with sealant, pill bottles with slot sealant, tape and coating, and other techniques. Any approach is acceptable as long as it provides a reliable environmental seal. Chapter 5 of the *Traffic Detector Handbook* contains further details concerning sealing techniques [1].

6.1.7 Grounding the loop

Grounding of the loop at the cabinet is governed by the recommendations of the equipment manufacturer and agency policy. Figure 6.5 illustrates the recommended method of grounding the lead-in cable if grounding is used [1]. This allows most electrical disturbances or interference to be safely grounded without affecting the performance of the lead-in cable and inductive loop detector. Some equipment manufacturers and agencies recommend that the shield of the cable not be connected to a ground terminal. The justification for not grounding is that the inductive loop detector system operates at low voltage and may, therefore, be sensitive to current flows induced by more than one grounding point. Such ground loops can be produced by grounding the shield at the cabinet since the cabinet and electronics unit are already connected to ground. Additional installation recommendations for inductive loops are found in the *Traffic Detector Handbook*, Chapter 5 and in the *Standard Practice for the Installation of Inductive Loop Detectors* published by ASTM [7].

6.2 LOOP SYSTEM SENSITIVITY

When a vehicle with metal content passes over an inductive loop or is stopped within the loop, the vehicle induces eddy currents in the wire loops that decrease their inductance.

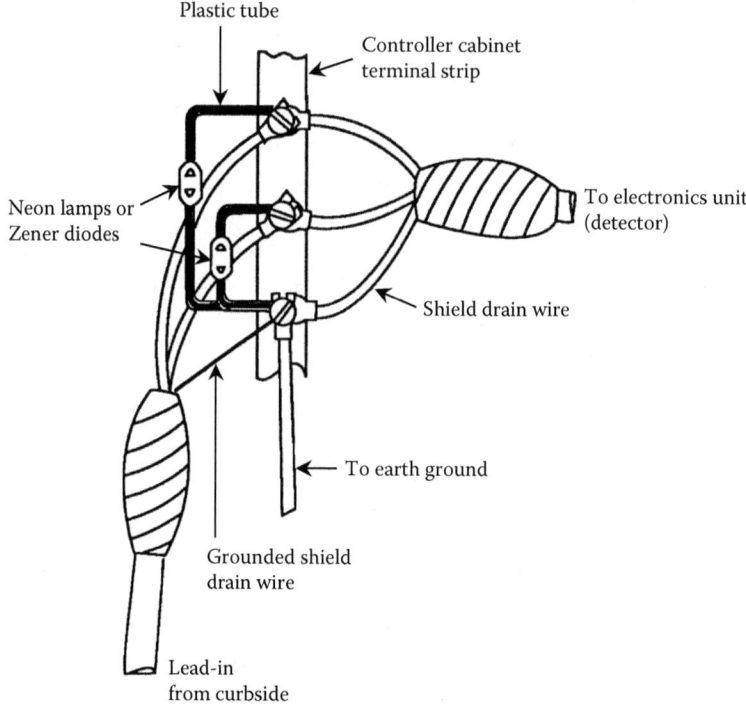

Figure 6.5 Lead-in cable grounding at the field terminal strip in the cabinet. (From L.A. Klein, D. Gibson, and M.K. Mills, *Traffic Detector Handbook: Third Edition*, FHWA-HRT-06-108 (Vol. I) and FHWA-HRT-06-139 (Vol. II), Federal Highway Administration, U.S. Department of Transportation, Washington, DC, October 2006. www.fhwa.dot.gov/publications/research/operations/its/06108/06108.pdf and www.fhwa.dot.gov/publications/research/operations/its/06139/06139.pdf.)

In order for the decrease in inductance to be sensed by the detector electronics unit, the size and the number of turns in the loop or combination of loops, together with the length of the lead-in cable, must produce an inductance value that is compatible with the tuning range of the electronics unit and with other requirements established by the traffic engineer. NEMA standards for inductive loop detectors specify that an electronics unit must be capable of operating satisfactorily over an inductance range of 50–700 µH [1]. Some units tolerate much larger inductance values, for example, from several loops wired in series.

Loop system sensitivity is defined as the smallest change of inductance ΔL at the electronics unit terminals that will cause the controller to actuate. The sensitivity of the electronics unit must be set equal to or greater than the calculated loop system threshold sensitivity at the electronics unit. Some electronic detectors specify sensitivity in terms of $\Delta L/L$ (in percent) and some in terms of simply ΔL as shown in Table 6.3. Corresponding loop system response times are also listed in the table.

NEMA specifies the sensitivity threshold for three classes of test vehicles centered on a 6-ft × 6-ft (1.8-m × 1.8-m) three-turn loop with 100 ft (30.5 m) of lead-in cable [8]. These classes are as follows:

- *Class 1—small motorcycles*: 0.13% $\Delta L/L$ inductance change (defined as the change in inductance in µH divided by the combined inductance of the loop, lead-in wire, and lead-in cable in µH) or 0.12 µH ΔL change in inductance.

Table 6.3 Electronics unit (detector) sensitivities, thresholds, and loop system response times as specified by three device manufacturers

Sensitivity $\Delta L/L$ (%)	Loop system response time (ms)	Sensitivity $\Delta L/L$ (%)	Loop system response time (ms)		Sensitivity threshold ΔL (μH)	Loop system response time (ms)
			Min.	Max.		
1.28	3.5 ± 2.5	1.28	1.4	3.4	1.024	<5
0.64	3.5 ± 2.5	0.64	1.4	3.4	0.512	<6
0.32	3.5 ± 2.5	0.32	1.4	3.4	0.256	<6
0.16	3.5 ± 2.5	0.16	1.4	3.5	0.128	<8
0.08	4.5 ± 3.5	0.08	1.4	5	0.064	<12
0.04	7.05 ± 6.0	0.04	2	9	0.032	<20
0.02	11.5 ± 10.5	0.02	5	18.5	0.016	<34
0.01	21.5 ± 20.5	0.01	9	32.5	0.008	<64

- *Class 2—large motorcycles*: 0.32% $\Delta L/L$ inductance change or 0.3 µH ΔL change in inductance.
- *Class 3—automobile*: 3.2% $\Delta L/L$ inductance change or 3.0 µH ΔL change in inductance.

When two loops are connected together, the sensitivity of the combination is reduced at the input to the electronics unit. The sensitivity of the combined loops depends on the way the loops are connected, in series or in parallel. The methods for calculating the sensitivity of a single loop, two loops connected in series, and two loops connected in parallel are described below.

6.3 SENSITIVITY OF A SINGLE LOOP

These two examples calculate the sensitivities of one three-turn and one four-turn loop at the pull box and then at the input terminals to the detector electronics unit located in the controller cabinet. Figure 6.6 depicts the plan view and equivalent electrical circuit for a single loop configuration.

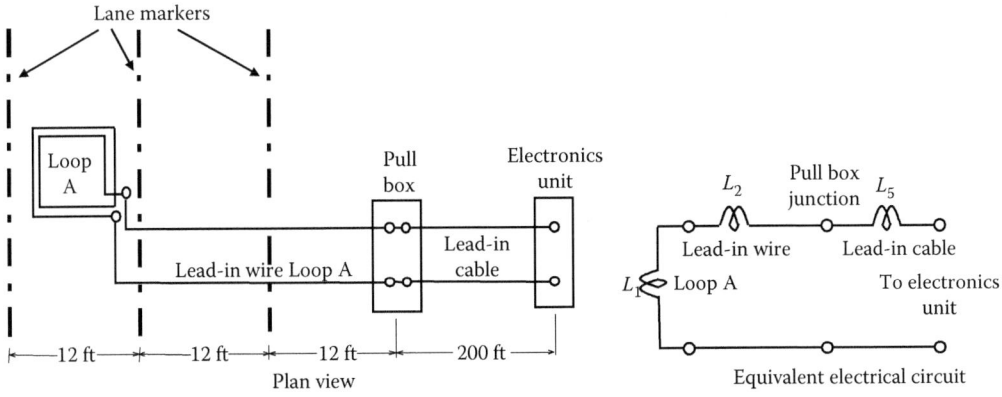

Figure 6.6 Lead-in wire and lead-in cable lengths and equivalent electrical circuit for single inductive loop.

6.3.1 Three-turn 6-ft × 6-ft (1.8-m × 1.8-m) loop

A high-bed vehicle with an undercarriage height of 4 ft (1.2 m) produces a sensitivity S_L of 0.1% when passing over a three-turn, 6-ft × 6-ft (1.8-m × 1.8-m) loop of #14 AWG wire [9]. The inductance for #14 AWG loop wire with 5 twists per foot (16 twists per meter) is 0.22 μH/ft (0.72 μH/m) [10]. If the lead-in wire length to the pull box is 24 ft (7.3 m) of twisted loop wire as shown in Figure 6.6, the lead-in wire inductance L_2 becomes

$$L_2 = 0.22 \, \mu\text{H/ft} \times 24 \, \text{ft} = 5.3 \, \mu\text{H}. \tag{6.5}$$

The self-inductance $L_L = L_1$ of a three-turn, 6-ft × 6-ft (1.8-m × 1.8-m) loop of #14 AWG wire at 20 kHz is 74 μH [11]. Therefore, the sensitivity S_P (in percent) at the pull box is

$$S_P = \frac{S_L}{1+(L_2/L_L)} = \frac{0.1\%}{1+(5.3 \, \mu\text{H}/74 \, \mu\text{H})} = 0.093\%. \tag{6.6}$$

The inductance of Type 8720 shielded lead-in cable that connects the pull box to the electronics unit is 0.21 μH/ft (0.69 μH/m) [12]. If 200 ft (61 m) of the cable is used, the total series inductance between the loop and the input terminals of the electronics unit becomes

$$L_S = L_2 + L_5 = 0.22 \times 24 + 0.21 \times 200 = 5.3 \, \mu\text{H} + 42 \, \mu\text{H} = 47.3 \, \mu\text{H}. \tag{6.7}$$

Then, the sensitivity S_D at the input terminals of the electronics unit is

$$S_D = \frac{S_L}{1+(L_S/L_L)} = \frac{0.1\%}{1+(47.3 \, \mu\text{H} / 74 \, \mu\text{H})} = 0.061\%. \tag{6.8}$$

The *Traffic Detector Handbook* contains values of S_L as a function of vehicle undercarriage height that apply to other sized loops and loops installed over reinforced steel pavements.

6.3.2 Four-turn 6-ft × 6-ft (1.8-m × 1.8-m) loop

For a four-turn, 6-ft × 6-ft (1.8-m × 1.8-m) loop, the sensitivity S_L for a 4-ft (1.2-m) high undercarriage vehicle is approximately 0.1%. The four-turn loop self-inductance is 125 μH at 20 kHz [11]. The series inductance is the same as in the previous example. Therefore,

$$S_D = \frac{S_L}{1+(L_S/L_L)} = \frac{0.1\%}{1+(47.3 \, \mu\text{H}/125 \, \mu\text{H})} = 0.073\%. \tag{6.9}$$

6.4 SENSITIVITY OF TWO LOOPS IN SERIES

Figure 6.7 illustrates the configuration used to compute the inductive loop system sensitivity at the input terminals of the electronics unit when a second identical loop (Loop B) is connected in series with the loop sensing the vehicle (Loop A). The series connection is

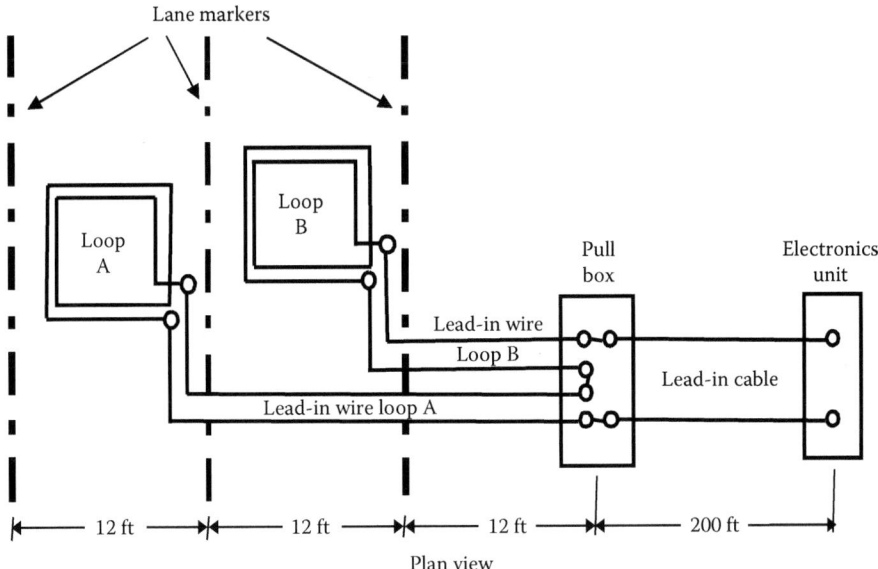

Figure 6.7 Lead-in wire and lead-in cable lengths for two inductive loops connected in series.

made in the pull box. Both the sensing loop (first loop) and the second loop are 6-ft × 6-ft (1.8-m × 1.8-m), three-turn loops of #14 AWG wire. Therefore, the self-inductance L_L of each loop is 74 µH at 20 kHz [11].

The lead-in wire inductances of Loop A and Loop B are part of the series electrical circuit shown in Figure 6.8, whose values are

$$L_S^A = (0.22\ \mu H/ft) \times (24\ ft) = 5.3\ \mu H \tag{6.10}$$

and

$$L_S^B = (0.22\ \mu H/ft) \times (12\ ft) = 2.6\ \mu H, \tag{6.11}$$

respectively, assuming #14 AWG loop wire with 5 twists per foot.

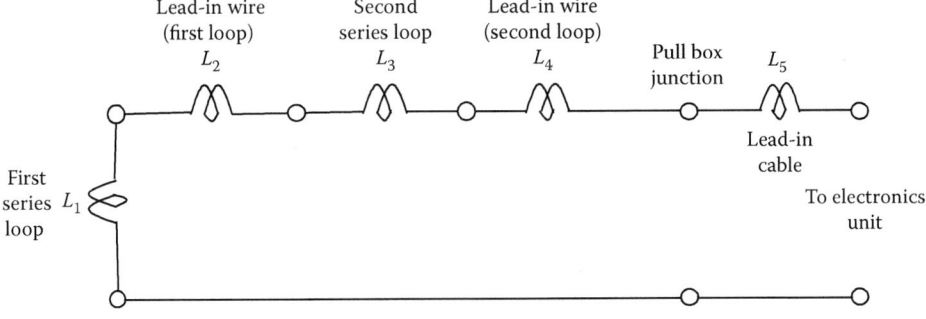

Figure 6.8 Equivalent electrical circuit for two inductive loops connected in series.

Therefore, the total series inductances of Loop A plus its lead-in wire and Loop B plus its lead-in wire to the pull box become

$$L_T^A = 74\ \mu H + 5.3\ \mu H = 79.3\ \mu H \tag{6.12}$$

and

$$L_T^B = 74\ \mu H + 2.6\ \mu H = 76.6\ \mu H. \tag{6.13}$$

The sensitivity of Loop A at the pull box is

$$S_P^A = \frac{S_L}{1 + ((L_T^A + L_T^B)/L_L)} = \frac{1.1\%}{1 + (155.9\ \mu H/74\ \mu H)} = 0.354\%, \tag{6.14}$$

where $S_L = 1.1\%$ for a vehicle with a 2-ft high undercarriage and a three-turn 6-ft × 6-ft loop of #14 AWG wire [9].

The total series inductance at the input terminals of the electronics unit becomes

$$L_T^S = L_T^A + L_T^B + L_S = 79.3\ \mu H + 76.6\ \mu H + 42\ \mu H = 197.9\ \mu H, \tag{6.15}$$

where L_S is 0.21 μH/ft for type 8720 cable multiplied by the lead-in cable length (0.21 μH/ft × 200 ft = 42 μH) [12].

Then, the sensitivity of Loop A at the input terminals of the electronics unit is given by

$$S_D^A = \frac{S_L}{1 + (L_T^S/L_L)} = \frac{1.1\%}{1 + (197.9\ \mu H/74\ \mu H)} = 0.299\%, \tag{6.16}$$

where S_L was set equal to 1.1%.

6.5 SENSITIVITY OF TWO LOOPS IN PARALLEL

Figure 6.9 illustrates the plan view and equivalent electrical circuit for calculating loop system sensitivity at the electronics unit terminals when two identical loops are connected in parallel. All parameters are the same as in the previous example for two inductive loops connected in series. The inductance and sensitivity of the first loop plus its lead-in wire are written as

$$L_{TS} = L_1 + L_2 \tag{6.17}$$

and

$$S_{TS} = \frac{S_L}{1 + (L_2/L_1)}, \tag{6.18}$$

where the self-inductance of the first loop is denoted by the subscript 1 and the self-inductance of its lead-in wire with a subscript 2.

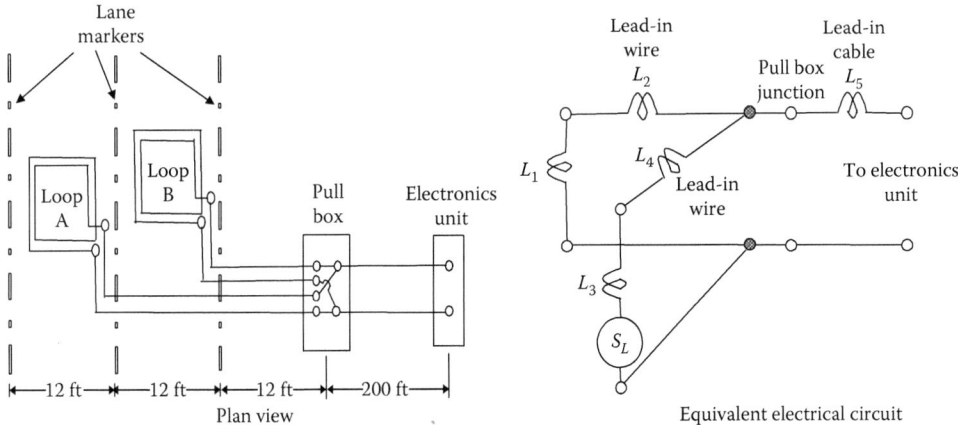

Figure 6.9 Lead-in wire and lead-in cable lengths and equivalent electrical circuit for two inductive loops connected.

Let

$$L_A = L_1 + L_2 \tag{6.19}$$

and

$$L_B = L_3 + L_4, \tag{6.20}$$

where L_3 is the self-inductance of the second loop and L_4 is the self-inductance of its lead-in wire.

Then, the inductance and sensitivity at the pull box are

$$L_{TP} = \frac{L_A \times L_B}{L_A + L_B} = \frac{(L_1 + L_2) \times (L_3 + L_4)}{L_1 + L_2 + L_3 + L_4} \tag{6.21}$$

and

$$S_{TP} = S_{TS} \frac{1}{1 + (L_A/L_B)} = S_{TS} \frac{1}{1 + ((L_1 + L_2)/(L_3 + L_4))}, \tag{6.22}$$

respectively.

The inductance at the input to the electronics unit is

$$L_D = L_{TP} + L_5 = \frac{L_A \times L_B}{L_A + L_B} + L_5 = \frac{(L_1 + L_2) \times (L_3 + L_4)}{L_1 + L_2 + L_3 + L_4} + L_5. \tag{6.23}$$

Therefore, the sensitivity at the input to the electronics unit is

$$S_D = S_L \frac{1}{1 + (L_2/L_1)} \times \frac{1}{1 + ((L_1 + L_2)/(L_3 + L_4))}$$
$$\times \frac{1}{1 + (L_D/(((L_1 + L_2) \times (L_3 + L_4))/(L_1 + L_2 + L_3 + L_4)))}, \tag{6.24}$$

where the first factor in Equation 6.24 is equal to S_{TS}.

Applying the inductance values from the three-turn series loop example,

$$S_D = 1.1\% \frac{1}{1 + (5.3\,\mu H/74\,\mu H)} \times \frac{1}{1 + (79.3\,\mu H/76.6\,\mu H)}$$
$$\times \frac{1}{1 + (81\,\mu H\,/\,((79.3\,\mu H \times 76.6\,\mu H)\,/\,(79.3\,\mu H + 76.6\,\mu H)))} = 0.164\%. \tag{6.25}$$

6.6 SUMMARY

Properly installed inductive loop detectors perform well for vehicle passage and presence detection if two important items are attended to, namely, following proper installation guidelines and maintaining the roadbed. Inductive loops are subject to intermittent performance and breaking if the roadbed deteriorates. Therefore, loops should be installed in a roadbed that is in good condition, that is, no cracks, pot holes, or crumbing pavement. Installation requires using a saw blade of sufficient width to cut the opening for the loop wires at the required depth, cleaning out the debris so that the wires are not cut or otherwise damaged by leftover rubble, laying the wires in the slot with the correct number of turns and winding direction, sealing the sawcuts with the proper sealant, and removing excess sealant from the road surface. The lead-in wires must be twisted with approximately 5 turns per foot and spliced in the pull box to the lead-in cable. Observations of the installation process indicate that sealing and splicing are critical steps in this procedure. When contractors are hired to install loops, it is imperative that an inspector from the responsible agency be on hand to ensure that the approved installation method is followed, especially cleaning of the sawcut before laying the loop wire, sealing the sawcut, and environmentally sealing the splice in the pull box.

Loops are used singularly or connected in series, parallel, and series-parallel combinations for a variety of traffic management applications. A calculation of loop system threshold sensitivity at the input to the electronics unit is required to determine the sensitivity setting of the detector electronics unit, which must be greater than the calculated threshold value. Consequently, if this condition is not satisfied and the inductive loop system design requires a vehicle to produce a large change in inductance when it passes over the wire loop, then the system may not be able to detect vehicles with high undercarriages or small metal content.

REFERENCES

1. L.A. Klein, D. Gibson, and M.K. Mills, *Traffic Detector Handbook: Third Edition*, FHWA-HRT-06-108 (Vol. I) and FHWA-HRT-06-139 (Vol. II), Federal Highway Administration, U.S. Department of Transportation, Washington, DC, October 2006. www.fhwa.dot.gov/publications/research/operations/its/06108/06108.pdf and www.fhwa.dot.gov/publications/research/operations/its/06139/06139.pdf. Accessed December 13, 2013.
2. *Loop Detector for Surveillance, Communication, and Control (SC&C)*, Special Specification 6574, Texas Department of Transportation, Austin District, 1993. ftp://ftp.dot.state.tx.us/pub/txdot-info/cmd/cserve/specs/1993/spec/es6574.pdf. Accessed December 27, 2015.
3. *Vehicle Detector Installation Details*, TC-82.10, Ohio Department of Transportation, Columbus, OH, Apr. 19, 2002. http://www.dot.state.oh.us/traffic/PublicationManuals/scds/SCD_PDF/tc8210.pdf. Accessed October 9, 2006.

4. *Electrical Systems (Detectors) Standard Plan RSP ES-5A*, California Department of Transportation, Sacramento, CA, May 1, 2006. http://www.dot.ca.gov/hq/esc/oe/project_plans/HTM/stdplns-US-customary-units-new06.htm#electrical. Accessed December 16, 2015.

5. *Standard Specification for Road and Bridge Construction*, Section 660, Florida Department of Transportation, Tallahassee, FL, 2007. ftp://ftp.dot.state.fl.us/LTS/CO/Specifications/SpecBook/2007Book/660.pdf. Accessed December 16, 2015.

6. *Electrical Systems (Detectors) Standard Plan ES-5B*, California Department of Transportation, Sacramento, CA, May 1, 2006. http://www.dot.ca.gov/hq/esc/oe/project_plans/HTM/stdplns-US-customary-units-new06.htm#electrical. Accessed December 16, 2015.

7. *Standard Practice for the Installation of Inductive Loop Detectors*, E 2561-07a, ASTM International, 100 Barr Harbor Drive, PO Box C700, West Conshohocken, PA, July 2007.

8. *Traffic Controller Assemblies with NTCIP Requirements*, NEMA Standards Publication TS 2-2003 (R2008) v02.06, National Electrical Manufacturers Association, Rosslyn, VA, 2012. www.peektraffic.com/portal/sites/default/files/NEMA%20TS2-1998.pdf. Accessed May 22, 2017.

9. L.A. Klein, D. Gibson, and M.K. Mills, FHWA-HRT-06-108 (Vol. I), op. cit., 2–17.

10. L.A. Klein, D. Gibson, and M.K. Mills, FHWA-HRT-06-108 (Vol. I), op. cit., 2–11.

11. L.A. Klein, D. Gibson, and M.K. Mills, FHWA-HRT-06-108 (Vol. I) or FHWA-HRT-06-139 (Vol. II), op. cit., 2–10 or C–4.

12. L.A. Klein, D. Gibson, and M.K. Mills, FHWA-HRT-06-108 (Vol. I), op. cit., 2–12.

Chapter 7

Overhead sensor installation and initialization

Sensor manufacturers make available step-by-step instructions for installation, software initialization, wiring connections, and detection zone configuration depending on the sensor model. Several vendors supply online tutorials and manuals that demonstrate a sensor's often unique installation and setup procedures. Some sensors have mounting height and setback requirements that must be satisfied for them to function according to their specifications. Among these are certain models of microwave presence-detecting radar sensors, acoustic and passive infrared sensors, and cameras used in video detection systems (VDSs). Higher-mounted cameras installed in conjunction with VDSs provide more accurate data and less occlusion from nearby vehicles, especially if they are positioned on the side of a roadway. This chapter contains typical installation and initialization procedures for a variety of traffic flow sensors in order to expose the reader to the different types of instruction that are generally available. Not all sensor types and models are discussed as this would be a daunting task. In all cases, the manufacturer should be contacted for information specific to a particular sensor.

7.1 VDS ARCHITECTURE, INSTALLATION, AND INITIALIZATION

VDS architectures vary according to the number of cameras and processors used. Figure 7.1 depicts single, dual, and quad camera architectures that might be encountered with VDSs. Table 7.1 lists the major steps for installing the camera and other sensor hardware, initializing the software, and positioning the detection zones for a VDS [1]. These procedures are illustrative of manufacturer recommendations for the installation and operation of their products. Before installing any sensor, the manufacturer or authorized representative should be contacted to ensure that the latest installation and user manuals and software have been obtained.

7.2 CAMERA MOUNTING AND FIELD-OF-VIEW FOR VIDEO DETECTION AT A SIGNALIZED INTERSECTION

Most signalized intersection control applications require detection at the stopline and, often, advance detection zones about 250 ft (76 m) upstream from the stopline. Typically, cameras used with VDSs are mounted on existing poles, luminaire arms, and signal mast arms often with the aid of an extension pole.

7.2.1 Mounting height

VDS detection range is a function of camera mounting height. A rule of thumb is that VDSs can detect the presence of vehicles approximately 10 ft (3 m) away from the camera for

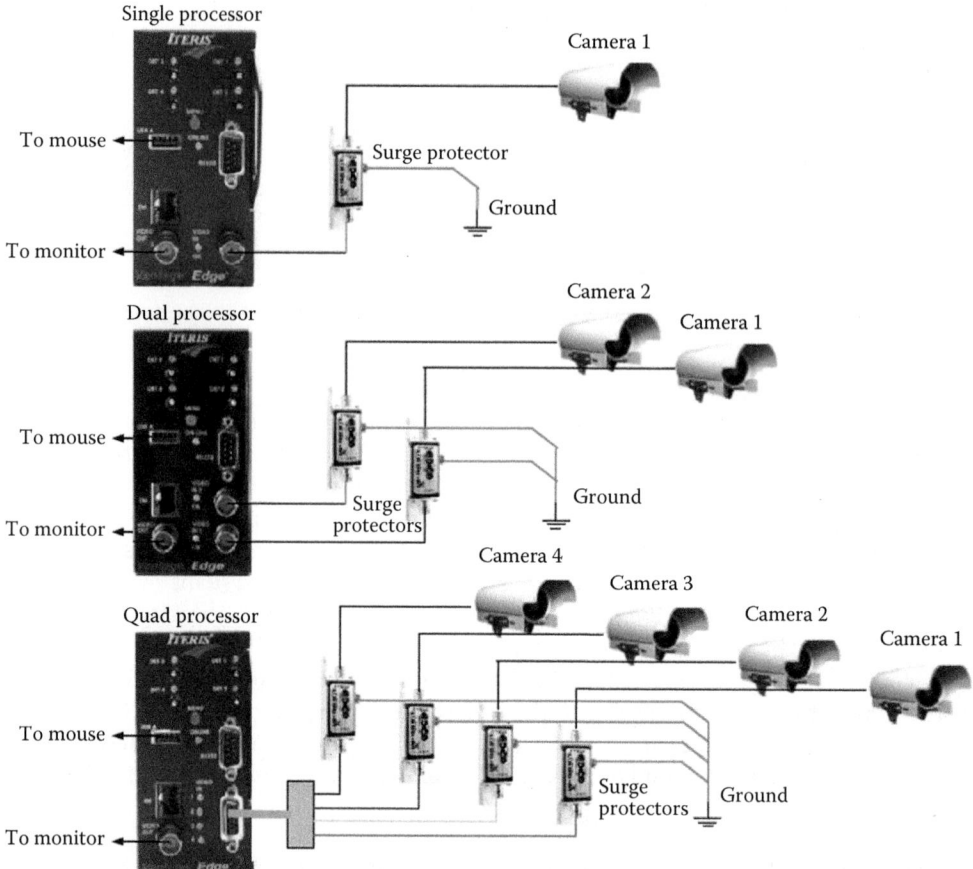

Figure 7.1 VDS camera and processor architectures. (Drawing courtesy of Iteris, Santa Ana, CA.)

every 1 ft (0.3 m) of camera height above the detection area [2]. Conservative design proce-dures may limit the range to smaller distances because of factors such as road configuration (e.g., elevation changes, curvature, and overhead or underpass structures), congestion level, vehicle mix, and inclement weather. Reduced vehicle headway can also decrease the effective surveillance range. The distance d (along the roadway from the base of the camera mounting structure to the vehicles in question) at which a VDS can distinguish between two closely spaced vehicles depends explicitly on camera mounting height, vehicle separation or gap, and vehicle height as illustrated in Figure 7.2. Its value is calculated from Equation 7.1 as

$$d = \frac{h \times \text{Intervehicle gap}}{\text{Vehicle height}}. \tag{7.1}$$

Figure 7.3 shows distance d as a function of the vehicle separation gap for 5-ft (1.5-m) and 13-ft (4-m) high vehicles and for camera mounting heights of 30, 45, and 60 ft (9, 14, and 18 m) [3,4].

The suggestions in Table 7.2 for optimal camera mounting to maximize vehicle detec-tion accuracy incorporate recommendations from Peek Traffic and Iteris [5,6]. Sometimes, the inherent characteristics of the installation site make it necessary to compromise these

Table 7.1 Installation and setup of a VDS (typical)

Major steps—Camera and hardware installation	Major steps—Software initialization and setup
1. Mount the camera. 2. Install video and power cables between the cabinet and the camera. 3. Install surge and lightning protection. 4. Mount the camera interface panel in the traffic control cabinet. 5. Wire the power and video cables to the camera. 6. Wire power and video cables to the interface panel. 7. Supply power to the interface panel from the cabinet. 8. Test the video output. 9. Install the detection card in the detector rack. 10. Connect the interface panel to the card. 11. Tilt and rotate camera as needed to optimize detection area. 12. Configure the detection zones. There may be more than one type of detection zone, e.g., count, presence, speed, bicycle. 13. Configure the card outputs. 14. Verify the operation of the detector outputs. 15. Switch the controller to accept input from the VDS.	1. Select a camera channel. 2. Assign detection zones to each camera channel. 3. Enable desired direction of vehicle travel. 4. Set detection zone sensitivity. 5. Enter zone and camera label information. 6. Enter zone delay. 7. Enter zone extend. 8. Enable conditional detection based on phases, overlaps, and other outputs for a specific zone. 9. Define detection zone fail-safe conditions when poor video or low contrast exist or under learning conditions. 10. Enable shadow filter. 11. Enable any additional VDS inputs required by the controller. 12. Edit detection zones as needed. 13. Save configuration.

guidelines. When this occurs, proper placement of the detection zones becomes even more critical. The higher the camera and more centered it is over the lanes of interest, the less likely that tall vehicles in lanes nearer the camera will occlude vehicles further away.

FLIR recommends a camera mounting height of 11–39 ft (3.5–12 m) for its TrafiCam VDS, leaving it up to the installer to make final adjustments [7]. Peek reports that FLIR

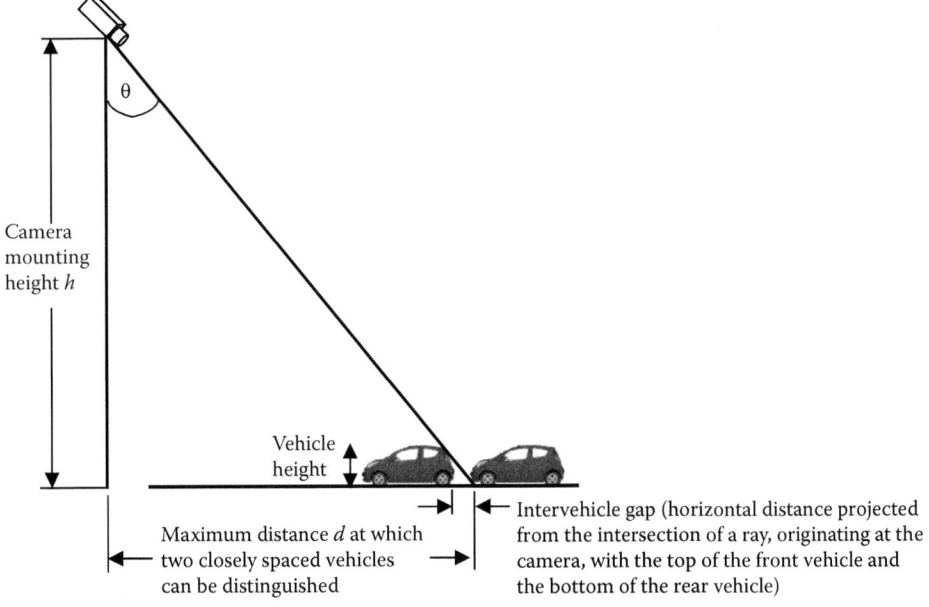

Figure 7.2 Distinguishing between two closely spaced vehicles as a function of camera mounting height and roadway distance between vehicle and camera location.

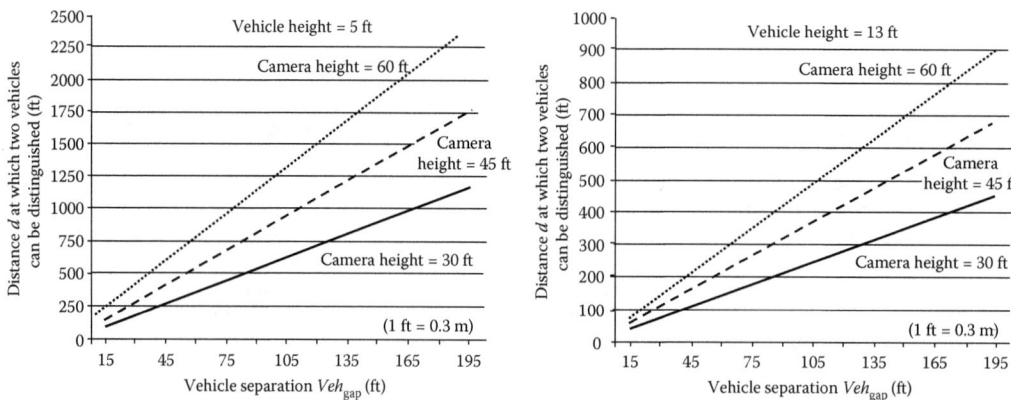

Figure 7.3 Distance *d* along the roadway at which a VDS can distinguish vehicles: (left) vehicle height = 5 ft (1.5 m); (right) vehicle height = 13 ft (3.9 m). These values may be further limited by road configuration, congestion level, vehicle mix, inclement weather, and pixel size.

Table 7.2 Recommendations for optimal camera mounting

Parameter	Description
Mounting height	25–40 ft (8–12 m).
Sighting for camera field-of-view (FOV)	Camera mounted no more than 15° off the line that separates the left-turn lane from the through lanes.
	Include stop bar and a 2–3 ft (0.6–0.9 m) section of roadway in front of the stop bar that extends into the intersection within the FOV.
	Stop bar in lower 1/3 of FOV.
	Car bumpers parallel to bottom of FOV.
	Area available for advance detection zones.
Cautions	Use a stable mounting location.
	Be aware of possible occlusion that will block view of traffic lanes.
	Avoid:
	• Power lines, tree limbs, and any other stationary objects in the detection zones.
	• Horizon or sky in the image FOV.
	• Background lighting, such as business signs or street lamps, appearing within the FOV.
	• Mounting camera directly beneath a street lamp.
	• Sun glint especially from east–west facing cameras.

thermal cameras can be utilized with the VideoTrak-IQ systems [1]. The Iteris height recommendation is approximately 30 ft (9 m) [6]. For lower mounting locations, such as a mast arm, Iteris recommends a suitable camera extension bracket be used to increase the camera height to a more ideal elevation. Camera mounting heights of 25 ft (8 m) or more can usually be obtained with the extension bracket. Low mounting heights can result in reduced system performance and vehicle occlusion.

7.2.2 Field-of-view

Figure 7.4 describes a typical field-of-view for the VideoTrak-IQ™ as it might appear in a video monitor after proper camera placement and adjustment of the camera's aim, zoom, and focus. For most intersections, cameras are aimed so the stopline detection zones are toward the top and centered left-to-right in the image. The traffic can also flow diagonally

Figure 7.4 Typical camera field-of-view with VideoTrak-IQ™ detection zones in place. (Photograph courtesy of Peek Traffic, Palmetto, FL.)

or across the image if this provides a better field-of-view [2,8], although not all VDS manufacturers recommend this option.

At the stopline, the camera should see an extra 1/4 or 1/2 lane on either side of the detection area. The image area should be filled with the detection targets, namely, the vehicles, bicycles, or pedestrians, while excluding extraneous objects or obstructing light sources that affect performance under some conditions. Adjustments such as rotating the barrel of the camera can assist in blocking unwanted light sources, for example, lighted signs, window glare, and signal heads, from the image. If zooming out for a wider view is necessary, the barrel should be moved forward only enough to see a little sunshield in the corners of the image.

When aiming the camera, it should be tilted down so the farthest detection area is toward the edge of the field-of-view. This increases the overall contrast of the image by excluding parts of the broader picture that have no detection purpose.

7.3 DETECTION ZONE PLACEMENT FOR VIDEO DETECTION AT A SIGNALIZED INTERSECTION

Figure 7.5 shows recommended field-of-views and detection zone placement for Iteris VDS. Rotation of the camera body is often required so that vehicles flow from the top of the monitor to the bottom to facilitate optimization of the detection zones. The setup of detection zones varies from manufacturer to manufacturer, for instance, in the shape of zones used for stopline detection. Therefore, the information that follows is for illustration only as the instructions from a specific sensor manufacturer should be used when configuring detection zones for their particular VDS.

Sections 7.3.1 through 7.3.3 below describe detection zone placement for VDS control of traffic signals as recommended in Autoscope® application notes [2,8]. These procedures are meant to introduce the reader to the criteria for configuring several types of detection zones utilized by VDS. In addition to the down-lane presence detectors at the stopline, stopline detectors, and speed detector zones that are described, many VDS also have detector configurations optimized for bicycle and pedestrian detection. Others add a variety of zones types that assist with signal control, such as extension, delay, and low contrast zones designed to determine if the VDS can adequately image the roadway under reduced visibility conditions. Section 7.3.4 describes the procedures recommended by Peek Traffic for configuring VDS presence detectors at a stopline.

Figure 7.5 Examples of recommended field-of-views and detection zone placement for Iteris VDS. (Photographs courtesy of Iteris, Santa Ana, CA.)

7.3.1 Autoscope® positioning and sizing down-lane presence detectors at stopline

The guidelines for placing down-lane presence detectors at a stopline, as depicted on the left side of Figure 7.6, are

1. In line with a typical vehicle's headlights and license plate or hood ornament.
2. 1.5–2 vehicle lengths long.
3. 0.5–0.9 ft (15–27 cm) wide (shoulder-width or about a side window size or large enough to cover the license plate to headlight area).
4. Close to detectors ahead of the stopline to help detect large dark vehicles and all vehicles at night.

7.3.2 Autoscope positioning and sizing stopline detectors

Procedures used for placing the stopline detectors illustrated in the middle of Figure 7.6 are as follows:

1. Configure the detection zone length equal to the length of 2–3 vehicles.
2. Position lane detection lines from the intersection side of the stopline to marker cones 60–100 ft (18–30 m) upstream from the intersection.
3. Adjacent detectors should share a common boundary.
4. Once the front and rear angles are drawn in this manner, the entire zone may be shifted as needed.
5. Enable the "Turn Off Shadow Processing at Midday" option.

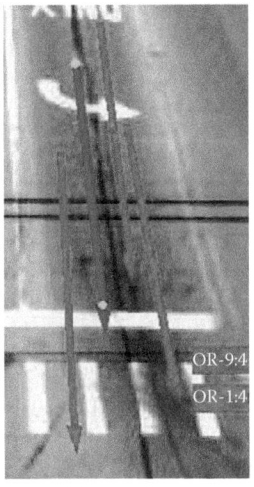 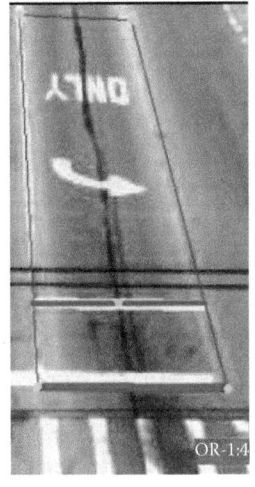 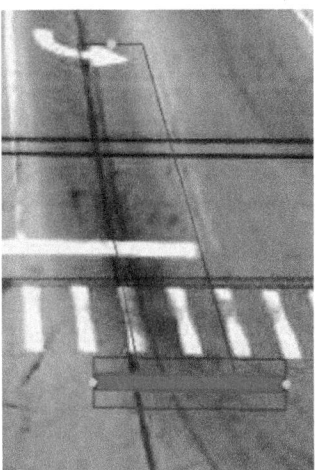

Down-lane detector   Stopline detector   Speed detector

Figure 7.6 Down-lane, stopline, and speed detectors as drawn for an Autoscope VDS. (Photographs courtesy of Econolite, Anaheim, CA.)

6. For best results, do not use the stopline detector with a black and white video camera or without phase colors.
7. Increase the detection zone length forward of the stopline and toward the intersection to assist in detecting large dark vehicles and all vehicles at night.

7.3.3 Autoscope positioning and sizing speed detectors

The methods for placing speed detectors, which are shown on the right side of Figure 7.6, are as follows:

1. Position speed detectors to maximize vehicle separation whenever possible, that is, closer to the camera or at the bottom of the field-of-view.
2. Place the detector where the hood ornament of a typical vehicle would be, or alternately in line with the lane or flow of traffic. Also adjust the count detector position.
3. Use the 50 ft (15 m) default length or 1.25–1.5 vehicle lengths.
4. Draw the sides parallel with each other or the lane marks.
5. Change Min Classification Speed to 6 mi/h.
6. Change Min Report Speed to 5 mi/h when traffic data are collected at the intersection.
7. Change Min Vehicle Length to 16 ft (5 m).
8. Change Occupancy Normalization when traffic data are collected at the intersection.

7.3.4 Peek Traffic guidelines for placing presence detectors at a stopline

Figure 7.7 depicts Peek Traffic guidelines for placing VideoTrak-IQ™ VDS presence detectors at a stopline. For intersection control, each zone at the stopline should be [1]

1. Located within the lane or slightly outside the lane, depending on the orientation of the vehicles from the camera's point of view. There are no limitations as far as placing the zone on a curb, painted lines and arrows, guard rails, or medians.

All zones are approximately 3–4 cars in length.

For proper night detection, place zones approximately 2–3 ft in front of stopline.

Figure 7.7 Guidelines for placing presence detectors at a stopline for a VideoTrak-IQ™ VDS. (Photograph courtesy of Peek Traffic, Palmetto, FL.)

2. 3–4 vehicle lengths long.
3. From side mirror to side mirror in width. Both headlights should be in the zone during normal traffic flow.
4. To improve night detection, zones should be placed 2–3 ft (0.6–0.9 m) in front of the stopline to maintain detection for vehicles that creep past the stopline at night. The rationale for this recommendation is that during night detection, the VideoTrak-IQ™ VDS detects headlights. Therefore, if the headlight beams extend beyond the detection zone, the call will be dropped if the controller is not set to lock calls.

7.4 INSTALLATION AND INITIALIZATION OF A PRESENCE-DETECTING MICROWAVE RADAR SENSOR

Table 7.3 gives the general procedure for sensor installation and software initialization and detection zone setup for an overhead presence-detecting microwave sensor. These steps are illustrative of those employed by manufacturers to instruct personnel in the use of modern presence-detecting microwave sensors. It is also instructive to examine detailed procedures for mounting and initializing presence-detecting microwave radar sensors.

Table 7.3 Installation and setup of an overhead microwave radar sensor (typical)

Major steps—Sensor installation	Major steps—Software initialization and setup
1. Ensure all necessary components and tools are available.	1. Install the sensor software.
2. Select the sensor's location.	2. Access the sensor from the computer used for setup.
3. Determine the height and setback and mount the sensor.	3. Enter the sensor settings, e.g., serial number, location, sensor height, RF channel associated with the sensor.
4. Align sensor to roadway.	
5. Attach data cable, surge protector and lightning arrestor if needed, and ground the sensor.	4. Select automatic or manual configuration.
6. Mount any specialized hardware in the traffic cabinet.	5. Configure size and shape of detection zones.
7. Wire power to cabinet hardware.	6. Map zones to channels.
8. Terminate the conductor cable at the hardware.	7. Verify and save configuration.
9. Connect to the detector rack cards and set rack card switches as required.	

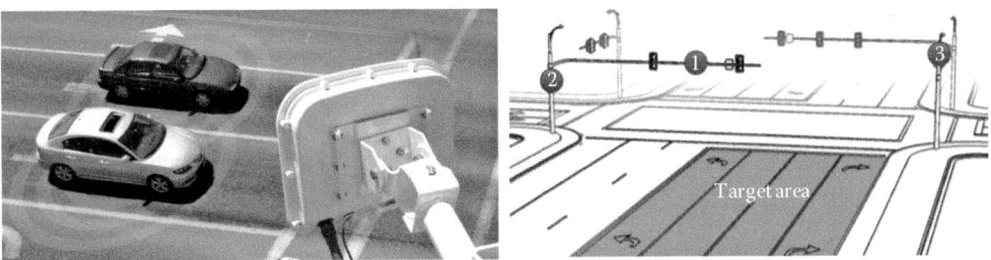

Figure 7.8 SmartSensor Matrix installation options for a signalized intersection. (Photographs courtesy of Wavetronix, Provo, UT; From *SmartSensor Matrix Installer Quick-Reference Guide*, Wavetronix, Provo, UT, 2012.)

This first example summarizes the eight-step mounting process for a Wavetronix™ SmartSensor™ Matrix installed at a signalized intersection as depicted on the left of Figure 7.8 [9].

7.4.1 Sensor mounting process

1. Ensure that all necessary components are available. These include sensors, mounting brackets, accessory cables, 6-conductor cable, and preassembled backplate.
2. Select the sensor's mounting location using the guidelines in Table 7.4. The three options on the right in Figure 7.8 are suggested mounting locations, although other mounting locations may be available depending on the specific configuration of the intersection.
   a. Back side of mast arm as indicated by the numeral 1 in Figure 7.8. This location allows sensor placement near the lanes of interest and may be the best option for wide approaches. This position functions well when the sensor is mounted near the end of the arm to reduce the possibility of the mast arm or departing traffic occluding approaching vehicles.
   b. Far side of approach as indicated by the numeral 2. Here, the sensor is usually mounted on a corner vertical mast pole or strain pole. Mounting on a vertical pole with a mast arm usually avoids occlusion because the sensor is located away from or below the mast arm.
   c. Near side of approach indicated by the numeral 3 is typically best if detecting the left-turn lane is less important. This location also allows the sensor to be high enough to avoid occlusion.
3. Determine the required height and offset and then mount and align the sensor by attaching the sensor mounting bracket to the pole and fastening the sensor to the mounting bracket. Alignment of the sensor to the roadway is performed as follows. The sensor's field-of-view fans out 45° to both sides of its viewing direction or boresight, as shown on the right side of Figure 7.9. Usually the radar beam is positioned so that the 90° detection or footprint area covers all lanes approaching the stop bar. The front edge of the field-of-view must be aligned to provide some coverage beyond the stop bar so that the sensor can detect vehicles that do not stop at or behind the stopline, as well as vehicles exiting queues [9,10]. This is accomplished as follows:
   a. Adjust the side-to-side angle so that the front edge of the field-of-view provides a view downstream of the stop bar.
   b. Tilt the sensor down to aim it at the center of the lanes of interest.

Table 7.4 Mounting guidelines for SmartSensor Matrix sensor

Parameter	Description
Field-of-view	Corner-shaped 90° coverage out to 140 ft (42.7 m) as illustrated in Figure 7.9.
Line of sight	Position so that the sensor can detect the entire area of interest. Avoid occlusion by installing the sensor away from trees, poles, signs, signal heads, and other roadside structures. Position it so that mast arms do not block the view of the detection area.
Mounting location	Select so that all stop bar detection zones on an approach are within a 6- to 140-ft (1.8- to 42.7-m) radial distance of the sensor.
Mast arm mounting	The mast arm is frequently a good place to mount the sensor.
Detection coverage	Position so that all specified stop bar detection zones are within the sensor's field-of-view. The sensor will often work better if it is positioned so that it tracks vehicles for several feet before the first zone in each lane. If the sensor has a view several feet beyond the stop bar, it is more likely to accurately detect queue dissipation.
Closest to lanes of primary interest	Mount the sensor on the side of the road closest to the principal lanes of interest. Table 7.5 assists in determining the mounting height as function of the distance to the closest monitored lane.
Minimum mounting height	12 ft (3.6 m). Mount the sensor high enough to prevent traffic from occluding approaching vehicles as allowed by mounting options at the installation site.
Maximum mounting height	60 ft (18.2 m).
Nominal mounting height	20 ± 5 ft (6.1 ± 1.5 m).
Preventing occlusion	Placing the sensor higher will result in less occlusion. Placing it lower could result in more occlusion. However, if the nearest detection area is less than about 20 ft (6.1 m) away, the sensor may perform better with a lower mounting position.
Minimum mounting setback	6 ft (1.8 m) to the first lane of interest is required. The farther the sensor is from the first lane of interest, the higher the sensor should be mounted.
Multiple sensors at an intersection	It is possible for multiple sensors to monitor the same approach. Multiple sensors are needed when zones are spread over more than 140 ft (42.7 m).
Interference from proximity to other sensors	When multiple sensors are mounted at the same intersection, interference can be avoided by configuring each sensor to operate on a unique RF channel.
Loss of extended range performance	Lanes that have stop bars or detection zones placed at extended range may show some loss in performance, even with a proper mounting height. This is more apparent at locations with many travel lanes or where detection zones are positioned near the far edges of the detection area.

    c. If necessary, rotate the sensor so that the bottom edge of the sensor is parallel with the roadway. This is necessary where the intersection approach has a significant grade.

4. Attach the 6-conductor cable and ground the sensor. To avoid undue movement from wind, strap the cable to the pole or run it through a conduit, leaving a small amount of slack at the top of the cable to reduce strain.

The sensor provides its own surge protection. Therefore, a pole-mount box on the sensor side of the cable is not needed and the cable should run directly to the main traffic cabinet. It is necessary, however, to ground the sensor using the following technique:

    a. Connect a grounding wire to the grounding lug on the bottom of the sensor.

    b. Connect the other end of the grounding wire to the earth ground for the pole on which the sensor is mounted. *Do not run the grounding wire back to the main traffic cabinet.*

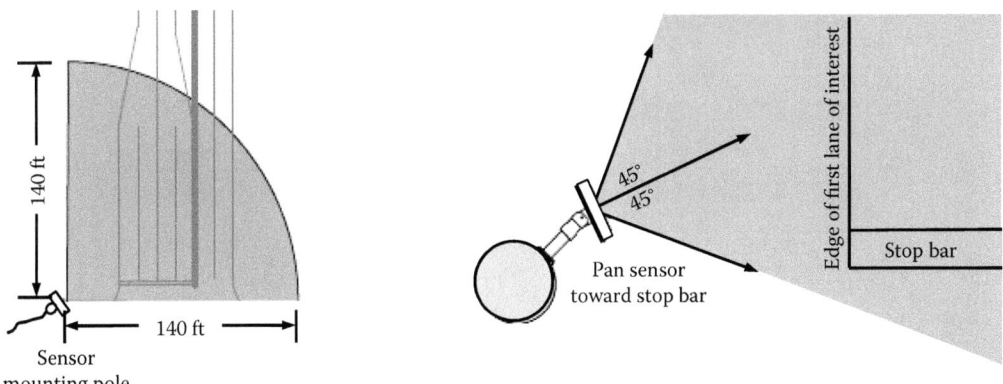

Figure 7.9 Field-of-view of SmartSensor Matrix radar sensor. (Drawings courtesy of Wavetronix, Provo, UT; From *SmartSensor Matrix User Guide*, Wavetronix, Provo, UT, 2015. http://www.wavetronix.com/en/support/downloads/316-smartsensor-matrix-user-guide; *SmartSensor Matrix Quick-Reference Guide*, Wavetronix, Provo, UT, 2014. http://www.wavetronix.com/en/support/downloads/545-smartsensor-matrix-quick-reference-guide.)

Table 7.5 Sensor height as a function of distance to closest monitored lane for SmartSensor Matrix sensor

Distance to closest monitored lane (ft)	Sensor height (ft)
6–15	12–25
15–50	15–25
>50	25–60

5. Mount the preassembled backplate in the main traffic cabinet once installation of the sensor is complete. To do so, locate the area planned for mounting the backplate such as the side panel of a NEMA-style cabinet. Then attach the backplate with the U-channel mounting screws.

6. Wire power to backplate. Use the steps below to connect power to the AC terminal block on the bottom DIN rail [11]:

   a. Connect a line wire (usually a black wire) to the bottom of the "L" terminal block shown on the left of Figure 7.10.

   b. Connect a neutral wire (usually a white wire) to the bottom side of the "N" terminal block.

   c. Connect a ground wire (usually a green wire) to the bottom of the "G" terminal block.

   d. Turn on AC main power.

   e. Press the circuit breaker switch on the left side of the top DIN rail to switch power to the backplate.

   f. Verify power is regulated by confirming that the DC OK LEDs are illuminated on the 100–240 VAC to 24 VDC power converters.

7. Terminate the 6-conductor cable by installing it into the terminal block as follows:

   a. After routing the 6-conductor cable into the cabinet, strip back the cable jacket and shielding on the service end of the cable.

   b. Open the insulation displacement connectors on the plug by inserting a small screwdriver into each square slot and rocking it back.

  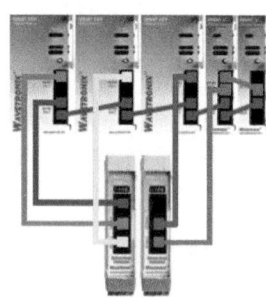

Wire power to backplate   Terminate cable in terminal block   Connect to detector rack cards

Figure 7.10 Connecting power and the sensor outputs to the detector cards. (Photographs courtesy of Wavetronix, Provo, UT; *SmartSensor Matrix Installer Quick-Reference Guide*, Wavetronix, Provo, UT, 2012.)

    c. Insert the wire leads into the bottom side of the plug-in terminal according to the colors of the wires and the labels on the plug. The wires should be completely inserted into the terminal as illustrated in the middle of Figure 7.10.

    d. Close the insulation displacement connector by reinserting the screwdriver into the square slot and rocking it forward. The plug-in terminals will automatically complete the electrical connection. There is no need to manually strip the insulation on the end of each wire.

    e. If the plug was removed to connect the cable, insert it back into the terminal block.

8. Connect to the detector rack cards using the procedure below:

    a. Confirm the DIP switches are set.

    b. Connect from the RS-485 A port (surge-protection device) to a bus 1 port on the appropriate rack card as shown on the right in Figure 7.10. Connect from the RS-485 B port to a bus 1 port on another rack card.

    c. If using file cards, use a patch cord to share bus 1 between cards dedicated to the same sensor. If there are more than two sensors in the system, repeat Steps 8a through 8c to connect bus 1 for all remaining rack cards.

    d. Connect from a bridge port to bus 2 of the rack cards.

    e. Daisy-chain between the bus 2 ports of all of the rack cards for device configuration.

7.4.2 Software initialization

The procedure for installing and initializing the SmartSensor Manager Matrix (SSMM) software on a laptop or PC is as follows.

1. Download the setup program from the Wavetronix website (http://www.wavetronix.com) under the Support link. Open the file and follow the steps in the install wizard.

2. Connect the target computer to the sensor.

    a. This is done with a communication module, such as a serial-to-Ethernet converter or a Bluetooth® module. The communication module must be mounted on the same T-bus as the system surge protector.

  b. Open SSMM software and select Communication on the main menu.
  c. Select the preferred type of connection (serial or Internet; the virtual connection is for training and demonstration purposes) by clicking on the tab at the top of the page.
  d. Change the timeout if desired.
  e. For an Internet connection, enter the IP address and port of the sensor. For a serial connection, enter the port (if needed).
  f. Select the type of search to perform, Quick or Full. The Full search option is used for a first-time sensor search. Subsequent searches may use the Quick search option.
  g. When the software is finished detecting sensors in the area, select the sensor of interest from the list. Then click Connect.
  h. The squares in the lower left of the main menu allow adjustment of the SSMM's display size.
3. Enter the sensor settings. These are found under Sensor Settings on the SSMM main menu.
  a. General tab:
      Serial number: The sensor's serial number; cannot be edited.
      Sensor ID: Used to uniquely identify all sensors on a multi-drop bus; by default, the sensor ID is the last seven digits of the sensor's serial number and cannot be edited.
      Description/location/approach: For identification and information. Description and location are limited to 64 characters, approach to 32 characters.
      RF channel: Assigning sensors to different channels prevents radars from interfering with each other.
      Sensor height: In feet; affects the display of data in SSMM.
      Wash-out time: Time the sensor has to see a constant power-level tracker before it washes out into the background.
      Units: Choose between standard or metric.
  b. Comm tab:
      Response delay: Configures how long the sensor will wait before responding to a received message. The green arrow points to the port servicing the current connection.
      Data push: Assigns the port on which data are being pushed.
      Source: Choose between Antenna and Diagnostic.
4. Lane configuration Option 1: Automatic configuration of the SmartSensor Matrix.
  a. Select Sensor Setup from the SSMM main menu. The Sensor Setup screen will appear. If the Lanes & Stop Bars screen is not already open, click on tab 1 to open it.
  b. Move the sensor to the desired orientation by clicking the Move Sensor button at the bottom of screen.
  c. Click the Erase button to clear the edit area.
  d. Start automatic lane configuration by clicking the button and selecting Restart Auto Lane Cfg from the window. Allow the intersection to cycle at least twice before proceeding. To see the automatically configured lanes, set the display to Automatic Configuration overlay by clicking the Auto Cfg tab at the bottom right of the screen.
  e. Once the automatically configured lanes have appeared, capture the lanes and stop bars. To capture, click once on a lane to highlight it, and then again to bring up the Capture Lane window. Select Capture Lane or Capture All. Any stop bars in the captured lanes will also be captured.
  f. Make manual adjustments, if necessary, as explained in Part 5 of the software initialization instructions.
  g. Save changes to the sensor by clicking the Save button.

5. Lane configuration Option 2: Manual configuration. The Lanes & Stop Bars tab allows manual adjustment of the automatically configured lanes, or complete manual configuration of the lanes.
    a. Add or delete a lane:
       To add a lane, click in the edit area where you would like to add a lane. The Edit Area window will appear; then click on the Add Lane button. A maximum of 10 lanes are allowed.
       To delete a lane, select the lane you want to delete, then click it again. The Edit Lane window will appear. Click on the Delete Lane.
    b. Insert, delete, or move a stop bar:
       To insert a stop bar, select the lane you want to insert it into and then click on it again. The Edit Lane window will appear. Click on the Insert Stop Bar button.
       To delete or move a stop bar, select the lane and then click on the stop bar. The Delete Stop Bar window will appear. Click Delete Stop Bar to delete the stop bar or use the arrow buttons to move it. Also, a stop bar may be grabbed and dragged in the edit area and moved to a new position.
    c. Insert, delete, adjust, or move a lane node:
       Lane nodes are used to change the trajectory, curve, or width of a lane.
       To insert a lane node, select the lane to insert it into and then click again. The Edit Lane window will appear. Click on the Insert Node button. A maximum of six nodes per lane are allowed.
       To delete or adjust the width of a lane node, select the lane and then click on the lane node. The Node Adjustment window will appear. Click Delete Node to delete the node. Use the arrow bumpers under Width to adjust the width of the node.
       To move a lane node, click on the lane and then the lane node. In the Node Adjustment window, use the arrows to move the node (or grab and drag the node in the edit area).
       Upon finishing, save changes to the sensor by clicking the Save button.
6. Set up detection zones.
    a. Click on tab 2 for the Zones & Channels screen. If there are stop bars and no zones are currently configured when the tab is opened, an option appears to automatically place 20-ft (6.1-m) detection zones at each stop bar. You can position these zones at any point by clicking the Place Auto Zones button.
    b. Manually set or adjust the zones.
       To change a zone's shape, select a zone and then drag its corners.
       To add or delete a zone, drag a green zone from the zone bank onto the edit area to add it. To delete a zone, drag it back out to the zone bank.
       To move a zone, select it and then click the zone button (the number will change based on the zone selected) for the Edit Zone window. Use the arrows to move the zone or click and drag the zone in the edit area.
7. Map zones to channels.
    a. When automatically placing zones, the first four are mapped to C1, C2, C3, and C4, respectively. Zones can also be manually mapped to channels in three different ways.
    b. Multiple zones can be mapped to the same output channel. In this case, the zone detections are "or"-ed together, meaning that if any of the zones associated with a channel is active, then the channel output will be active.
    c. Edit Zone window: To move a zone, select it and click on the Edit Zone button.
    d. Edit Channel window: To map a specific channel, click on the Edit Channel button. Click on one of the gray indicators marked Z to map the zone to the channel.
       To select a different channel, click on the Edit Channel button to cycle through

until the desired channel appears. To add Delay, Extend, and Phase information, click anywhere on the right side of the window.

e. Zones/Channel Map: To see all channels and zones, click the Zone/Channel Map button. Click on the indicators in the table to map or un-map zones and channels. A zone is mapped to a channel if the corresponding indicator is green; it is not mapped if the indicator is gray.

f. When finished, click to save changes to the sensor.

8. Verify the configuration.

a. Select tab 3 to open the Verification screen.

b. Verify the sensor is configured and working properly: the blue rectangles indicate detections; the indicators at the top turn red when the associated channel is active. To see the zones associated with a channel, click on that channel's indicator.

7.5 RTMS® SX-300 PRESENCE-DETECTING MICROWAVE RADAR SENSOR

The RTMS® Sx-300 is another presence-detecting radar sensor with multilane vehicle detection capability. Once mounted, multiple operating modes may be accessed to optimize internal parameter settings for highway (mainly free-flowing traffic) and midblock urban street (mainly congested traffic) applications from the Setup Utility. The sensor's detection area extends out to 76 m (250 ft), capable of detecting up to 12 lanes of traffic as indicated in Figure 7.11 [12]. The length of the detection zone is determined by the antenna beam's footprint, that is, detection area.

7.5.1 Mounting options

Figure 7.12 illustrates five cases that are typical of side-fired highway mounting options for the RTMS Sx-300 sensor on existing poles and road structures.

1. Case 1: Maximal utilization of the sensor's zone capability. Cautions to be observed with this mounting configuration are as follows:

a. The requirement for 12-lane coverage implies a larger setback distance to the first lane. If setback is insufficient, two sensors may be required, one for each travel direction.

b. Limitations in mapping range slices to lanes will cause decreased accuracy. This requires the site designer to trade-off level of accuracy with cost.

c. In almost all cases, the sensor can resolve the return signal from the barrier that separates opposite direction travel lanes from that of the vehicles in the lane immediately behind it as long as 50% of vehicles can be seen.

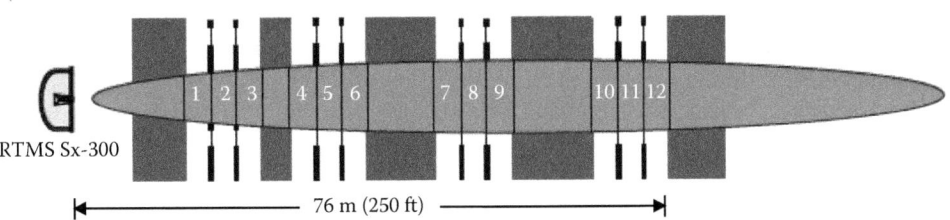

Figure 7.11 RTMS Sx-300 footprint. (Drawing courtesy of ISS, St. Paul, MN.)

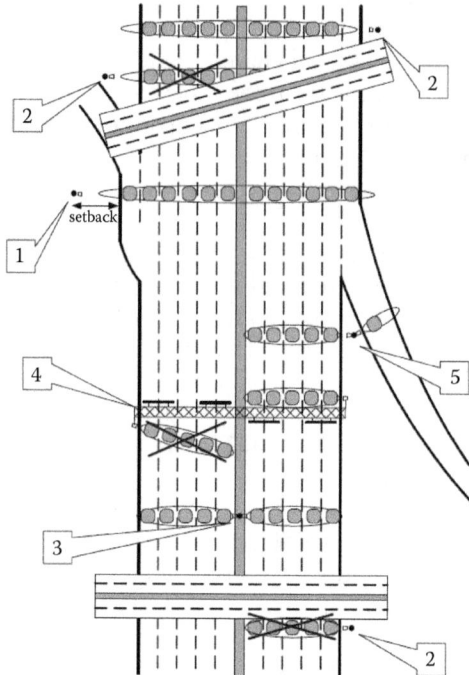

Figure 7.12 RTMS Sx-300 side-fired highway mounting options. (Drawing courtesy of ISS, St. Paul, MN.)

2. Case 2: Overpass installations. Do not mount the sensor on a perpendicular overpass. Instead, use poles located at least 5 m (16 ft) from the overpass to avoid multipath where reflected signals from vehicles can also be detected by the sensor through secondary reflection from a large flat surface (such as a sign or overpass). If the overpass is at an angle to the road, take advantage of the angle to point the sensor at the monitored roadway and away from the overpass. Do not aim the beam under it.

3. Case 3: Using median poles to mount two sensor units, one per direction, may save poles but the designer should verify if the setback is satisfactory.

4. Case 4: Sign-structure installations.
   a. Installation on message sign structures is acceptable only if the sensor is offset from the overhead span of the structure as it can reflect the microwave signal and produce false or missed detections. Some structures such as dynamic message signs (DMS) have wide, flat metal bottoms that are similar to those found in bridges. These can cause more interference than lattice-work structures and may require consultation with RTMS Sx-300 technical support.
   b. The best way to mount the sensor is on a horizontal mast arm or pipe located approximately 1.3 m (4 ft) from the structure, 1.8–2.4 m (6–8 ft) if a DMS, ideally on the back of the structure away from any lighting or signs. The sensor should be aimed perpendicular to the traffic flow direction.

5. Case 5: Typical ramp metering site. The viewing direction should be perpendicular to the traffic flow direction.

Placement for midblock applications is similar to that for side-fired highway use. If sufficient setback is available, up to 12 zones of traffic can be configured. If the sensor is mounted in a zero-setback configuration, only the nearest four zones are available for detection. However, the nearest zones can be excluded from detection (creating adequate sensor setback) if data from zones farther away are desired.

Mounting and aiming of the sensor proceeds as follows:

1. Attach the bracket to the roadside pole (or another specified location) using bolts or stainless steel bands.
2. Secure the sensor to the mounting bracket using the washer, lock washer, and nut. The cable connector should be at the bottom of the unit when it is mounted.
3. Adjust the sensor to be perpendicular to the travel lanes and level side to side.
4. Look from behind the unit and use the top sight-ridge as a guide to align the boresight.
5. Tilt the sensor so that the top is aimed at the first 1/3 of the monitored lanes as in Figure 7.13.

   Steps 3 through 5 are general guidelines. Mounting and tilt may need to be adjusted based on other factors such as obstacles and number of lanes.

6. Secure the mounting position by tightening the nuts.
7. Connect the cables for power, communications, and surge suppression. It is recommended that power and communication cables connected to sensor have surge protection and each sensor be properly grounded.
8. Configure the sensor using the Setup Utility.

7.5.2 Setup Utility

The following procedure is utilized to connect the Setup Utility to the RTMS Sx-300 sensor [12].

1. Using a serial cable, connect the sensor to the serial port of the computer that has the Setup Utility installed.
2. Power up the sensor.
3. Select Start > All Programs > ISS > RTMS Sx-300 Setup Utility > RTMS Sx-300 Setup Utility or double-click the shortcut icon on the desktop.
4. Select language in which screens will be displayed.
5. Select region of the world in which the sensor is physically located.
6. If communication is established with the sensor, the Main Screen illustrated on the left of Figure 7.14 will appear. If communication is not established or if multiple units are located, the Start Screen on the right of the figure will appear.

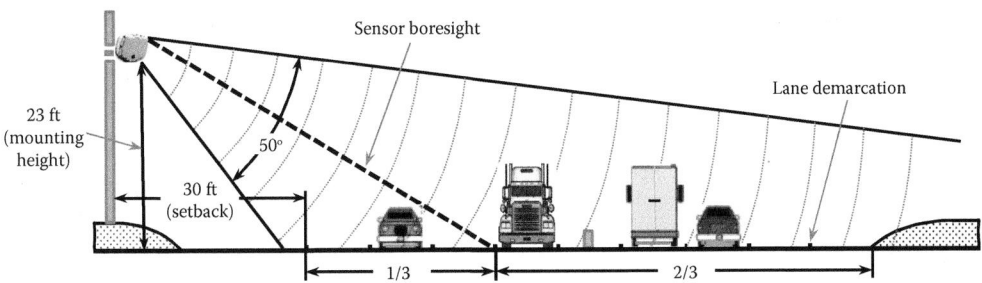

Figure 7.13 RTMS Sx-300 initial aiming. (Drawing courtesy of ISS, St. Paul, MN.)

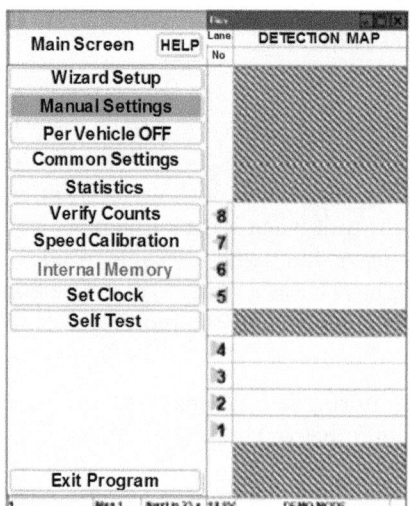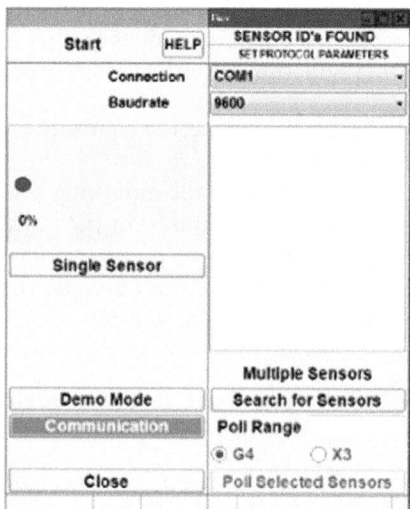

Figure 7.14 Setup Utility main and start screens for RTMS Sx-300. (Pictures courtesy of ISS, St. Paul, MN.)

The Main Screen is divided into two vertical panels. The left panel contains the following menu buttons: Wizard Setup that provides an automated zone setup process; Manual Settings used to configure or reconfigure the sensor; Per Vehicle that adds a time stamp, lane number, classification, speed, and dwell time of the detected vehicles; Common Settings that saves the sensor configuration for loading into other sensor units as setup parameters are often common to all sensors at a single site; Statistics that monitors data on the user interface; Verify Counts to compare manual vehicle counts with the sensor vehicle counts; Speed Calibration that matches actual speeds with the speed calculated by the sensor; Internal Memory to store data inside the sensor; Set Clock to synchronize the sensor clock with the computer clock; and Self Test to initiate an internal diagnostic test of the sensor. The status bar at the bottom of the screen shows, from left to right, the serial number of the sensor and firmware version, message number, time to next message, voltage at sensor unit, COM port and speed, and COM indicator. The firmware version as of January 2017 was 8.0.4.0.

The right panel of the Main Screen displays the detection map with the current detection zones and the real-time detections indicated by moving vehicles. The Start Screen is most often used to run the demonstration mode and to test sensors in a polled mode.

7.5.2.1 Configuration process

The Setup Utility allows the user to configure each RTMS Sx-300 sensor once the hardware and software are installed. This process requires physical connection of the computer that contains the Setup Utility to each sensor in the system. Each sensor is configured according to the following process:

1. Set the region (if not already done). The region setting indicates where the sensor is installed and ensures proper operation of the sensor. The setting applies to both the computer in which the Setup Utility is running and the sensor itself. If there is a mismatch between the computer and connected sensor, a warning message is displayed when the Setup Utility is started.

2. Set the application mode to Side-Fired or Midblock through the Manual Settings screen. Side-Fired Highway applies, in general, when traffic is primarily free-flowing, farther detection distance is needed, and lanes are wide. Midblock applies, in general, when traffic is mainly urban (can be congested), shorter detection distance is needed, and lanes are narrow.

3. Run the Wizard. The automated zone setup process requires free-flowing traffic in all lanes of interest. The Wizard scans the detection range of the sensor's microwave beam and positions up to 12 detection zones, representing lanes where vehicles are detected. The Setup Utility then configures lane parameters utilizing a two-stage process:
   - Initial setup where the Wizard finds zones that match up to the lanes of traffic.
   - Final setup where the Wizard fine-tunes the zone boundaries and detection parameters.

4. Adjust the zones. A zone created by the sensor ideally represents a detected lane of traffic. Zone adjustment occurs by monitoring the vehicle icons in each zone on the detection map and comparing them with what is physically seen on the road. Pressing the space bar on the computer produces a beep sound every time the sensor detects a vehicle in the zone. This allows the user to watch the road and listen for the beep to determine if what is being seen on the physical road is the same as what the system is detecting. If the vehicle icons on the screen match the physical vehicles on the road, then the zones are adjusted properly. If the system is not detecting smaller vehicles or double counting trucks, increase or decrease the sensitivity of the sensor accordingly.

5. Verify vehicle counts. Vehicle count verification ensures that the detection zones are set up properly. This process compares a sensor's volume counts over a period of time to a manual (visual) count for the same interval. The recommended procedure is to use a handheld counter to perform the verification individually on each zone. However, if a sufficient number of personnel are available, all zones can be verified simultaneously. This alternate process requires at least one person per zone to gather the manual counts. A minimum of 50 vehicles should be counted in each zone being verified. If the difference between the sensor count and manual count is greater than 5%, modify the detection zone setup to improve the accuracy. Section 7.5.2.2 contains further suggestions for improving vehicle count accuracy.

6. Calibrate speed. Speed calibration is a three-step process that includes (1) entering the reference speed into the Wizard, which is the average speed that most vehicles travel through a zone, (2) running the automatic calibration, and finally (3) checking the calibration. The reference speed is confirmed with a lidar radar gun or similar device. Speed calibration should be performed when traffic is moving at the posted speed limit, and not during periods of fluctuation or congestion.

7. Define Message Composition. Message Composition identifies the content (e.g., volume, occupancy, speed, vehicle class, gap or headway, 85% speed, and time stamp) and format of the statistical messages that are sent from the sensor to connected hardware (e.g., computer, smart phone, or tablet). Messages are automatically sent every message period when the data mode is set to Stat or Normal. A Polled data mode is also available that transmits statistical data currently stored in the sensor buffer when a matching sensor ID is received by the sensor.

8. Define vehicle classifications. The correct classification of vehicles by length requires the definition of minimum and maximum lengths for up to six vehicle classes. These are entered into the Setup Utility by the user. The maximum length that can be specified is 25.5 m (83.7 ft). The default size for each class is shown below:
   - Small: 0–5 m (0–16.4 ft).
   - Regular: 5–7 m (16.4–23.0 ft).

- Medium: 7–10 m (23.0–32.8 ft).
- Large: 10–15 m (32.8–49.2 ft).
- Truck: 15–20 m (49.2–65.6 ft).
- Extra large: Greater than 20 m (65.6 ft).

9. For best results, the differences between minimum-to-maximum lengths should be greater than 3 m (≈10 ft), especially for larger vehicles. Small minimum-to-maximum differences increase the potential for merging of classes and result in vehicle class counting errors.

10. Save the configuration file. After completing the configuration setup, save it to a file on the computer's hard drive.

7.5.2.2 Optimizing volume count accuracy

The most common reasons for vehicle count discrepancies are the following:

- Zone boundaries overlap or are too close: Occurs when vehicles in one zone are shown as being detected in an adjacent zone. This is referred to as splashing. Splashing is eliminated by changing the zone dimensions by increasing or decreasing the boundary by one or more micro-slices.
- Improper sensor aiming: Occurs when vehicle counts are below some expected value. Vehicles may not be detected if the sensor is aimed too low or high, or is not perpendicular to the zone.
- Sensitivity too high or low: If the sensor is correctly aimed, incorrect counts could be caused by an improper sensitivity setting.
- Obstruction between the sensor and zone: An obstruction, such as a concrete lane divider, may cause smaller vehicles to be missed.
- Occlusion: Occurs when a vehicle is hidden from view by another vehicle or object such as by a large truck masking a small car located behind it.

7.5.2.3 Communications options

Communications options are accessed through the Manual Settings screen and include serial, dial-up, Bluetooth, and TCP/IP, as illustrated in Figure 7.15.

7.6 SETBACK AND MOUNTING HEIGHT IN GRAPHICAL OR STATEMENT FORM

Sensor manufacturers supply setback and mounting height information in several ways. These include graphs, figures, and statements of what is permitted and what is not as illustrated in the following examples.

7.6.1 Wavetronix radar sensors

The Wavetronix Matrix sensor specifies a minimum setback distance of 6 ft (1.8 m) and a mounting height that depends on the distance to the closest monitored lane as was described in Tables 7.4 and 7.5.

The Wavetronix SmartSensor Advance™ presence-detecting microwave sensor offers mounting height and setback instructions as follows [13]:

- Mounting height: Mounting the sensor as high as possible is recommended to reduce same lane occlusion. A maximum of 40 ft (12 m) and minimum of 17 ft (5 m) are

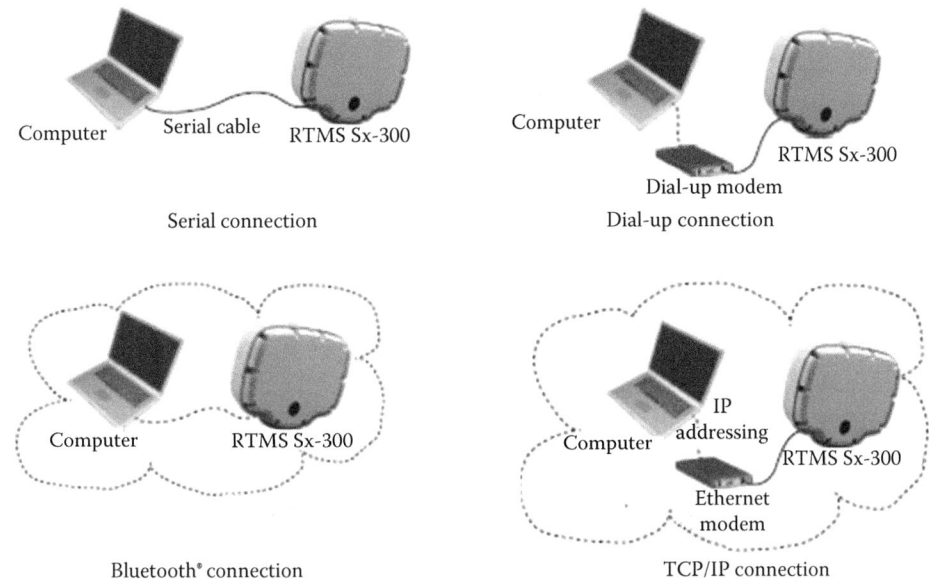

Figure 7.15 RTMS Sx-300 communications options. (Drawing courtesy of ISS, St. Paul, MN.)

recommended. If the sensor is higher than 30 ft (9 m), the offset should be less than 50 ft (15 m) to increase accuracy.
• Setback: Mounting the sensor closer to the lanes of interest will usually increase detection accuracies. A maximum offset within 50 ft (15 m) of the center of the monitored lanes is recommended, but the sensor will still reliably track vehicles at larger offsets. Mounting with a smaller offset will generally increase the line of sight.

7.6.2 RTMS Sx-300 microwave radar sensor

Setback is a limiting installation parameter of the RTMS Sx-300 as it is with most overhead sensors. The sensor's setback requirements vary with the number of lanes monitored. All lanes of traffic must be within 76 m (250 ft) of the sensor for vehicles to be detected. Setback is measured as the distance between the nearest edge of the first lane of monitored traffic to the front of the structure on which the sensor is mounted. More lanes can be monitored with a larger setback.

7.6.2.1 Standard setback

The RTMS Sx-300 sensor has a minimum mounting height of 5 m (17 ft) to minimize occlusion of vehicles even by the tallest trucks. The sensor must be set back from the first monitored lane to ensure it includes all required lanes within its field-of-view. The amount of setback varies with the width of the monitored road as illustrated by the graph in Figure 7.16. The correct installation height can be determined once the setback is known using Figure 7.17. Height is measured relative to the road surface at the detection area and not from the bottom of the mounting pole [12]. Other considerations for sensor mounting are as follows:

• It is almost always better to be further back from the first monitored lane than the required minimum distance.

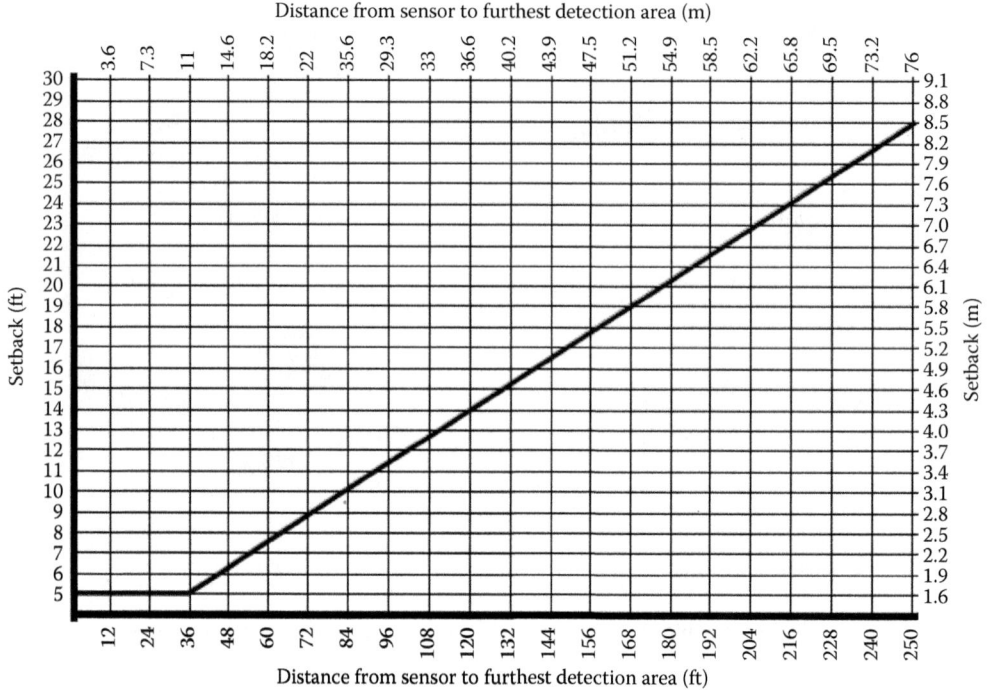

Figure 7.16 RTMS Sx-300 setback as a function of distance from sensor to furthest point in detection area. (Drawing courtesy of ISS, St. Paul, MN.)

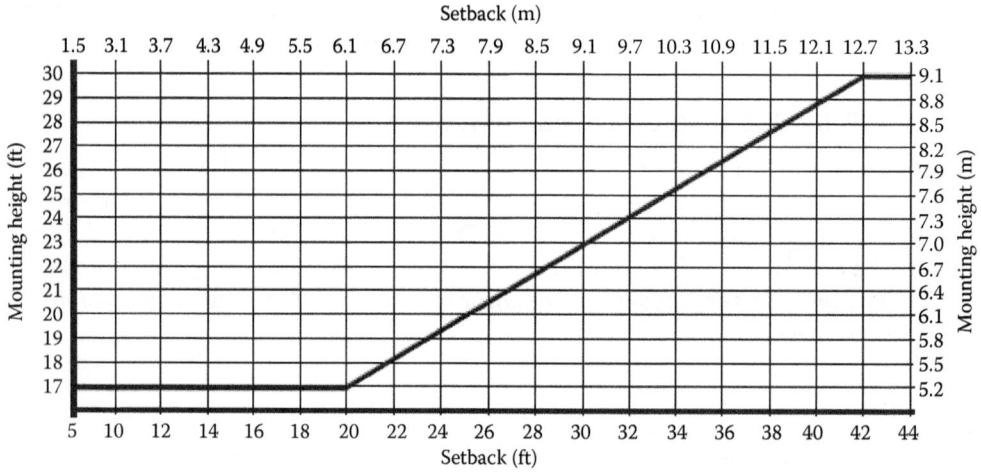

Figure 7.17 RTMS Sx-300 mounting height as a function of setback. (Drawing courtesy of ISS, St. Paul, MN.)

- Using the recommended mounting-height value allows the sensor to be aimed so that it receives maximum return signal while monitoring all required lanes. Mounting the sensor at an incorrect height reduces accuracy.
- Widths of roadway medians must be included in the total detection area. For example, the installer may be able to set up 12 zones, but all zones must be within 76 m (250 ft) from the sensor.

7.6.2.2 Zero setback

The RTMS Sx-300 has the ability to detect vehicles in lanes with zero setback, that is, pole location immediately beside the first lane of detection. This feature can accommodate many midblock detection sites and bridges that have limited setback. Zero-setback operation is limited to a maximum of four lanes and requires a mounting height of approximately 4 m (14 ft). The zero-setback feature should be used only if the situation requires this option as adequate setback improves the sensor's vehicle detection ability.

7.6.2.3 Guidelines for mounting, configuring, and transmitting data

Table 7.6 lists procedures that contribute to the successful setup and use of the RTMS Sx-300 sensor.

7.6.3 SmartTek SAS-1 sensor

The SmartTek acoustic SAS-1 sensor provides mounting instructions as shown in Figure 7.18 [14]. They also supply software installation and setup instructions similar to those offered by the VDS and radar sensor manufacturers.

7.6.4 Xtralis ASIM passive infrared sensors

Xtralis ASIM passive infrared sensors provide the field-of-view versus mounting height information in Figure 7.19 for the 300 series of devices. These sensors are typically mounted on signal head, gantry, or bridge structures and aimed to face approaching traffic or

Table 7.6 Guidelines for mounting an RTMS Sx-300 sensor

Do	Don't
Ensure setback is sufficient and height is not greater than manufacturer's recommendations.	Attempt to mount a side-fired RTMS Sx-300 closer than 1.5 m (≈5 ft) to the first monitored lane without reviewing the restrictions for zero setback.
Use extension arms where needed to improve sensor location on existing structures.	Install sensors where overhead structures can interfere with the microwave beam, e.g., under overpass bridges and heavy structures.
Aim the sensor perpendicular to the direction of traffic flow in the monitored lanes.	Aim the sensor at an angle exceeding five degrees from the perpendicular to the monitored lane.
Aim the sensor according to the 1/3 rule, and then verify aiming interactively with the sensor's Setup Utility by checking detection in all lanes.	Separate installation and aiming from the setup stage. Increase sensitivity to offset poor aiming.
Pay attention to site cabling design. Ensure serial port access is available for setup. If necessary, add a pole-mounted junction box.	Run the sensor cable directly to cabinets out of visual range of the sensor's detection area.
When powering with a low voltage input AC transformer, design for 16 VAC.	Specify use of controllers in new applications requiring data only.
Evaluate power arrangements vs. distance. Use either heavy gauge power wires to reduce voltage drop, or a higher supply voltage.	Use thin wires to carry power with a low voltage supply. Attempt connecting wires thicker than #18 with the sensor's MS connector.
Use wireless communication for • Long distances to offset trenching cost. • Quick deployment and portability.	Specify dial-up communication in applications requiring real-time data. It is applicable to infrequent downloads of traffic counting data.

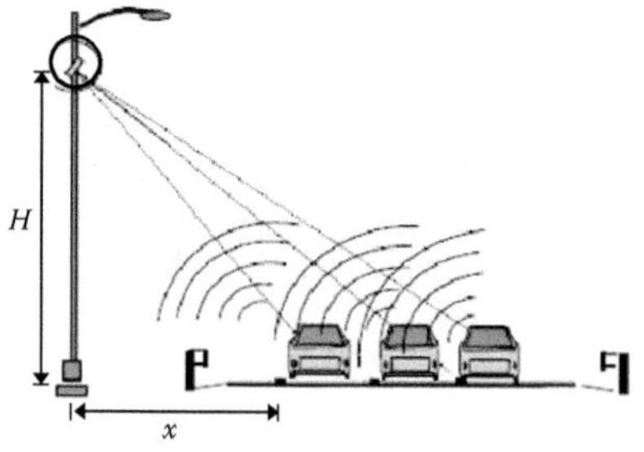

Recommended installation height *H* for SAS-1

Distance *x* from nearest lane	Number of lanes monitored				
	5	4	3	2	1
6 ft	34 ft	30 ft	26 ft	24 ft	20 ft
12 ft	36 ft	32 ft	28 ft	26 ft	22 ft
18 ft	38 ft	34 ft	30 ft	28 ft	24 ft
24 ft	40 ft	36 ft	32 ft	28 ft	24 ft
30 ft	44 ft	38 ft	34 ft	30 ft	24 ft

Figure 7.18 Sensor mounting information for SAS-1 acoustic sensor. (Drawing courtesy of SmarTek, Woodbridge, VA.)

Model IR 301/303 narrow beam

	Width *W* (m)			
Distance *D* (m)	*H* = 3.5 m	*H* = 4.5 m	*H* = 5.5 m	*H* = 6.5 m
10	0.75	0.80	0.80	0.85
20	1.35	1.40	1.40	1.45
30	2.10	2.10	2.10	2.15
40	2.70	2.75	2.80	2.80
50	3.40	3.40	3.40	3.40

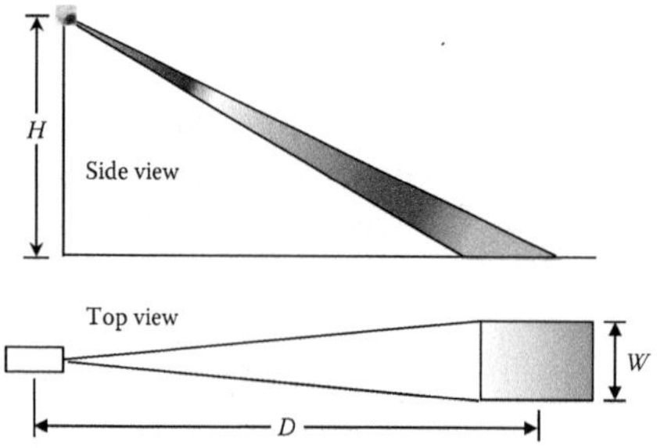

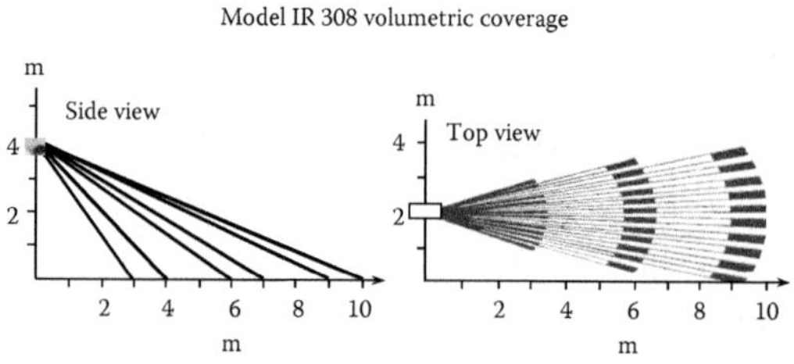

Model IR 308 volumetric coverage

Figure 7.19 Fields-of-view as a function of mounting height for ASIM 300 Series passive IR sensors. (Drawings courtesy of ASIM Xtralis, Kiel, DE.)

pedestrians. They detect objects by their movement and positive or negative temperature contrast against the background [15]. The recommended applications and maximum ranges of these sensors are listed below:

- IR 301/303 narrow beam model (recommended for green phase request and extension). Side mounting gives up to 50 m range.
- IR 308 volumetric coverage model (recommended for pedestrian detection for extension of green phase). Side mounting gives up to 10 m range.

7.7 SUMMARY

The overhead sensor installation, software initialization, and detection zone configuration procedures that have been described are meant only to illustrate the variety of methods employed by manufacturers to instruct personnel in the use of their products. Before installing any sensor, the manufacturer or authorized representative should be contacted to ensure that the latest installation and user manuals and software have been obtained. The procedures given in this chapter are not prescriptive and should not be relied on to contain all of the information required to successfully install and operate a sensor.

REFERENCES

1. *Installation and Setup of a VideoTrak-IQ (Typical)*, email received from Peek Traffic Corporation, January 4, 2016.
2. *Econolite application note—Aiming Video Sensors for Intersection Applications*, Reference: AN2104, January 30, 2009.
3. L.A. Klein and M.R. Kelley, *Detection Technology for IVHS, Final Report*, FHWA-RD-95-100, U.S. Federal Highway Administration, McLean, VA, December 1996. http://ntl.bts.gov/lib/jpodocs/repts_te/6184.pdf. Accessed December 10, 2015.
4. L.A. Klein, D. Gibson, and M.K. Mills, *Traffic Detector Handbook: Third Edition*, FHWA-HRT-06-108 (Vol. I), Federal Highway Administration, U.S. Department of Transportation, Washington, DC, October 2006. http://www.fhwa.dot.gov/publications/research/operations/its/06108/06108.pdf.
5. *Operating Manual, VideoTrak-IQ™ Multi-Channel Video Traffic Detection System*, Part Number 99-541 Rev. 2, Peek Traffic Corporation, Palmetto, FL, July 1, 2014.
6. *Vantage Edge 2 User Guide*, Part Number 493241401 Rev. H, Iteris, Santa Ana, CA, 2015.
7. *FLIR TrafiCam Series Vehicle Presence Sensor*, FLIR Systems, Inc., Wilsonville, OR, 2014.
8. *Econolite Application Note—Tips for Drawing Detectors Autoscope® Software Suite*, Document Number: AN2161, June 16, 2015.
9. *SmartSensor Matrix User Guide*, Wavetronix, Provo, UT, 2015. http://www.wavetronix.com/en/support/downloads/316-smartsensor-matrix-user-guide. Accessed December 18, 2015.
10. *SmartSensor Matrix Quick-Reference Guide*, Wavetronix, Provo, UT, 2014. http://www.wavetronix.com/en/support/downloads/545-smartsensor-matrix-quick-reference-guide. Accessed December 18, 2015.
11. *SmartSensor Matrix Installer Quick-Reference Guide*, Wavetronix, Provo, UT, 2012.
12. *RTMS Sx-300 User Guide*, PN A900-1155-1 Rev. B, Image Sensing Systems Inc., St. Paul, MN, 2015.
13. *SmartSensor Advance User Guide*, Wavetronix, Provo, UT, 2014. http://www.wavetronix.com/attachments/bbd1c11624a0554fe7dba6f4973606ec7311d2ec/store/fc244097ca15c318aa09a-9128b06ee99f367ac47db4898d00ae839e24b48/smartsensor-matrix-user-guide-user-guide-en.pdf. Accessed December 21, 2015.
14. *SmarTek Acoustic Sensor—Version 1 (SAS-1) Installation and Setup Guide Part-E, Installation at the Traffic Monitoring Site*, Woodbridge, VA, Mar. 28, 2003.
15. *ASIM by Xtralis Traffic Detectors—IR 30X Passive Infrared Vehicle Detector*, Document Number: 18261_04.

Chapter 8

Sensor field tests

Testing of candidate sensors under actual operating conditions is essential before large scale purchases occur in order to validate their performance under operational traffic flow environments that often vary with season, vehicle mix, unique road configurations, lighting, and weather. The documentation of the test conditions and results for future reference will provide a valuable resource if additional sensor purchases are required at some later time.

This chapter describes several evaluations of traffic flow sensors, some containing multiple sensor models and technologies and some one type of sensor technology or only one sensor model. The purpose is to introduce the reader to the process of conducting sensor testing, preparing test site documentation, and to the types of data usually sought and obtained. The investigations described are FHWA's Detection Technology for Intelligent Vehicle-Highway Systems (IVHS), Evaluation of Non-Intrusive Technologies for Traffic Detection—Phase 3, Evaluation of Video Detection Systems, and Queue Estimation Using Magnetometers. Although the tests were conducted on models or software versions that are no longer available or have been superseded by newer versions, they are still valuable in that they alert the user to potential issues that may arise when these sensor technologies are installed and used to monitor traffic flow.

8.1 DETECTION TECHNOLOGY FOR IVHS OVERVIEW

The purpose of this pioneering project, which occurred from 1992 to 1995, was to comprehensively measure the laboratory and field performance of commercial vehicle sensors that apply technologies compatible with above-the-road, surface, and subsurface mounting to measure traffic parameters on freeways and arterial streets with acceptable accuracy, precision, and repeatability. The sensors were installed in three states having diverse traffic, climate, and weather ranging from cold winter and snow in Minneapolis, Minnesota; humidity, rain, lightning, and heat in Orlando, Florida; warm, dry weather in Phoenix and Tucson, Arizona; and hot summer temperatures with thunderstorms in Phoenix. As part of this FHWA IVHS Program, the forerunner of Intelligent Transportation Systems (ITS), traffic parameter specifications, for example, vehicle count, presence, speed, and lane occupancy, were developed for interconnected intersection signal control, isolated intersection signal control, freeway incident detection, traffic data collection, real-time adaptive signal control, and vehicle-to-roadway communications. This project also assessed the best performing sensor technologies that were available at the time by application and examined the need for a national sensor test facility. A series of 10 reports were issued during the course of the project that described traffic parameter specifications, the field sites selected for the testing, laboratory test specifications and test plans, the sensor selection process, results of the laboratory tests, field-test specifications and test plans, evaluation results from the field tests, and the final reports [1–10].

Table 8.1 Sensor technologies evaluated during the FHWA Detection
Technology for IVHS Program

Technology	Number of models[a]
Ultrasonic	3
Microwave presence-detecting radar	1
Microwave Doppler sensor	4
Laser radar (lidar)	1
Passive infrared	2
Imaging infrared	1
Passive acoustic array	1
Video detection systems	5
Magnetometer	2
Magnetic (passive)	1
Inductive loop[b]	6

[a] Not all models of all sensors were available at each test location as new sensors became available as the program continued.
[b] The number of models for the inductive loop corresponds to the number of electronics units or detectors that were on hand.

Manufacturers and vendors of the sensor technologies listed in Table 8.1 were contacted to obtain representative samples of their products. The sensors were installed, calibrated, and then operated for a period of time to ensure they were functioning properly. If a sensor was not operating correctly, it was returned to the manufacturer for repair or replacement before the actual performance tests were begun. This procedure assured that all sensors were capable of performing within their specifications.

Each state in which sensor testing and evaluation occurred had two test sites: a freeway site and a surface street site. Table 8.2 summarizes the test locations, weather conditions, and direction of traffic flow with respect to sensor orientation at each test site. These field tests began after the laboratory tests were completed and competence with the sensors mounting requirements and performance characteristics was achieved. The first field tests were conducted in downtown Los Angeles, which was near the prime contractor's facilities. This enabled the

Table 8.2 Detection technology evaluation sites

Location	Evaluation period	Weather	Traffic direction
Minneapolis freeway: I-394 at Penn Avenue	Winter 1993	Cold, snow, sleet, fog	Departing in a.m.
			Departing and approaching in p.m.
Minneapolis surface street: Olson Hwy at East Lyndale Avenue	Winter 1993	Cold, snow, sleet, fog	Departing
Orlando freeway: I-4 at SR 436	Summer 1993	Hot, humid, heavy rain, lightning	Approaching
Orlando surface street: SR 436 at I-4	Summer 1993	Hot, humid, heavy rain, lightning	Departing
Phoenix freeway: I-10 at 13th Street	Autumn 1993	Warm, rain	Approaching
Tucson surface street: Oracle Road at Auto Mall Drive	Winter 1994	Warm	Departing
Phoenix freeway: I-10 at 13th Street	Summer 1994	Hot, low humidity, thunder storms, lightning	Approaching

testing team to familiarize themselves with street installation procedures and the transmission of data and video to a central location via several types of communications media.

The following descriptions highlight the techniques utilized to install the sensors, calibrate the test sites, and emphasize the more critical results. Although the sensor models tested likely have been superseded by devices that contain improved hardware and data processing and hence offer improved performance, the discussions offer insights into issues that may arise when similar technologies are used to gather traffic flow data.

8.2 MINNEAPOLIS TESTS

The Minneapolis, Minnesota tests took place from January through March 1993. For the freeway portion of the tests, sensors were installed above a section of I-394 along the outside of the Penn Avenue overpass as illustrated in Figure 8.1. Most sensors viewed eastbound traffic into Minneapolis, but several detected traffic in the middle reversible HOV lanes (eastbound during the a.m. peak, westbound during the p.m. peak). Figure 8.2 depicts the sensors as they were mounted on the overpass. At most sites, the overhead sensors were mounted on 1.5 in. (3.8 cm) galvanized iron pipe that was attached to an overhead structure such as an overpass, sign support, or traffic signal mast arm. Data and power cables ran from the sensors to a data acquisition trailer parked in a grassy area beyond the shoulder of the eastbound freeway lanes.

The left portion of Figure 8.3 shows the areas or detection zones where vehicles were detected by each sensor in each lane. The inside-most lane was designated Lane 1. The table on the right lists the specific detection zones for each sensor along with the symbol used to identify the sensor. This information was recreated for each test site so that the acquired data could be correlated as best as possible with traffic flow patterns. Detection zone 0 is located directly under the sensor mounting brackets. The detection zones were confirmed by moving emitters (for passive mode sensors) or reflectors (for active mode sensors) of energy along the lanes until a detection event was recorded in the trailer.

Figure 8.1 Sensor view of eastbound I-394 and the two middle reversible lanes.

Figure 8.2 Sensors in place overlooking I-394.

Video detection systems (VDSs) are denoted by a VP symbol in this study and are indicated by VIP as part of the model number designation that appears in the figures. The symbol U represents an ultrasonic sensor, M a microwave sensor, IR a lidar or passive infrared (PIR) sensor, and IL an inductive loop. Multiple sensors of the same model were differentiated by an auxiliary letter designation. For example, M-5 represented a particular model of a microwave sensor. If two of these models were deployed over different lanes, one would be labeled M-5A and the other M-5B. One exception was the inductive loops where one dual detector card was used for the two loops in each monitored lane. Hence, the detector card for Lane 1 was designated with an A, Lane 2 with a B, and Lane 3 with a C. Another was the microwave presence-detecting radar when it was used to monitor multiple lanes. In this case, the detection zone for each lane was designated with a different letter.

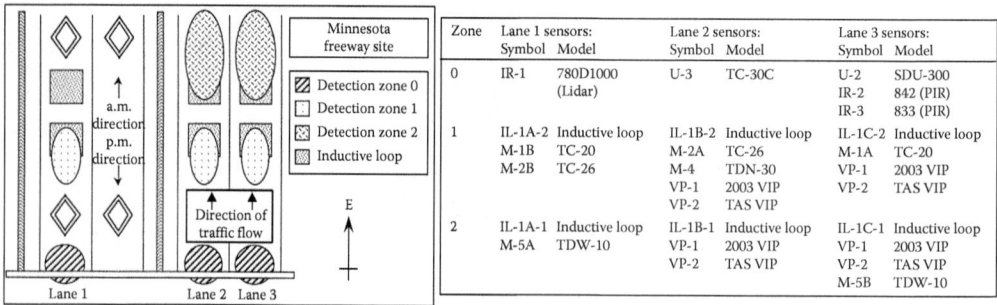

Figure 8.3 Sensor detection zones along I-394.

Figure 8.4 Olson Highway surface street evaluation site and with sensors on back of sign bridge.

Figure 8.4 shows the Olson Highway surface street test site and the mounting of the sensors and galvanized pipe structure to the back side of the sign bridge facing departing traffic. A table similar to the one in Figure 8.3 was created to show the sensor detection zones on each lane of Olson Highway.

Vehicle counts are a popular metric for evaluating sensor performance. Figure 8.5 plots the sensor counts on I-394 for the a.m. traffic peak during cold weather and light snow flurries. Because of time and money constraints, this project could not analyze data by each detection event as recommended in Chapter 9, which discusses sensor requirements specification,

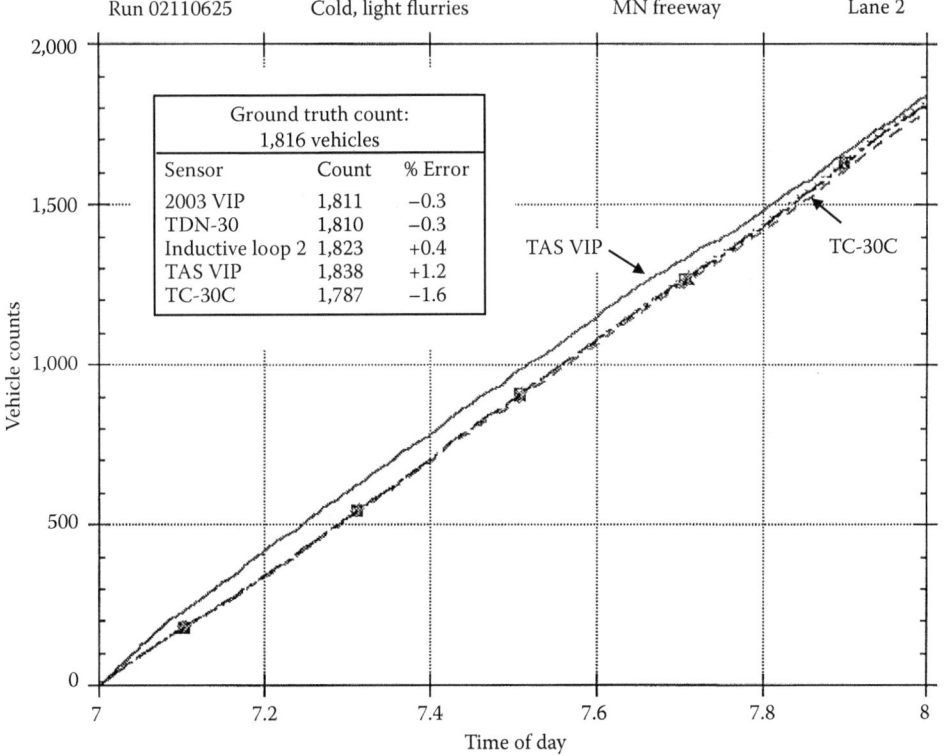

Sensor	Count	% Error
2003 VIP	1,811	−0.3
TDN-30	1,810	−0.3
Inductive loop 2	1,823	+0.4
TAS VIP	1,838	+1.2
TC-30C	1,787	−1.6

Figure 8.5 Vehicle count comparisons on I-394.

testing procedures, and data evaluation. Instead, aggregated vehicle counts, here over a 1-h time period, were used to obtain a course measure of sensor performance. Aggregated measurements, such as vehicle count over a time interval, can obscure the actual vehicle detection accuracy of a sensor since failures to detect are often canceled by false detections. The preferred and recommended metrics for evaluating sensors for vehicle count accuracy are by

- *Correct detection*—indication by a sensor that a vehicle passing over the detection area of the sensor is detected by the sensor.
- *False detection*—indication by a sensor that a vehicle *not* passing over the detection area of the sensor is detected by the sensor.
- *Missed detection*—indication by a sensor that a vehicle passing over the detection area of the sensor is *not* detected by the sensor.

The sensor models listed in Figure 8.5 are no longer manufactured and have been replaced by improved devices with better performing hardware and algorithms. Therefore, the information in the figure should be used only to judge the relative performance of the sensors available at the time the tests were conducted. The data may also be useful for determining which types of sensor technologies are better suited to a particular application and where their weaknesses lie (such as in inclement weather and poor lighting). Ground truth vehicle counts were obtained by counting the number of vehicles in video recordings of the traffic for the time period in question.

Figure 8.6 shows the decrease in vehicle speed with increase in traffic flow rate during the a.m. traffic peak on I-394 as measured with a Doppler microwave sensor. The Doppler sensor was suitable for this application because the vehicles were moving.

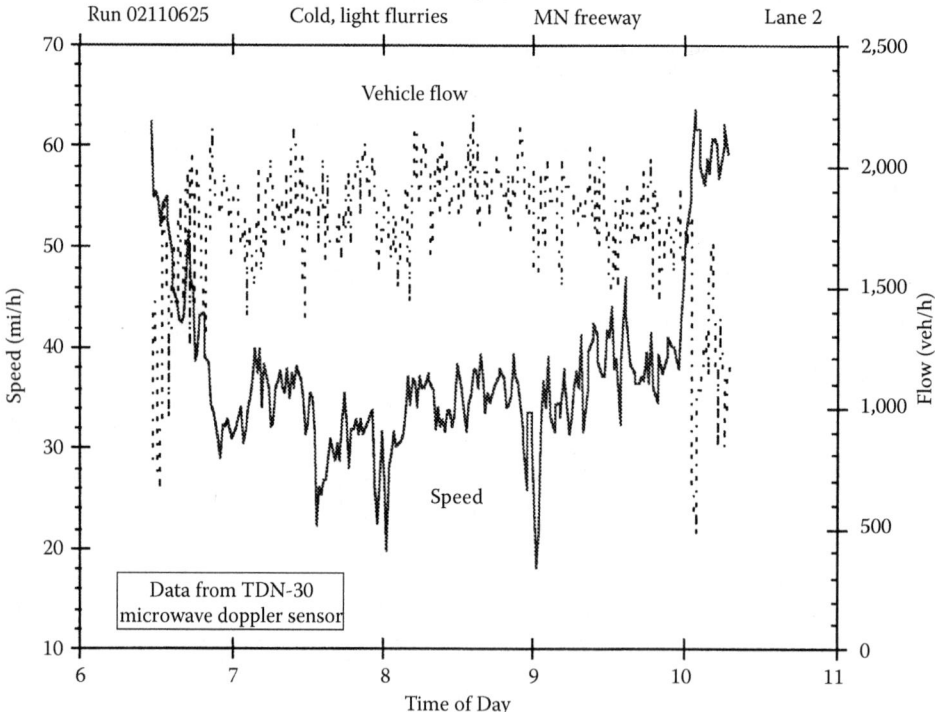

Figure 8.6 Vehicle speed and flow rate on I-394.

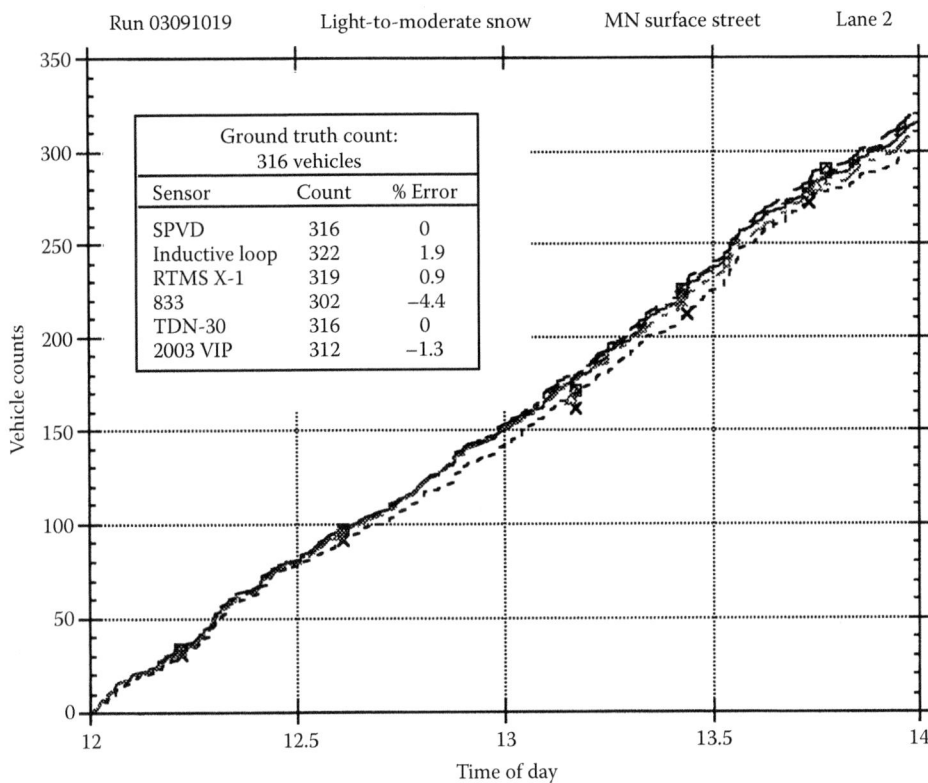

Figure 8.7 Vehicles count comparisons on Olson Highway.

The vehicle counts contained in Figure 8.7 are for a two-hour mid-day time period on the Olson Highway surface street test site where volumes were relatively low. All sensors, except for a passive infrared model, were within 2% of the ground truth value.

8.3 ORLANDO TESTS

Sensor performance evaluation in Florida occurred in Altamonte Springs near Orlando during the summer of 1993. The two test locations were at the intersection of the I-4 freeway and the SR-436 surface street arterial. The surface street test site was conveniently located above the freeway as illustrated in Figure 8.8. For the freeway evaluation, the sensors were attached to the SR-436 overpass facing approaching traffic. A Doppler microwave sensor was also mounted on a pole on the right shoulder of the I-4 freeway as indicated in Figure 8.9. For the SR-436 evaluation, they were attached to the back of the sign bridge that extended across the arterial and faced departing traffic.

To assist in calibrating the field-of-view for later video analysis, lines were painted along the shoulder of the freeway at 25-ft (7.6-m) intervals and traffic cones placed on the lines as a viewing aid as represented in Figure 8.10. The lane-by-lane vehicle count and speed detection zones for the 2003 VIP VDS were recorded as drawn in Figure 8.11 to assist with later video analysis of the data. Mounting of the overhead sensors at the SR-436 site is illustrated in Figure 8.12.

Figure 8.8 Sensors mounted on SR-436 overpass above I-4 at Altamonte Springs evaluation site.

Figure 8.13 depicts sensor occupancy data. During rush-hour periods (6:00 to 8:00 a.m.), the occupancy is higher than at off-peak hours (approximately 8:30 to 11:00 a.m.). Depending on the application of the data, either the raw and fluctuating occupancy data or a smoother polynomial curve fit to the data may be utilized. An interesting result from the analysis of these data is the observation that the polynomial curve fit for data acquired at 30-s integration intervals is identical to that for data acquired at 5-min integration intervals. Thus, if a smoothed curve fit is adequate to determine occupancy for a particular application, then the data can be aggregated over larger intervals to reduce the quantity of data acquired.

Figure 8.9 Side-looking Doppler microwave sensor at I-4 evaluation site.

Traffic cones at 25-ft
(7.6-m) intervals

Figure 8.10 I-4 distance measures for VDS calibration.

The purpose of Figure 8.14 is to point out that lane occupancy values are dependent on the sensor used to acquire the data. The dependency is due to the different size detection zones of each sensor and different length of times sensors hold their output. Thus, a vehicle detected by a sensor with a larger detection zone will be in that sensor's detection area longer than that of the same vehicle detected by a sensor with a smaller detection zone. Therefore, the occupancy values will be different even though the vehicle was traveling at the same speed through the detection areas of both sensors.

The implication for sensor specification is that the same sensor should be installed along a section of roadway if occupancies are to be compared along that section in accordance with some traffic management application and algorithm. An example is the implementation of an automatic incident detection algorithm that evaluates occupancies upstream and downstream of a potential incident, such as in the California-type algorithms. If different sensor types were to be used to acquire data for such an incident detection algorithm, a false declaration of an incident could be made if the upstream sensor occupancy measurement were to increase because a sensor with a larger detection zone was being used.

Figure 8.15 indicates the percent difference in vehicle counts registered by an inductive loop detector and a presence-detecting microwave radar over several traffic signal cycles along SR-436. The average value of the count difference was -0.75 counts per interval and its standard deviation was 1.73 counts. The standard deviation of the percent difference values was 5.35%. These results show that the loop did not consistently overcount or under-count with respect to the microwave radar.

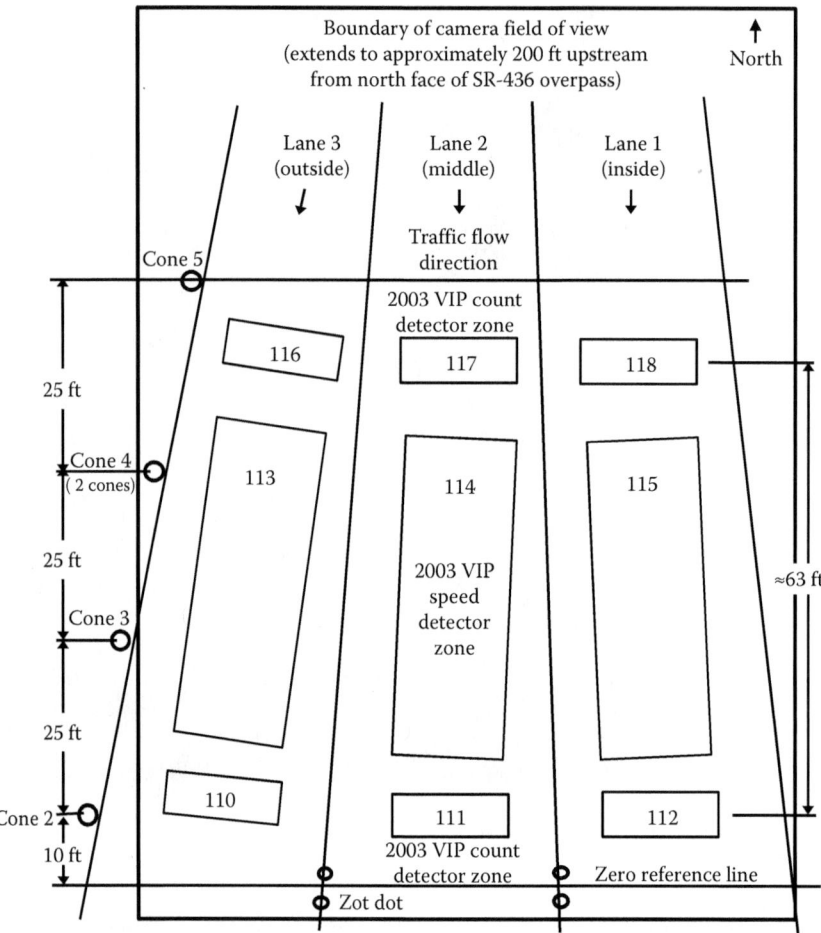

Figure 8.11 2003 VIP detection zones along I-4. The smaller rectangles superimposed on the lanes represent vehicle count detector zones, while the larger rectangles represent vehicle speed detector zones.

Figure 8.12 SR-436 surface street evaluation site.

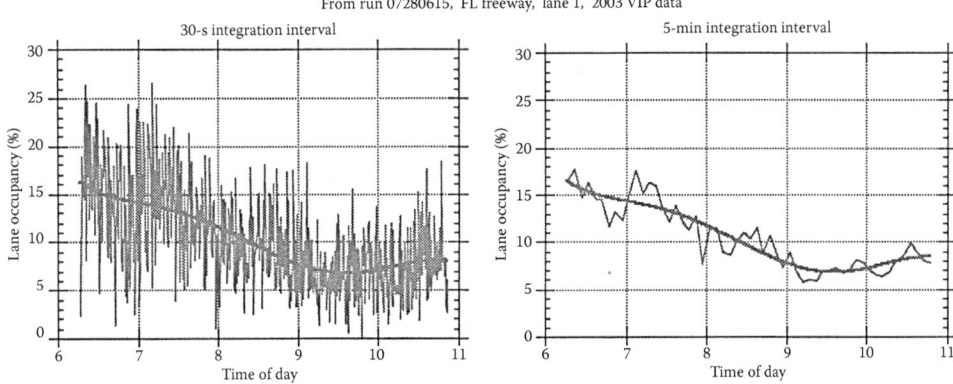

Figure 8.13 Lane occupancy raw data with the corresponding polynomial curve fit. Analysis showed that integrating the raw data for longer intervals did not change the shape of the curve fitted to the data.

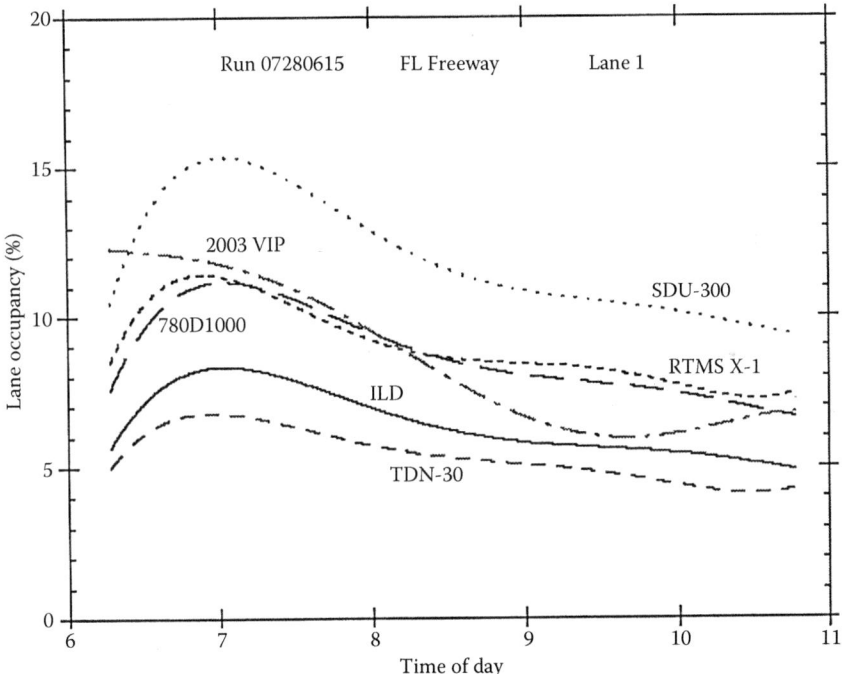

Figure 8.14 Lane occupancy value is dependent on the size of the sensor's detection area and data hold time.

8.4 PHOENIX TESTS

The only test site in Phoenix was a freeway site on I-10. The corresponding surface street site for this climate condition was in Tucson, Arizona. Phoenix testing occurred during November through December 1993 and then again during July through August 1994. The sensors were installed on a sign bridge facing approaching freeway traffic as shown in the photographs in Figure 8.16. The sensor types and mounting locations are indicated in the

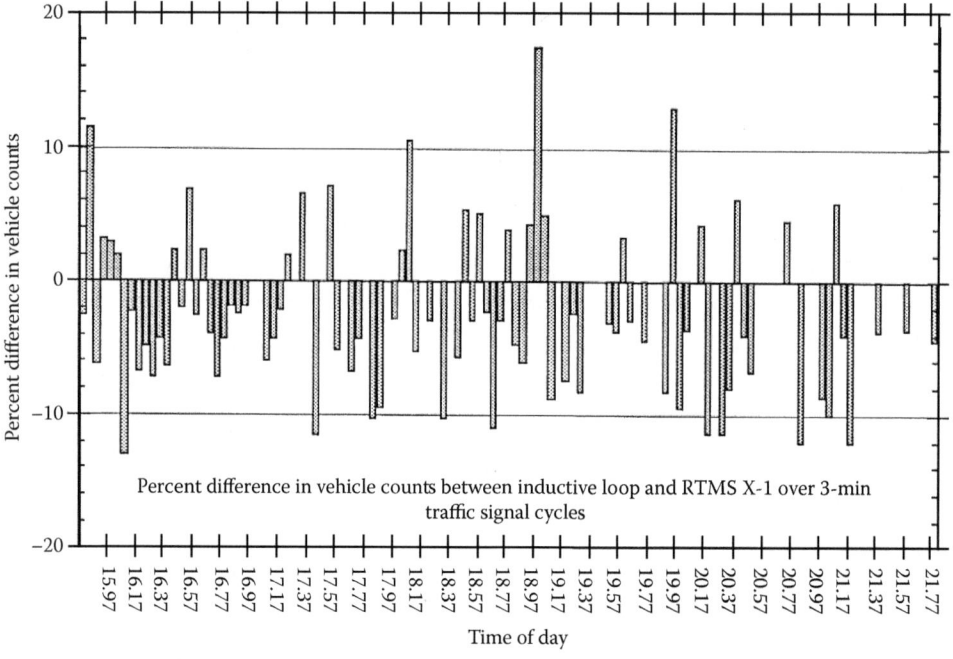

Figure 8.15 Vehicle count comparisons over many signal cycles at SR-436 intersection.

U-1	SDU-200	M-5	TDW-10
U-2	SDU-300	M-6	RTMS X-1
U-3	TC-30C	IR-1	780D1000
M-1	TC-20	IR-2	842
M-2	TC-26	IR-3	833
M-4	TDN-30	A-1	TSS-1

Figure 8.16 I-10 freeway sensor evaluation site near 13th Street in downtown Phoenix. Overhead sensors were mounted on the sign bridge facing approaching traffic.

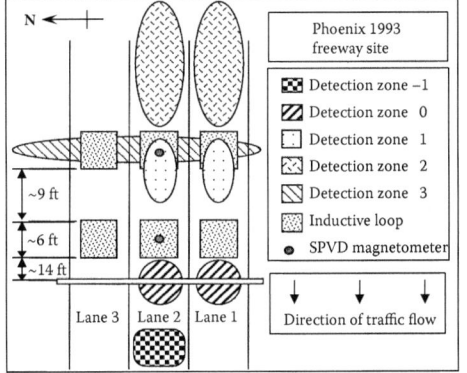

Zone	Lane 1 Sensors Symbol	Model	Lane 2 Sensors Symbol	Model	Lane 3 Sensors Symbol	Model
−1			A-1	TSS-1		
			IR-3	833		
0	U-3A	TC-30C	U-3B	TC-30C	IL.-1C-2	Inductive loop
	U-2A	SDU-300	U-2B	SDU-300		
	IL.-1A-2	Inductive loop	IL-1B-2	Inductive loop		
	IR-1	780D1000	MG-2A-2	SPVD		
	IR-2	842				
1	IL-1A-1	Inductive loop	IL-1B-1	Inductive loop	IL-1C-1	Inductive loop
	M-4B	TDN-30	M-4A	TDN-30		
	M-2B	TC-26	M-1A	TC-20		
	M-6A	RTMS X-1	MG-2A-1	SPVD		
	U-1	SDU-200				
2	VP-1	2003 VIP	VP-1	2003 VIP	VP-1	2003 VIP
	M-6B	RTMS X-1	M-5A	TDW-10		
3	M-6C	RTMS X-1	M-6D	RTMS X-1	M-6E	RTMS X-1

RTMS X-1 units in zones 1 and 2 were forward looking. Those in zone 3 were side looking.

Figure 8.17 Sensor detection zones along I-10.

bottom, right of the figure. Detection zones where vehicles were detected by each sensor in each lane were located and recorded as with the other test sites as indicated in Figure 8.17. Shoulder paint markings and traffic cones were used to help locate the vehicles in the video recordings that established the ground truth vehicle counts. The symbol A denotes a passive acoustic sensor, MG a magnetometer, while the other sensor symbols remain as defined before.

Figure 8.18 illustrates the vehicle flow in Lane 2 on I-10 using data obtained from a Doppler microwave sensor, passive infrared sensor, and VDS. Vehicle flows in the upper leftmost plots were computed from 1-min data integration intervals for the 1-h period between 4:00 and 5:00 p.m. for which ground truth vehicle counts were calculated from video imagery. The result of using a 1-min integration interval produces a discrete and "spiky" curve. Shorter integration intervals, such as 30 s, produce flow values that reflect the microscopic movement of individual vehicles and thus may generate a curve that shows higher instantaneous flow rates. Short integration intervals may be required when the maximum peak flow on a roadway is needed and for some software simulation models.

Plots in the upper right of the figure show vehicle flow versus time of day for the same three sensors over the entire 3.5-hour run. These vehicle flows were computed using a 5-min data integration interval. If peak traffic flow information is required, then a shorter integration interval that does not average the flow data as much should be applied.

The flow reported by the three sensors generally coincides except for a pronounced spike exhibited by the VDS from about 3:00 to 3:10 p.m. Upon examining the database file from which the plots were drawn, it was found that the counts recorded from the VDS detection zone 3 (downstream detection zone) on the data recorder suspended temporarily. The rising edge of a pulse was received by the data recorder (corresponding to a vehicle entering the VDS detection zone), but the falling edge of the pulse was not received until 51 s later.

The curves in the bottom center of the figure present a fifth-order polynomial fit to the flow data in the upper right plots. The discrete and spiky flow characteristics evident in the upper right curves are smoothed out by the polynomial fit. Fitting the data in this manner shows long-term traffic trends as opposed to the instantaneous flow.

Figure 8.19 contains the fifth-order polynomial fit to the actual lane occupancy data for the same Doppler microwave, passive infrared, and VDS sensors over approximately the same 3.5-hour period. Occupancy was computed as the sum of the vehicle presence times collected over a 5-min integration interval divided by the integration interval of 5 min. This

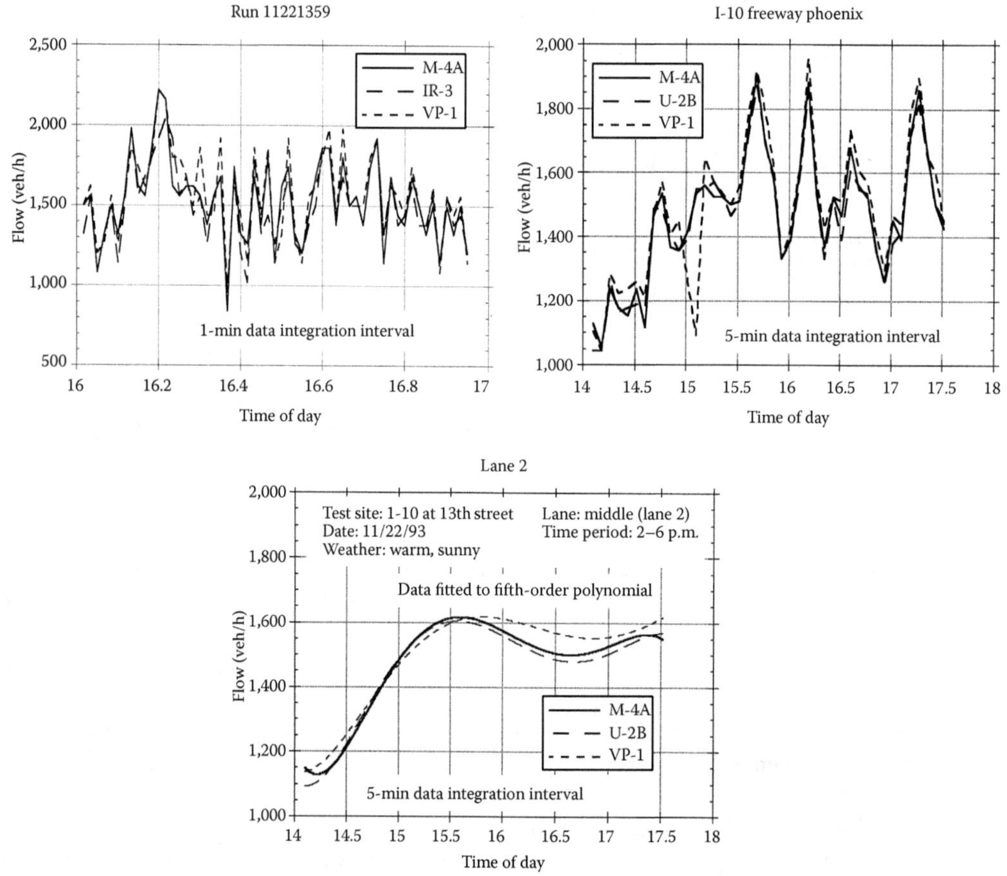

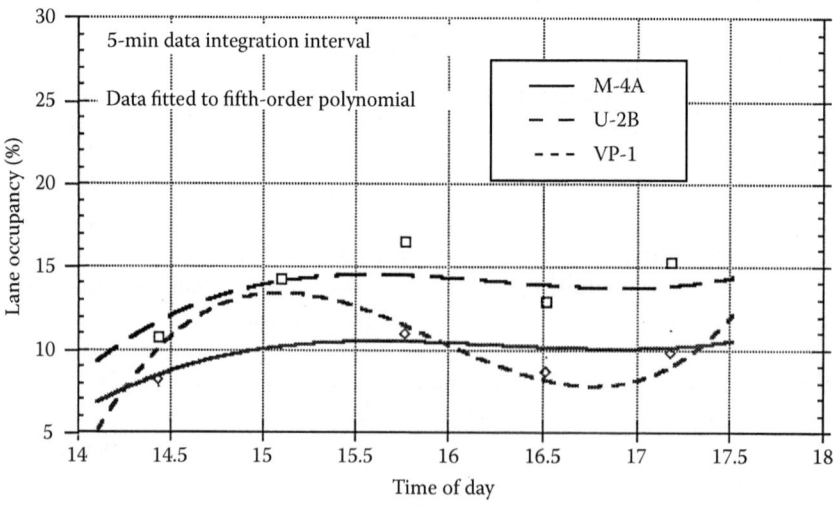

Figure 8.18 Vehicle flow rate comparisons for different data integration times.

Figure 8.19 Lane occupancy comparisons on I-10.

yielded the percentage of time that the sensor was actively detecting vehicle presence during the 5-min interval.

Once again, the differences in the occupancies measured by the sensors are due to different sensor detection areas and hold times. Therefore, the same sensor should be used along a section of roadway if occupancies are to be compared along that section in accordance with some traffic management application and algorithm.

8.5 TUCSON TESTS

Tucson testing occurred during March through April 1994 during warm, dry weather on Oracle Road near Auto Mall Drive. The photograph in Figure 8.20 shows the test area and the traffic signal mast arm on which the sensors were mounted to face departing traffic on a three-lane arterial. Before the sensors were installed, the city performed a load analysis to make sure the mast arm could safely carry the weight of the sensors. The data recording equipment was located in a trailer situated in an area to the left of the street (not shown in this photograph).

The top of Figure 8.21 displays the sensors attached to the mast arm, while the bottom identifies the sensor types and models. The symbol IIR represents an imaging passive infrared sensor, while the other sensor symbols remain as defined before. Compare the size of the imaging passive infrared sensor (IIR-1) in Figure 8.21 mounted between Lanes 2 and 3 with the modern embodiment shown in Figure 5.15. Detection zones where vehicles were detected by each sensor in each lane were located and recorded as with the other test sites. Four groups of detection zones in two lanes of traffic were used. Zone 0 was dedicated to sensors oriented at or near nadir. Zones 1–3 were located progressively downstream from the mast arm and appeared similar to those in Figure 8.17.

Figure 8.22 illustrates the test site layout showing the locations of the data acquisition trailer, trench where the data and power cables were laid, the overhead mast arm, and locations of the surface and subsurface sensors. Figure 8.23 shows the surface and subsurface sensors and the right-lane paint markings that helped calibrate distances in the video recordings used for ground truth vehicle counts.

The imaging infrared sensor used in Tucson was designed to detect vehicle presence, rather than provide vehicle counts or other vehicle data. Vehicle presence was indicated

Figure 8.20 Tucson sensor evaluation site at Oracle Road and Auto Mall Drive.

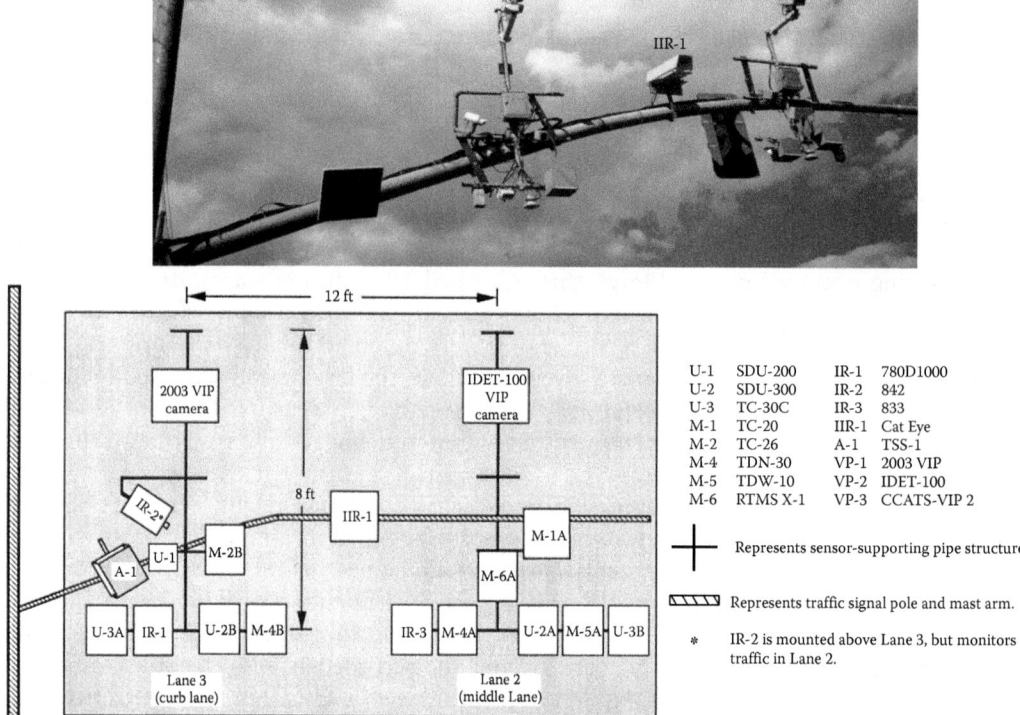

Figure 8.21 Oracle Road sensor array mounted on traffic signal mast arm over right and middle lanes.

at 1-s intervals if one or more vehicles were within its detection zone in the preceding 1-s interval. An example of the infrared imagery appears in Figure 8.24. The hottest areas on the vehicles and road surface appear white.

An interesting experiment performed in Tucson was motivated by the absence of large accumulations of snow during the Minneapolis tests. Therefore, it was decided to simulate dry snow conditions by attaching 1- and 2-in. (25- and 51-mm) thick sheets of Styrofoam to the top of a probe vehicle and drive it repeatedly through one of the instrumented lanes that had been closed to normal traffic. The combination of the 1- and 2-in. sheets provided a 3-in. (76-mm) thick option as well.

Of particular interest was the response of the ultrasonic, lidar, and microwave Doppler sensors to the Styrofoam. It was postulated that the irregular surface of the Styrofoam layer may scatter or absorb a portion of the transmitted energy or modify the emitted energy, causing the sensor to miss the vehicle detection opportunity. The results indicated otherwise. The ultrasonic sensor detected the Styrofoam-covered probe vehicle in all of the runs with all thicknesses of Styrofoam, as did the other two overhead sensors examined, namely, the lidar and the microwave Doppler sensor.

8.6 CONCLUSIONS FROM THE DETECTION TECHNOLOGY FOR IVHS PROGRAM

More than 20 sensors representing about eleven technologies were evaluated at nine locations having different climates and weather patterns. The sensors appeared to have satisfactory

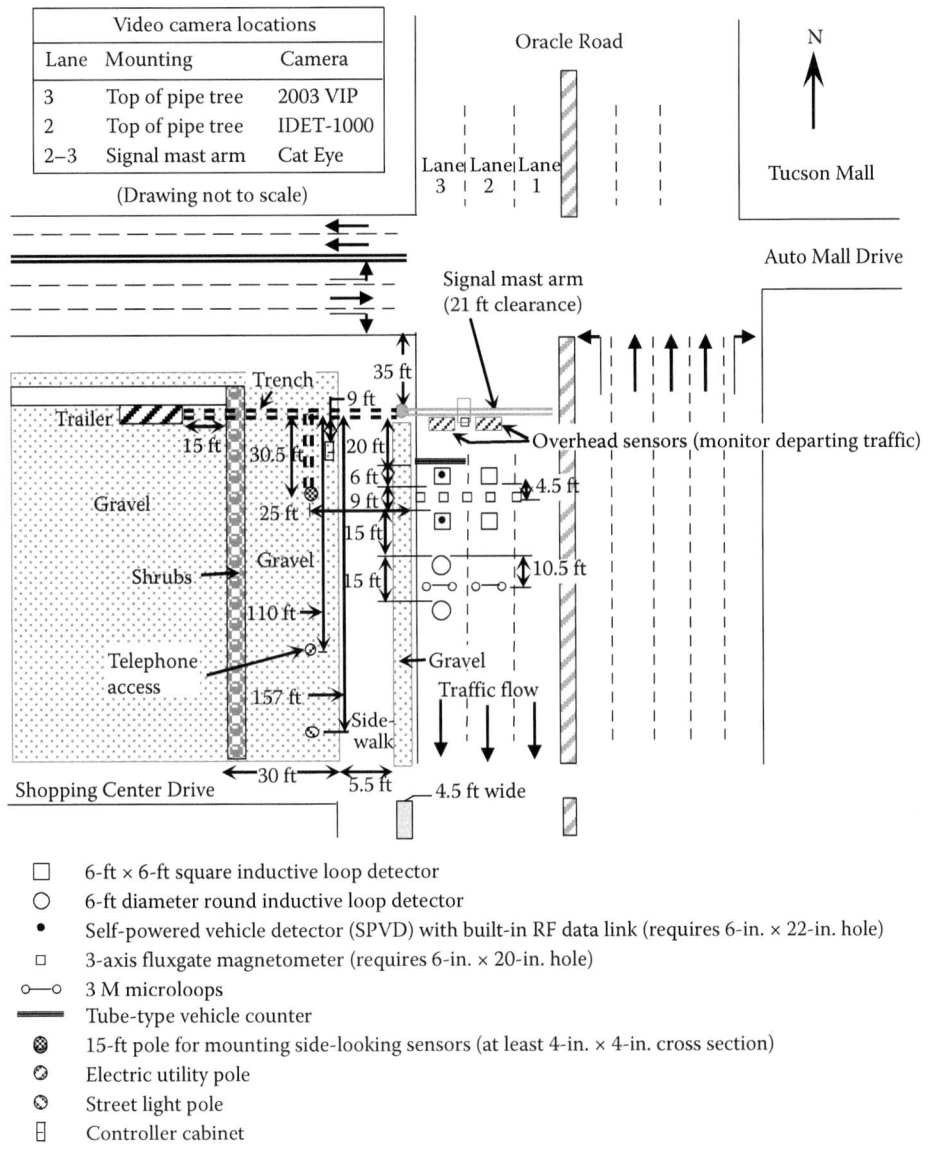

Video camera locations		
Lane	Mounting	Camera
3	Top of pipe tree	2003 VIP
2	Top of pipe tree	IDET-1000
2–3	Signal mast arm	Cat Eye

(Drawing not to scale)

□ 6-ft × 6-ft square inductive loop detector
○ 6-ft diameter round inductive loop detector
• Self-powered vehicle detector (SPVD) with built-in RF data link (requires 6-in. × 22-in. hole)
▫ 3-axis fluxgate magnetometer (requires 6-in. × 20-in. hole)
o—o 3 M microloops
▬ Tube-type vehicle counter
⊗ 15-ft pole for mounting side-looking sensors (at least 4-in. × 4-in. cross section)
⊘ Electric utility pole
⊙ Street light pole
⊟ Controller cabinet

Figure 8.22 Oracle Road test site layout.

performance to meet the requirements of traffic management applications of their time. Further development was required to meet more demanding future ITS applications. Among these are queue length measurement for interconnected intersection control, vehicle tracking to detect lane changing and weaving vehicles, and higher accuracy vehicle speed measurement for freeway congestion prediction.

Both quantitative and qualitative observations from the field-test evaluation sites are reflected in the Table 8.3 assessments of how well a particular technology performed relative to others in providing different traffic flow parameters. The conclusions are based on results from the limited number of runs analyzed and the general qualitative opinions gained from using these devices over an 18-month evaluation period. Many of the qualitative results were

Figure 8.23 Oracle Road showing surface and subsurface sensors and paint for calibrating the VDS field of view. The inductive loops and magnetometers at the center of the loops have also been outlined.

gained from the familiarity that came with utilizing these sensors day in and day out in a number of different weather and traffic flow environments. The dynamic nature of the field tests and the interest displayed by the sensor manufacturers to participate in them caused the number of devices under evaluation to grow steadily during the project.

The assessments of sensor performance were made with respect to only sensor operation and not cost. Cost considerations must be traded off by the procuring organization. The cost-effectiveness of a particular sensor or type of technology can only be judged when applied to a specific application and should include total life-cycle costs (i.e., purchase price,

Figure 8.24 Thermal images of vehicles on Oracle Road obtained with the imaging infrared camera system.

Table 8.3 Qualitative assessment of best performing technologies by application from the Detection Technology for IVHS Program

Technology	Count in low volume	Count in high volume	Speed in low volume	Speed in high volume	Best in inclement weather
Ultrasonic	•	•	•	•	•
Microwave Doppler[a]	√	√	√	√	√
Microwave true presence	√	√			√
Passive infrared	•	•	•	•	•
Active infrared (lidar)	•	•	•	•	•
Visible VDS	√	√			
Infrared VDS					
Acoustic array	•	•			
SPVD magnetometer	√	•	•	•	√
Inductive loop	√	√	•	•	√

Note: √ indicates the best performing technologies; • indicates performance not among the best, but may still be adequate for the application; no entry indicates not enough data reduced to make a judgment.

[a] Does not detect stopped vehicles. Therefore, although this technology is among the best performing, it is not suitable in applications where vehicle presence detection is required.

installation, data interface preparation, and maintenance over an extended time period of 10–20 years) and the equivalent number of lower cost sensors (e.g., inductive loops) that it replaces.

At the conclusion of the project, five CDs were produced that contain all the data gathered during the project and the reports that were issued.

8.7 EVALUATION OF NON-INTRUSIVE TECHNOLOGIES FOR TRAFFIC DETECTION—PHASE 3

The site for this sensor evaluation was the same as that used for the Minneapolis portion of the Detection Technology for IVHS freeway tests, namely, the I-394 and Penn Avenue intersection in Minneapolis. The sensor tested was a side-mounted, multilane Wavetronix SmartSensor HD microwave radar capable of detecting vehicle presence. Ground truth was provided by piezo-loop-piezo sensors and manual counts from video imagery [11]. In addition, the Infrared Traffic Logger (TIRTL) was capable of providing vehicle counts, classification, and lane-by-lane speed of passing vehicles. The test site configuration is shown in the top photograph in Figure 8.25.

Traffic volume results for the SmartSensor HD radar were consistent over the course of multiple months of data collection and generally were not affected by weather and traffic volume levels. An exception was when occlusion reduced the sensor's detection ability in the lanes further from the sensor under heavily congested conditions. This occurred because large trucks in the lane nearest the sensor blocked its view of vehicles in adjacent lanes. Trucks only comprise 5% of the traffic, but slow-moving trucks occlude the other lanes for a significant amount of time.

The plots in the lower portion of Figure 8.25 demonstrate that slow traffic in Lane 3 (the right-most lane and the lane nearest the sensor) affected the cumulative volume measurement accuracies in Lanes 2 and 1 (the far two lanes). Traffic was unusually slow due to light snow conditions. Traffic undercounting in Lanes 2 and 1 is attributed to occlusion from

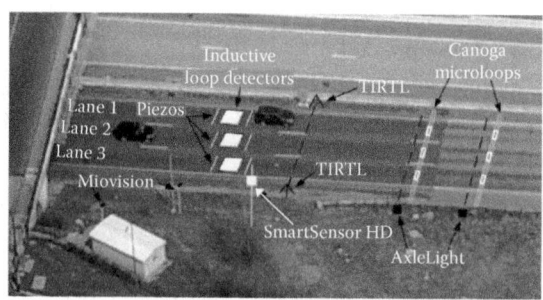

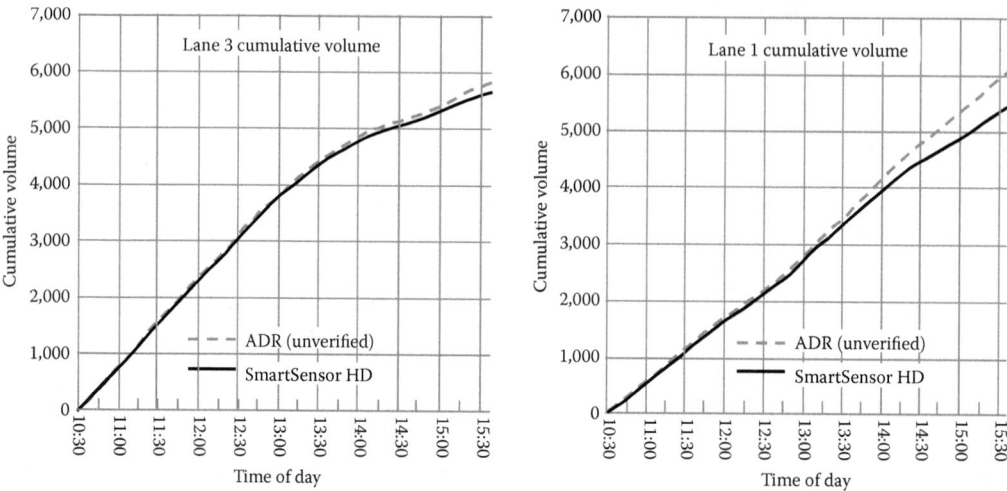

Figure 8.25 I-394 test site configuration and traffic volume data from Evaluation of Non-Intrusive Technologies for Traffic Detection—Phase 3. (Adapted from E. Minge et al., Evaluation of non-intrusive technologies for traffic detection—Phase 3, Presented at 90th Annual Meeting, Transportation Research Board, Washington, DC, 2011.)

Lane 3. While Lane 3 volumes are consistently accurate, even when speeds decreased, Lanes 2 and 1 volumes were reduced due to occlusion from Lane 3. From 2:30 to 3:30 p.m., the sensor missed 20.0% of the vehicles in Lane 2 and 22.6% of vehicles in Lane 1. During the same period of slow-moving traffic, it only undercounted 12.2% of vehicles in Lane 3. During this period, vehicles in Lane 3 were traveling 0–10 mi/h (0–16 km/h). In free-flow conditions, occlusion is not a factor and volume accuracy is typically within 2% error.

In summary, the tests found that (1) occlusion affects the SmartSensor HD count accuracy in far lanes in heavy congestion; (2) small numbers of slow-moving trucks traveling 0–10 mi/h can occlude more distant lanes for significant time; (3) vehicle undercounting is approximately 10% higher in the furthest lane as compared with the closest lane to the sensor (22% vs. 12%); and (4) during free flow occlusion is not a factor as traffic flow volume measurements are typically within 2% error. This test did not evaluate sensor performance in terms of correct detection, false detection, and missed detection.

8.8 EVALUATION OF VDSs IN INCLEMENT WEATHER

This weather-related evaluation of VDSs occurred in the urban area of Rantoul, Illinois. It studied the effects of fog, rain, and snow under six conditions, namely, light fog in daytime,

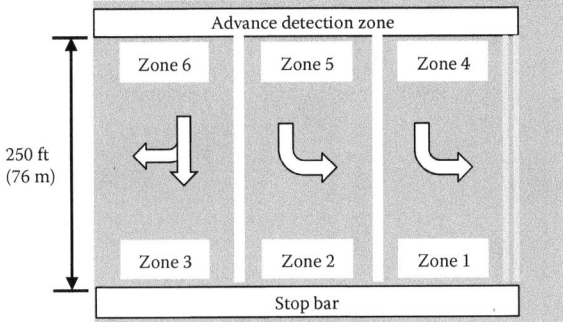

Cameras from the three manufacturers were installed on a luminaire arm located past the intersection at a height of approximately 40 ft above the projection of the center of the through lane. Six 6-ft × 6-ft inductive loops were also installed at the stop bar and advance detection zone locations.

Figure 8.26 VDS test site. (Adapted from J.C. Medina, R.F. Benekohal, and M.V. Chitturi, *Evaluation of Video Detection Systems Vol. 4—Effects of Adverse Weather Conditions in the Performance of Video Detection Systems*, Research Report FHWA-ICT-09-039, UILU-ENG-2009-2010, Illinois Center for Transportation, Department of Civil and Environmental Engineering, University of Illinois at Urbana-Champaign, Urbana, IL, March 2009.)

dense fog in daytime, rain in daytime, snow in daytime, rain in nighttime, and snow in nighttime on the performance of VDSs [12].

The intersection approach at the test site has two left-turn lanes and a shared right-turn, through lane as shown in Figure 8.26. The speed limit is 35 mi/h (56 km/h). Cameras from three VDS manufacturers (Autoscope® SoloPro with v. 8.13 firmware, Peek Unitrak with v. 2.2 firmware, and Iteris Edge 2 with v. 1.08 firmware) were installed on a luminaire arm located past the intersection at a height of approximately 40 ft (12 m) [12]. The cameras were placed above the projection of the center of the through lane, and not above the projection of the center of the approach lane, because the luminaire arm did not extend out that far. The field of vision from camera location to both stop bar and advance detection zones was clear of obstacles.

In addition to the VDS sensors, six inductive loops, each 6 ft × 6 ft (1.8 m × 1.8 m), were installed at the stop bar and advance locations, which were 250 ft (76 m) apart. The inductive loops, located at Zones 1 through 6, served as pointers to potential errors in detection. VDSs were configured by the manufacturers or the distributors to detect vehicles at the locations where the loop detectors were installed. Thus, each camera had three advance and three stop bar detection zones as explained in the figure. A representative from one of the manufacturers was present at the evaluation site during the setup. Distributors were present for the other two systems and received technical support from the manufacturers via telephone.

Four types of detection errors were recorded (false, missed, stuck-on, and dropped calls) at stop bar and advance detection zones. The results reflect the performance of the sensors after two rounds of configuration modifications were made by the manufacturers or their distributors and after the preliminary data were analyzed. The reported sensor product performance is a snapshot in time, that is, it reflects the hardware and algorithm technology available at the time of the tests. Newer VDS models and products from these manufacturers may function differently from the ones evaluated in Illinois. For example, the version of Autoscope firmware as of May 2015 was 10.5.0, while the version used in these tests was 8.13. As of August 2015, Peek was supplying firmware version 2.21, whereas the version in the tests was 2.2. As of August 2016, the version of Iteris firmware for its Vantage products was 06.01.20, while version 1.08 was used in the Illinois tests.

(a)   (b)

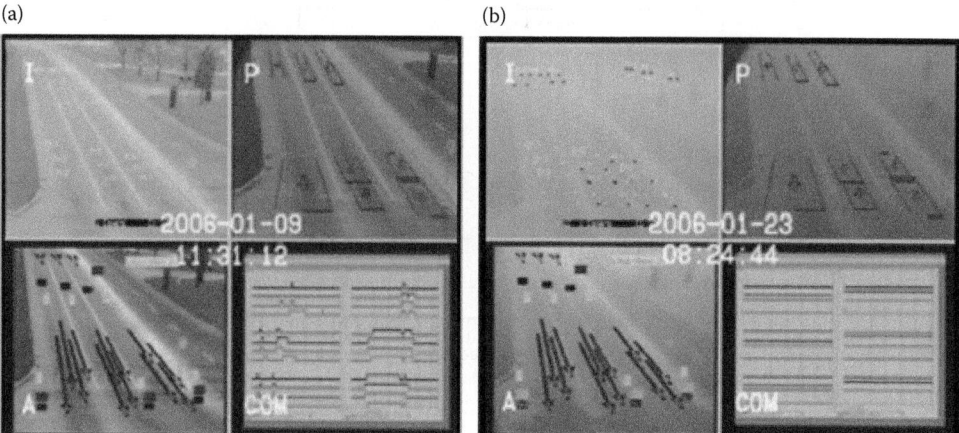

I = Iteris, P = Peek, and A = Autoscope in the pictures

Figure 8.27 Sample images of base and dense fog conditions encountered during the VDS evaluation. (a) Base condition and (b) Dense fog condition. (From J.C. Medina, R.F. Benekohal, and M.V. Chitturi, *Evaluation of Video Detection Systems Vol. 4—Effects of Adverse Weather Conditions in the Performance of Video Detection Systems*, Research Report FHWA-ICT-09-039, UILU-ENG-2009-2010, Illinois Center for Transportation, Department of Civil and Environmental Engineering, University of Illinois at Urbana-Champaign, Urbana, IL, March 2009.)

8.8.1 Light fog in daytime

The effects of light fog on VDS performance at stop bar and advance zones were limited, although there was a moderate increase in false calls (less than 10%) for the Autoscope and Iteris products.

8.8.2 Dense fog in daytime

Figure 8.27 includes sample images of the base and dense fog conditions. The Iteris VDS changed its operating mode and placed constant calls in all its zones during over 75% of the analyzed time period due to contrast loss in periods of heavy fog, while Autoscope placed constant calls in the front zones for about 13% of the time. Peek continued its operating mode without any apparent change. The constant calls led to inefficiencies in the operation of the intersection but avoided any missed calls. The Peek VDS, having no fail-safe mode, did not increase false calls at stop bar zones, but increased the number of missed calls in Zone 3, from 0.1% in conditions with favorable weather to 13.8% in dense fog.

At the advance zones, missed calls were as high as 50% or more, indicating that most of the vehicles were not detected. False calls and stuck-on calls were not negatively affected by the dense fog conditions during normal operating mode.

8.8.3 Rain in daytime

At stop bar locations, false calls increased on the average between 9.5% and 11.7% because of headlight reflection from approaching vehicles on the adjacent lanes. This was mainly due to the wet pavement, and was not observed in favorable weather when most vehicles had their headlights off. Rain did not affect the missed calls at the stop bar locations.

At advance locations, false calls also increased due to the reflection of headlights on the adjacent lanes (with averages between 6.3% and 12.2%). The reflections generated a better contrast to detect vehicles and reduced the missed calls in Peak and Iteris systems (<1%).

8.8.4 Snow in daytime

At stop bar zones, false calls in snow were high for all three products and constituted 64%, 88%, and 91% of the calls for Iteris, Autoscope, and Peek, respectively. No generalized effect was observed in missed calls, except for a particular zone (Zone 3) in Peek that increased missed vehicles by close to 6%. Stuck-on calls significantly increased in snow conditions, but to levels lower than 2.5%.

At the advance zones, the false calls were high for all three systems and constituted 43%, 69%, and 79% of the calls for Iteris, Autoscope, and Peek, respectively. Missed calls also increased in all three zones and with all systems except Zone 4 in Autoscope, ranging from 3.5% to 34.5%. Stuck-on calls were not affected at the advance zones and were nonexistent during periods of snow in daytime.

8.8.5 Rain in nighttime

During nighttime, rain significantly increased false calls to between 24% and 47% for the three stop bar zones combined. The reflection of headlights from vehicles approaching in the adjacent lane was the main cause of this increase. Similarly, stuck-on calls increased for Autoscope to 1.2%, for Iteris to 4.4%, and for Peek remained at zero. Missed calls were not affected during the rain condition in nighttime at the stop bar zones.

At the advance zones, false calls were high mainly in Zones 5 and 6, ranging from 9% to 50%, and mostly due to the reflection of headlights from vehicles approaching in the adjacent lane. Missed calls were slightly affected and remained lower than 1% for all three advance zones combined, similar to stuck-on calls, which also had averages lower than 1%.

8.8.6 Snow in nighttime

At stop bar zones, false calls increased mostly when pavement was partially covered with snow and when wind was present. False calls constituted 65%, 68%, and 83% of the calls for Autoscope, Iteris, and Peek, respectively. Missed calls and stuck-on calls did not show great variation in snow conditions at nighttime, limiting the snow effects to false calls only.

At advance zones, the increases in false calls mostly occurred in periods of partially snow-covered pavement and wind. False calls constituted 57%, 68%, and 87% of the calls for Autoscope, Iteris, and Peek, respectively. No significant increase was observed in missed calls, except in Zone 5 of Autoscope, where 6.2% of the calls were missed. The number of stuck-on calls did not change at any advance zone for any VDS.

8.8.7 Conclusions from VDS evaluation

In summary, the tests of these three video systems found that (1) performance was not greatly impacted in daytime with light fog or in daytime with rain but no wind, and (2) significant changes were observed under dense fog and snow during the day, and snow and rain during night. Another conclusion of note is that each VDS performed differently from the others with respect to the numbers of false, missed, stuck-on, and dropped calls they experienced. This observation reinforces the need for on-site testing of a detection system before purchase.

8.9 QUEUE ESTIMATION USING MAGNETOMETERS

Magnetometer sensors were installed on a freeway on-ramp to measure the effectiveness of four queue estimation techniques based on (1) occupancy measurements at the ramp entrance, (2) vehicle counts at the on-ramp entrance and exit, (3) speed measurements at the ramp entrance, or (4) vehicle re-identification [13]. As a result of this study, information was also obtained about the effectiveness of magnetometer sensors in detecting congested vehicles on curved roadways.

8.9.1 Sensor configuration

Figure 8.28 illustrates the sensor configuration at the Hegenberger Road on-ramp to I-880 southbound in Oakland, California. Two arrays of seven sensors, separated by 1 ft (30 cm) and centered on the lane width, were located at the entrance and at the exit of the on-ramp for vehicle identification and re-identification. These were Mode F sensors that transmit x, y, z data samples only while a vehicle is present. Four additional sensors were installed at the entrance of the on-ramp and arranged in a speed trap configuration. These operated as Mode B sensors that determine when a vehicle arrives at and departs from the sensor detection zone. Two of these sensors provided leading vehicle detection (SL1 and SL2), and another two, trailing vehicle detection (ST1 and ST2). Utilizing two sensors side-by-side increases the lateral detection zone to capture vehicles that may not be traveling down the center of the lane. The speed trap sensors were used to study the queue estimation methods based on speed and occupancy and to compare the vehicle counting performance of a single sensor and a sensor array.

8.9.2 Ground truth

Ground truth queue information was obtained from imagery captured by three video cameras. The first camera recorded vehicles waiting at the metering light and leaving the on-ramp and passing the exit sensor array. The second camera recorded vehicles entering the on-ramp and passing through the speed trap and the entrance sensor array. The third camera was not fixed at any specific location, as it was used to capture queue dynamics during the ground truth data collection period. All cameras were synchronized with a common clock so that data extracted from different videos could be merged, or meaningfully compared with the magnetometer detection system data.

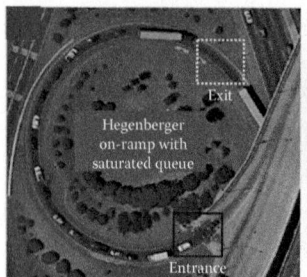 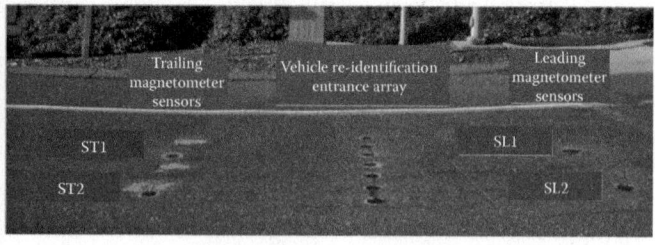

Figure 8.28 Magnetometer sensor configuration at Hegenberger Road on-ramp to I-880 in Oakland, CA. (From R.O. Sanchez, R. Horowitz, and P. Varaiya, Analysis of queue estimation methods using wireless magnetic sensors, Paper 11-3173, 90th Annual Meeting, Transportation Research Board, Washington, DC, 2011.)

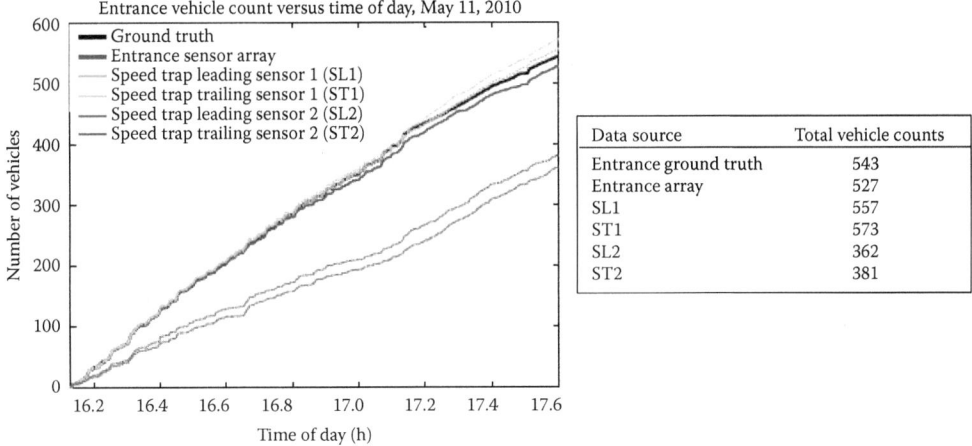

Figure 8.29 Entrance vehicle counts from ground truth, entrance array, and SLI, SL2, STI, and ST2 sensors. (Adapted from R.O. Sanchez, R. Horowitz, and P. Varaiya, Analysis of queue estimation methods using wireless magnetic sensors, Paper II-3173, Presented at 90th Annual Meeting, Transportation Research Board, Washington, DC, 2011.)

8.9.3 Results

Figure 8.29 contains a comparison of the entrance vehicle counts from speed trap sensors, the entrance array, and ground truth. If cars traveled through the middle of the lane, a similar count would be expected from all speed trap sensors. However, it was observed that vehicles tend to travel on the right side of the lane, which is reflected in the data of Figure 8.29 since sensors SL2 and ST2, located on the left side of the lane, register significantly lower vehicle counts over the data collection time interval. The difference in vehicle count from sensors SL1 and ST1 and the ground truth values becomes evident only after hour 17.1, which corresponds to the time the ramp goes into saturation. It appears from these data that congestion affects the counting performance of both speed trap (Mode B) and re-identification array (Mode F) sensors.

As with the sensor array data, single sensors sometimes register multiple detections for the same vehicle and a single detection event for multiple vehicles. Nevertheless, the total count for sensors SL1 and ST1 exceeded the ground truth value, which is the opposite of the total entrance array vehicle count. This suggests that the speed trap Mode B sensors may be more likely to generate multiple detections for the same vehicle, which was not observed for the re-identification array Mode F sensors.

Leading and trailing sensors on the same side of the lane were expected to have similar vehicle counts. However, trailing sensors registered higher total vehicle counts than leading sensors. This suggests that vehicle counting by speed trap sensors is dependent on the lateral as well as the longitudinal location of the sensor in the ramp lane.

Long and slow-moving maneuvering trucks at the on-ramp entrance may introduce counting errors of one or two vehicles. Curved on-ramps have wide lanes that allow drivers to maneuver as they go around. The extra lateral space results in vehicles traveling highly off-centered, completely missing the vehicle detection station.

The worst performance of the queue estimation methods was observed when the on-ramp was heavily congested, that is, the queue extended beyond the ramp entrance. Under this condition, two adjacent cars close to each other are likely to stop on top or very close to the detection zone of the sensors. This creates undercounting problems, since two cars may be registered as one.

8.9.4 Conclusions from queue estimation tests with magnetometer sensors

Each of the tested queue estimation methods had its limitations under either unsaturated, saturated, or transition conditions. Occupancy queue estimation may be used to determine if the ramp is either empty or full, but it does not estimate queue length accurately. This approach is highly dependent on the time calculation interval over which occupancy is calculated and, when on-ramp saturation occurs, may yield misleading results due to a vehicle's tendency to miss the sensor detection zone while at rest.

Queue length based on vehicle counts from sensors is not an acceptable method to estimate the queue due to its inability to correct for errors such as sensor miscounts and offsets resulting from initial conditions.

Speed-based queue estimation appears capable of instantaneously determining the state of the ramp, unsaturated, saturated, or in transition. However, the results obtained for this queue estimation approach do not match the results from traffic simulations.

Finally, queue estimation from vehicle re-identification performed better than the other methods when the ramp was not saturated, but it underperformed during saturated conditions. The low vehicle re-identification match rate during ramp saturation occurred because the algorithm's ability to correct for errors was degraded by the inaccuracy of the vehicle counts under these conditions. In order to make this method reliable for queue estimation and regulation, it would be necessary to develop a vehicle re-identification algorithm that takes into account on-ramp specific factors such as ramp curvature, slope, vehicle headway, and sensor location.

With respect to the use of magnetometers to detect the vehicles on the ramp, their relatively restricted detection zone makes it necessary to use several across a lane to ensure that a vehicle not traveling down the center of the lane will be detected. It was also observed that speed trap Mode B sensors may be more likely to generate multiple detections for the same vehicle under congested conditions.

8.10 SUMMARY

A rationale has been presented for conducting testing and evaluation of traffic flow monitoring sensors that are being considered for purchase. The reasons include ensuring they provide the required data and information under the often unique conditions encountered at a particular venue. Conditions that frequently vary from location to location that may affect sensor performance include seasonal variations in traffic volumes and vehicle mix, congestion level, time-of-day and lane-to-lane variations in vehicle mix, unique road configurations, lighting, bridge and tunnel deployment, and weather. The test and evaluation results that were presented show that sensor performance does indeed vary as these conditions change. The chapter also described techniques and documentation examples that serve as a resource for conducting these types of tests. A critical observation is that new devices that incorporate modern sensor technologies appear to be less susceptible to several of the factors that degraded their performance in the past. Nonetheless, testing remains a critical part of a sensor selection process.

REFERENCES

1. L.A. Klein and M.S. MacCalden Jr., *Development of IVHS Traffic Parameter Specifications, Task A Report for Detection Technology for IVHS*, U.S. Federal Highway Administration, McLean, VA, April 1995.
2. L.A. Klein, *Select Field Sites for Detector Field Tests, Task B Report for Detection Technology for IVHS*, U.S. Federal Highway Administration, McLean, VA, March 1994.
3. L.A. Klein, *Vehicle Detector Laboratory Test Specifications and Test Plan, Task C Report for Detection Technology for IVHS*, U.S. Federal Highway Administration, McLean, VA, March 1995.
4. L.A. Klein, *Select and Obtain Vehicle Detectors, Task D Report for Detection Technology for IVHS*, U.S. Federal Highway Administration, McLean, VA, Rev. December 1994.
5. L.A. Klein, *Results of Laboratory Detector Tests, Part I, Task E Report for Detection Technology for IVHS*, U.S. Federal Highway Administration, McLean, VA, Rev. January 1994.
6. L.A. Klein, *Results of Laboratory Detector Tests, Part II, Task E Report for Detection Technology for IVHS*, U.S. Federal Highway Administration, McLean, VA, October 1993.
7. L.A. Klein, *Results of Laboratory Detector Tests, Part III, Task E Report for Detection Technology for IVHS*, U.S. Federal Highway Administration, McLean, VA, August 1993.
8. L.A. Klein, *Vehicle Detector Field Test Specifications and Field Test Plan, Task F Report for Detection Technology for IVHS*, U.S. Federal Highway Administration, McLean, VA, March 1995.
9. L.A. Klein and M.R. Kelley, *Detection Technology for IVHS, Final Report*, FHWA-RD-95-100, U.S. Federal Highway Administration, McLean, VA, December 1996. http://ntl.bts.gov/lib/jpodocs/repts_te/6184.pdf. Accessed December 10, 2015.
10. L.A. Klein, *Task L Report for Detection Technology for IVHS*, Final Report Addendum, U.S. Federal Highway Administration, McLean, VA, 1996.
11. E. Minge, S. Peterson, and J. Kotzenmacher, Evaluation of non-intrusive technologies for traffic detection—Phase 3, Presented at 90th Annual Meeting, Transportation Research Board, Washington, DC, 2011.
12. J.C. Medina, R.F. Benekohal, and M.V. Chitturi, *Evaluation of Video Detection Systems Vol. 4—Effects of Adverse Weather Conditions in the Performance of Video Detection Systems*, Research Report FHWA-ICT-09-039, UILU-ENG-2009–2010, Illinois Center for Transportation, Department of Civil and Environmental Engineering, University of Illinois at Urbana-Champaign, Urbana, IL, March 2009.
13. R.O. Sanchez, R. Horowitz, and P. Varaiya, Analysis of queue estimation methods using wireless magnetic sensors, Paper 11-3173, Presented at 90th Annual Meeting, Transportation Research Board, Washington, DC, 2011.

Chapter 9

Sensor specification and testing tools

Presence detection is the most ubiquitous application of vehicle detection systems deployed on both freeways and surface street arterials. Its uses include arterial traffic signal control, freeway ramp metering, incident detection, wrong-way vehicle detection, and toll collection.

Given that a vehicle is actually located within a specified detection zone, a sensor can either correctly detect the vehicle or fail to detect the vehicle. Given the non-presence of a vehicle, a sensor can either correctly not detect a vehicle or incorrectly report the presence of a vehicle. In summary, there are three possible outcomes covering both situations:

1. *Correct detection (of an actual vehicle)*: Indication by a sensor that a vehicle passing over the detection area of the sensor is detected by the sensor.
2. *False detection (when no vehicle is present)*: Indication by a sensor that a vehicle *not* passing over the detection area of the sensor is detected by the sensor.
3. *Failure to detect (an actual vehicle) or missed detection*: Indication by a sensor that a vehicle passing over the detection area of the sensor is *not* detected by the sensor.

Aggregated measurements such as total vehicle count within a specified time period can obscure the actual accuracy of a sensor, since failures to detect are canceled by false detections. Therefore, evaluation methods based upon aggregated metrics can provide misleading conclusions. A common example is the case of a loop detector connected incorrectly that reports data from another lane. Reasonable results are generated since adjacent lane counts, average speeds, and occupancies are similar, and such a sensor would be classified by a sensor condition monitoring system as "good," despite the fact that the sensor was not measuring the intended phenomena in the correct lane. Therefore, the metrics listed above, rather than the aggregated vehicle count, are recommended for determining the ability of a sensor to accurately detect and count vehicles.

In this chapter, we discuss three topics related to testing and evaluating sensor performance. The first is a review of the information available in testing standards such as those developed through ASTM International. The second examines the concepts of confidence intervals and confidence levels that should be included in any standard or specification that is prepared for sensor accuracy measurement and purchase. The third topic concerns interoperability as it relates to institutions, policies and procedures, and technical concerns such as interfacing with other components and data transfer among devices.

9.1 ASTM STANDARDS OVERVIEW

ASTM International is an organization that relies on volunteers working in industry, as consultants, and in the teaching profession to develop specifications and standards that specify the performance and compliance testing procedures for a wide range of products and materials. The standards are available for purchase from ASTM International through their website http://www.astm.org/Standard/. The descriptions that follow include the salient features of two specifications that were developed to assist in the purchase and testing of traffic flow sensors. They contain the purpose of the standard, critical definitions, sensor accuracy definitions, requests for the purchaser to state the types of sensor data that are required and the conditions under which the sensor will operate, and the testing protocols that will be used to verify that the purchased product satisfies the requirements of the purchase order. These specifications may be treated as advisory if the purchaser and seller so wish and may be modified by the purchaser and seller to meet their unique needs.

9.2 STANDARD SPECIFICATION FOR HIGHWAY TRAFFIC MONITORING DEVICES E 2300-09

This first specification describes the recommended procedure for identifying the performance requirements and operating conditions to be included in a purchase order for traffic monitoring devices (TMDs), more commonly known as sensors or detectors [1]. It also defines terminology so that the purchaser and seller can understand what both want and offer. Thus, the specification can be referenced by each to determine compliance with the specified requirements.

9.2.1 Definitions

Although this specification makes an effort to harmonize the terminology associated with traffic flow sensors, other terms are sometimes encountered. For example, accepted reference value (ARV), a term used in the standard, is commonly referred to as ground truth elsewhere and electronics unit as used with inductive loop detectors is commonly referred to as the detector. Below are several of the more pertinent definitions contained in E 2300 and, where applicable, additional explanatory notes.

Accepted reference value: A particular quantity, (e.g., number of vehicles in a particular class defined by number of axles and interaxle spacings, vehicle count, lane occupancy, or vehicle speed) produced by a method agreed upon by the purchaser and seller in advance of testing of a TMD, which has an accuracy associated with its value that is appropriate for the given purpose. ARV is often referred to as ground truth value.

Accuracy: Closeness of agreement between a value indicated by a TMD and an ARV.

Electronics unit: Device that provides power to one or more sensors, filters and amplifies the signals produced by the sensors, and may perform other functions such as sensitivity adjustment, failure indication, and delayed and extended actuation of traffic control signals. Often referred to as a detector.

Lane occupancy: Percent of selected time interval that vehicles are detected in the detection area of a sensor; the time interval during which the lane occupancy is measured is usually 20–30 s. Different sensor models or technologies used to measure lane occupancy may have different detection area sizes and, hence, produce different occupancy values, although all devices are operating properly.

Sensor: Device for acquiring a signal that provides data to indicate the presence or passage of a vehicle or of a vehicle component over the detection area, often with respect to time, (e.g., flow or number of axles and their spacing), or one or more distinctive features of the vehicle such as height or mass. Also referred to in this standard as a TMD. Some literature uses the term detector for sensor, although this can cause confusion as the same word is often applied to the electronics unit defined above.

Tolerance: Allowable deviation of a value indicated by the device under test or a device in service from an ARV, that is, ground truth value.

Traffic monitoring device: Equipment that counts and classifies vehicles and measures vehicle flow characteristics such as vehicle speed, lane occupancy, turning movements, and other parameters typically used to portray traffic movement. Frequently called a sensor or detector.

9.2.2 Device ordering information

When purchasing a TMD, that is, a sensor, several pieces of information must be included in the purchase specification and order. They are the device type, the data accuracy or tolerance error that can be accepted, and the conditions under which the device will be used. It is also vital to specify a confidence level for the measurement accuracy. The tolerance may be specified in several ways depending on the application of the data as explained below.

9.2.2.1 Device type

E 2300 suggests a scheme to associate the function of a TMD and the vehicle characteristics it detects with a device type, which is used in the purchase specification. Accordingly, the standard requires the purchaser to specify a TMD by (1) a type identifier shown in Table 9.1 that relates to the TMD's function, detected vehicle characteristics, and specific data measured or recorded; (2) a tolerance for each required data item; and (3) the conditions under which the device will be operated. Table 9.2 contains a list of installation and operating conditions that should be considered for inclusion in any purchase specification.

9.2.2.2 Tolerance

A tolerance is required for each data item output by the TMD. A TMD that records or outputs multiple data items may have different tolerances specified for each data item.

The tolerance should be specified in a manner consistent with the application supported by the TMD output data. Accordingly, the tolerance of the TMD may be specified in three ways: (1) percent difference; (2) single-interval absolute value difference (SAVD); and (3) multiple-interval absolute value difference (MAVD).

1. *Percent difference*: Percent difference is defined as an absolute value given by

$$\text{Percent difference} = \frac{|\text{TMD output value} - \text{ARV}|}{\text{ARV}} \times 100. \tag{9.1}$$

When vehicle presence is the data item of interest, ARV or ground truth value may be defined as the actual time record of the presence of all vehicles or the nonpresence of vehicles on a particular facility. One of the greatest challenges in the evaluation of sensor accuracy is the generation of this reference data set [2,3]. The recommended metrics for evaluating the ability of a TMD (i.e., sensor) to detect

Table 9.1   TMD functions, types, detected vehicle characteristics, and data recorded or data collection interval

Function	Type	Detected vehicle characteristic	Data recorded[a]
A—Traffic counting	A-1	Axle passage	Number of axles
	A-2	Vehicle passage	Number of vehicles
	A-3	Vehicle presence	Number of vehicles
B—Traffic counting/ classifying	B-1 (classification by number of axles and interaxle spacings)	Vehicle passage, number of axles and interaxle spacings during vehicle passage	Number of axles, number of vehicles per class, vehicle speed, vehicle class by number of axels and interaxle spacings
	B-2 (classification by length)	Vehicle passage and speed	Number of vehicles, vehicle speed, vehicle length and class, vehicle presence, lane occupancy
C—Incident detection data	C-1	Vehicle passage, presence, and speed	Number of vehicles, vehicle speed, vehicle presence, or lane occupancy
D—Speed monitoring	D-1	Speed	Number of vehicles, vehicle speed
E—Metering data (ramp, mainline, or freeway-to-freeway)	E-1	Vehicle presence	Number of vehicles, vehicle presence, or lane occupancy
F—Signal control data	F-1	Vehicle presence	Number of vehicles, vehicle presence, or lane occupancy
G—Enforcement aid	G-1 (speed)	Speed	Vehicle speed
	G-2 (red signal)	Location of front of vehicle, red signal indication	Number of vehicles and violations
	G-3 (dimension)	Vehicle location and specified overall dimensions	Vehicle presence, specified overall dimension

[a]   The purchaser may specify the recording of a device identifier by the TMD and data time stamp when needed.

the presence of a vehicle are correct detection, false detection, and failure to detect as defined above.

The percent difference for the number of *correctly detected* vehicles is given by

$$\text{Percent difference} = \frac{|\text{TMD output value for correctly detected vehicles} - \text{ARV}|}{\text{ARV}} \times 100.$$

(9.2)

Thus, a TMD that correctly detects 1539 vehicles when the ARV is 1600 is said to have correctly measured the number of vehicles to within a ±3.8% tolerance.

The percent difference for the number of *falsely detected* vehicles is given by

$$\text{Percent difference} = \frac{|\text{TMD output value for falsely detected vehicles} - \text{ARV}|}{\text{ARV}} \times 100.$$

(9.3)

Table 9.2 Installation, operating, and maintenance requirements to be included in TMD purchase specifications

• *Environment*	• *Installation*
Ambient temperature	Weight and size limitations
Humidity	Mounting or other installation constraints
Lighting	Power availability
Sun position and angle	Power surge and lightning protection
Precipitation types (e.g., rain, snow, hail)	Input power interface
Other atmospheric obscurants (e.g., fog,	Special cables and connectors
dust, smoke)	• *Setup and calibration*
Vibration and shock	Operating and calibration software
Wind	Operating, installation, and repair manuals
• *Vehicle characteristics*	• *Miscellaneous*
Vehicle class mix	Fail safe operation if device fails
Vehicle-to-vehicle gaps required to define	Warranty
vehicle flow rate and evaluate TMD	Software upgrades and product maintenance
detection accuracy	Other pertinent items affecting installation,
• *Output data items*	operation, maintenance, and storage
Data recording interval	
Data communication link	
Data interface	
Data display	

For example, if the number of falsely detected vehicles is 40 and the ARV is 1600, the TMD is said to have falsely detected ±2.5% of the vehicles.

The percent difference for the number of *missed detections* is given by

$$\text{Percent difference} = \frac{|\text{TMD output value for missed detections} - \text{ARV}|}{\text{ARV}} \times 100. \quad (9.4)$$

For example, if the number of missed vehicle detections is 15 and the ARV is 1600, the TMD is said to have missed the detection of ±0.9% of the vehicles.

2. *Single-interval absolute value difference*: SAVD specifies a single maximum allowable deviation of the TMD output with respect to the comparable ARV. Thus,

$$\text{SAVD} = |\text{TMD output value} - \text{ARV}|. \quad (9.5)$$

The SAVD is stated in units that correspond to the data item indicated. For example, a maximum difference of 3 mi/h (5 km/h) with respect to the ARV is specified for the measurement of vehicle speed within a single user-defined speed interval, say 10–80 mi/h (16–130 km/h), inclusive.

3. *Multiple-interval absolute value difference*: MAVD specifies a different allowable deviation in TMD output with respect to the comparable ARV for each interval of data item values included in the TMD specification. Thus, the MAVD permits different deviations to be established for distinct intervals of the measured data item.

The MAVD is calculated using Equation 9.5. For example, a maximum difference of 3 mi/h (5 km/h) is required when measuring the speed of vehicles traveling at or above

55 mi/h (88 km/h), but a maximum difference of 1 mi/h (2 km/h) is required for vehicles traveling below 55 mi/h (88 km/h).

9.2.3 Acceptance tests

There are two kinds of acceptance tests, the Type-Approval Test and the Onsite Verification Test, which can be required by E 2300 before the device will be accepted by the purchaser. The Type-Approval Test is the more rigorous of the two. It is specified when a sensor has never passed this kind of test before, for example, when purchasing a new or improved model of a device and installing it for the first time. This test may last for several weeks. The thoroughness of the test is meant to verify the functionality of all features of the TMD and the accuracy of the data item outputs when monitoring vehicle flows consisting of a mix of all anticipated and specified vehicle classes under the specified operating conditions.

The Onsite Verification Test is a shortened version of the acceptance test and is intended for sensors that have previously passed a Type-Approval Test. It is applied when additional sensors of a type previously purchased are repurchased and installed perhaps at a new location. This test determines whether the production version of a TMD installed at a particular site meets the performance and user requirements identified in the purchase order and sensor specification. Similar to the Type-Approval Test, the Onsite Verification Test defines the required tests for evaluating the performance of a TMD according to the functions it performs, the data it provides, and the required accuracy of the data for the conditions under which the device operates.

9.3 STANDARD TEST METHODS FOR EVALUATING PERFORMANCE OF HIGHWAY TRAFFIC MONITORING DEVICES E 2532-09

This standard defines the acceptance test conditions, specifies the procedures for performing both the Type-Approval Test and the Onsite Verification Test, and suggests methods to obtain reference value or ground truth data against which the outputs of the TMD under test are compared [4].

9.3.1 Test conditions

In addition to reviewing the definitions of terms used during the test, the standard lists the conditions under which the acceptance test will be performed. They should match the conditions that were included in the purchase specification and order. Typical testing conditions and other items that should be identified are given in Table 9.3.

9.3.2 Accuracy required of ARV measuring equipment

The data measuring accuracy requirements for all equipment used to obtain ARV (ground truth) data shall be agreed upon by the purchaser and seller before testing begins. When possible, it is recommended that such equipment have an accuracy at least an order of magnitude greater than the accuracy specified for the TMD under test.

9.3.3 Summary of procedure for conducting Type-Approval Test

The Type-Approval Test provides performance evaluation of an untested TMD brand and model in a field environment under operational conditions. The test determines whether

Table 9.3 Acceptance test conditions to be included in a test procedure document

1. Installation requirements
2. Vehicle flow rates and vehicle classes
3. Lighting
4. Temperature
5. Other environmental conditions: Rain and rain rates, fog and visibility range, snow and snowfall rate, wind-borne dust, movement caused by wind and vibration, and any other conditions that are perceived by the user to affect the performance of the TMD
6. Seller-provided evidence that the TMD can operate under the specified environmental conditions
7. Power requirements
8. Data and video communication link
9. Options, exceptions, and added features
10. Other operating conditions that affect sensor performance including road geometry, structures that impede sensor line of sight, lane dropping or adding, restrictive lane widths, operation on a metal deck or with a metal superstructure, tunnel operation

the TMD meets the requirements in the TMD specification developed in accordance with E 2300.

9.3.3.1 Approval of site and test conditions

It is recommended that both the purchaser and the seller approve the type-approval test site and TMD installation prior to conducting the Type-Approval Test. TMD settings and other test conditions must be documented to verify compliance with the test conditions described in Table 9.3. Pictures often assist in recalling how the equipment was set up and configured.

9.3.3.2 Calibration and preliminary testing

The TMD under test is calibrated by the seller and approved by the purchaser. The calibration procedures are documented and made available to the purchaser. Often, a preliminary test is performed to confirm that the device under test is operating properly. If an obvious defect is present, the manufacturer should be contacted for repairs before beginning the contractual Type-Approval Test.

9.3.3.3 Duration of Type-Approval Test

Type-Approval Test duration continues until the required data are recorded to verify correct operation of the TMD under all of the environmental and other operating conditions specified by the purchaser and, therefore, may last several days.

9.3.3.4 Type-Approval Test method

1. Install the TMD according to the seller's instructions or according to another procedure mutually agreed upon by the purchaser and seller or their designated representatives.
2. Adjust variable TMD operating parameters to values agreed upon by the purchaser and seller or their designated representatives and record these values in the documentation.
3. Record all data output by the TMD under test along with ARV data using a device capable of time stamping the data. Each vehicle detection event shall be output by the TMD in a format that can be directly correlated with the video record of the test. The digitizing of data from the TMD and the reference value equipment shall occur at a sampling frequency that prevents compromising of data quality by aliasing.

4. Document the test and test conditions with the time of day, TMD identifier, vehicle class, ambient lighting, weather, and other items listed in Table 9.3.

5. Evaluate TMD performance for the vehicle flow rates and mix of vehicle classes specified in accordance with Table 9.3.

6. TMD testing is also performed under various lighting, temperature, weather, other local environmental conditions, and distinctive road geometry and features when the performance of the TMD is deemed by the purchaser to possibly vary under these conditions.

7. For the purposes of verifying TMD performance, lane-straddling vehicles are eliminated from consideration by identifying them from the video recordings made while obtaining ARV data.

9.3.3.5 Generating ARV data

Suggested methods for obtaining ARV or ground truth data for axle count, vehicle count, vehicle speed, vehicle classification, vehicle presence, and lane occupancy are described in this section. These methods rely on two human observers analyzing video recordings for acquiring the pertinent data. It is recommended that the detection area of the TMD under test be marked with tape, paint, or other means so that it is visible in the recorded imagery. Alternatively, the detection area may be indicated by a digital overlay on the digitized video. An automated method of comparing the data from the TMD under test to reference values is described in Appendix X2 of E 2532-09 and is summarized in Section 9.3.3.7.

The seller shall have primary responsibility for supplying the equipment and personnel for obtaining the ARV data needed for interpreting the results of the Type-Approval Test. The purchaser or a third party may conduct the test or provide other assistance.

1. *Axle count reference values*: Axle count reference values shall be found by analyzing imagery recorded by a video camera installed to have an unimpeded view of the vehicle axles as they pass over the effective detection area of the TMD under test. Two or more human observers shall each record the reference number of axles by viewing the video imagery. Each observer shall view the imagery for no longer than a 15-min interval before taking a rest of at least 5 min to help assure accurate determination of the reference value.

   If the difference in axle counts reported by any observer exceeds the largest of the observer-reported values by 10% of the specified device tolerance (calculated as a percentage of the largest observer-reported value and rounded up to the nearest whole integer), repeat the observations. For example, if the tolerance is 10%, the axle count obtained by two observers shall not differ by more than 1% (10% of 10%). When satisfactory agreement among observed axle counts is achieved, use the average of the reported counts as the reference value against which to compare the device under test.

2. *Vehicle count reference values*: Vehicle count reference values for the number of correct detections shall be determined on a vehicle-by-vehicle basis by human observer analysis of the vehicle images recorded by one or more video cameras installed to give an unimpeded view of the vehicles as they pass over the effective detection area of the TMD under test. The observers shall also calculate the numbers of missed detections and false detections by comparing their recorded observations with the output of the TMD under test when the TMD reports detections on an individual vehicle basis. When the TMD reports detections aggregated over a known time interval, the

observers shall calculate the numbers of missed detections and false detections by aggregating their recorded observations over the same interval and then comparing that value with the output of the TMD under test.

3. *Vehicle speed reference values*: Vehicle speed reference values shall be obtained using two or more matched axle-detecting sensors at known distances from each other, which are installed on or in the pavement as near as feasible to midway within the detection area of the TMD under test. The vehicle speed shall be calculated as the distance between any two axle sensors divided by the time difference between actuation of the second and first axle sensors. A microwave radar or lidar speed gun operated by trained personnel may be used as an alternative device for acquiring speed reference values.

4. *Vehicle classification reference values*: Vehicle classification reference values shall consist of the number of vehicles of a particular class as displayed on imagery recorded by one or more video cameras installed to give an unimpeded view of the vehicles as they pass over the effective detection area of the TMD under test.

5. *Vehicle presence reference values*: Vehicle presence reference values shall consist of the appearance of a vehicle as displayed on imagery recorded by one or more video cameras installed to give an unimpeded view of the vehicles as they pass over the effective detection area of the TMD under test. The presence of a vehicle in the effective detection area of the TMD shall be noted from the recorded imagery and shall be sufficient for establishing a vehicle presence reference.

6. *Lane occupancy reference values*: Lane occupancy reference values for a vehicle shall consist of the percent of a selected time interval the vehicle is in the effective detection area of a video camera installed in a manner that provides an unimpeded view of the vehicle as it passes over the effective detection area of the TMD under test.

9.3.3.6 *Tolerance compliance calculation*

The tolerance compliance computation is key to accepting the device under test. As E 2532 is now written, it does not incorporate a confidence interval or confidence level along with the accuracy specification. Therefore, the standard contains a clause that says even if only one piece of data does not fall within the accuracy or tolerance specification, you must fail the device. The notion of a confidence interval avoids this ill-advised trap by recognizing that TMD and ARV data measurements contain a random error component that should be accounted for. In any event, it is important to document the test conditions and the test results. Remember, this standard can be used as guidance. Any user of the standard is free to modify it to include features such as a confidence interval.

Tolerance compliance is performed by calculating the difference between the ARV and the TMD output for each data item using the percent difference, SAVD, or MAVD defined in Specification E 2300 and shown in Section 9.2.2.2. The calculated tolerance value is then compared with the tolerance specified in the purchase specification. If the calculated value is within the specified limits, the device has met the accuracy specification.

9.3.3.7 *Automated methods of comparing data from the TMD under test to reference values*

Sensor testing and evaluation methods requiring comparison against human-verified ARVs are not practical when many sensors of different types are concurrently tested on as many as six lanes, or when large data recording intervals generate thousands of records. In these cases, an automated test process becomes an important adjunct to the testing protocol. Such a system, for example, the Video Vehicle Detector Verification System (V²DVS) developed by MacCarley [5], validates the detection output quantities from individual sensors based on the fusion of data from each of the sensors under test with data from a reference image processing system. This creates a reliable composite ARV record. This approach reduces the human labor required for sensor validation by replacing the single reference sensor or human observer that generates the ARV with an automatic technique and compares it with the output of the sensors under test.

Figure 9.1 displays the V²DVS as deployed in Irvine, CA on the I-405 Freeway. The system contains reference video cameras above each freeway lane and side-viewing, multilane sensors and other devices that can be mounted on one of two roadside poles. Each of the six traffic lanes is equipped with duplex inductive loops and provision for other interchangeable roadway sensors. Other equipment includes a cluster of data acquisition computers (field machines), one per lane, and a central server for archiving and automated processing of data. These are housed in a roadside Caltrans Type 334C cabinet as shown in Figure 9.2. A PC-based client program facilitates remote monitoring and control of all field machines, manual verification of ARVs, and generation of test results through the central server. The field machines are 2-U industrial rack-mount Linux/PC platforms, each interfaced to a video camera located on an overcrossing above an assigned lane. The collected raw data consist of JPEG-compressed images and a database containing the time of arrival, speed, other metrics of every detected vehicle in each lane, and a reference record created by the V²DVS based on real-time image analysis. The system supports multiple test sites, with a maximum of eight hardwired sensors with contact closure pairs and an unlimited number of network or serial-communicating sensors for each lane at each site. At maximum traffic capacity, as many as 96,000 records per hour per site are generated.

The most significant labor-saving feature of V²DVS is its ability to automatically generate an accurate ARV record, against which all individual detection events are compared to

Figure 9.1 Over-lane video cameras and roadside sensor-mounting pole at I-405 V²DVS sensor evaluation location.

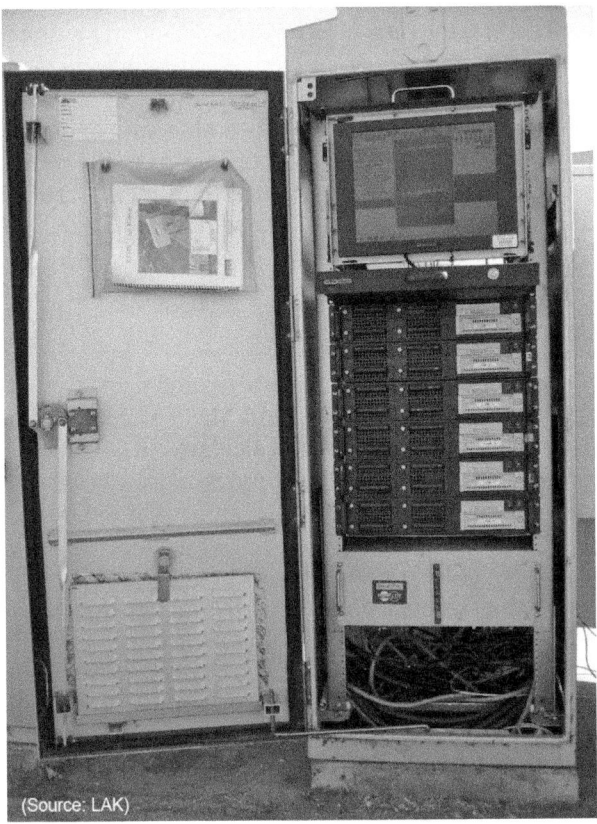

(Source: LAK)

Figure 9.2 V²DVS data acquisition computers (field machines) in Type 334C cabinet. A rack-mounted fold-down monitor and keyboard are also installed at the top of the equipment rack.

determine and report the accuracy of the individual sensors under test. A biased voting process is used for each detection event, in which the conclusion of the weighted majority of the sensors for each lane is believed to be the truth, (i.e., if the majority of the sensors saw it, it must have been there). An adaptive learning algorithm continuously optimizes the weighting coefficients to maximize the accuracy of the synthesized ARV data set.

A type of recursive filter gradually adapts the weighting coefficient $a_i(k)$ for the ith sensor based upon its agreement or disagreement with the consensus for each detection event k. If every detection is correct, (i.e., agrees with the consensus), $a_i(k)$ asymptotically approaches unity. If the sensor consistently fails to detect or falsely detects, $a_i(k)$ asymptotically approaches zero. Therefore, sensors that are frequently incorrect, (i.e., in disagreement with the weighted majority) are devalued in the consensus voting, while accurate sensors (those in agreement with the weighted majority) are more and more strongly weighted.

Automated data reduction greatly reduces the workload associated with ARV generation, since it requires human verification only for detection records that cannot be automatically correlated. In the final analysis, vehicle detections are classified as either correct, false, or failure to detect by this automated approach. Errors are most commonly due to ambiguous vehicle lane position. Accuracy is dependent upon the size of the admissible time/distance aperture, with more conservative settings tending to reject valid detections and less conservative settings admitting incorrect matches that sometimes cause alignment errors that propagate to other proximate vehicles in the ARV data set. Additional details concerning the data fusion architecture, data fusion algorithms, computer vision detection methods, and automated data reduction and reporting methods are found in [5].

The implementation of the automated data comparison process described above is not the only embodiment possible. Rather, it is provided as an example of how the data from the TMD under test can be compared with ARV data in a partially or fully automated manner by applying modern technology. In fact, other embodiments have been developed using laser sensors, a relational database, and a video acquisition system to generate the ARV data [6–8].

9.3.3.8 Interpretation of test results and report

All specified data collection, data processing features, and options for the TMD under test shall be demonstrated to function properly before the TMD is accepted. If any specified TMD data item is not output or its difference as calculated in Section 9.3.3.6 (*with the addition of a confidence interval if so desired by the purchaser and seller*) exceeds the specified tolerance, declare the TMD nonfunctional or inaccurate and record that it failed the Type-Approval Test.

Whether or not the TMD fails or passes the Type-Approval Test, the purchaser or his representative shall prepare a written report, which documents the test result, all device settings, test conditions and duration, drawings and photographs that illustrate the location of the TMD under test with respect to the traffic flow direction and devices used to acquire ARV data, detection areas of the TMD and the devices used to acquire ARV data overlaid on the road surface, ARV data, and TMD output data used to determine the test result. A copy of the report shall be furnished to the purchaser and seller.

9.3.4 Summary of procedure for conducting Onsite Verification Test

The Onsite Verification Test is intended for TMDs that have previously passed the more rigorous Type-Approval Test. It determines whether the production version of a TMD installed at a particular site meets the performance and purchaser requirements identified in the purchase order and sensor specification. It usually takes less time to conduct than the Type-Approval Test. The conditions under which the Onsite Verification Test is performed are different than those in the Type-Approval Test.

9.3.4.1 Approval of site and test conditions

Both the purchaser and the seller approve the onsite verification test site and the TMD installation prior to the start of the Onsite Verification Test.

9.3.4.2 Duration of Onsite Verification Test

The Onsite Verification Test continues until the required numbers of measurements defined in Table 9.4 are obtained.

9.3.4.3 Onsite Verification Test method

The Onsite Verification Test is conducted by the purchaser in cooperation with the seller or their designated representatives. The following steps are required for each instrumented lane.

Table 9.4 Summary of procedures for obtaining ARV data during an Onsite Verification Test

Data item	Procedure[a]
Axle count	Relies on two or more human observers to record the number of axles on a data sheet prepared by the user.[b] A minimum of 50 axles shall be counted.
Vehicle count	Relies on two or more human observers to record the number of vehicles on a data sheet prepared by the user.[b] It is preferable to use correct detection, false detection, and missed detection rather than aggregate vehicle count over a time interval when evaluating vehicle count accuracy. A minimum of 50 vehicles shall be counted.
Vehicle speed	Utilizes a microwave radar or lidar speed gun operated by trained personnel to measure the speed of a vehicle as it passes through the effective detection area of the TMD under test. Speed gun values are entered on a data sheet prepared by the user. A minimum of 50 vehicles shall have their speeds measured.
Vehicle classification	Relies on two or more human observers to record the class of vehicles on a data sheet prepared by the user.[b] A minimum of 50 vehicles among all observed classes shall be included in the test.
Vehicle presence	Relies on two or more human observers to record the presence of vehicles on a data sheet prepared by the user while they observe the mix of vehicles passing through the effective detection area of the TMD under test.[b] A minimum of 50 vehicles shall be included in the test.
Lane occupancy	Use same procedures as in 9.3.3.5, Step 6. A minimum of 50 vehicles shall be included.

[a] The detection area of the TMD under test is marked as in 9.3.3.5 when acquiring ARV data. Data sheets used to record reference value data contain, as a minimum, the following information: TMD identifier; type of data acquired; test date; test start and end times; weather and lighting conditions; road description (number of lanes and their widths, road surface type and condition, grades, pertinent bridge and tunnel information); location of TMD under test and its detection area with respect to the roadway; pertinent TMD installation criteria; ARV data; names, affiliation, and contact information for data recorders; signature of data recorders at conclusion of test.

[b] When axles or vehicles are counted or when vehicles are classified or their presence noted by human observers, the observers shall record data in no more than 15-min intervals before taking a rest of at least 5 min.

1. The seller calibrates the TMD under test using the procedure developed for meeting the requirements of 9.3.3.2.
2. The purchaser or his representative installs the TMD according to the procedures identified in 9.3.3.4, Step 1.
3. The purchaser configures the TMD in accordance with the manufacturer's requirements.
4. The purchaser records TMD output data for vehicle flow rates, vehicle classes, and applicable environmental factors that apply at the selected test site.
5. While acquiring the TMD output data referred to above, the purchaser simultaneously acquires reference value data according to the procedures described in Table 9.4.

9.3.4.4 Tolerance compliance calculation

Same requirements as in 9.3.3.6.

9.3.4.5 Interpretation of test results and report

Same requirements as in 9.3.3.8.

9.4 SUMMARY OF ASTM SPECIFICATIONS

The ASTM specifications and standards are a worthwhile tool for ensuring that a traffic management agency obtains sensors that will meet its needs. The testing procedures are complex, and they require time, personnel, and funds to execute. Any shortcomings in them can be remedied by mutual agreement between buyer and seller. Methods other than those described in Section 9.3.3.5 may be utilized to obtain ARV data. Of course, it is possible for one agency to learn from another's test results. Therefore, consulting published sensor evaluation reports and journal articles is a valuable effort in which agencies should engage when they are considering sensor purchases.

9.5 BRIEF TUTORIAL ABOUT CONFIDENCE INTERVALS

The material in this section is intended to provide the reader with a basic knowledge of confidence intervals and confidence levels. The topics discussed are unbiased estimators, the normal distribution, and confidence intervals.

9.5.1 Estimating statistics of a population

Suppose we have a population and we want to draw conclusions about it from a random sampling of members of the population. The sample mean $\bar{x}$ is an unbiased estimator of an unknown population mean μ if the samples are random and represent the entire population. Under these circumstances, the standard deviation of the sample mean σ_x (sometimes called the standard error) is given by

$$\sigma_x = \frac{\sigma}{\sqrt{n}}, \tag{9.6}$$

where σ is the standard deviation of the entire population and n is the sample size.

The standard deviation of the sample mean is smaller than the standard deviation of the entire population σ since the standard deviation of the sample mean is obtained by dividing the standard deviation of the population by the square root of the number of observations in the sample.

If the random variables that characterize the population are normally distributed, then there is approximately a 68% probability that the sample mean is within ± 1 standard deviations of the population mean, approximately a 95% probability that the sample mean is within ± 2 standard deviations of the population mean, and approximately a 99.7% probability that the sample mean is within ± 3 standard deviations of the population mean as illustrated in Figure 9.3 [9].

Now let us discuss the notions of sample mean, confidence interval, and margin of error if we have random variables that are normally distributed. Suppose the mean score of a "standardization group" on an aptitude test is 500 and the standard deviation is 100. The scale is maintained from year to year, but the mean in any year can be different than 500.

We want to estimate the mean test score for more than 250,000 students using a sample of test scores from 500 students. Accordingly, the test is given to a random sample of 500 students, who get a mean score of 461. What can be said about the mean score of the total population of 250,000?

The sample mean $\bar{x}$ is equal to 461 and the standard deviation of the sample mean is equal to $100/\sqrt{500} \approx 4.5$. Thus, we can say we are 95% confident that the unknown mean score for the 250,000 students lies between $\bar{x} - 9 = 461 - 9 = 452$ and $\bar{x} + 9 = 461 + 9 = 470$.

The interval of numbers $\bar{x} \pm 9$ is the 95% *confidence interval* for μ where μ is population mean. The *margin of error* is ± 9.

Figure 9.4 describes the interpretation of a 95% confidence interval in repeated sampling. The center of each interval is marked by a dot and the arrows span the confidence interval.

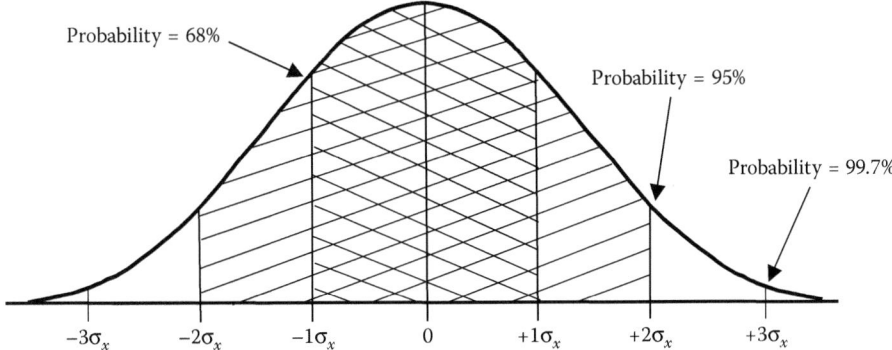

Figure 9.3 Normal distribution.

For a large number of samples, 95% of the confidence intervals will contain μ. Hence, one can be 95% confident that an interval built around a specific sample mean will contain the population mean [10]. In the example in Figure 9.4, all except 2 of the 30 intervals include the true value of the population mean μ.

9.5.2 Confidence intervals

Confidence intervals have two aspects: the interval computed from the data and the confidence level that gives the probability that the method produces an interval that includes

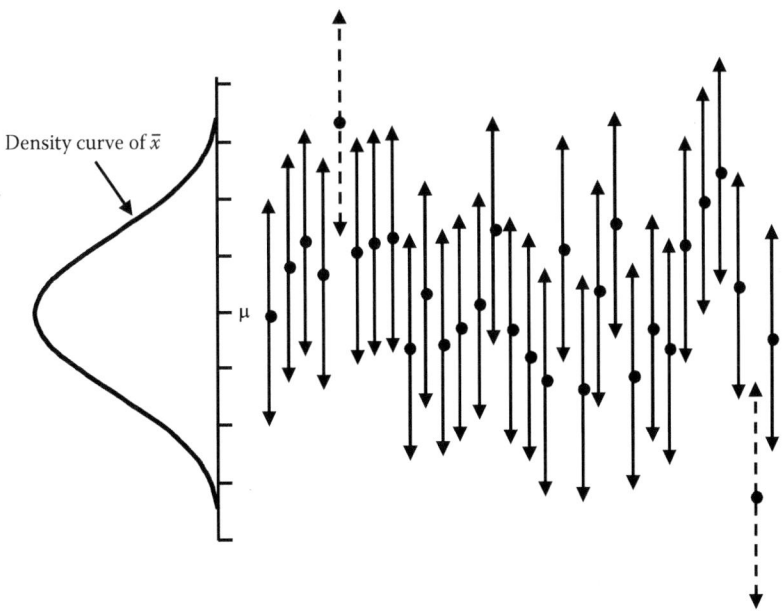

Figure 9.4 Interpretation of a 95% confidence interval when 30 samples are obtained from the same population. For a large number of samples, 95% of the confidence intervals will contain the population mean μ. In this illustration, all except 2 of the 30 intervals (indicated by dashed lines) include the true value of the population mean.

the parameter of interest. Most often, a confidence level greater than or equal to 90% is selected. If C is the confidence level in decimal form, then a level C confidence interval for a parameter θ is an interval computed from sample data by a method that has probability C of producing an interval containing the true value of θ.

For example, suppose it is desired to find a level C confidence interval for the mean μ of a population from an unbiased random data sample of size n. The confidence interval is based on the sampling distribution for the sample mean $\bar{x}$, which is equal to $N(\mu, \sigma/\sqrt{n})$ when the sample is obtained from a population having the $N(\mu, \sigma)$ distribution. In this notation, N represents a normal distribution, μ the mean of the entire population, and σ the standard deviation of the entire population. The central limit theorem confirms that a normal distribution is a valid representation of the sampling distribution of the sample mean when the sample size is sufficiently large regardless of the probability density function that describes the statistics of the entire population.

The construction of a 95% confidence interval is based on the observation that any normal distribution has probability 0.95 that the true value of the population mean lies within two standard deviations of the sample mean. A confidence level C (where C is expressed in decimal form) must include the central area C under the normal curve. To ensure that this area is captured by the confidence level, a number z^* is found such that there is a probability C that a sample from any normal distribution falls within z^* standard deviations of the distribution's mean. The number z^* is listed in tables of standard normal probabilities such as the summary given in Table 9.5 [11].

The value z^* for confidence C encompasses the central area C between $-z^*$ and $+z^*$, thus omitting the area $1 - C$. Half the omitted area lies in each tail. Because the area corresponding to z^* has area $(1 - C)/2$ to its right under the standard normal curve, it is called the upper $(1 - C)/2$ or p critical value of the standard normal distribution. For example, if $C = 0.95$, there is a $(1 - 0.95)/2$ or 2.5% chance that the true population mean is more than two standard deviations larger than the sample mean and an equal probability that it is more than two standard deviations lower than the sample mean. In this case, z^* equal to 1.960 is the upper 2.5% critical value for the standard normal distribution.

9.5.3 Confidence interval for a population mean

If the samples are randomly selected and unbiased, come from a normally distributed unstratified population, and contain no outliers (i.e., no individual observations that fall well outside the overall pattern of the data), then the confidence interval is found as follows.

Table 9.5 z^* and p critical values for selected confidence levels

Confidence level (%)	p critical value: $(1 - C)/2$	z^*
90	0.05	1.645
95	0.025	1.960
96	0.02	2.054
98	0.01	2.326
99	0.005	2.576
99.5	0.0025	2.807
99.8	0.001	3.091
99.9	0.0005	3.291

Under the stated conditions, the sample mean $\bar{x}$ has a normal distribution $N(\mu, \sigma/\sqrt{n})$, and the probability is C that $\bar{x}$ lies between

$$\mu - z^* \frac{\sigma}{\sqrt{n}} \quad \text{and} \quad \mu + z^* \frac{\sigma}{\sqrt{n}}, \tag{9.7}$$

or equivalently that the unknown population mean μ lies between

$$\bar{x} - z^* \frac{\sigma}{\sqrt{n}} \quad \text{and} \quad \bar{x} + z^* \frac{\sigma}{\sqrt{n}}. \tag{9.8}$$

Restated, there is a probability C that the interval $\bar{x} \pm z^* \sigma/\sqrt{n}$ contains μ. Therefore, the desired confidence interval is $\bar{x} \pm z^* \sigma/\sqrt{n}$. The estimator of the unknown μ is $\bar{x}$ and the margin of error M is

$$M = z^* \frac{\sigma}{\sqrt{n}}. \tag{9.9}$$

Thus, the sample size n needed to obtain a confidence interval with a specified margin of error M is

$$n = \left(\frac{z^* \sigma}{M} \right)^2, \tag{9.10}$$

assuming randomly selected and unbiased samples, a normally distributed unstratified population, and no outliers. The requisite sample size increases as the desired level of confidence increases, dispersion of the sample data increases, and the allowable error decreases. The size of the entire population does not influence the sample size as long as the population is much larger than the sample. The confidence interval is exact when the population distribution is normal and is approximately correct for large n for other distributions by application of the central limit theorem.

There is a trade-off between the confidence level and the margin of error. To obtain higher confidence from the same data requires acceptance of a larger margin of error. Thus, it is more difficult to arrive at the exact value of the mean μ of a highly variable population, which is why the margin of error of a confidence interval increases with σ. The selected confidence interval depends on the usage of the data (e.g., vehicle detection and tracking, incident detection, traffic signal actuation, vehicle counting, average vehicle speed measurement, or historical data collection).

The margin of error in a confidence interval indicates the error expected from chance variation in randomized data production. When random samples are not obtained because of omission of some affected groups from the sample data or nonresponse from some groups, additional errors are introduced that may be larger than the random sampling error. If the population is not normal and contains extreme outliers or is strongly skewed, the confidence level will be different from C.

9.5.4 The *n*-sigma dilemma

Now that the concepts of a confidence interval and confidence level have been reviewed, let us examine their impact on a user of a TMD or sensor and find out why the user must specify not only the accuracy but also the confidence level of the measured data. Often, the user desires a measurement to $\pm3\sigma$ accuracy (99.7%), but the manufacturer of the device specifies the accuracy at only a $\pm1\sigma$ or $\pm2\sigma$ level, and often does not include that bit of information on the specification sheet for the device. Thus, it is critical to let a vendor know the confidence level that is associated with the specified sensor accuracy. For example, if a sensor accuracy of 98% is required at a 99.7% confidence level, then the specification must include either that statement or its equivalent, 98% accuracy with a $\pm3\sigma$ confidence interval.

9.6 INTEROPERABILITY

Sometimes, devices are purchased without considering whether they have the interfaces needed to operate properly in the system. These interfaces take many forms, for example, software, connectors and other hardware, power, environmental, and data transfer protocols and standards that are supported.

One definition of interoperability is the "Ability of systems to provide services to and accept services from other systems and to use the services so exchanged to enable them to operate effectively together" [12]. Interoperability addresses technical, procedural, and institutional risks and barriers to the successful deployment of interoperable systems, including intelligent transportation systems, and engages in activities to mitigate risks and remove barriers. It does this by determining potential solutions and provides advice for users and others to achieve efficiencies and economies through the incorporation of selected standardized interfaces, while encouraging a collaborative process to address institutional and procedural issues. The discussions concerning collocated TMC operations, automated and connected vehicle development, the systems engineering process, and National ITS Architectures in Chapters 2, 11, 12, 13, and 14, respectively, address many of these issues. It is often institutional barriers rather than technical issues that limit deployment or impede development of interagency projects. Mutually beneficial discussions among agencies are often needed to address policies, funding mechanisms, and processes that will ensure effective implementation, operation, and maintenance of the system. Some specific tools that prove helpful are agreements and memoranda of understanding to establish policies concerning joint operations and information sharing, identification of funding for initial and sustained operations, an identified champion, executive buy-in and commitment, a documented organizational structure, defined roles and responsibilities, involvement of all stakeholders, and external and internal marketing, outreach, and education.

Technical interoperability deals with the capability of hardware elements to communicate. Procedural interoperability concerns the adoption of common procedures and common data element definitions to facilitate the exchange of meaningful information. Systems and organizations must work effectively at all three levels to be truly interoperable.

Consider the technical interoperability among sensors having different data reporting intervals as described in Table 9.6. If it is required to compare the vehicle count outputs of the presence-detecting microwave radar with that of video detection system (VDS) 3, it is necessary to calculate the average value of two data samples acquired from the VDS over 10 s to compare with the one sample from the radar that is obtained after 10 s. Similarly, if the vehicle count output of VDS 2 is to be compared with the output from VDS 3, it is necessary to compute the average value of 12 samples from VDS 3 to compare with one sample from VDS 2 after 1 min. A related consideration is the requirement that the software used by an operator of a traffic management system be able to read the serial data output of the sensors. Therefore, having a standard data transfer protocol for all sensor manufacturers to adhere to is beneficial.

Table 9.6 Technical interoperability example using data and update intervals of sensors with RS-232 interfaces

Sensor	Update interval	Count	Lane occupancy	Speed	Vehicle type[a]
Presence-detecting microwave radar	10 s to 10 min[b]	•	•	•	
Microwave Doppler sensor	Per vehicle	•		•	
VDS 1	10 s to 1 h	•	•	•	•
VDS 2	1 min	•	•	•	•
VDS 3	5 s	•		•	•
VDS 4	Per vehicle	•		•	•
VDS 5	Per vehicle	•	•	•	•
Infrared VDS	1 s	•		•	

[a] Based on user-selected vehicle lengths.
[b] User selected in 10-s increments.

Another example concerns early DSRC standards developed for tolling operations where several methods were used by manufacturers of in-vehicle toll tags to communicate with roadside readers. The differences were in the active or passive nature of the toll tags and the data transfer protocols. Currently, there are efforts to standardize the tag types and data transfer protocols in the United States to enable vehicles from different tolling jurisdictions to use their tags elsewhere and in connected vehicle applications. Similar efforts exist in the European Union.

9.7 SUMMARY

Methods for specifying, testing, and evaluating the performance of traffic flow sensors have been described. It was emphasized that using aggregate vehicle counts over an extended time period as a metric to verify sensor accuracy is not the best approach. It is far better to employ a technique that allows the identification of the numbers of correct detections, false detections, and missed detections. Two specifications, one for sensor performance as could be referenced in a purchase order and another for testing the delivered product, were discussed. A shortcoming in the testing specification, namely, the lack of inclusion of the concept of confidence level, was pointed out and its importance was illustrated. Finally, the notion of interoperability was introduced at three levels: technical, procedural, and institutional. This is an area that is addressed by a variety of agencies, disciplines, and applications such as traffic management and emergency operations centers that have collocated operations with different organizations; systems engineering; and the Connected Vehicle Program in particular, and is critical to the successful deployment of automated and connected vehicles.

REFERENCES

1. *Standard Specification for Highway Traffic Monitoring Devices*, E 2300-09, ASTM International, 100 Barr Harbor Drive, PO Box C700, West Conshohocken, PA, July 2009.
2. L.A. Klein and M.R. Kelley, *Detection Technology for IVHS, Volume I: Final Report*, FHWA-RD-95-100, Federal Highway Administration, U.S. Department of Transportation, Washington, DC, December 1996. http://ntl.bts.gov/lib/jpodocs/repts_te/6184.pdf.
3. Cambridge Systematics, Inc. and C.A. MacCarley, *Caltrans Statewide Detection Plan— Detection System Testing*, California Department of Transportation, Sacramento, CA, September 2008.

4. *Standard Test Methods for Evaluating Performance of Highway Traffic Monitoring Devices*, E2532-09, ASTM International, 100 Barr Harbor Drive, PO Box C700, West Conshohocken, PA, July 2009.
5. C.A. MacCarley, *Video Vehicle Detector Verification System (V2DVS) Operators Manual—Revision 9*, California Polytechnic State University, San Luis Obispo, CA, March 2012.
6. *Performance Evaluation on ITS Devices in Korea*, Korea Institute of Construction Technology, Goyang, South Korea, 2008.
7. J. Jang and S. Kim, *Evaluation of Traffic Data Accuracy in the Korea Detector Testbed*, Korea Institute of Construction Technology, Goyang, South Korea, June 2009.
8. J. Jang and S. Byun, Evaluation of traffic data accuracy using Korea Detector Testbed, *IET Intelligent Transport Systems*, 5(4):286–293, December 2011.
9. L.A. Klein, *Sensor and Data Fusion: A Tool for Information Assessment and Decision Making*, Second Edition, Press Monograph 222, SPIE, Bellingham, WA, 2012.
10. Confidence Intervals and Sample Size. Chapter 7. https://www.scribd.com/doc/31682662/ch07. Accessed January 13, 2017.
11. D.S. Moore and G.P. McCabe, *Introduction to the Practice of Statistics*, Fourth Edition, W.H. Freeman and Company, New York, NY, August 2002.
12. ISO/TC204, ITS-America 1997 Interoperability Workshop, George Mason University, Fairfax, VA.

Chapter 10

Alternative sources of navigation and traffic flow data

Most travelers, whether in a vehicle, walking, or bicycling, have a device, such as a cellular telephone, with them that contains a global navigation satellite system (GNSS). Installing roadside readers that receive the Bluetooth®-transmitted signals from the device makes it possible to retrieve the media access control (MAC) hardware number or address of the device and, hence, enables the anonymous tracking of automobiles, trucks, buses and other transit vehicles, cyclists, and pedestrians. Such a rich data source can provide link traffic volumes, travel times and speeds, and origin–destination pairs useful for determining the need and locations for future roads and possibly even transit routes, incident detection, traffic signal timing adjustments, and travel route and mode advisories.

GNSS applications are evolving and becoming more prevalent in many countries as the types and numbers of mobile devices with GNSS functionality increase. In the United States, for example, the Global Positioning System (GPS) supports tracking of transit vehicles; taxis; hazmat, police, fire, and paramedic service vehicles; street and highway work zone vehicles and personnel; tree harvesters in forests; snow plows; and commercial vehicles. GPS is also utilized by package delivery services to ensure timely delivery of merchandise and efficient operations, and city, county, state, and national agencies to track service vehicles, monitor search and rescue efforts, and enable future air traffic control systems. Tracking of vehicles and pedestrians in real time is synergistic with the U.S. Department of Transportation (DOT) Connected Vehicle Program and the European Union Cooperative Intelligent Transport Systems Program (described in Chapter 12), enabling such functions as green signal phase and walk time extensions, approaching vehicle warnings, and work zone advisories. In urban areas, where tall buildings often make GPS reception difficult, devices that use combinations of GPS and inertial navigation system (INS) technology may be beneficial.

Following descriptions of GNSS developed by or under development by other countries, this chapter describes the U.S. GPS, INS, and Bluetooth technologies and examines several of their applications.

10.1 GLOBAL NAVIGATION SATELLITE SYSTEMS

Russia, the European Union countries, China, India, and Japan (discussed in Section 10.4) are developing and launching GNSS constellations similar to those in the U.S. The Russian Global Navigation Satellite System (GLONASS) consists of 27 satellites, 24 operational that are more useful in northern latitudes. Smartphone manufacturers are increasingly exploiting a combination of U.S. GPS and GLONASS to improve the tracking accuracy of their devices.

The European Union is constructing the Galileo GPS system to provide a civilian-controlled global positioning service. Galileo contains 30 satellites (24 operational with six in-orbit spares) positioned in three circular medium Earth orbit (MEO) planes at a 23,222-km altitude above the Earth, with an orbital plane inclination of 56° to the equator. Initial services began in December 2016, with system completion scheduled for 2020. At completion, the Galileo navigation signals will offer good coverage at latitudes up to 75° north, which corresponds to Norway's North Cape, the most northerly tip of Europe, and beyond. Galileo is interoperable with GPS and GLONASS. By offering dual frequencies as standard, Galileo promises real-time positioning accuracy down to the meter range.

China's constellation of 35 satellites, called the BeiDou-2 Navigation Satellite System (formerly known as COMPASS), has been under construction since January 2015. It is currently operational in China and the Asia-Pacific region with 22 satellites in use as of March 2016. The satellite constellation includes five geostationary orbit satellites for backward compatibility with BeiDou-1, and 30 non-geostationary satellites (27 in MEO and 3 in inclined geosynchronous orbit), which will offer complete coverage of the globe by 2020. The precision of the system is 10 m public and 0.1 m encrypted.

The Indian Regional Navigation Satellite System (IRNSS) was developed by the Indian Space Research Organization (ISRO). It provides accurate position information to users in India and to a primary region extending up to 1500 km from its boundary. An extended service area lies between the primary service area and an area enclosed by the rectangle from latitude 30° south to 50° north and longitude 30° east to 130° east. IRNSS offers a standard position service to the public and a restricted service to authorized users such as the military. The system consists of a constellation of seven satellites, three located in suitable geostationary orbital slots and four in geosynchronous orbits, with inclination and equatorial crossings in two different planes. In April 2016, with the last launch of the constellation's satellite, IRNSS was renamed the Navigation Indian Constellation (NAVIC).

10.2 U.S. GPS ARCHITECTURE

Figure 10.1 illustrates the three GPS segments, space, control, and user, for the U.S. GPS. The space segment consists of a constellation of NAVSTAR GPS satellites transmitting radio

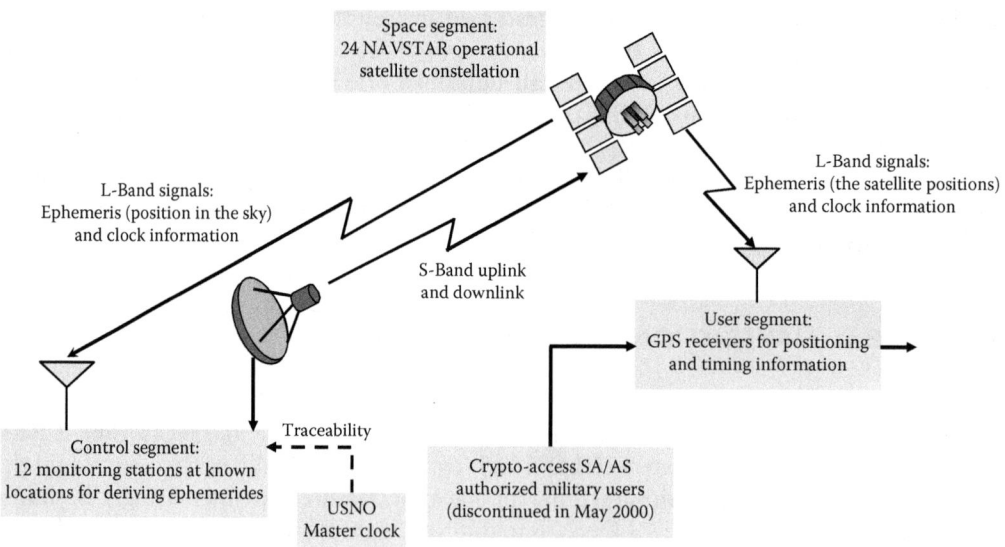

Figure 10.1 U.S. GPS architecture.

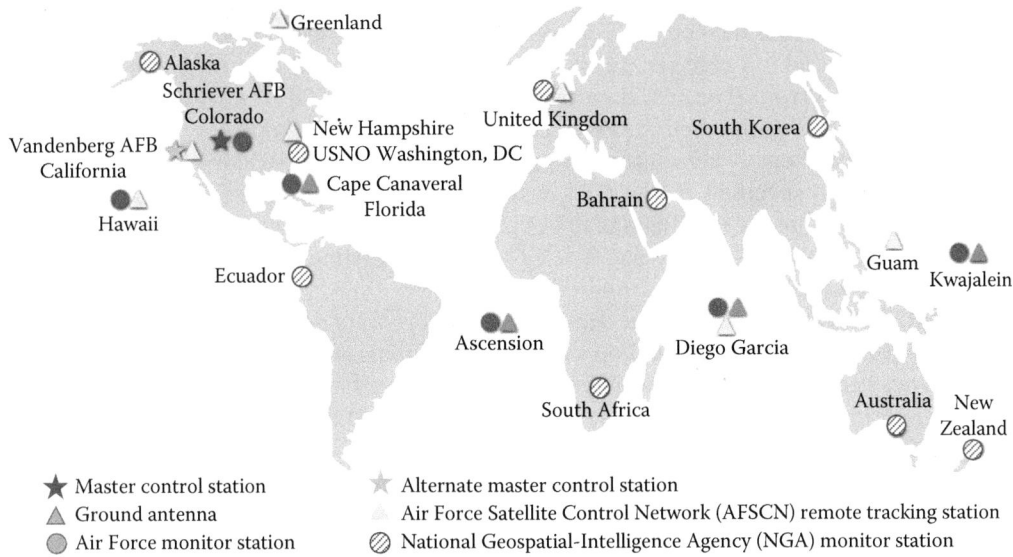

Figure 10.2 GPS control segment. (From National Coordination Office for Space-Based Positioning, Navigation, and Timing. http://www.gps.gov/systems/gps/control/)

signals to users. The system is committed to maintaining the availability of at least 24 operational satellites, 95% of the time. To ensure this commitment, the U.S. Air Force now flies 31 operational GPS satellites in MEO at an altitude of approximately 20,200 km (12,550 mi). Each satellite circles the Earth twice a day. The satellites in the GPS constellation are arranged into six equally spaced orbital planes surrounding the Earth. Each plane contains four slots occupied by baseline satellites. This 24-slot arrangement ensures users can view at least four satellites from virtually any point on the planet. Signals from the four satellites permit the x, y, and z coordinates of the GPS user device and the receiver clock biases to be calculated.

The GPS control segment consists of a global network of ground facilities that track the GPS satellites, monitor their transmissions, perform analyses, and send commands and data to the constellation. As of April 2016, the operational control segment that appears in Figure 10.2 includes a master control station in Colorado Springs, Colorado, an alternate master control station at Vandenberg Air Force Base in California, four command and control ground antennas spaced around the globe, Air Force Satellite Control Network (AFSCN) remote tracking stations, and a combination of 15 Air Force and National Geospatial-Intelligence Agency (NGA) monitoring sites around the Earth, including one at the Naval Observatory (NO) in Washington, DC [1]. A monitoring station can track up to 11 satellites at a time. The known location of the monitoring stations is exploited to correct errors in the satellite's orbit and clock that would otherwise degrade the position calculated by GPS receivers in the area served by the monitoring station. The user segment is simply the totality of GPS devices carried or otherwise utilized by the consumers of the service.

10.3 GPS ACCURACY

A standard, inexpensive (~U.S. $200–$400) single-frequency GPS receiver tracks the code signal of the NAVSTAR constellation at the nominal L1 coarse/acquisition (C/A) frequency of 1575 MHz. The transmission time from the known satellite locations provides a pseudo-range measurement that depends on receiver location, difference in clock time between the satellite and the tracking receiver caused by special and general relativity effects [2], and atmospheric delay errors. The actual measurement given by the GPS receiver is an integrated phase from the first epoch when the receiver began tracking the signal.

The U.S. government is committed to providing GPS to the civilian community at the performance levels specified in the GPS Standard Positioning Service (SPS) Performance Standard. For example, the GPS signal in space will provide a worst case pseudo-range accuracy of 7.8 m at a 95% confidence level. This is not the same as user accuracy as pseudo-range is the distance from a GPS satellite to a receiver.

The actual accuracy users attain depends on factors outside the government's control, including diffraction-induced bending of the radio signals as they propagate through the atmospheric, sky blockage, and receiver quality. Real-world data from the Federal Aviation Administration (FAA) show that their high-quality GPS SPS receivers afford better than 3.5-m horizontal accuracy.

10.3.1 GPS operation

To calculate a receiver's x, y, and z position coordinates, the system automatically measures the distances to the satellites, obtains satellite positions (ephemerides), performs triangulation calculations, and compensates for local clock bias. The satellite position data are broadcast in a fine and coarse resolution mode, namely, ephemeris and almanac. The ephemeris data measure the precise distance to the satellite, while the almanac data contain course orbital parameters for all satellites in the constellation. Ephemeris data are exact orbital and clock corrections for each satellite and are required to calculate the precise position of the satellite. Each satellite broadcasts only its own ephemeris data every 30 s. Ephemeris data are only valid for about 30 min.

Each satellite also broadcasts almanac data for all satellites. Almanac data are not precise and are valid for up to several months. Each frame contains a part of the almanac and the complete almanac is transmitted by each satellite in 25 frames total, requiring 12.5 min.

The almanac serves several purposes. The first applies the almanac's coarse orbit and status information for each satellite in the constellation to assist in acquiring satellites at power-up. The list of visible satellites generated by the GPS receiver is based on stored position and time data, while an ephemeris from each satellite is needed to compute position fixes using that satellite. In older hardware, lack of an almanac in a receiver operating for the first time would cause long delays before providing a valid position because the search for each satellite was a slow process. Advances in hardware design have made the acquisition process much faster, so not having an almanac is no longer an issue. The second purpose utilizes the ionospheric model to correct a single-frequency receiver for ionospheric delay error by using a global ionospheric model. The corrections are not as accurate as augmentation systems or dual-frequency receivers. However, it is often better than no correction, since ionospheric error is the largest error source for a single-frequency GPS receiver. Finally, the almanac contains information that relates GPS-derived time to coordinated universal time (UTC), the primary time standard through which the world regulates clocks and time.

10.3.2 GPS error sources

Several error sources that affect GPS accuracy are displayed in Figure 10.3. Inaccuracies in reporting position occur with GPS since the position calculation assumes the radio

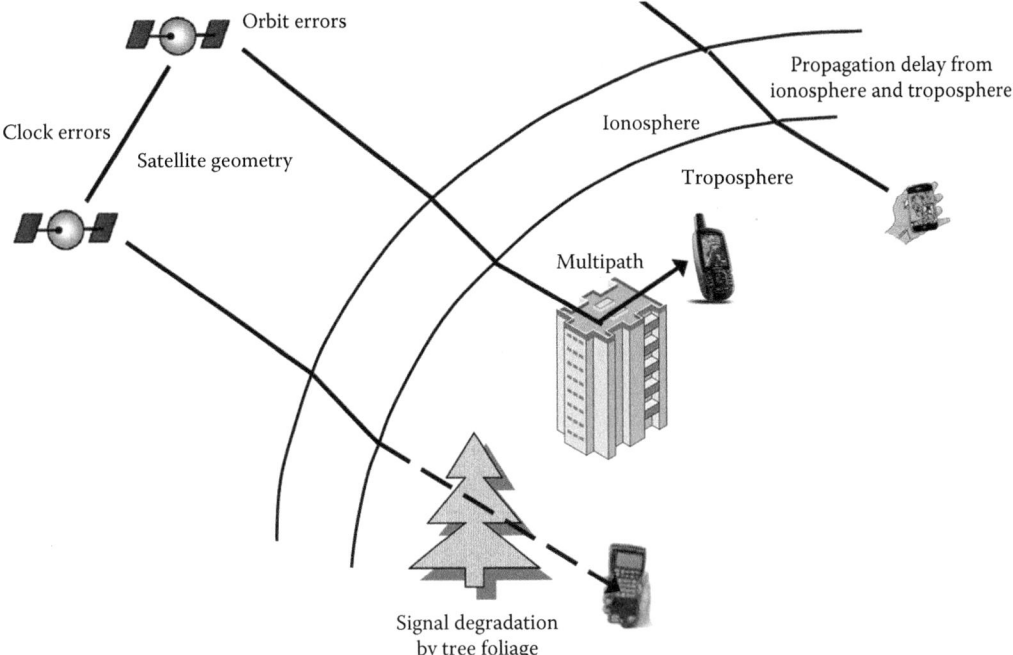

Figure 10.3 GPS error sources.

signals make their way through the atmosphere at a constant speed (the speed of light). In fact, the Earth's atmosphere slows the electromagnetic energy down somewhat, particularly as it goes through the ionosphere and troposphere. The delay varies depending on where you are on Earth, which means it is difficult to accurately factor this into the distance calculations. Problems can also occur when radio signals bounce off large objects, such as tall buildings, and hence travel to the receiver along a path that is longer than a direct path from satellite to receiver. This gives a receiver the impression that a satellite is farther away than it actually is. Additionally, satellites sometimes just send out bad almanac data, misreporting their own position. Other errors are caused by attenuation and scattering of the satellite signal, such as by tree foliage, buildings, or any other object in the line of sight.

10.4 GPS AUGMENTATION SYSTEMS

Higher accuracy is attainable by using GPS in combination with augmentation systems such as those described below [3]. These systems enable real-time positioning to within a few centimeters, and post-mission measurements at the millimeter level.

- Nationwide Differential GPS System (NDGPS) is a ground-based augmentation system that provides increased accuracy and integrity of GPS information to users on U.S. land and waterways. The system consists of the Maritime Differential GPS System operated by the U.S. Coast Guard and an inland component funded by the DOT. NDGPS is built to international standards, and similar systems have been implemented by 50 countries around the world.

- Wide Area Augmentation System (WAAS), a satellite-based augmentation system operated by the FAA, supports aircraft navigation across North America. Although designed primarily for aviation users, WAAS is widely available in receivers used by other positioning, navigation, and timing communities. The FAA is committed to providing WAAS service at the performance levels specified in the GPS WAAS Performance Standard [4] and is improving WAAS to incorporate the future GPS safety-of-life signal for even better performance. Other similar space-based augmentation systems include Japan's Multi-Functional Transport Satellite (MTSAT)-Based Satellite Augmentation System (MSAS) and Quasi-Zenith Satellite System (QZSS), the European Geostationary Navigation Overlay Service (EGNOS), and India's GPS And Geo-Augmented Navigation (GAGAN) system.
- The U.S. Continuously Operating Reference Stations (CORS) network, managed by the National Oceanic and Atmospheric Administration, archives and distributes GPS data for precise positioning tied to the National Spatial Reference System. Over 200 private, public, and academic organizations contribute data from over 1800 GPS tracking stations to CORS. The Web-based Online Positioning User Service (OPUS) offers free post-processing of GPS data sets to the centimeter level using CORS information. CORS is also being modernized to support real-time users.
- Global Differential GPS (GDGPS) is a high-accuracy GPS augmentation system, developed by the NASA Jet Propulsion Laboratory (JPL) to support the real-time positioning, timing, and determination requirements of NASA science missions. NASA plans to use the Tracking and Data Relay Satellite System (TDRSS) to disseminate via satellite a real-time differential correction message. This system is referred to as the TDRSS Augmentation Service Satellites (TASS).
- International GNSS Service (IGS) is a network of over 350 GPS monitoring stations from 200 contributing organizations in 80 countries. Its mission is to provide the highest quality data and products as the standard for GNSSs in support of Earth science research, multidisciplinary applications, education, and other applications benefiting society. Approximately 100 IGS stations transmit their tracking data within 1 h of collection.
- Similar in its goals to WAAS, the Quasi-Zenith Satellite System is a centimeter-scale GNSS being deployed by the Japanese to augment the U.S.-operated GPS. Under development by Mitsubishi Electric, QZSS addresses the degraded performance of current differential GPS (DGPS) in urban canyons where satellite views may be blocked and where resolution is not adequate for some applications [5]. The system will be capable of centimeter-scale horizontal and vertical position accuracies of about 1.3 cm horizontally and 2.9 cm vertically. It will provide navigation signals at L1 C/A, L1C, L2C, and L5 frequencies (defined in Section 10.5) that are predicted to improve the time percentage of positioning availability from 90% (GPS only) to 99.8% (GPS + QZSS).

    Applications for the QZSS include navigation and position data for connected and conventional vehicles, precision farming and construction equipment, unmanned aerial vehicles, and autonomous vehicles in general. The Japanese government and the European Union intend to connect their GPSs to speed up development of autonomous driving technologies as early as 2018. The link will be a common digital language that the systems will use to transmit information. This will allow driverless cars and automobile parts developed for the Japanese market to be shipped and used outside Japan [6].

    Four QZSS satellites are scheduled to be in place by end of 2017 with a total of seven satellites planned to furnish redundancy. The four-satellite orbit shown in Figure 10.4 traces an asymmetrical figure eight in the sky from the perspective of a person

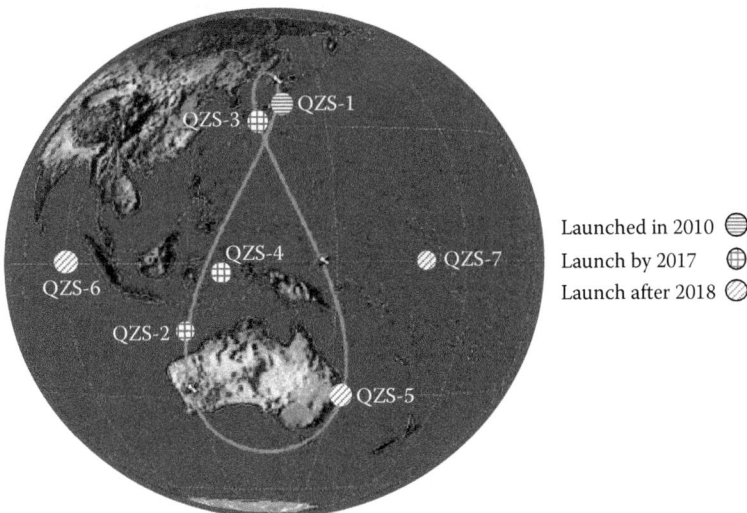

Figure 10.4 Quasi-Zenith Satellite System orbits. (Adapted from Quasi-Zenith Satellite System, Office of National Space Policy, Cabinet Office, Government of Japan, 2012. http://www.unoosa.org/pdf/icg/2012/icg-7/5.pdf.)

in Japan. Although the orbit extends as far south as Australia at its widest arc, it will narrow its path over Japan so that at least one satellite is always in view high in sky.

Errors are corrected using a master control center that compares the satellite's signals received by the reference stations with the distance between stations and the satellite's predicted location. The corrected signals are compressed from an overall 2 Mb/s data rate to 2 kb/s and transmitted to the satellite, which then broadcasts them to users' receivers in real time.

- Additional GPS augmentation systems are available worldwide, both governmental and commercial. These systems use differential, static, or real-time techniques. There are also systems that augment other GNSSs. The United States and other nations are cooperating to ensure the interoperability of international augmentation systems with GPS and U.S. GPS augmentations.

10.5 GPS MODERNIZATION PROGRAM

The GPS modernization program added new civilian signals and frequencies to GPS satellites, enabling ionospheric correction for all users and providing the technology needed to eliminate the accuracy difference between military and civilian systems [7]. Furthermore, dual-frequency receivers, those that receive the original L1 C/A 1575 MHz frequency and the more recent L2C 1227 MHz frequency, deliver faster signal acquisition, enhanced reliability, and greater operating range. L2C signals became available in 2005 and will be accessible on 24 GPS satellites around 2018. These transmissions broadcast at a higher effective power than the legacy L1 C/A signal, making it easier to receive under trees and indoors.

The third GPS signal L5 (1176 MHz) began launching in GPS satellites in 2010 and will be available on 24 GPS satellites around 2021. It will be used by aircraft in combination with L1 C/A to improve accuracy (via ionospheric correction) and robustness (via signal redundancy). In addition to enhancing safety, L5 use will increase capacity and fuel efficiency within U.S. airspace, railroads, waterways, and highways. Beyond transportation, L5 will provide users worldwide with the most advanced civilian GPS signal. When used in combination with L1 C/A and L2C, L5 will deliver a highly robust service. Through a technique called trilaning, the three GPS frequencies may enable sub-meter accuracy without augmentations, and very long-range operations with augmentations.

The fourth GPS signal L1C (1575 MHz) begins launching in 2017 with GPS III and is expected to be available on 24 GPS satellites in the late 2020s. It is designed to enable interoperability between GPS and international satellite navigation systems, and improve mobile GPS reception in cities and other challenging environments.

10.6 DIFFERENTIAL GPS

DGPS helps correct errors in the position calculation. The concept, depicted in Figure 10.5, is to compare the GPS-calculated location for a stationary base station with its known location. Since the DGPS hardware at the station already knows its own position, it can easily calculate its receiver's inaccuracy. The station then broadcasts a radio signal that provides signal correction information to all remote DGPS receivers in the area served by the base station. In general, access to the correction information makes DGPS receivers much more accurate than ordinary receivers.

The signal correction contains two pieces of information, the pseudo-range correction (PRC) and range-rate correction (RRC), which are transmitted to the remote receivers in near real time. The remote receivers apply the corrections to the measured pseudo-ranges and perform point positioning with the corrected pseudo-ranges.

10.7 GPS SPOOFING

In normal operation, GPS receivers deduce their position by calculating their distance from several satellites at once. Each satellite carries an atomic clock and broadcasts its location,

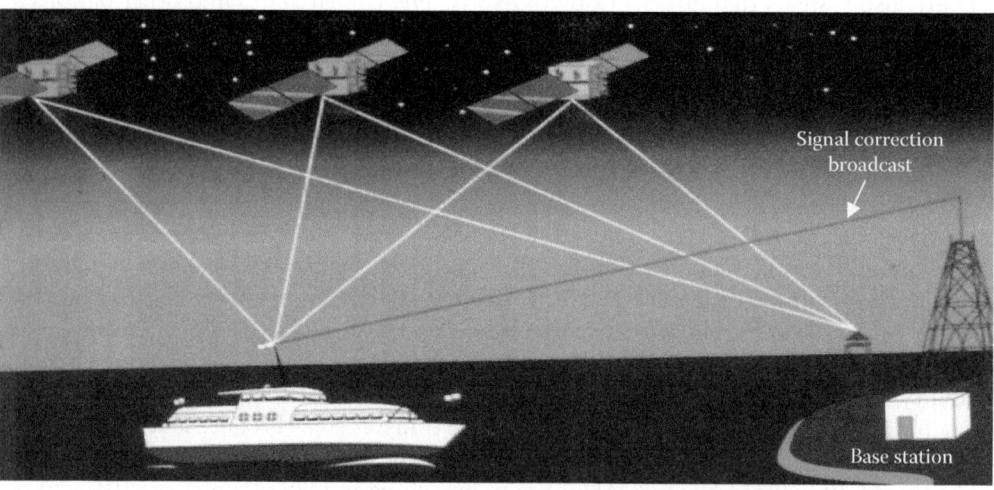

Figure 10.5 Three satellite pattern as received by a ship with DGPS. The monitoring station on the right side of the figure knows its location and can broadcast signal correction information to users in its vicinity.

the time, and a signature pattern of 1023 plus and minus signs known as a pseudorandom noise (PRN) code. These codes identify a signal as originating from, for instance, satellite A versus satellite B, which is necessary because all GPS satellites broadcast civilian signals on the same frequency. The PRN code patterns also repeat over time, and their distinctive arrangements of pluses and minuses enable GPS receivers to use them to determine the signal transmission delay between a satellite and the receiver. A receiver uses these delays, along with the satellite positions and time stamps, to triangulate its precise location. To get a good fix, a receiver must receive signals from four or more satellites at a time—it can figure coordinates based on just three, but it needs the fourth to synchronize its inexpensive, drift-prone clock with the constellation's precise atomic clocks.

10.7.1 Attack initiation

A GPS spoofer transmits false GPS signals, which to a navigation system are indistinguishable from real ones. To attack civilian receivers, a spoofer's operator determines which GPS satellites will be in the vicinity of the target at a given time based on the satellites' orbits. The spoofer then fabricates the PRN code for each satellite using formulas available in the public database. Next, the spoofer broadcasts faint signals carrying the same codes as all of the nearby satellites at once. The GPS receiver registers these weak signals as though they were part of the stronger, true signals transmitted by those satellites [8].

What follows is the delicate art of the "drag-off," in which attackers must gently override the true signals. To do this, the spoofer's operator gradually increases the power of the false GPS signals until the receiver latches onto these new signals. If the signal increase is too abrupt, the receiver or even the ship's human navigators might detect something amiss. Once the receiver has latched onto the false signals, the operator can adjust the spoofer and receiver to a new set of coordinates and leave the true signals behind.

10.7.2 Protection methods

Psiaki and Humphreys [8] report that there are three main ways to protect against GPS spoofing: cryptography, signal distortion detection, and direction-of-arrival sensing. No single method can stop every spoof, but a combination of strategies can provide a reasonably secure countermeasure that could be commercially deployed.

10.7.2.1 Cryptographic methods

Cryptographic methods present an approach for users to authenticate signals on the fly. In one method, civilian receivers use PRN codes that are totally or partially unpredictable, similar to those used by the U.S. military, so a spoofer cannot synthesize the codes ahead of time. But to verify each new signal, every civilian receiver would have to carry an encryption key similar to those held by military receivers, and it would be difficult to keep attackers from obtaining such widely distributed keys.

Alternatively, a receiver could simply record the unpredictable part of the signal and wait for its sender to broadcast a digitally signed encryption key to verify its origin. However, this approach would require the U.S. Air Force to revise the way GPS signals are broadcast and manufacturers of civilian receivers to change how those devices are built. It would also require a slight delay, which would mean that navigation updates would not be verified instantaneously.

An easier way to protect civilians is to have them piggyback off of the encrypted U.S. military signals. Military signals can already be received and recorded by a civilian receiver, although they cannot be decrypted and utilized for navigation. Once they record the signals, civilian receivers can observe the noisy trace of a PRN code even if they cannot decipher the actual code. That means these receivers could authenticate a civilian signal by looking for the trace of an encrypted military signal behind it. This strategy relies on a second civilian receiver at a secure location to verify what the trace should look like within the signal. Otherwise, a spoofer could generate a fake trace to accompany any civilian signal the operator wished to spoof.

The downside of cryptographic techniques is that they are all vulnerable to attacks by specialized systems that can intercept any signal, delay it, and rebroadcast it with more power, persuading a receiver to switch from the legitimate signal to the delayed one. Such gear, which is called a meacon, can use multiple antennas to add delays of different lengths. By tuning the lengths, the spoofer's operator can choose how he or she subverts a GPS receiver.

10.7.2.2 Signal distortion detection

Distortion detection can alert users to suspicious activity based on a brief but observable blip that occurs when a GPS signal is spoofed. Typically, a GPS receiver uses a few different strategies to track the spike of an incoming signal's amplitude. When a copycat signal is transmitted, the receiver sees a combination of the original signal and the false one, and this combination causes a blip in the amplitude profile during drag-off.

Distortion detection requires additional signal processing channels and, possibly, a modest amount of hardware so that users can track a signal's amplitude profile with greater precision. This technique looks for unnatural features—an amplitude spike beyond a certain height or width, for example. However, a distortion detector works only if it catches the signal between the beginning of the attack and the end of drag-off—a process that may last just a few minutes.

10.7.2.3 Direction-of-arrival sensing

Direction-of-arrival sensing was demonstrated at White Sands, New Mexico, but it required hours of off-line data processing to detect the spoof [8]. Direction-of-arrival sensing exploits the fact that a practical spoofer can be in only one place at a time. However, a spoofer transmits a false signal for each GPS satellite the operator wishes to imitate by fabricating the PRN codes for every satellite in the vicinity of a target. The catch is that the spoofer sends all those signals from a single antenna, and they arrive from the same direction. Authentic GPS signals, on the other hand, come from several satellites, and therefore from several angles. Hence, if you independently sense the direction from which each signal arrives, you can easily determine whether you are being spoofed.

To test this idea, Psiaki and Humphreys built a system that uses software and two antennas to apply interferometry principles to spoofing detection. It measures the carrier phase to discern how a signal varies from one antenna to the next, and then determines what that variation implies about the signal's angle of arrival. If the difference in carrier phase as measured between the detector's two antennas varied widely from satellite to satellite, the detection system knew the signals had arrived from multiple directions. But if the system detected little or no variance among carrier-phase differences, that meant it was receiving a set of signals coming from a single spoofer.

Using a GPS software radio, whose key components such as mixers, filters, modulators, and demodulators are implemented with software rather than hardware, enabled the real-time use of Psiaki's off-line code with only 6 s of delay. This approach proved effective in detecting and alerting the crew aboard a yacht to a simulated spoofing attack [8].

10.8 INERTIAL NAVIGATION SYSTEMS

Several studies have examined the viability of using GPS to track vehicles. As you may expect, vehicle tracking in urban areas degrades from multipath, signal degradation from foliage, and the inability to receive signals from the required three or four satellites because of blockage from tall buildings. Therefore, investigators explored the pairing of a GPS with an INS to augment the GPS when its signal is weak or unavailable. On the other hand, the GPS can be used to calibrate the location of the INS device, which is subject to errors caused by the integration of INS acceleration information into a position estimate. Before we examine the performance of the combined system, let us explore how an INS functions.

Inertial navigation is a self-contained navigation technique in which measurements provided by accelerometers and gyroscopes are used to track the position and orientation of an object relative to a known starting point, orientation, and velocity. An INS consists of an inertial measurement unit (IMU) and navigation computers. IMUs typically contain three orthogonal rate-gyroscopes and three orthogonal accelerometers, measuring angular velocity and linear acceleration, respectively. Position and orientation information is obtained by processing signals from these devices.

An INS applies Newton's first law of motion: an object at rest stays at rest and an object in motion stays in motion with the same speed and in the same direction unless acted upon by an unbalanced force. An INS estimates the object's position, velocity, and attitude based on the known initial states and the sensor measurements. The acceleration measurement a from the accelerometers must be converted into the required value of position or displacement x. Therefore, the acceleration measurement is integrated twice with respect to time to first obtain the magnitude of the velocity and then the displacement x as

$$v = \int_0^{t_1} a \, dt \tag{10.1}$$

and

$$x = \int_0^{t_1} v \, dt. \tag{10.2}$$

The integration process increases the effect of any errors present in measuring a on the calculated value of position x. Here's why.

If an object is moving under constant acceleration, its motion as a function of time is expressed as

$$a(t) = c_1, \tag{10.3}$$

where c_1 is a constant.

Since the acceleration is the derivative of a velocity, the velocity function can be found by integration as

$$v(t) = c_1 t + c_2, \tag{10.4}$$

where c_2 is another constant.

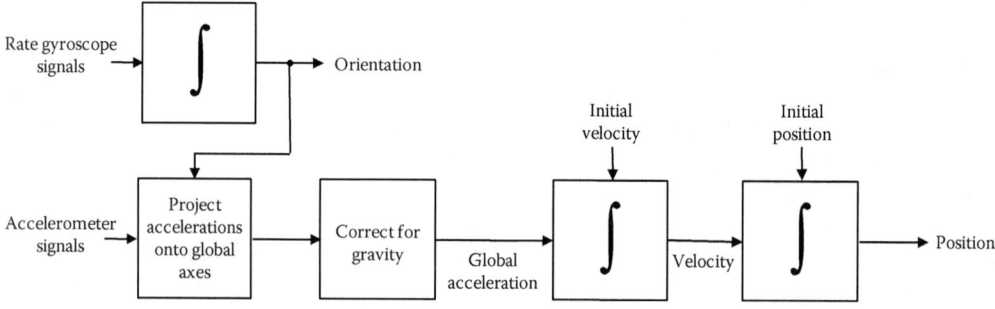

Figure 10.6 Strapdown INS.

Thus, any errors c_1 present in the measurement of a are multiplied by time t in the equation for velocity.

Another integration gives the displacement as

$$x(t) = c_1 t^2 + c_2 t + c_3, \tag{10.5}$$

where c_3 is another constant.

Here, the error in the measurement of a is multiplied by t^2 in the equation for displacement x. Therefore, the effect of any measurement error in a or drift of the accelerometer output is greatly magnified in the reported value of position. The strapdown INS algorithm in Figure 10.6 illustrates how an INS implements the position, velocity, and acceleration equations derived above.

10.9 GPS AND COMBINED GPS–INS POSITIONING SYSTEMS AND TEST PROCEDURE

GPS and INS operation complement each other and combine to create a robust system for detecting and tracking vehicles. GPS does not work well under tree canopies, near high-rise buildings, in tunnels, and in naturally occurring canyon areas, for example. On the other hand, an INS does not rely on an external reference, but its errors increase without upper limit as the INS device operates over time without periodic calibration inputs. Furthermore, an INS cannot be spoofed or interfered with (jammed) by outside sources. Hence, the GPS–INS integrated system removes the limitations of the individual GPS and INS by virtue of their combined operational strengths. The integrated system relies on the INS when the GPS is blocked, while the GPS helps maintain the INS alignment and calibration.

Integrated GPS–INS approaches to navigation are used by the military [9] and civilian communities. The following discussion relates to the application of an integrated system to vehicle location and tracking and the determination of the tracking accuracy of the individual and combined systems.

The GPS and combined GPS–INS devices shown in Figure 10.7 were placed in a vehicle driven around downtown Seattle and Bellevue, Washington [10]. The Trimble Pro XR is a 12-channel, real-time DGPS receiver operating on single frequency. The GPS–INS integrated system is the POS/LS model from the Applanix Corporation. The equipment was chosen based on popularity and availability of the models. These two devices were placed in a probe vehicle for the field test. The antennae for the two systems were attached to the roof of the test vehicle above the POS/LS device.

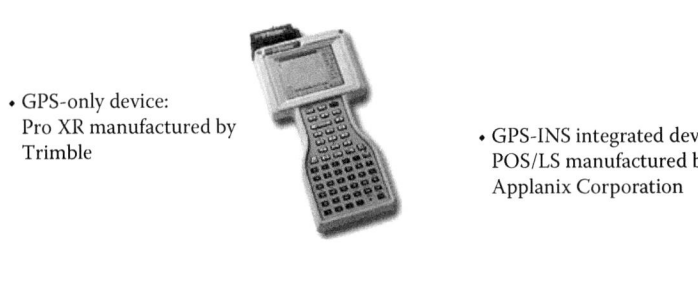

• GPS-only device:
Pro XR manufactured by
Trimble

• GPS-INS integrated device:
POS/LS manufactured by
Applanix Corporation

Figure 10.7 GPS-only and combined GPS–INS devices used in University of Washington tests. (Photographs courtesy of Trimble.)

10.9.1 Test routes

Three different types of test routes were selected to demonstrate the capabilities and limitations of the GPS and combined GPS–INS devices. The first contained a freeway loop through downtown Seattle and downtown Bellevue, the second a local street loop that crosses an urban canyon area and non-canyon area, and the third a local street loop with significant elevation change in an urban canyon area.

Maps in Figure 10.8 depict the three test routes. The first route was a closed loop of about 34 km composed of freeway sections of I-5, SR-520, I-405, and I-90. This route was selected because it passes through downtown Seattle and downtown Bellevue, and contains typical kinds of freeway canopies, including tunnels, overpasses, and bridges. Routes 2 and 3 were selected to analyze the effects of road surface altitude changes and high-rise buildings, respectively. They contain the local streets found in downtown Seattle. Route 2 is completely located in an urban canyon area with the road surface elevation increasing significantly from southwest to northeast. This route was used to evaluate the impact of road surface elevation on positioning accuracy. Route 3 traverses both an urban canyon

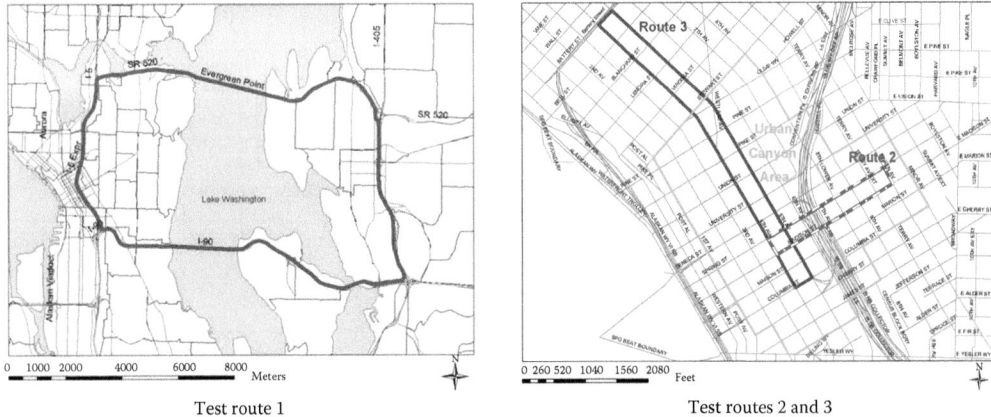

Test route 1   Test routes 2 and 3

Figure 10.8 GPS and GPS–INS test routes through Seattle and Bellevue, Washington. (From Smart Transportation Applications and Research [STAR] Laboratory, University of Washington, Seattle, WA.)

area and a non-canyon area. In the non-canyon area, there are fewer tall buildings than in the urban canyon area. Data from Route 3 were used to show the performance difference in the devices while in an urban canyon area and an urban non-canyon area.

10.9.2 Test method

The test method consisted of collecting vehicle location data at 1 Hz from both devices (Pro XR and POS/LS), mapping the collected location data into geographic information system (GIS) maps and visually examining the location accuracy, and finally statistically analyzing and comparing the performance of the two systems for location accuracy and location update interval.

10.9.3 Position accuracy analysis

The GPS and combined GPS–INS devices give the logged position of the vehicle at any time t. However, the vehicle may actually be elsewhere along the road because the devices are not 100% accurate as indicated in Figure 10.9.

To analyze the positioning accuracy of the devices, the error of each logged position had to be calculated. Since the exact location of the test vehicle at a particular time was unknown, calculating the exact position error was difficult. However, decomposing the tracking error into across-track error and along-track error allowed easy calculation of the across-track error, which was of interest for vehicle positioning.

Figure 10.10 illustrates the definitions of across-track error and along-track error. The across-track error was defined as the perpendicular distance obtained from the GPS or GPS–INS device's observed position to the corresponding street along which the test vehicle was traveling. The along-track error was defined as the distance from the projected point to the real position of the vehicle on the street. As the true position of the test vehicle at a particular time is unknown, along-track position error cannot be calculated. Therefore, the across-track error was used to estimate the position of the vehicle on the road.

When turning corners, a logged position may correspond to a true vehicle position on either of the roads. In such cases, the across-track error for each condition was calculated and the smaller one was chosen. The across-track error and the along-track error should

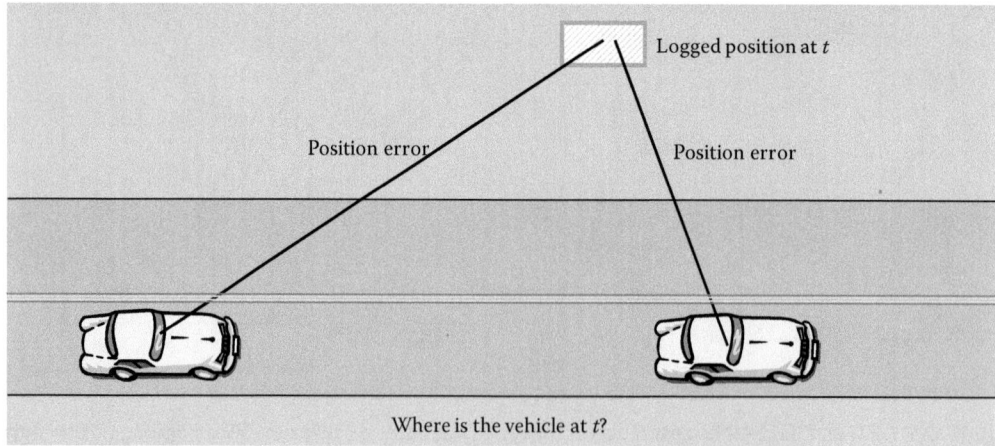

Figure 10.9 Along-track vehicle location error.

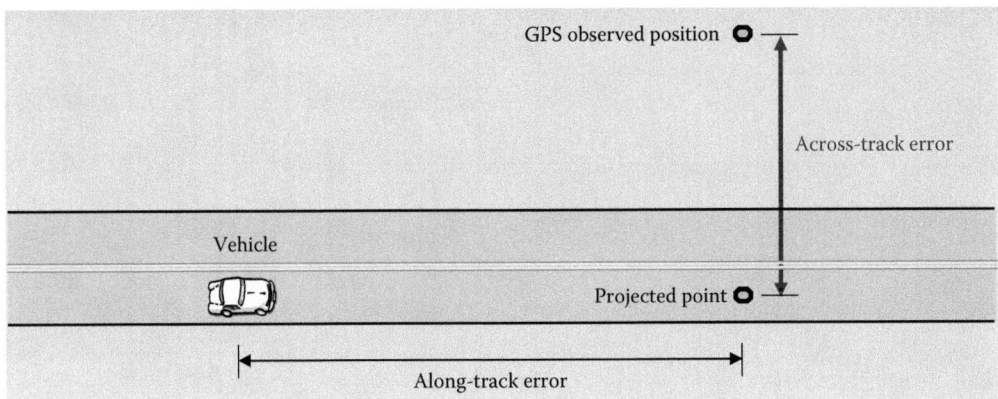

Figure 10.10 Across-track and along-track errors as used to estimate vehicle position on the road.

be of the same order of magnitude. Therefore, the across-track error can be used to represent the true tracking error at each logged point. However, the across-track error is always smaller than or equal to the true tracking error. Therefore, the evaluated positioning error in this study should be considered an optimistic value.

The across-track errors were calculated from both the logged position data and the street position data from the GIS. The geographic information for the roadways in the test area was obtained from the Washington State Geospatial Data Archive in the University of Washington Libraries. The coverage files were provided by the King County Street Network (KCSN). The ArcGIS^SM System was used to display the roadway maps.

10.10 GPS AND GPS–INS SYSTEMS TEST RESULTS

10.10.1 GPS Route 1 results

Most of the GPS tracking results on the freeway are accurate and of high precision. Although the test vehicle traveled under the bridge shown in Figure 10.11, the position errors were not prohibitive. This is because GPS signals could be quickly re-obtained after the test vehicle passed the bridge.

However, if the test vehicle traveled in a long tunnel, GPS signals would be totally lost and the position error would become larger. The longest location update interval in a tunnel was less than 38 s during the freeway testing along Route 1.

10.10.2 GPS Route 2 and 3 results

As expected, the GPS had difficulties accurately measuring the vehicle's position in the urban canyon areas displayed in Figure 10.12 where GPS signals are difficult to receive. This is demonstrated along Routes 2 and 3 by the erratic vehicle positions reported by the GPS. Table 10.1 lists the errors experienced by the GPS on these routes based on the logged position data. The road surface elevation changes more on Route 2 than on Route 3. The average slope rate of Route 2 is approximately 10.6:1, and that of Route 3 is approximately 42.8:1. Since GPS receivers in the 2D mode assume position elevation is constant when calculating their new position coordinates, the elevation difference along Route 2 should decrease the positioning accuracy of the GPS device. However, even in 3D–2D mode,

Figure 10.11 GPS Route 1 test scenario and results. (From Smart Transportation Applications and Research [STAR] Laboratory, University of Washington, Seattle, WA.)

the mean position error of the device for Route 2 was 6.89 m, which was much less than the 30.00 m error for Route 3 in the urban canyon area.

This result implies that the effect of road surface elevation on GPS position error may not be dominant in certain urban environments. For this particular case, the lower error for Route 2 is probably due to the lower density of high-rise buildings along it. During our test runs along Route 2, the signal availability was good and about 60% of positioning calculations were done in the 3D mode even though the GPS device was set to automatically use the 3D–2D mode. This indicates that the constant elevation assumption required for the positioning calculation in the 2D mode was not frequently used. Hence, the error, possibly caused by the large elevation change along Route 2, was largely avoided.

For each route, the positioning accuracies of the GPS device were significantly different between the 3D mode and the 3D–2D mode. For Route 2, the mean error for the 3D mode was 4.36 m, which was significantly lower than the mean error of 6.89 m for the 3D–2D mode at the $p = 0.01$ significance level.

10.10.3 GPS–INS Route 2 and 3 results

Test results confirm that the GPS–INS integrated system is more accurate than GPS alone. This is demonstrated by the absence of random vehicle tracks in Figure 10.13 as compared with those that appear in Figure 10.12.

The GPS–INS integrated system errors were based on the device's logged position data for Routes 2 and 3. The performance of the integrated device was more consistent and

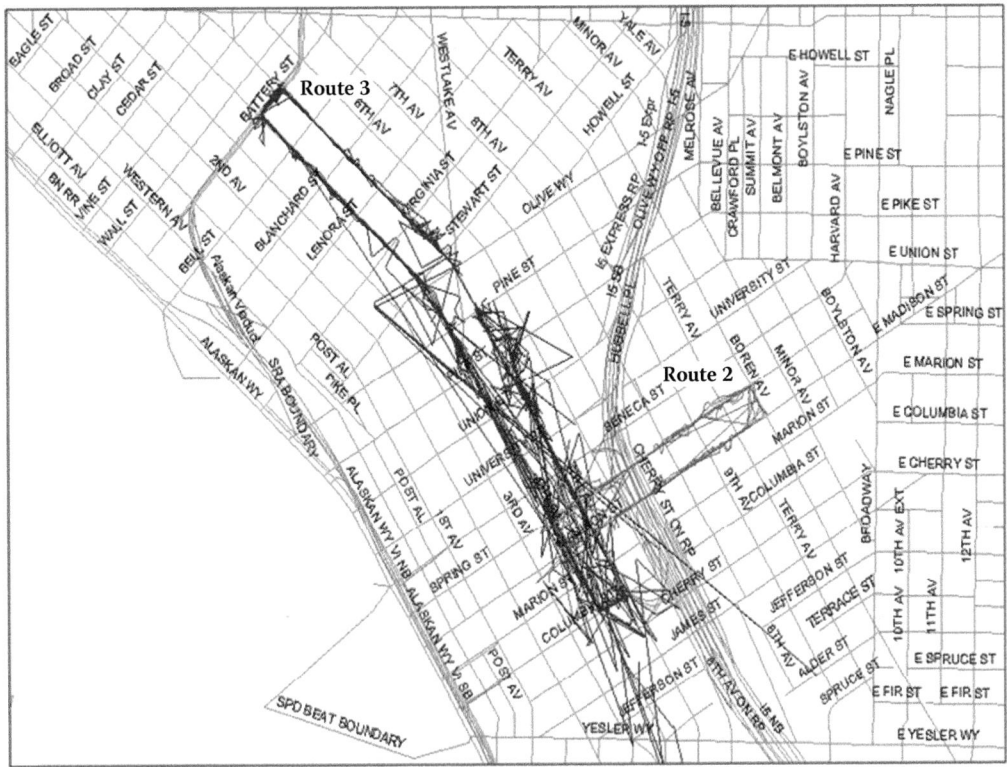

Figure 10.12 GPS Route 2 and 3 test scenarios and results. (From Smart Transportation Applications and Research [STAR] Laboratory, University of Washington, Seattle, WA.)

significantly better than that of the GPS alone regardless of the road surface slope and high-rise building density. The data in Table 10.2 collected from both urban canyon and urban non-canyon areas show the mean error of the integrated system as 4.23 m and the standard deviation as 3.44 m. Thus, the integrated system provided greater positioning accuracy in the test than the GPS functioning alone.

10.10.4 GPS data update interval

Table 10.3 corroborates that the GPS location update intervals in urban canyons are longer than those in urban non-canyons. When operating in the 3D position mode, 20 update

Table 10.1 GPS errors for Routes 2 and 3

Route	Area	Position mode	Positions logged by GPS	Error (m)			
				Mean	Standard deviation	Maximum	Minimum
2	Urban canyon	3D	511	4.36	4.48	41.76	0.04
		3D–2D	821	6.89	13.74	247.42	0.00
3	Urban canyon	3D	850	14.51	20.43	178.06	0.01
		3D–2D	495	30.00	46.56	663.09	0.01
3	Urban non-canyon	3D	1324	4.27	4.06	43.94	0.01
		3D–2D	545	6.14	8.21	66.61	0.02

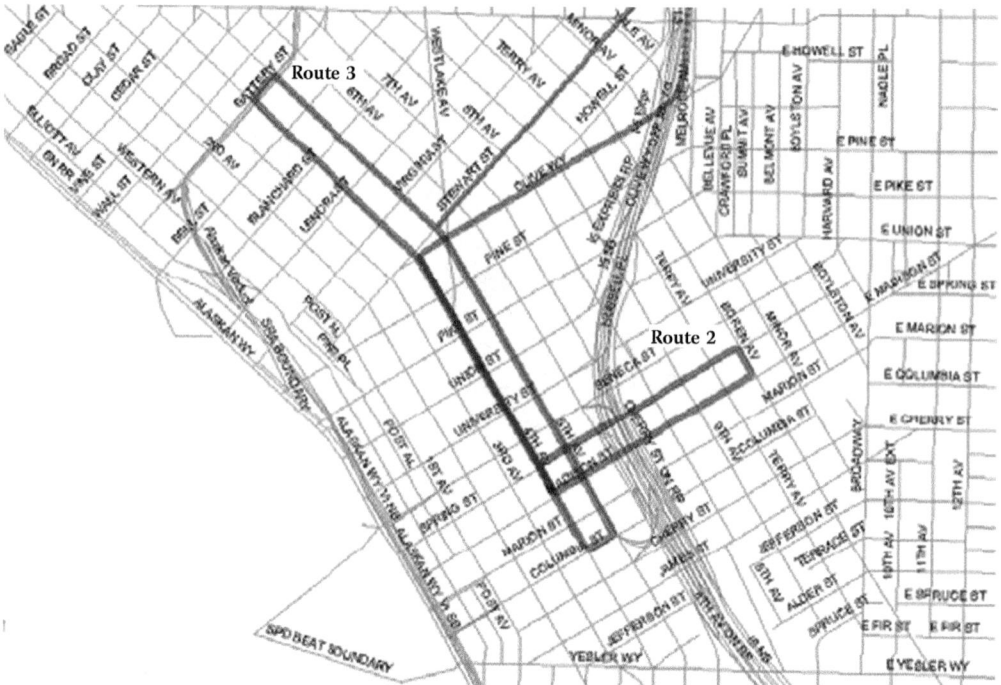

Figure 10.13 GPS–INS Route 2 and 3 test results. (From Smart Transportation Applications and Research [STAR] Laboratory, University of Washington, Seattle, WA.)

intervals were observed to be longer than 1 min in the urban canyon area, but only 2 intervals were observed in the urban non-canyon area to be longer than 1 min. When operating in the 3D–2D mode, the difference between the two areas, 8 in the urban canyon area versus 0 in the urban non-canyon area, is also significant.

In addition, since the 3D–2D mode reduces the required number of visible satellites from four to three, it should have fewer long update intervals than the 3D mode. The data for both Route 2 and Route 3 confirm that observed update intervals longer than 1 min are more frequent in the 3D mode than in the 3D–2D mode.

Unlike the GPS device, whose position data update always depends on satellite signal availability, the GPS–INS device can update position at a constant frequency because the INS takes over the role of positioning from the GPS when satellite signal availability is poor since the INS does not rely on external inputs. Throughout the test, the integrated GPS–INS device consistently provided position updates at 1 Hz as planned. If necessary, the location data update frequency of the GPS–INS device can be set even higher as the system supports position data update rates to 200 Hz.

Table 10.2 GPS–INS errors for Routes 2 and 3

	Positions logged by integrated system	Error (m)			
Area		Mean	Standard deviation	Maximum	Minimum
Urban canyon	12,355	4.22	3.48	14.47	0.001
Urban non-canyon	2844	4.27	3.22	12.79	0.001
Both areas	15,199	4.23	3.44	14.47	0.001

Table 10.3 GPS data update interval

Route	Area	Position mode	Total travel time (min)	Number of long update intervals			
				1–2 (min)	2–3 (min)	3–4 (min)	>4 (min)
2	Urban canyon	3D	22.72	5	2	0	0
		3D–2D	24.72	1	0	0	0
3	Urban canyon	3D	64.75	15	2	2	1
		3D–2D	24.38	7	0	1	0
3	Urban non-canyon	3D	28.07	1	0	1	0
		3D–2D	10.30	0	0	0	0

10.10.5 GPS-alone and combined GPS–INS accuracy

When operating in 3D mode, the GPS device produced significantly more accurate position data than in 3D–2D mode. Its mean across-track error varied from 4.36 to 30.00 m in the test runs in urban canyon areas and from 4.27 to 6.14 m in urban non-canyon areas. The standard deviation of GPS across-track error varied from 4.06 to 46.56 m for different modes and locations in urban areas. The maximum GPS positioning error identified in this study was more than 663 m, which was enough to misplace the vehicle several blocks away. If such misplacement errors occur frequently, users will definitely question the accuracy of the tracking system.

The GPS–INS integrated system, on the other hand, worked consistently well whether in urban canyon areas or in urban non-canyon areas. The mean across-track error of the integrated system varied from 4.22 to 4.27 m, and the standard deviation varied from 3.22 to 3.48 m.

The GPS–INS integrated system surpassed the GPS-alone device in both positioning accuracy and data update frequency in the test. The mean positioning error of the integrated system was less than 4.27 m, and the largest positioning error was 14.47 m in urban areas. Even the largest positioning error was still not enough to cause misplacement issues. At the 95% confidence level, the GPS–INS integrated system position error in the urban area is within 14 ft (4.2 m). The GPS-alone error is as large as 2175 ft (663 m). On the other hand, the largest error for the GPS–INS integrated system is 47 ft (14 m).

10.11 CONCLUSIONS FROM GPS–INS STUDY

Position accuracy of the GPS–INS integrated system is superior to the GPS-alone system in an urban area for the following reasons:

- The GPS–INS integrated system reliably updates its location every 1 s, but the GPS-alone system may not be able to update its location for several minutes in an urban area.
- When the GPS operates in 3D–2D mode, it has fewer long update intervals, but worse location accuracy.
- GPS is an effective solution for tracking vehicles on freeways.

Although the integrated GPS–INS system is capable of tracking vehicles in an urban canyon area, the systems were too expensive at the time of the test for use by the general public. Therefore, further research is needed to find alternative solutions for tracking vehicles accurately in urban canyon areas, lower cost GPS–INS devices need to be developed, or the public needs to wait until the additional frequencies that are part of the GPS modernization program become operational as discussed in Section 10.5.

An example of a newer, low-cost approach to combining GPS and INS is the application of micro-electro-mechanical systems (MEMS) technology with its reduced size, weight, power consumption, and cost. VectorNav manufactures one such commercial product as a dual-antenna GPS-aided INS. Its VN-300 model packages three-axis accelerometers, three-axis gyros, three-axis magnetometers, a barometric pressure sensor, two GPS receivers, and a low-power microprocessor in a rugged aluminum enclosure about the size of a matchbox [11]. When in motion, the device couples the position and velocity measurements from the onboard GPS receivers with measurements from the onboard inertial sensors to provide position, velocity, and attitude estimates with greater accuracy and better dynamic response than a standalone GPS receiver or attitude heading reference system. The dual GPS receivers also provide accurate true north heading measurements when the sensor is stationary through GPS interferometry techniques. These utilize the raw pseudo-range and carrier-phase measurements from two separate internal GPS receivers and a known baseline to directly measure the heading of the vehicle or platform without any assumptions regarding the vehicle dynamics.

This product is suited for a wide variety of industrial and military applications that have size, weight, power, and cost (SWAP-C) constraints such as unmanned vehicle systems; antenna, camera, and platform stabilization; heavy machinery monitoring; robotics; and primary or secondary flight navigation. Its manufacturer claims it is also ideal for applications that require highly accurate inertial navigation measurements (position, velocity, and attitude), especially in environments with unreliable magnetic heading and GPS visibility. With development kits at several thousand dollars (U.S.), the product is made for commercial and industrial applications as it does not have the user-friendly interface required by the general public.

10.12 BLUETOOTH FOR TRAVEL TIME ESTIMATION

Devices transmitting Bluetooth signals facilitate the calculation of travel time estimates over a road segment. This concept is based on identifying a vehicle carrying a Bluetooth device by reading its unique 48-bit MAC hardware number or address. The MAC address can be captured by roadside equipment if the Bluetooth device is in discover mode. Most cell phones have Bluetooth capability and allow the owner of the phone to activate this mode of operation. The difference in the times at which the address was first read at one location and then again at a later time by reidentifying the device at another location allows the travel time to be measured and reported to travelers.

Bluetooth vehicle tracking applications are many and include congestion reporting on bridges, freeways, and even arterials with appropriate data processing; street network analysis to find the shortest or quickest path between two destinations; bus stop waiting time; bicycle and pedestrian travel times; comparison of toll-free lane and toll lane travel times; before and after studies of traffic signal timing plans; and rural travel time reporting. In addition to Bluetooth MAC address readers, other roadside equipment can access toll-tag devices and DSRC devices in connected and other vehicles. The MAC addresses of these devices can be used with the same reidentification process to estimate travel time along a road segment.

10.12.1 Bluetooth technology

The key features of Bluetooth technology are its ubiquitous nature, low transmitted power, and low cost. The Bluetooth specification [12] defines a uniform structure for a wide range of devices to connect and communicate with each other. Pairing occurs when two

Bluetooth-enabled devices connect to each other. The connections allow short-range wireless communications through ad hoc networks known as piconets. Piconets are established dynamically and automatically as these devices enter and leave radio proximity.

Bluetooth technology operates in the unlicensed industrial, scientific, and medical (ISM) band at 2.4–2.485 GHz utilizing a spread spectrum, frequency hopping, full-duplex signal at a nominal rate of 1600 hops/s. The 2.4 GHz ISM band is available and unlicensed in most countries.

Its adaptive frequency hopping (AFH) capability reduces interference between wireless technologies sharing the 2.4 GHz spectrum. AFH operates by detecting other devices in the spectrum and avoiding the frequencies they are using. This adaptive hopping among 79 frequencies at 1 MHz intervals gives a high degree of interference immunity and also allows for more efficient transmission within the spectrum.

Bluetooth purposely broadcasts a low-power signal to prevent interference. The Bluetooth specification lists three power classes as shown in Table 10.4. By contrast, a cell phone's output power can vary from 251 mW to 3 W depending on the phone generation and its power class.

10.12.2 Locating a Bluetooth device through its MAC address

A Bluetooth device is detected by scanning the full Bluetooth transmission spectrum, randomly jumping from frequency to frequency. The Bluetooth specification states that "The inquiry substate may have to last for 10.24 seconds unless the inquirer collects enough responses and determines to abort the inquiry substate earlier" [13]. Therefore, the mathematical relation between the time a Bluetooth-enabled device spends in a location and the chance of detecting it is given by

$$\text{Chance of obtaining MAC address} = \frac{\text{Time spent in detection zone}}{10.24 \text{ s}}. \tag{10.6}$$

The length of the detection zone for a vehicle is a function of its speed S. Hence, the distance d it travels in 10.24 s is

$$d = (S) \times (10.24 \text{ s}). \tag{10.7}$$

Therefore, a vehicle traveling at 60 mi/h (97 km/h) would cover

$$d = \frac{60 \text{ mi/h}}{3600 \text{ s/h}} \times 10.24 \text{ s} = 0.17 \text{ mi} \quad \text{or} \quad 897 \text{ ft } (273 \text{ m}) \text{ in } 10.24 \text{ s}. \tag{10.8}$$

An antenna is connected to the roadside MAC address reader to amplify the power received from the cell phone or other Bluetooth device that is being tracked. Antennas can be

Table 10.4 Bluetooth power and range classes

Class	Maximum power (mW)	Nominal power (mW)	Minimum power (mW)[a]	Range (m)
I	100	n/a	1	≈100
II	2.5	1	0.25	≈10
III	1	n/a	n/a	≈1

[a] At maximum power setting.

designed to detect signals coming from anywhere within a 360° sphere or from a particular direction. Antenna gain, calculated in dBi, is a measure of the amount of focus or directivity that an antenna can apply to the incoming signal relative to an isotropic radiator (i.e., a dispersion pattern that radiates the energy equally in all directions onto an imaginary sphere surrounding a point source). Thus, an antenna with 5 dBi of gain focuses the energy so that some areas on an imaginary sphere surrounding the antenna will have 5 dB more signal strength than the strength of the strongest spot on the sphere around an isotropic radiator.

10.12.3 Travel time estimation using Bluetooth device reidentification

Point-to-point travel times on both freeways and arterials can be collected using MAC address readers since a Bluetooth device in discover mode will remain visible for at least 10.24 s. In the tests reported below, antennas with several gains were connected to the roadside readers, namely, 7- and 9-dBi omnidirectional antennas and a 12-dBi directional antenna. The reidentification tests reported here were conducted at three sites: SR-522 and SR-520 in Seattle, WA and a site near Yreka, CA along the I-5 freeway [14,15].

10.12.3.1 SR-522 test segment

On the SR-522 study segment in Figure 10.14, automatic license plate readers (ALPRs) were already in place at the NE 170th Street and 61st Avenue NE sites and could be used to obtain ground truth data to compare against the nearby MAC address readers. The directional gain patterns of the antennas and the ALPR detection zones are illustrated in Figure 10.15 for the 61st Avenue NE site. Traffic volumes were moderate with between 20,000 and 40,000 annual average daily traffic (AADT). Vehicle speeds were between 40 and 50 mi/h (64–80 km/h).

The tests were conducted over a 24-h period from October 8th and October 9th, 2009. The 12-dBi directional antenna was used on October 8th and the 7-dBi omnidirectional antenna on October 9th. The 12-dBi directional antenna recorded 1595 readings for both directions at the 61st Avenue site, yielding a detection rate of 10%. At the NE 170th Street site, there were 1375 readings for both directions and a detection rate of 8%. There were 792 matches (0.55 matches/min) giving a matching rate of 58% (792/1375). The 7-dBi

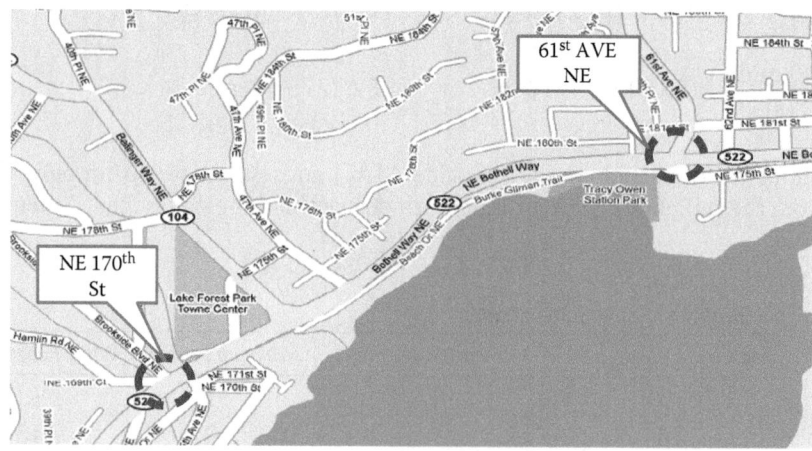

Figure 10.14 SR-522 arterial test segment. (From Smart Transportation Applications and Research [STAR] Laboratory, University of Washington, Seattle, WA.)

Figure 10.15 Antenna gain patterns and ALPR detection zones at the 61st Avenue NE test site. The dotted line represents the 7-dBi omnidirectional antenna pattern, the solid line the 12-dBi directional antenna range pattern, and the rectangles the ALPR detection zones. The X marks the location of the Bluetooth MAC address reader. (From Smart Transportation Applications and Research [STAR] Laboratory, University of Washington, Seattle, WA.)

omnidirectional antenna recorded 1926 readings for both directions at the 61st Avenue site, yielding a detection rate of 11%. At the NE 170th Street site, there were 2124 readings for both directions and a detection rate of 12%. There were 1340 matches (0.93 matches/min) giving a matching rate of 70% (1340/1926).

These Bluetooth reader matching rates are certainly adequate to get valid estimates of travel times along a road segment. One of the lessons learned during this test was to avoid setting up equipment near bus stops or other places with easy public access to avoid vandalism and theft of equipment.

Travel time ground truth was obtained from the ALPR. Figure 10.16 is a sample of the collected data that shows that Bluetooth-measured travel times were always greater than the ALPR travel times. This trend was present with both directional and omnidirectional antennas and occurs because Bluetooth exhibits a bias toward slower vehicles and hence overestimates travel time on a road segment. The bias arises from the increased probability of obtaining a device's MAC address from a slower-moving vehicle because more time is spent in the detection zone. Obvious errors in travel time have been replaced with default travel time values calculated by assuming the vehicle was traveling in a free-flow condition at the posted speed limit.

10.12.3.2 SR-520 test segment

Bluetooth travel time estimation was evaluated for higher-speed vehicles using the segment of the SR-520 freeway shown in Figure 10.17 that crosses Lake Washington. Roadside readers recorded MAC addresses of the vehicles, while ALPRs were again used to collect ground truth data. The locations of the ALPR and the 7-dBi omnidirectional antenna for the MAC address reader are described in the figure. Data collection occurred on February 22nd at the 24th Avenue and 76th Avenue intersections with SR-520 from 8:00 a.m. to 9:00 a.m. The antennas recorded 432 readings for both directions at the 24th Avenue intersection

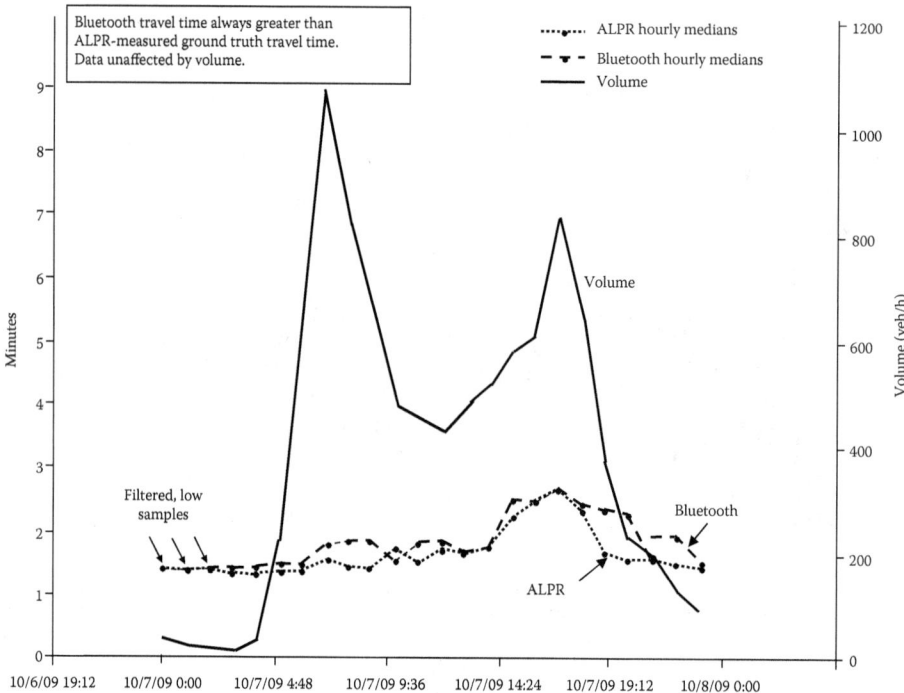

Figure 10.16 ALPR and directional antenna travel time data comparison on westbound SR-522. (From Smart Transportation Applications and Research [STAR] Laboratory, University of Washington, Seattle, WA.)

and 190 readings for both directions at the 76th Avenue intersection. This longer corridor provided a MAC address matching rate of 61% (116 out of 190), which is ample for travel time estimation.

10.12.3.3 I-5 Yreka test segment

Another test at Yreka along 7.6 mi (12.2 km) of the I-5 freeway between Walter's Road and Anderson Grade was performed to verify the ability of roadside readers to read

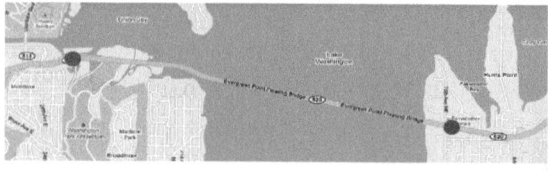

- ALPR and Bluetooth antenna mounted to the inside of the overpass.
- 7-dBi omnidirectional antenna mounted next to rear-viewing ALPR.

Figure 10.17 SR-520 freeway test site. (From Smart Transportation Applications and Research [STAR] Laboratory, University of Washington, Seattle, WA.)

Figure 10.18 I-5 Yreka freeway test site. (From Smart Transportation Applications and Research [STAR] Laboratory, University of Washington, Seattle, WA.)

the MAC addresses of Bluetooth devices on very high-speed vehicles. This scenario is illustrated in Figure 10.18. The 7- and 9-dBi omnidirectional antennas for the MAC address readers were deployed at this location and were mounted to signs at each end of the test route. ALPRs provided ground truth data. AADT for the corridor was approximately 20,000 vehicles. Communication occurred via cellular GSM™ (Global System for Mobile Communications) that incorporated narrowband time-division multiple access (TDMA) that allowed eight simultaneous calls on the same frequency. During the 24-h tests from 6:00 p.m. to 6:00 p.m. on June 15th and 16th, the antenna recorded 1118 readings for both directions at Anderson Grade and 336 readings for both directions at Walter's Road. The matching rate was 68% (228 out of 336), which again is sufficient for travel time estimation.

10.12.4 Bluetooth travel time test conclusions

Table 10.5 contains a summary of the test location parameters and collected data. Bluetooth travel time data collection using roadside MAC address readers produces reasonably accurate

Table 10.5 Test location characteristics and data summary for Bluetooth travel time estimation

Location	Road type	Duration of test (h)	Antenna gain	Volume (AADT)	Number detections/ detection rate	Number matches/ matching rate
SR-522	Arterial	24	12 dBi directional	20,000–40,000	1595/10% 1375/8%	792/58%
			7 dBi omnidirectional	20,000–40,000	1926/11% 2124/12%	1340/70%
SR-520	Freeway	1	7 dBi omnidirectional	Higher speed, longer corridor	432 190	116/61%
I-5	Freeway	24	7 dBi omnidirectional	20,000	1118 336	228/68%

travel time measurements. The high matching rate (60%–70%) implies that the majority of the Bluetooth devices were captured by the readers. Bluetooth travel times are generally overestimates because there is bias toward capturing data from slower vehicles. The percent error rate varies with distance, with longer corridors experiencing a lower error rate. When measuring travel time along arterials, intersection delay will manifest itself in a decrease in estimated travel speed along that road segment.

On arterials, band-mounting devices to poles was acceptable. There was no vandalism when the hardware was left alone for a week. Avoid bus stops and other locations where theft may be an issue. On freeways, the recommended mounting procedure includes band-mounting ALPRs and antennas to overpass railings, informing local DOTs and police about the devices that are mounted, and mounting the devices on the inside of overpasses for safety reasons.

10.12.5 Issues in applying MAC address reading to travel time estimation

Remaining concerns affecting the use of MAC address readers to estimate travel time include filtering data that are not representative of the true travel times, overestimating travel times from MAC address reidentification data, adequate probability of reidentifying a Bluetooth device, rural challenges, rural applicability, noise, privacy, functionality, and communications and data processing costs. These issues are discussed below.

10.12.5.1 Filtering of Bluetooth data

Travel time estimates obtained from reidentification of Bluetooth devices is subject to errors produced by vehicles traveling faster or slower than the prevailing traffic flow or by devices that may not be in vehicles at all [16,17]. For example, motorcycles traveling in between lanes can travel faster than the prevailing speed, while vehicles stopping at rest areas that are later reidentified on the main roadway may appear to be traveling much slower than the prevailing speed. Filters derived from multiples of the free-flow speed (based on the speed limit of the facility) or some percentile of predicted facility speeds (based on recent or historical speeds) can be employed to remove these outliers.

Filtering may also be implemented at the MAC address reader to remove addresses that are not representative of Bluetooth devices in vehicles traveling past the reader. As an example of data filtering at the address reader level, consider a MAC address that was read seven times over a 26-s period (elapsed time between the first and last reads) [17]. The time difference between successive reads is either 4 or 5 s, which corresponds to the length of the inquiry mode programmed into the address reader. Should this MAC address data be filtered? If it represents a traveling vehicle, it is traveling at a slow speed through the antenna coverage area of the reader. In this example, other MAC addresses are read just before and after the address in question, and these addresses may generate data that are more typical for a Bluetooth-enabled device in a vehicle moving at a speed close to the speed limit. Therefore, the questionable address may not be in a moving vehicle and can be removed. The fact that the address was read seven times in 26 s is not enough information to indicate that the data should be filtered since congestion causing slower vehicle speeds could also generate such data. In this case, the address would not be filtered. However, there are instances where a MAC address is detected many times over a long period of time indicating that the address was not from a Bluetooth device in a passing vehicle. Data of this type should be filtered at the MAC address reader level.

Another example of filtering at the MAC address reader level occurs when an address is read five times in 7 min. Unlike the previous example, the longer time interval between reads indicates that it may not represent a Bluetooth-enabled device in a passing vehicle, for example, it could originate from a pedestrian carrying a cell phone. The minimum time between reads is 16 s, but four of the five time intervals are close to 1 min. It would appear that this MAC address should be filtered at the address reader level.

Other travel time data filtering techniques discussed in Porter et al. [17] to eliminate non-vehicle travel time samples, or vehicle travel times that are outliers with respect to the corresponding traffic flow conditions are as follows:

- A moving standard deviation algorithm where the travel time mean and standard deviation are calculated for a fixed number of neighboring (in time) travel time data samples. If a travel time is a pre-established number of standard deviations above the mean, the obtained travel time is determined to be an outlier [18].
- A two-step travel time filtering mechanism. The first step is based on a histogram of speeds (rather than travel times) generated from collected MAC addresses. The histogram is smoothed by computing the average of eight consecutive histogram frequencies. The smoothed histogram represents an estimate of the distribution of speeds that occur along a particular road segment, which includes outlier travel times. The outliers are identified as those speeds that occur outside of lower and upper limits [19].
- Comparison of each new calculated travel time sample to the most recent average for the road segment. If the new travel time differs by more than a certain percentage (e.g., 25%), the new travel time would be labeled invalid and would be discarded. This filtering method was particularly successful on high-volume freeways that do not have much variance in speed. However, it tends to discard many potentially valid records on arterials where the travel speeds were more varied [20].
- A lower threshold for the free-flow speed estimated from previous traffic studies and modifying it based on the aggregated average speed of the vehicles. This lower bound removes abnormal travel speeds. Kalman filtering prediction of the next speed sample value can also be applied to identify outliers [21].
- Box plot filtering and z-score filtering distribution-based techniques [22]. Box plot filtering identifies outliers based on quartile values of a distribution, while z-score filtering identifies outliers based on the standardized sample mean test statistic [23]:

$$z = \frac{(\bar{x} - \mu)}{(\sigma/\sqrt{n})},$$

(10.9)

where $\bar{x}$ is sample mean, which is an unbiased estimator of the unknown population mean μ if the samples are random and represent the entire population, $\sigma_{\bar{x}} = \sigma/\sqrt{n}$ is the standard deviation of the sample mean (sometimes called the standard error), σ is the standard deviation of the entire population, and n is sample size.

10.12.5.2 Bluetooth data bias

As noted in Section 10.12.3.1, Bluetooth exhibits a bias toward slower vehicles and hence overestimates travel time on a road segment. Other biases arise from multiple devices in a vehicle such as a bus and devices on slower-moving transportation modes such as bicycles. It is interesting to consider that the latter biases can be exploited to identify these transportation modes. For example, if some number of detected Bluetooth-enabled devices moving along a roadway appears grouped together and is perhaps traveling slower than other vehicles in a lane where buses are expected, then a bus is likely to be the transportation vehicle and appropriate data analysis techniques can be applied. Similarly, if one or two Bluetooth-enabled devices are found to be moving slowly on a conveyance traveling near the curb or in the right-most lane, then a bicycle is likely to be the transportation vehicle. A related factor in gathering travel time or speed data from Bluetooth devices is to ensure that the antenna attached to the address reader has appropriate beamwidth to cover and gather data from all the travel lanes of interest.

10.12.5.3 Probability of reidentifying a Bluetooth device

Computation of detection rates requires contemporaneous collection of both Bluetooth and traffic volume data at each site. Furthermore, the complexity of the detection rate computation increases with the number of sites. The probability of detecting a single vehicle at a single location is the joint probability of a vehicle having a Bluetooth device in discovery mode and the probability that the Bluetooth device's MAC address can be read. Effinger et al. [24] state that these probabilities should be independent of each other. It is also reasonable to assume that a vehicle has a Bluetooth device at location B if it had a Bluetooth device at location A, but the same cannot be said about the probabilities of *reading* a MAC address for any device. The conditional probability of reading a MAC address at location B given that it was read at location A is considerably less than 1. Consequently, the probability of detecting a given triple (i.e., reading the same MAC address at three different reader locations) in a traffic stream is considerably less than detecting the device at two reader stations. Similarly, the probability of detecting a quadruple is likely to be even less. It is also necessary to assure that there are no site conditions that result in unusually high (or low) concentrations of Bluetooth devices in the traffic stream, for example, by placing a MAC address reader at a new car dealership unless, of course, that is the defined purpose of an origin–destination pair study.

10.12.5.4 Potential rural challenges

About 8%–10% of traffic was detected at each location in the University of Washington study [14]. Of that, 50% is matched to obtain travel time. To get one reading every minute, you need to have at least 120 veh/h. However, low-volume roads may experience a lower occurrence of vehicles. There is also some bias toward recording the travel time of buses because there are a lot of devices on one vehicle. You then have to determine if buses are the dominant source of data on the road where travel time is needed and, if necessary, devise an approach to filter out these travel time estimates perhaps by noting that there is a lot of data coming from multiple devices at one location. Two additional considerations need to be addressed in rural areas: how to account for slower-moving trucks being more likely to be detected by Bluetooth readers and how to account for travelers stopping at rest areas. The rest area question is frequently solved by filtering out travel times that are not consistent with the majority of travelers on the highway [16,17]. However, this approach may not be sufficient when the number of vehicles is small and the average speed has a large variance.

10.12.5.5 Rural applicability

Consider the hypothetical case where volume is approximately 500 veh/h and penetration is 10% or 50 veh/h. Some vehicles may even divert off the highway on which they are first detected, making the potential field of available vehicles even lower. Assume 50% matching or 25 veh/h. This will vary with vehicle speed. Furthermore, assume 80% capture rate giving 20 veh/h. Is this capture of a vehicle every 3 min good enough to get a valid estimate of travel time or, in other terms, will capturing 10 vehicles in 30 min be sufficient to obtain a useful measure of travel time?

10.12.5.6 Noise

Wi-Fi networks in the vicinity of a Bluetooth MAC address reader can potentially cause interference as can other nearby Bluetooth readers. However, tests show there is little difference between using two MAC address readers versus one at the same location in terms of interference.

10.12.5.7 Privacy

Privacy can be addressed by maintaining the trust of the public through measures such as not maintaining a central database, not tying the MAC address to an individual, encouraging the use of technology that scrambles the MAC address in each device, and deleting expired addresses.

10.12.5.8 Functionality

Functionality concerns include synchronization of MAC address readers by using identical timestamps at each reader station, spatial organization of data, and using sensor networks. An antenna is needed on each MAC address reader. This requires an enclosure made of plastic or glass if the antenna is internal or a sealed outside port.

10.12.5.9 Communications and data processing costs

GSM is able to communicate via http. Costs among cellular service providers vary. However, they appear to be decreasing with time. In 2010, the cost was about 1 cent per update with AT&T.

A low-cost database option that saves power and money is a MySQL™ server that updates every minute and only if data are present [14]. According to the *MySQL 5.7 Reference Manual*, MySQL software provides a very fast, multithreaded, multiuser, and robust SQL (structured query language) database server [25].

10.12.6 Bluetooth versus ALPR for vehicle counting and travel time measurement

The advantages and limitations of Bluetooth and ALPR for vehicle counting and travel time estimation are summarized in Table 10.6. A Transport of London study shows that Bluetooth has considerable practical and cost advantages over ALPR and that its data correlated well with vehicle movement [26].

Table 10.6 Advantages and limitations of Bluetooth and ALPR for vehicle counting and travel time estimation

Technology	Advantages	Limitations
Bluetooth	One MAC address reader can cover a multilane highway with a range of 200 m Not disrupted by inclement weather Low installation and maintenance costs Bluetooth readers recorded fast-moving cyclists and motorcyclists Low security (hacking) concerns	Overcounting when large numbers of passengers on a transit vehicle have smart phones Undercounting when a vehicle does not contain a Bluetooth device or if the device is in non-discover or hidden mode Siting of Bluetooth readers is critical as they pick up signals over a long distance and at junctions, and may collect data from vehicles on the wrong roads
ALPR	ALPR captures a larger sample of the traffic as not all vehicles emit Bluetooth signals	Cost

REFERENCES

1. *GPS Control Segment*, Federal Aviation Administration, U.S. Department of Transportation, Washington, DC, November 13, 2014. http://www.faa.gov/about/office_org/headquarters_offices/ato/service_units/techops/navservices/gnss/gps/controlsegments/. Accessed December 24, 2015. Also see Official U.S. Government Global Positioning System (GPS) and related topics Web site, National Coordination Office for Space-Based Positioning, Navigation, and Timing, Washington, DC, October 4, 2016. http://www.gps.gov/systems/gps/control/. Accessed October 4, 2016.
2. N. Ashby, *Relativistic Effects in the Global Positioning System*, July 18, 2006. https://www.aapt.org/doorway/TGRU/articles/Ashbyarticle.pdf. Accessed September 24, 2016.
3. *Augmentation Systems*, http://www.gps.gov/systems/augmentations/. Accessed November 26, 2015.
4. *Global Positioning System Wide Area Augmentation System (WAAS) Performance Standard*, First Edition, Federal Aviation Administration, U.S. Department of Transportation, Washington, DC, October 31, 2008. http://www.gps.gov/technical/ps/2008-WAAS-performance-standard.pdf. Accessed December 24, 2015.
5. J. Boyd, Centimeter-scale GPS coming to Japan, *IEEE Spectrum*, 14–15, May 2014.
6. *EU Galileo and Japanese QZSS combination pushes driverless cars*, http://galileognss.eu/-eu-galileo-and-japanese-qzss-combination-pushes-driverless-cars/. Accessed November 10, 2016.
7. *New Civil Signals*, http://www.gps.gov/systems/gps/modernization/civilsignals/. Accessed November 26, 2015.
8. M.L. Psiaki and T.E. Humphreys, GPS lies, *IEEE Spectrum*, 26–32:52–53, August 2016. http://spectrum.ieee.org/telecom/security/protecting-gps-from-spoofers-is-critical-to-the-future-of-navigation. Accessed August 3, 2016.
9. G.T. Schmidt, Navigation sensors and systems in GNSS degraded and denied environments, *Chinese Journal of Aeronautics*, Elsevier ScienceDirect, 2015. http://dx.doi.org/10.1016/j.cja.2014.12.001. Accessed September 26, 2016.
10. J. Zheng, *Performance Evaluation of the GPS/INS Integrated System in Forest Environment and Urban Area*, Master's thesis, Department of Civil Engineering, University of Washington, Seattle, WA, 2002.
11. VectorNav VN-300 Dual Antenna GPS/INS, http://www.unmannedsystemstechnology.com/wp-content/uploads/2013/12/VN-300-Product-Brief.pdf. Accessed December 24, 2015.
12. *Master Table of Contents & Compliance Requirements——Bluetooth Core Specification 4.2*, December 2, 2014. https://www.bluetooth.org/en-us/specification/adopted-specifications?_ga=1.144607541.1884128135.1447794149. Accessed December 3, 2015.

13. *Ibid.*, 404.
14. Y. Wang, Y. Malinovskiy, Y.-J. Wu, and U.-K. Lee, *Field Experiments with Bluetooth Sensors*, STAR Laboratory, Department of Civil Engineering, University of Washington, Seattle, WA, 2010.
15. Y. Malinovskiy, Y. Wu, Y. Wang, and U.-K. Lee, Field experiments on Bluetooth-based travel time data collection, Paper 10-3134, Presented at the 89th Annual Meeting of the Transportation Research Board, National Research Council, Washington, DC, 2010.
16. N.-E. El Faouzi, L.A. Klein, and O. De Mouzon, Improving travel time estimates from inductive loop and toll collection data with Dempster-Shafer data fusion, *Transportation Research Record*, Journal of the Transportation Research Board, No. 2129: Intelligent Transportation Systems and Vehicle–Highway Automation 2009, TRB, National Research Council, Washington, DC, 73–80, 2009.
17. J.D. Porter, D.S. Kim, S.-J. Park, A. Saeedi, and M.E. Magaña, *Wireless Data Collection System for Travel Time Estimation and Traffic Performance Evaluation*, Chapter 2—Literature Review, FHWA-OR-RD-12-13, SPR 737, OTREC-RR012-06, Oregon State University, Corvallis, OR, May 2012. http://www.oregon.gov/odot/td/tp_res/docs/reports/2012/spr737_wireless.pdf. Accessed January 20, 2016.
18. S.M. Quayle, P. Koonce, D. DePencier, and D. Bullock, Arterial performance measures using MAC readers: Portland pilot study, Paper 10-3417, Presented at the 89th Annual Meeting of the Transportation Research Board, National Research Council, Washington, DC, 2010.
19. A. Haghani, H. Masoud, F.S. Kaveh, S. Young, and P. Tarnoff, Freeway travel time ground truth data collection using Bluetooth Sensors, Paper 10-0729, Presented at the 89th Annual Meeting of the Transportation Research Board, National Research Council, Washington, DC, 2010.
20. D.D. Puckett and M.J. Vickich, *Bluetooth-Based Travel Time/Speed Measuring Systems Development*, Final Report, Project Report #09-00-17, University Transportation Center for Mobility, Texas Transportation Institute, 2010.
21. J. Barceló, L. Montero, L. Marques, and C. Carmona, Travel time forecasting and dynamic OD estimation in freeways based on Bluetooth traffic monitoring, Paper 10-3123, Presented at the 89th Annual Meeting of the Transportation Research Board, National Research Council, Washington, DC, 2010.
22. M. Schneider, M. Linauer, N. Hainitz, and H. Koller, Traveller information service based on real-time toll data in Austria, *IET Intelligent Transport Systems*, 3(2):124–137, 2009.
23. D.S. Moore and G.P. McCabe, *Introduction to the Practice of Statistics*, Second Edition, W.H. Freeman and Company, New York, NY, 1993.
24. J. Effinger, A. Horowitz, Y. Liu, and J.W. Shaw, Bluetooth Vehicle Re-identification for Analysis of Work Zone Diversion, Paper 13-2159, Presented at the 92nd Annual Meeting of the Transportation Research Board, National Research Council, Washington, DC, 2013.
25. *MySQL 5.7 Reference Manual*, http://dev.mysql.com/doc/refman/5.7/en/introduction.html. Accessed December 3, 2015.
26. M. Glaskin, The numbers game, *Traffic Technology International*, 22–28, August/September 2015.

Chapter 11

Automated vehicles

Availability of advanced automated features in passenger, freight, and transit vehicles is increasing rapidly as manufacturers add safety-related and self-driving elements to their products continually. Just when fully autonomous vehicles will be available for general use is difficult to predict with certainty, but at a minimum, vehicles are increasingly being equipped with a variety of attributes that improve their safety, energy efficiency, and performance along with traveling comfort and convenience for the public. Partial automation, especially for long-distance limited-access highway portions of longer trips, will be available earlier than the full automation that is the popular public perception [1]. In fact, vehicles that enable an automated driving system (ADS) to assume several aspects of the dynamic driving task (DDT) with the expectation that the human driver will respond appropriately to a request to intervene are already making appearances. This chapter examines methods of classifying levels of vehicle automation that have been and are currently in use, driving environments in which automated vehicles must operate, and international conventions that regulate the safe operation of all vehicles. It discusses the U.S. National Highway Traffic Safety Administration's (NHTSA) 2016 automated vehicles policy that sets forth an approach to accelerate the development, testing, and operation of highly automated vehicles (HAVs) and describes driver assist and automation options made available by a variety of vehicle manufacturers. The chapter concludes with a summary of the Mobility as a Service concept that exposes the traveler to all of the available travel mode options for a given trip. Technical, security, policy, legal, and institutional issues are explored as they relate to the above topics. The latter subjects are investigated more fully in the next chapter.

11.1 DEFINITIONS OF AUTOMATION-RELATED TERMS

Several expressions are commonly used to describe vehicles with self-driving features [2]. Among these is the term automated vehicle, which describes a vehicle that contains one or more driving functions designed to relieve the driver of a particular task. In the limit, an array of these automated driving functions will make autonomous vehicle operation possible. Autonomous vehicle refers to any vehicle equipped with technology capable of operating the vehicle without the active physical control or monitoring of a driver, whether or not the technology is engaged. Autonomous driving mode occurs when an autonomous vehicle is operating or driving in autonomous mode, that is, with the autonomous technology engaged. Cooperative or connected vehicles are those that have telematics to engage vehicle-to-vehicle (V2V), vehicle-to-infrastructure (V2I), or infrastructure-to-vehicle (I2V) exchange of information that warns the driver of a potential crash or other unsafe driving condition, for example, curve ahead, slippery road surface, work zone, or pedestrian crossing.

Table 11.1 lists examples of automated driving functions that are implemented solely within the vehicle, solely within the infrastructure such as in traffic signals and roadside cabinets, or in both a vehicle and the infrastructure. Another way of categorizing automated functions is by tactical and strategic. Tactical automated functions are those that are activated in a matter of seconds (or less) and are usually supported by telematics installed in the vehicle or in devices carried by people in the vehicle. Strategic functions are implemented over several seconds or minutes and are generally implemented in the infrastructure and are available to all vehicles and pedestrians equipped with devices that can access and use the infrastructure's communications media.

11.2 VEHICLE AUTOMATION CLASSIFICATION TAXONOMIES

Several schemes exist for classifying vehicle automation levels. The four discussed below are the human–computer interaction, the Society of Automotive Engineers (SAE) six-level, the German Federal Highway Research Institute five-level, and the U.S. NHTSA's five-level automation taxonomies.

11.2.1 Human–computer interaction model

Research examining human interactions with automated systems demonstrates that automation does not simply supplant human activity, but rather changes it often in ways unintended and unanticipated by the designers of the automation [3]. This poses new coordination demands on the human operator and may have implications for autonomous vehicle operation. The human–computer interaction research of Sheridan and others [3,4] led to the 10-level scale for decision automation identified in Table 11.2. Automation was defined as a device or system that accomplishes (partially or fully) a function that was previously, or conceivably could be, carried out (partially or fully) by a human operator. Thus, automation is not an all or none process, but rather a continuum that progresses from the lowest level of fully manual performance to the highest level of full automation.

Table 11.3 lists several automated driving functions at various stages of development and the decision automation levels, also referred to as technology readiness levels, into which

Table 11.1 Tactical and strategic automated driving function examples

Function	Location	Type	Example
Actuation of steering, engine, or brakes (transparent to driver)	Vehicle	Tactical	Antilock braking system
Powertrain and chassis control (transparent to driver)	Vehicle	Tactical	Stability control
Real-time information collection	Vehicle and infrastructure	Tactical and strategic	Driving environment perception from sensing and communication
Hazard assessment	Vehicle and infrastructure	Tactical and strategic	Curve, work zone, and road surface condition warnings
Decision-making	Vehicle and infrastructure	Tactical and strategic	Tactical microscopic maneuvering to strategic route planning
Management of vehicle flows	Infrastructure (traffic management center or signal controller cabinet)	Strategic	Traffic management through incident detection or adaptive signal control

Table 11.2 Decision and action automation levels based on human–computer interaction research

Level	Characterization
1	Computer offers no assistance: human must take all decisions and actions
2	Computer offers a complete set of decisions or action alternatives to human (navigation system, cruise control), but has no further say in which decision is selected
3	Computer narrows selection to a few options
4	Computer suggests one alternative (route guidance or collision warning) with the human retaining authority for executing that alternative or choosing another one
5	Computer executes suggestion if human approves
6	Computer allows human a limited time to veto before it commences automatic execution
7	Computer acts automatically, then informs human, e.g., CACC
8	Computer informs human only if asked
9	Computer informs human if the computer wants to
10	Computer decides everything, acting autonomously and ignoring the human

they fit. Several level 3, 7, and 9 automated functions are already offered by car manufacturers. Among them are cooperative adaptive cruise control (CACC), which automatically adjusts a vehicle's velocity to that of the surrounding vehicles to improve traffic throughput and assist in dissolving shockwaves in a safe way [5], lane merge assist, and driverless parking in a garage using a smart phone application.

Table 11.3 Technology readiness levels for automated driving applications

Application	Level	Development phase
Safety pull over	3	Proof-of-concept validation
Automated lane keeping	9	Already available
Automated steering assist in case of road blocks	3	Proof-of-concept validation
Platooning by means of CACC	7	Proof-of-concept validation
Forward collision warning	9	Blind spot monitoring and collision warning systems are already available
Lane merge assist	7	Already available
Driver warning in case of drowsiness	9	Already available
Emergency braking for vulnerable road users	9	Already available
Automated parking and parking spot reservations	3	Automated parking is already available, but parking spot reservation is in the proof-of-concept phase
Contextual speed limit	7	Proven to be technically possible, but there is currently no system that sets the speed of a vehicle based on communication from the infrastructure
Cooperative adaptive cruise control	7	Already available
Road pricing	9	Already available
Fuel use optimization	9	Already available, e.g., a navigation system that calculates a fuel-saving route
Driverless parking in a garage	7	Already available through a smart phone application interface
Automatic emergency braking	10	To be standard on all light vehicles manufactured in the United States by 2022

11.2.2 SAE six-level taxonomy for motor vehicle driving automation systems

The SAE J3016™ taxonomy applies to motor vehicle driving automation systems that perform part or all of the DDT on a sustained basis. The amount of driving automation ranges from none (Level 0) to full (Level 5) [6–10]. The levels relate to the driving automation features that are engaged in any given instance of on-road operation of an equipped vehicle. As such, although a given vehicle may be equipped with a driving automation system that is capable of delivering multiple driving automation features that perform at different levels, the level of driving automation exhibited in any given instance is determined by the features that are engaged [8].

The DDT includes all of the real-time operational and tactical functions required to operate a vehicle in on-road traffic, excluding the strategic functions such as trip scheduling and selection of destinations and waypoints. The DDT incorporates the following:

1. Lateral vehicle motion control via steering (operational).
2. Longitudinal vehicle motion control via acceleration and deceleration (operational).
3. Monitoring the driving environment via object and event detection, recognition, classification, and response preparation (operational and tactical).
4. Object and event response execution (operational and tactical).
5. Maneuver planning (tactical).
6. Enhancing conspicuity via lighting, signaling, gesturing, and so on (tactical).

Subtasks 3 and 4 are referred to collectively as Object and Event Detection and Response (OEDR).

The terms operational, tactical, and strategic are defined in the J3016 Recommended Practice as follows:

- Operational tasks involve split-second reactions that can be considered precognitive or innate, such as making micro-corrections to steering, braking, and accelerating to maintain lane position in traffic or to avoid a sudden obstacle or hazardous event in the vehicle's pathway.
- Tactical tasks involve maneuvering the vehicle in traffic during a trip, including deciding whether and when to overtake another vehicle or change lanes, selecting an appropriate speed, checking mirrors, etc.
- Strategic tasks involve trip planning, such as deciding whether, when, and where to go; travel mode selection; best routes to take; etc.

SAE J3016 references three primary participants in the driving task: the (human) driver, the driving automation system, and other vehicle systems and components. The other vehicle systems (or the vehicle) do not include the driving automation system, even though a driving automation system may actually share hardware and software components with other vehicle systems such as a processing modules or operating code. Active safety systems, such as electronic stability control and automated emergency braking, and certain types of driver assistance systems, such as lane keeping assistance, are excluded from the scope of the SAE driving automation taxonomy because they do not perform part or all of the DDT on a sustained basis. Rather they provide momentary intervention during potentially hazardous situations and their intervention does not change or eliminate the role of the driver in performing part or all of the DDT. Thus, they are not considered to be driving automation. However, crash avoidance features, including intervention-type active safety systems, may be included in vehicles equipped with driving automation systems at any level. For ADS-equipped vehicles (i.e., Levels 3–5) that perform the complete DDT, crash avoidance capability is part of ADS functionality. An ADS encompasses the hardware and software that are collectively capable of performing the entire DDT on a sustained basis, regardless of whether it is limited to a specific operational design domain (ODD). The ODD is discussed further in Section 11.4.

The SAE driving system automation model in Table 11.4 divides the automation levels into two general categories—those where a human driver monitors and is responsible for OEDR and those where the system is responsible for OEDR. The heavy line in the table delineates the levels in which the system, and not the driver, is responsible for OEDR. The automation levels in the table are descriptive rather than normative and technical rather than legal. They imply no particular order of market introduction. The descriptions indicate minimum rather than maximum system capabilities for each level.

Level 0 represents no automation. Here, the human driver is responsible for performing all aspects of the DDT, even when enhanced by warning or intervention systems.

Level 1 is characterized by automation of a single driving function, such as speed control (acceleration and deceleration) or steering control. The driver at all times performs the remainder of the DDT and supervises the driving automation system, intervening as necessary to maintain safe operation of the vehicle. The driver also determines whether and when engagement or disengagement of the driving automation system is appropriate and immediately performs the entire DDT whenever required or desired. This level of assistance is already available, including adaptive cruise control systems that adjust the speed in response to that of the vehicle directly ahead of it, parallel parking assist, and obstacle warning.

Level 2 automation performs part of the DDT by executing both lateral and longitudinal vehicle motion control subtasks, that is, combines automated functions for both speed and steering control such as CACC with autonomous lane guidance and steering control. The driver's duties are the same as in Level 1 driver assistance. Level 2 features are available on some car models now, while other manufacturers will make these offerings available in the near future.

Level 3 and higher level automated vehicles let the driver disengage from the driving task for periods that range from short times to the full extent of a journey. The driver's role (while the ADS is not engaged) with Level 3 automation is to verify operational readiness of the ADS-equipped vehicle, determine when engagement of ADS is appropriate, and become the DDT fallback-ready user when the ADS is engaged. When the ADS is engaged, the driver's role is that of DDT fallback-ready user. Now the driver's duties are to be receptive to a request to intervene and respond by performing DDT fallback in a timely manner; be receptive to DDT performance-relevant system failures in vehicle systems and, upon occurrence, perform DDT fallback in a timely manner; determine whether and how to achieve a minimal risk condition; and finally to become the driver upon requesting disengagement of the ADS.

As Figure 11.1 implies, a Level 3 vehicle system allows the driver to do something else, but still be prepared to take control when manual operation of the vehicle is necessary such as if the system fails. For example, a Level 3 system may recognize lane markings and other vehicles, but may not recognize flaggers or cones in a work zone. Another challenging situation might be a rainy night when lane markings are not very clear and there is glare off the water on the road. In this situation, the system would not see the lane markings. If the driver cannot see them, neither can one of these systems. Drivers must be made aware when a mode change is about to occur and that their attention is required. Mercedes and Tesla offer cars that automate speed, steering, and lane keeping control. These systems are limited in that they are not designed to function under all driving conditions.

Level 4 adds a system capability to bring the vehicle to a safe state if the driver fails to re-engage. The role of the driver or dispatcher (while the ADS is not engaged) is to verify

Table 11.4 SAE J3016 definitions of driving automation levels

Automation level	Definition	Dynamic driving task			Operational design domain
		Sustained lateral and longitudinal vehicle motion control	Object and event detection and response	Dynamic driving task fallback	
Driver performs part or all of the DDT					
Level 0: No driving automation	The performance by the driver of the entire DDT, even when enhanced by active safety systems.	Driver	Driver	Driver	Not applicable
Level 1: Driver assistance	The sustained and ODD-specific execution by a driving automation system of either the lateral or the longitudinal vehicle motion control subtask of the DDT (but not both simultaneously) with the expectation that the driver performs the remainder of the DDT.	Driver and system	Driver	Driver	Limited to some driving modes
Level 2: Partial driving automation	The sustained and ODD-specific execution by a driving automation system of both the lateral and longitudinal vehicle motion control subtasks of the DDT with the expectation that the driver completes the OEDR subtask and supervises the driving automation system.	System	Driver	Driver	Limited to some driving modes
ADS performs the entire DDT (while engaged)					
Level 3: Conditional driving automation	The sustained and ODD-specific performance by an ADS of the entire DDT with the expectation that the DDT fallback-ready user is receptive to ADS-issued requests to intervene, as well as to DDT performance-relevant system failures in other vehicle systems, and will respond appropriately.	System	System	Fallback-ready (user becomes the driver during fallback)	Limited to some driving modes
Level 4: High driving automation	The sustained and ODD-specific performance by an ADS of the entire DDT and DDT fallback without any expectation that a user will respond to a request to intervene.	System	System	System	Limited to some driving modes
Level 5: Full driving automation	The sustained and unconditional (i.e., not ODD-specific) performance by an ADS of the entire DDT and DDT fallback without any expectation that a user will respond to a request to intervene.	System	System	System	Unlimited

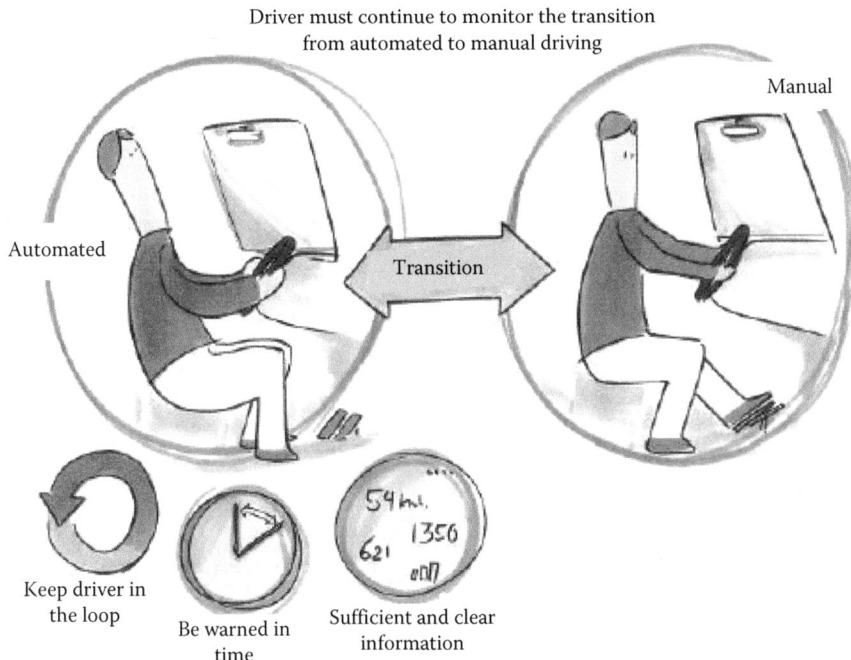

Driver must continue to monitor the transition
from automated to manual driving

Manual

Automated

Transition

Keep driver in
the loop

Be warned in
time

Sufficient and clear
information

Figure 11.1 Driver prepared to transition between automatic and manual driving modes. (From M. van Schijndel-de Nooij et al. *Definition of Necessary Vehicle and Infrastructure Systems for Automated Driving,* European Commission SMART 2010/0064 Study Report, Brussels, BE, June 2011, Version 1.2.)

the operational readiness of the ADS-equipped vehicle, to determine whether to engage the ADS, and to become a passenger when the ADS is engaged only if the person is physically present in the vehicle. While the ADS is engaged, the passenger or dispatcher need not perform the DDT or DDT fallback, need not determine whether and how to achieve a minimal risk condition, may perform the DDT fallback following a request to intervene, may request that the ADS disengage and perhaps achieve a minimal risk condition after it is disengaged, and finally may become the driver after a requested disengagement.

For instance, a Level 3 system is not designed for a situation where the driver is reading a tablet or a book, falls asleep, and does not hear the vehicle's alarms. But a Level 4 system will be able to bring the vehicle to a safe state, even if the driver does not respond. That might mean maneuvering a vehicle over to a shoulder and stopping as displayed in Figure 11.2. In other driving scenarios where there is no shoulder, perhaps in a tunnel, the ability of the automated system to result in a safe state for the vehicle may be compromised. In the tunnel example, does bringing the vehicle to a safe state mean the vehicle needs to drive you all the way to the end of the tunnel until it gets to a shoulder? Or does it stop in a traffic lane inside the tunnel, which is not generally a safe thing to do? Those are still unanswered questions. Therefore, what it means to go to a so-called minimal risk condition is still largely undefined. It is a concept whose details are still being explored and developed.

Level 5 is full vehicle automation that allows the vehicle to go anywhere and do anything under automatic control, without driver involvement. The driver's or dispatcher's role (while the ADS is not engaged) is to verify operational readiness of the ADS-equipped vehicle, to determine whether to engage the ADS, and to become a passenger when the ADS is engaged only if physically present in the vehicle. While the ADS is engaged, the passenger or dispatcher need not perform the DDT or DDT fallback, need not determine whether and how to achieve

Figure 11.2 Level 4 automation concept to bring a vehicle to a safe stop on the shoulder of a highway. (From M. van Schijndel-de Nooij et al. *Definition of Necessary Vehicle and Infrastructure Systems for Automated Driving*, European Commission SMART 2010/0064 Study Report, Brussels, BE, June 2011, Version 1.2.)

a minimal risk condition, may perform the DDT fallback following a request to intervene, may request that the ADS disengage and perhaps achieve a minimal risk condition after it is disengaged, and may become the driver after a requested disengagement. This type of system may be a decade or more away before becoming a common occurrence on city roadways. Table 11.5 lists applications and corresponding driver responsibilities for Levels 1–5 of the SAE model.

Table 11.5 Example applications and driver responsibilities using SAE automation levels

Level	Example applications	Driver roles
1	Adaptive cruise control OR Lane keeping assistance	Must drive and control all functions other than the one that is functioning automatically at the time, and monitor driving environment
2	Adaptive cruise control AND Lane keeping assistance	Must continue to monitor driving environment (system reminds driver to try to ensure this occurs)
3	Following of vehicles in heavy traffic congestion (pilot)	May read a book, text, or web surf, but must be prepared to intervene when needed
	Cooperative adaptive cruise control AND automatic lane keeping	Driver may take hands off steering wheel and foot off accelerator, but must be prepared to intervene when needed
	Automated parking on a street without driver assistance	May watch in awe as this application executes
4	Highway driving pilot	May sleep—system reverts to minimum risk condition if driver fails to re-engage
	Closed campus driverless shuttle	No driver needed
	Driverless parking in a garage (Audi, Volvo)	Requires a smart phone with the appropriate application interface
	Automated buses on special transitways or appropriately marked or instrumented roads	No driver needed
5	Automated taxi, even for children	No driver needed
	Car-share repositioning system	No driver needed

11.2.3 German Federal Highway Research Institute vehicle automation classification

The German Bundesanstalt für Strassenwesen or Federal Highway Research Institute (BASt) developed a five-level classification concept, described in Table 11.6, that harmonizes and guides vehicle automation to achieve legal certainty as to whether the driver or the vehicle's automation system is responsible for executing a function, such as longitudinal and lateral vehicle control, in a safe manner and to bring regulatory law into compliance with these decisions [5,11,12].

Level 1, the driver only level, assigns all driving responsibility to the driver such that the human driver executes the manual driving tasks.

In Level 2, the driver assistance level, the driver is in permanent command of either longitudinal or lateral control. The other task can be automated to a certain extent by a vehicle assistance system.

Level 3, partial automation, lets vehicle systems take over longitudinal and lateral control, while the driver permanently monitors the system and is prepared to assume control at any time.

Level 4, high automation, finds the vehicle systems in control of longitudinal and lateral motion, and where the driver is no longer required to permanently monitor the system. In case of a takeover request, the driver must assume control within a certain time buffer.

Level 5, full automation, allows vehicle systems to assume longitudinal and lateral control completely and permanently. In case of a takeover request that is not followed, the system will return to the minimal risk condition by itself, for example, by automatically braking and reducing speed until the vehicle comes to a complete stop. The driver need not monitor the system.

The first three levels in the BASt model are available today in some vehicles. The last two levels represent future concepts not available today to the general driving public or to commercial fleets.

11.2.4 U.S. NHTSA vehicle automation levels

The USDOT, through NHTSA, is concerned with facilitating the safe introduction and deployment of HAVs as a transportation mode. The agency has broad enforcement authority to protect the safety of the driving public against unreasonable risks of harm that may occur because of the design, construction, or performance of a motor vehicle or motor vehicle equipment, and to mitigate risks of harm, including risks that may be emerging or contingent. This authority and responsibility extends to cover defects and unreasonable risks to safety that may arise in connection with HAVs.

In 2013, NHTSA defined a preliminary statement of policy concerning automated vehicles that included a five-level vehicle automation taxonomy [13,14] summarized in Table 11.7, which is similar to the BASt model. However, in later guidance released in September 2016, it deferred to the SAE model and its six-level characterization of vehicle automation. The original NHTSA taxonomy is included as it is referred to in Chapter 12 where the 2013 NHTSA recommendations for licensing drivers for self-driving vehicle testing are discussed.

11.3 AUTOMATED VEHICLE DRIVING ENVIRONMENTS

Automated driving functions, at whatever level, must be designed to operate under variable traffic flow conditions, changing roadway configurations and surface types, and changeable

Table 11.6 BASt automation levels

Automation level	Driver's expectations	Exemplary systems
Level 1: Driver only	Driver continuously (throughout the compete trip) performs longitudinal (accelerating and braking) and lateral (steering) control.	No active driver assistance intervenes in the longitudinal and lateral control functions.
Level 2: Assisted	Driver continuously performs either lateral or longitudinal control. The other or remaining task is performed by the automating system to a certain level only. Driver's duties include • Permanently monitoring the system. • Must be prepared at any time to assume complete control of the vehicle.	Cooperative adaptive cruise control: • Longitudinal control with adaptive distance and speed control. Parking assistance: • Lateral control by the parking assistance system that provides automatic steering into a parking space. The driver performs longitudinal control.
Level 3: Partially automated	System takes over lateral and longitudinal control for a certain amount of time and/or in specific situations. Driver's duties include • Permanently monitoring the system. • Must be prepared at any time to assume complete control of the vehicle.	Motorway assistant: • Automatic longitudinal and lateral control on motorways up to an upper speed limit.
Level 4: Highly automated	System assumes lateral and longitudinal control for a certain amount of time in specific situations. • Driver need not permanently monitor the system as long as it is active. • If necessary, the system requests the driver to assume control within a certain time buffer. • All system limits are detected by the system. The system is not capable of reestablishing the minimal risk condition from every initial state.	Motorway chauffeur: • Automatic longitudinal and lateral control on motorways up to an upper speed limit.
Level 5: Fully automated	System assumes lateral and longitudinal control completely within the individual specification of the application. • Driver need not monitor the system. • Before the specified limits of the application are reached, the system requests the driver to assume control within a sufficient time buffer. • In absence of a takeover request, the system will return to the minimal risk condition by itself. • All system limits are detected by the system and the system is capable of returning to the minimum risk condition in all situations.	Motorway pilot: • Automatic longitudinal and lateral control on motorways up to an upper speed limit.

Source: T.M. Gasser and D. Westhoff, *BASt-study: Definitions of Automation and Legal Issues in Germany*, German Federal Highway Research Institute, July 25, 2012.

Table 11.7 U.S. NHTSA 2013 automation levels

Automation level	Definition	Exemplary systems
Level 0: No automation	Driver is in complete and sole control of the primary vehicle controls (namely, brake, steering, throttle, and motive power) at all times, and is solely responsible for monitoring the roadway and for safe operation of all vehicle controls. Level 0 technology could augment and facilitate the full implementation of many applications typical of other levels of automation.	Provide only warnings (e.g., forward collision warning, lane departure warning, blind spot monitoring) and automated secondary controls such as wipers, headlights, turn signals, and hazard lights.
Level 1: Function-specific automation	Automation of one or more specific control functions. If multiple functions are automated, they operate independently from each other. Driver has overall control and is solely responsible for safe operation. • Drivers are not disengaged from physically operating the vehicle by having their hands off the steering wheel and feet off the pedals at the same time. • Driver can choose to cede limited authority over a primary control (as in adaptive cruise control), or the vehicle can automatically assume limited authority over a primary control (as in electronic stability control), or the automated system can provide added control to aid the driver in certain normal driving or crash-imminent situations (as with dynamic brake support in emergencies).	Electronic stability control, cruise control, lane keeping, or precharged brakes where the vehicle automatically assists with braking to enable the driver to regain control of the vehicle or stop faster than possible by acting alone.
Level 2: Combined function automation	Automation of at least two primary control functions that operate in unison to relieve the driver of control of those functions. Driver is still responsible for monitoring the roadway and safe operation and is expected to be available for control at all times and on short notice. Drivers are disengaged from physically operating the vehicle by having their hands off the steering wheel AND feet off the pedals at the same time. System can relinquish control with no advance warning and the driver must be ready to control the vehicle safely. The major distinction between Level 1 and Level 2 is in the specific operating conditions for which the Level 2 system is designed.	Adaptive cruise control in combination with lane centering
Level 3: Limited self-driving automation	Enables the driver to cede full control of all safety-critical functions under certain traffic or environmental conditions and, in those conditions, to rely heavily on the vehicle to monitor for changes that require transition back to driver control. Driver is not expected to constantly monitor the roadway while driving. Driver is expected to be available for occasional control, but with sufficient transition time. The major distinction between Level 2 and Level 3 is that a Level 3 vehicle is designed with the expectation that the driver is not constantly monitoring the roadway while driving.	Automated or self-driving vehicle that determines when the system is no longer able to support automation, such as when approaching a construction area, and then signals to the driver to reengage in the driving task and regain manual control.

(Continued)

Table 11.7 (Continued) U.S. NHTSA 2013 automation levels

Automation level	Definition	Exemplary systems
Level 4: Full self-driving automation	Vehicle performs all safety-critical driving functions and monitors roadway conditions for an entire trip. • Safe operation rests solely on the automated vehicle system. Driver provides destination or navigation input, but is not expected to be available for control at any time during the trip. This includes both occupied and unoccupied vehicles.	No examples currently available to the driving public, although several automobile manufacturers are developing these vehicles.

Source: U.S. Department of Transportation Releases Policy on Automated Vehicle Development, May 30, 2013. http://www.nhtsa.gov/About+NHTSA/Press+Releases/U.S.+Department+of+Transportation+Releases+Policy+on+A utomated+Vehicle+Development; Preliminary Statement of Policy Concerning Automated Vehicles, National Highway Traffic Safety Administration, U.S. Department of Transportation, National Highway Traffic Safety Administration, Washington, DC, May 30, 2013. http://www.nhtsa.gov/staticfiles/rulemaking/pdf/Automated_ Vehicles_Policy.pdf.

weather. Furthermore, their designs may have to accommodate different regulatory statutes as they travel from state-to-state, province-to-province, or country-to-country. The scope of these challenges is discussed below.

11.3.1 Roadway

There are three types of roadway infrastructures on which an automated vehicle may find itself: the existing infrastructure unchanged, the existing infrastructure modified for automated vehicles, and an entirely new and separate infrastructure designed to service only automated vehicles. The automated systems built into a vehicle must adapt to all of these circumstances for the owner and driver of the vehicle to maximize their driving options with the assurance of safe operation on each of them.

An unchanged existing infrastructure has several renditions that include all existing roads, off-roads (usually unpaved), all paved roads, well-marked paved roads, urban and suburban arterials, rural highways, residential streets, limited-access highways (freeways and toll roads including bridges and tunnels), parking facilities, and parks or low-speed pedestrian zones. The class of existing infrastructure augmented for automation includes dedicated lanes within limited-access highways and those with special markings or electronics added. A separate new infrastructure may present itself in several versions, namely, as dedicated, protected lanes on limited-access highways; fully automated parking facilities; or physically separated guideways using, for example, personal rapid transit (PRT).

PRT vehicles are types of automated vehicles that travel on a separate and new infrastructure. Typical PRT vehicles (sometimes called pods) under development are golf cart–sized, motorized enclosures that normally hold up to four people and bicycles or luggage. There are a number of schemes that utilize a guideway from which the pods get their power. Figure 11.3 illustrates one such concept [15]. Guideways can be miniature rails, monorails, or paved surfaces with buried wires for tracking. Generally, the vehicles run on rubber tires. Some have sliding power connections below the surface that the wheels ride upon or alongside the guideway.

PRT vehicles are small; two could pass in the width of a typical driveway. The ratio of vehicle to passenger weight is much lower than that for buses, cars, or trains. They also can change elevation quickly. This allows a three-dimensional right-of-way to be carved out where nothing else would fit. PRT vehicles do not stop at every station along the way; they only stop at the selected destinations. This means that stations are miniature sidings along the route. It also permits routes to be interconnected in multiple ways to allow loops.

Figure 11.3 PRT concept. (Vectus PRT Test Track Video, Uppsala Sweden. http://faculty.washington.edu/jbs/itrans/prtquick.htm.)

There is no waiting time as a PRT does not run on a schedule. Riders just enter the next available vehicle, and select their destination and go. The overall PRT system is controlled by an intelligent computer system. Thus, passenger traffic is monitored continuously and empty vehicles are routed where they are needed based on historic and actual demand. PRT vehicles do not run empty unless the central computer is redistributing them.

11.3.2 Traffic flow

Driving environments vary from location to location, season to season, and time-of-day to time-of-day. Hence, automated vehicles must be able to operate under a diverse set of conditions, which are delineated in Table 11.8. Units of density are vehicles per lane per mile when speed is measured in miles per hour.

11.3.3 Combinations of road type and traffic flow

Automated vehicles must not only cope with the temporal and spatial variations of traffic densities, but also with different road construction and road types as described by the matrix in Table 11.9 [16].

Table 11.8 Traffic flow conditions under which automated driving functions must operate

Level of service	Density	Speed	Characterization and typical location
A	Low	Low	Mixed (residential)
B	High	Low	Well-behaved (urban)
C	Low	High	Well-behaved (rural highway)
D	High	High	Well-behaved (urban highway)
E	High	Low	Chaotic (Bangkok, Moscow)
F	High	High	Chaotic (rural, developing countries)

Table 11.9 Road and traffic flow environments for automated vehicle operation

Road type	Speed:	Low	Low	High	High	Low	High
	Density:	Low	High	Low	High	High	High
	Characterization:	Mixed	Behaved	Behaved	Behaved	Chaotic	Chaotic
	Level of service:	A	B	C	D	E	F
Existing infrastructure—unchanged							
Off road							
All roads							
All paved roads							
Well-marked paved roads							
Urban and suburban arterials							
Rural highways							
Residential streets							
Limited-access highways							
Parking facility							
Parks or low-speed pedestrian zones							
Existing infrastructure—augmented for automation							
Dedicated lanes within limited-access highway							
Special markings or electronics added							
Separate new infrastructure							
Dedicated, protected lanes on limited-access highway							
Fully automated parking facilities							
Physically separated guideways, e.g., PRT							

Source: S.E. Shladover, Automated Vehicles: Terminology and Taxonomy, Taxonomy Working Group, University of California PATH Program.

11.3.4 Weather

Weather sensors and the incorporation of the means to respond to weather changes are essential if self-driving vehicles are to become a ubiquitous part of our landscape. Level 4 and 5 automated vehicles (SAE taxonomy), in particular, must be able to operate safely under a wide range of weather and lighting conditions. These include fair weather (baseline); variable lighting conditions that change with daylight, nighttime, and sun angle, for example, glare from low sun angle driving; precipitation in the form of light rain, heavy rain, snow, sleet, or drizzle; other visibility challenges such as fog, dust, sand, and smoke; wind; and finally pavement surface condition due to precipitation (e.g., dry, wet, snow, and ice) and maintenance level (e.g., pot holes, debris, buckling, and condition of lane markings).

11.3.5 Road configurations

Automated vehicles must have the capability to operate safely on different types of road geometries and grades, on bridges and through tunnels, and when encountering planned special events and unscheduled incidents. Road conditions may be static but still challenging from curves of various radii and super elevation, grades and abrupt grade changes, line of sight restrictions from the built environment, road surface roughness, roadway and lane delineation markings, and roadway signage condition. Scheduled event traffic control by officers may interrupt travel on well-delineated lanes and known roads as may work zones. Similarly, dynamic or unscheduled incidents, incident responders blocking traffic, and law enforcement actions may require the automated vehicle to change lanes or roadways in a manner that was not anticipated when travel began.

11.3.6 Legal evaluation

Although the BASt effort is addressing legal issues related to autonomous vehicle operation, each country must ensure that its laws have been updated to deal fairly with this new transportation mode. Regulatory law is synonymous with national road traffic codes. These address drivers' duties with respect to national and international regulations and conventions [17,18]. The European Transport Safety Council (ETSC) is concerned with challenges that include the need for autonomous cars sold in Europe to be capable of following national road rules in 28 EU countries and updating driving license regulations to ensure that drivers learn to safely resume control from ADSs [19].

Figure 11.4 shows the origins of a national road traffic safety law and an international convention designed to promote road safety by standardizing traffic rules among the contracting parties. The convention that currently affects the introduction of automated vehicles in over 70 ratifying nations, but not the United States and United Kingdom, is the Vienna Convention on Road Traffic [5]. Its history and purpose are shown in the figure. It is interesting to note that in 1968 (the year the convention was created), animals were still used to

National Road Traffic Codes	Vienna Convention
The Road Traffic Safety Law of the People's Republic of China (中华人民共和国道路交通安全法) was passed by the National People's Congress of the People's Republic of China on October 28, 2003, promulgated by Decree No. 8 of the President of the PRC Hu Jintao, and took effect on May 1, 2004 on all parts of mainland China (but not in Hong Kong and Macau which have their own judicial systems.) It is the PRC's first-ever road traffic safety law, and is intended to address an alarmingly high traffic fatality rate, which is four or five times greater than that of other nations.	The Vienna Convention on Road Traffic is an international treaty designed to facilitate international road traffic and to increase road safety by standardizing the uniform traffic rules among the contracting parties. Agreed upon at the United Nations Economic and Social Council's Conference on Road Traffic (October. 7, 1968–November. 8, 1968). It came into force on May 21 1977. Ireland, Spain and UK have not ratified the treaty.

Figure 11.4 National road traffic code and international convention examples for promoting safety when traveling in vehicles.

move vehicles and the concept of autonomous driving was considered to be science fiction. This is important when interpreting the text of the treaty, whether in a strict interpretation to the letter of the text, or an interpretation of what is meant (at that time). Thus, regulatory law as it applies to automated vehicles may not be clear or may require legal evaluation as to its interpretation for automated driving scenarios [20].

Some of the questions still to be answered are, Is it the driver's obligation to permanently monitor surrounding traffic and status of the vehicle and be ready to override and oversteer in the event automated system control appears inadequate? In this context, override means the driver must be able to assume control of the vehicle from the automatic system and act to maneuver the vehicle in a different manner than that commanded by the automatic system. Oversteer refers to the need for the driver to turn the car by more than the amount specified by the automated controls to perform a given action or maneuver. Or are these driver responsibilities only in Level 1 and 2 (BASt automation levels) automated systems? Can drivers rely on the vehicle to alert them to instances when they are required to assume manual control of the vehicle as prescribed by the definitions of Level 3, 4, and 5 automation?

Another view on the legal issues is provided by Smith. In his 2014 Law Review article he states the following [17]:

> The Geneva Convention, to which the United States is a party, probably does not prohibit automated driving. The treaty promotes road safety by establishing uniform rules, one of which requires every vehicle or combination thereof to have a driver who is "at all times … able to control" it. However, this requirement is likely satisfied if a human is able to intervene in the automated vehicle's operation.
>
> NHTSA's regulations, which include the Federal Motor Vehicle Safety Standards to which new vehicles must be certified, do not generally prohibit or uniquely burden automated vehicles, with the possible exception of one rule regarding emergency flashers.
>
> State vehicle codes probably do not prohibit—but may complicate—automated driving. These codes assume the presence of licensed human drivers who are able to exercise human judgment, and particular rules may functionally require that presence. New York somewhat uniquely directs a driver to keep one hand on the wheel at all times. In addition, far more common rules mandating reasonable, prudent, practicable, and safe driving have uncertain application to automated vehicles and their users. Following distance requirements may also restrict the lawful operation of tightly spaced vehicle platoons. Many of these issues arise even in the several states that expressly regulate automated vehicles.

Smith proceeds to recommend five near-term measures that may help increase legal certainty without producing premature regulation.

1. Regulators and standards organizations should develop common vocabularies and definitions that are useful in the legal, technical, and public realms.
2. The United States should closely monitor efforts to amend or interpret the 1969 Vienna Convention, which contains language similar to the Geneva Convention but does not bind the United States.
3. NHTSA should indicate the likely scope and schedule of potential regulatory action. (*Author's note*: In 2016, NHTSA unveiled two policy statements. The first addresses automated vehicle development and operation as discussed in Section 11.4. The second is a Notice of Proposed Rulemaking to establish a new Federal Motor Vehicle Safety Standard (FMVSS) that mandates V2V communications for new light vehicles and standardizes the message and format of V2V transmissions as discussed in Section 12.10.)

4. States in the United States should analyze how their vehicle codes would or should apply to automated vehicles, including those that have an identifiable human operator and those that do not.

5. Additional research on laws applicable to trucks, buses, taxis, low-speed vehicles, and other specialty vehicles may be useful. This is in addition to ongoing research into the other legal aspects of vehicle automation.

In the EU, the ETSC admits that it is far from answering the many research and regulatory questions that must be considered. The challenges include ensuring that autonomous cars sold in Europe are capable of following national road rules in 28 EU countries, updating EU driving license regulations to reflect the need for drivers to learn how to safely resume control from ADSs, promoting standards for infrastructure changes for automated and semiautomated vehicles such as clear road markings, and determining how autonomous vehicles will interact with human-driven vehicles, pedestrians, and cyclists. The ETSC initially wants the EU to mandate the installation of effective and proven driver assistance systems in all new cars and to develop an EU framework for approving automated technologies and autonomous vehicles. Furthermore, it requests car makers to be fully open and transparent in disclosing automated vehicle collision data to help prevent future collisions [21].

11.4 2016 NHTSA AUTOMATED VEHICLES POLICY

NHTSA's automated vehicles policy, unveiled in September 2016, recognizes the complex driving environments and legal issues that beset the widespread deployment of automated vehicles and "sets out an ambitious approach to accelerate the highly automated vehicle revolution" [22]. In this policy, NHTSA acknowledges "the remarkable speed with which increasingly complex HAVs are evolving" and challenging the USDOT "to take new approaches that ensure these technologies are safely introduced (i.e., do not introduce significant new safety risks), provide safety benefits today, and achieve their full safety potential in the future." The DOT expects and intends the policy and its guidance to be iterative, changing based on public comment; the experience of the agency, manufacturers, suppliers, consumers, and others; and further technological innovation.

The policy adopts the SAE International definitions for levels of automation based on whether a human driver or the ADS monitors the driving environment and performs other tasks [8]. NHTSA refers the public to "Key Considerations in the Development of Driving Automation Systems" [23] for examples and applications of classifying HAV systems to the SAE levels of automation. This document was prepared by the Crash Avoidance Metrics Partnership (CAMP) Automated Vehicle Research Consortium consisting of representatives from Nissan North America, Volkswagen Group of America, Ford Motor Company, General Motors, Mercedes-Benz, CAMP, Toyota Motor Engineering and Manufacturing North America, Inc., and NHTSA.

Throughout the NHTSA policy, the term HAV represents SAE Level 3–5 vehicles with automated systems responsible for monitoring the driving environment. NHTSA defines an automated vehicle system as a combination of hardware and software (both remote and onboard) that performs a driving function, with or without a human actively monitoring the driving environment. A vehicle has a separate automated vehicle system for each ODD. The ODD defines the conditions in which that function is intended to operate with respect to geographical location, roadway types, speed range, lighting (day, night, or both), weather, and other design constraints. Accordingly, SAE Level 2, 3, or 4 vehicles could have one or multiple automated systems, one for each ODD (e.g., freeway driving with automatic acceleration and deceleration to maintain a preset intervehicle separation distance, self-parking, automatic lane changing, and restricted-area urban driving). SAE Level 5 vehicles have a single automated vehicle system that performs under all conditions. If a vehicle can perform freeway and nonfreeway driving, the ODD would contain the appropriate scenarios for safe vehicle operation on both types of roadways and the system would be considered one system.

The 2016 policy contains four areas that encompass

- Vehicle Performance Guidance for Automated Vehicles.
- A Model for HAV State Policy.
- Current NHTSA Regulatory Tools.
- New Tools and Authorities.

Below are brief explanations of the content of each of the areas followed by more detailed descriptions of the Vehicle Performance Guidance and the NHTSA Model for HAV State Policy. The reader is referred to [22] for additional information concerning current and new regulatory tools and authorities.

11.4.1 Vehicle Performance Guidance for Automated Vehicles

Vehicle Performance Guidance for Automated Vehicles outlines best practices for the safe predeployment design, development, and testing of HAVs prior to commercial sale or operation on public roads. The guidance is intended as an initial step to further influence the safe testing and deployment of HAVs. It establishes DOT's expectations of industry by providing reasonable practices and procedures that manufacturers, suppliers, and other entities should follow in the immediate short term to test and deploy HAVs. The data generated from these activities should be shared in a way that allows government, industry, and the public to increase their learning and understanding as technology evolves, but protects legitimate privacy and competitive interests.

Testing refers to analyses and evaluations of HAV systems and vehicles conducted by a researcher, manufacturer, entity, or expert third party at the request of one of those entities. Deployment refers to use of HAV systems and vehicles by members of the public who are not employees or agents of research or design organizations, manufacturers, or other entities. A manufacturer is an individual or company that manufactures HAVs for testing and deployment on public roadways. Manufacturers include original equipment manufacturers, multiple- and final-stage manufacturers, and alterers (individuals or companies making changes to a complete vehicle prior to first retail sale or deployment) and modifiers (individuals or companies making changes to existing vehicles after first retail sale or deployment) of HAVs. Other entities are individuals or companies that are not a manufacturer, and are involved with designing, supplying, testing, selling, operating, deploying, or helping to manufacture HAVs.

11.4.2 A Model for HAV State Policy

Integration of HAVs should not change the ability of a motorist to drive across state lines without a concern more complicated than "did the speed limit change?" Similarly, a manufacturer should be able to focus on developing a single HAV fleet rather than 50 different versions to meet individual state requirements. The Model State Policy confirms that states retain their traditional responsibilities for vehicle licensing and registration, traffic laws and enforcement, and motor vehicle insurance and liability regimes. Furthermore, it encourages states to evaluate current laws and regulations to identify unnecessary impediments to the safe testing, deployment, and operation of HAVs, and update references to a human driver to take into account changes that occur when higher levels of vehicle automation transition from a human to the automated vehicle as the agent for conducting the driving task and monitoring the driving environment.

11.4.3 Current NHTSA Regulatory Tools

NHTSA will continue to exercise its available regulatory authority over HAVs using its existing regulatory tools: interpretations, exemptions, notice-and-comment rulemaking, and defects and enforcement authority. This authority allows the agency to identify safety defects and recall vehicles or equipment that poses an unreasonable risk to safety even when there is no applicable FMVSS.

To aid regulated entities and the public in understanding the use of these tools (including the introduction of new HAVs), NHTSA prepared a new information and guidance document to be published in the Federal Register. This document provides instructions, practical guidance, and assistance to entities seeking to employ those tools. Furthermore, NHTSA has streamlined its review process and promises to issue simple HAV-related interpretations in 60 days, and rule on simple HAV-related exemption requests in 6 months.

11.4.4 New Tools and Authorities

Because today's governing statutes and regulations were developed when HAVs were only a notion, and because of the speed with which complex and novel HAV innovation is advancing, existing NHTSA regulatory tools may not be sufficient to ensure the full safety promise of the new technologies. Therefore, in addition to more effective use of existing regulatory tools, the initiative identifies potential new tools, authorities and regulatory structures that could aid the safe and appropriately speedy deployment of new technologies.

New tools focus on measures that support effective risk mitigation, safety performance, and premarket safety assurance. These tools include variable test procedures to ensure behavioral competence and avoid the gaming of tests designed to assure the safety of the public, functional and system safety reporting to identify possible safety-related defects and ensure that manufacturers are satisfying their duties with respect to such defects, use of an iterative and forward-looking process for establishing and updating testing protocols for HAVs, and enhanced event data recorders to allow reconstruction of the circumstances of crashes and gain an understanding of how a vehicle involved in a crash or incident sensed and responded to its driving environment immediately before and during the crash or near crash. New authorities include premarket approval, cease and desist to require manufacturers to take immediate action to mitigate safety risks that are serious and imminent, expansion of the existing exemption authority to exceed the current limit of 2500 vehicles per year for a 2-year period on the basis of equivalent safety, and post-sale authority to regulate software changes.

11.4.5 Vehicle Performance Guidance

Under current U.S. law, manufacturers bear the responsibility to self-certify that all of the vehicles they manufacture for use on public roadways comply with all applicable FMVSS. Therefore, if a vehicle is compliant within the existing FMVSS regulatory framework and maintains a conventional vehicle design, there is no specific federal legal barrier to an HAV being offered for sale.

However, manufacturers and other entities designing new automated vehicle systems are subject to NHTSA's defects, recall, and enforcement authority. The DOT anticipates that manufacturers and other entities planning to test and deploy HAVs will use this guidance, industry standards, and best practices to ensure that their systems will be reasonably safe under real-world conditions. Furthermore, NHTSA expects to pursue follow-on actions to this guidance, such as holding public workshops, obtaining public comment, and performing additional research in areas such as benefits assessment, human factors, cybersecurity, performance metrics, objective testing, and others as they are identified in the future.

The NHTSA guidance is not currently mandatory and is not intended for states to codify as legal requirements for the development, design, manufacture, testing, and operation of automated vehicles. However, NHTSA may consider, in the future, proposing to make some elements of this guidance mandatory and binding through regulatory actions.

11.4.5.1 Scope

The Vehicle Performance Guidance is intended for all individuals and companies manufacturing, designing, testing, or planning to sell automated light-, medium-, and heavy-duty vehicles and vehicle systems in the United States. These groups include, but are not limited to, equipment designers and suppliers; entities that outfit any vehicle with automation capabilities or HAV equipment for testing, commercial sale, or operation on public roadways; transit companies; automated fleet operators; driverless taxi companies; and any other individual or entity that offers services utilizing HAVs. The guidance targets vehicles that incorporate HAV systems, such as those for which there is no human driver at all, or for which the human driver can give control to the HAV system and is not be expected to perform any driving-related tasks for a period of time.

The guidance is applicable to both test- and production-level vehicles. If a vehicle is operated by members of the public who are not employees or agents of the manufacturer or other testing and production entities, the guidance considers that operation to be deployment (not testing). For use on public roadways, automated vehicles must meet all applicable FMVSS. If a manufacturer or other entity wishes to test or operate a vehicle that would not meet applicable safety standards, "[t]he Agency encourages manufacturers to, when appropriate, seek use of NHTSA's exemption authority to field test fleets that can demonstrate the safety benefits of fully autonomous vehicles" [24]. This statement also applies to entities that traditionally may not be considered manufacturers under NHTSA's regulations, for example, alterers, modifiers, transit companies, fleet owners, and others who may test or operate HAV systems.

In addition to safety, automated vehicles can provide significant, life-changing mobility benefits for persons with disabilities, older persons, and others who may not be the target of conventional vehicle design programs. Accordingly, the DOT encourages manufacturers and other entities to consider the full array of users and their specific needs during the development process.

11.4.5.2 Vehicle Performance Guidance framework

Figure 11.5 presents the framework for the DOT's Vehicle Performance Guidance. The framework identifies the key areas to be addressed by manufacturers and other entities prior to testing or deploying automated vehicles on public roadways. The manufacturer or other entity is responsible for determining their system's level of automation in conformity with

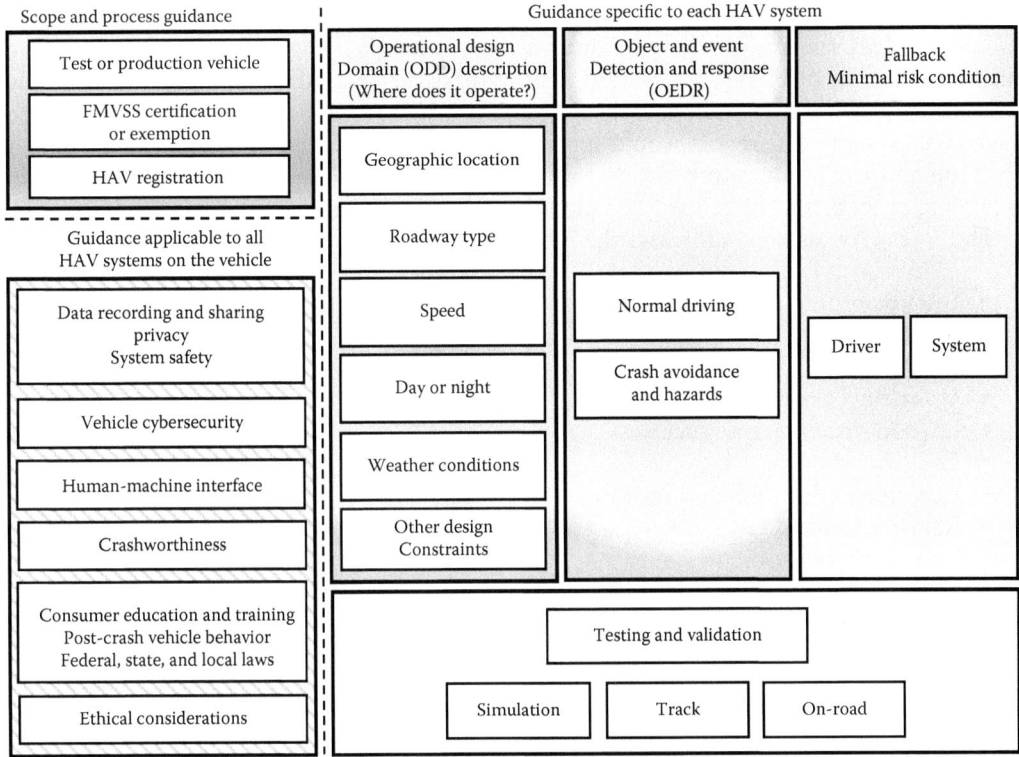

Scope and process guidance

Test or production vehicle

FMVSS certification or exemption

HAV registration

Guidance applicable to all HAV systems on the vehicle

Data recording and sharing privacy
System safety

Vehicle cybersecurity

Human-machine interface

Crashworthiness

Consumer education and training
Post-crash vehicle behavior
Federal, state, and local laws

Ethical considerations

Guidance specific to each HAV system

Operational design Domain (ODD) description (Where does it operate?)

Geographic location

Roadway type

Speed

Day or night

Weather conditions

Other design Constraints

Object and event Detection and response (OEDR)

Normal driving

Crash avoidance and hazards

Fallback Minimal risk condition

Driver

System

Testing and validation

Simulation

Track

On-road

Figure 11.5 USDOT framework for Automated Vehicle Performance Guidance.

SAE International's published definitions. NHTSA will review manufacturers' automation level designations and advise the manufacturer if it disagrees with the assigned level.

The framework pertains to test and production vehicles and to original equipment and replacement equipment or updates (including software updates and upgrades) used in automated vehicle systems. It applies to capabilities and systems that are cross-cutting, that is, those that apply to all automation functions on the vehicle, and those that apply to a specific automation function on the vehicle. Cross-cutting areas include data recording and sharing, privacy, system safety, cybersecurity, human–machine interface (HMI), crashworthiness, and consumer education and training. Areas specific to each vehicle automation function are the description of the ODD, OEDR, and fallback minimum risk condition. The framework incorporates testing to evaluate and validate that the HAV system can operate safely in the defined ODD and has the capability to return to a minimal risk condition when needed. Evaluation is performed through a combination of simulation, test track driving, or roadway driving.

Furthermore, manufacturers and other entities should place significant emphasis on assessing the risk of driver complacency and misuse of Level 2 systems, and develop effective countermeasures to assist drivers in properly using the system as the manufacturer expects. Manufacturers and other entities should develop tests, validation, and verification methods to assess their systems for effective complacency and misuse countermeasures. For example, a Level 2 vehicle might have a system to monitor human driver engagement, and take the vehicle to a safe fallback condition if the monitor determines the driver is not sufficiently engaged.

11.4.5.3 Safety assessment report to NHTSA

To aid NHTSA and the public in monitoring how safety is being addressed in the development and testing of HAVs, the Agency will request that manufacturers and other entities voluntarily provide Safety Assessment Reports to NHTSA's Office of the Chief Counsel for each HAV system. The report would specify how they are meeting the guidance at the time they intend their product to be ready for use (testing or deployment) on public roads. This reporting process may be refined and made mandatory through a future rulemaking.

The Safety Assessment addresses the following 15 areas:

- Data recording and sharing.
- Privacy.
- System safety.
- Vehicle cybersecurity.
- Human–machine interface.
- Crashworthiness.
- Consumer education and training.
- Registration and certification.
- Post-crash behavior.
- Federal, state, and local laws.
- Ethical considerations.
- Operational design domain.
- Object and event detection and response.
- Fallback (minimal risk condition).
- Validation methods.

If software or hardware updates materially change the way in which the vehicle complies (or takes it out of compliance) with any of the 15 elements in the Safety Assessment, the agency would require an update to the Safety Assessment that summarizes the particular changes.

11.4.6 Model HAV State Policy regulations

States in the United States are presently charged with reducing traffic crashes and the resulting deaths, injuries, and property damage [25]. They may use their authority to establish and maintain highway safety programs addressing issues including driver education and testing; licensing; pedestrian safety; law enforcement; vehicle registration and inspection; traffic control; highway design and maintenance; crash prevention, investigation, and record keeping; and emergency services.

States' responsibilities include motor vehicle regulations that should remain largely unchanged for HAVs, such as licensing (human) drivers and registering motor vehicles in their jurisdictions, enacting and enforcing traffic laws and regulations, conducting safety inspections where states choose to do so, and regulating motor vehicle insurance and liability. Nonetheless, other concerns appear when HAVs are introduced.

Table 11.10 summarizes the model framework proposed by NHTSA for states wishing to regulate procedures and conditions for testing, deployment, and operation of HAVs [22]. Evaluation of current laws and regulations by the states can address unnecessary impediments to the safe testing, deployment, and operation of HAVs, and update references to a human driver as appropriate. For example, for purposes of state traffic laws that apply to drivers of vehicles (e.g., speed limits and traffic signs), states may wish to deem an HAV system that

Table 11.10 NHTSA 2016 model state policy framework for test, deployment, and operation of HAVs

Function	State and manufacturer (or other entity) responsibilities
Administrative	States should identify a lead agency responsible for considering and approving testing of HAVs. Lead agency should create a jurisdictional automated safety technology committee that includes representatives from the governor's office, motor vehicle administration, state department of transportation, state law enforcement agency, state highway safety office, office of information technology, state insurance regulator, state office(s) representing the aging and disabled communities, toll authorities, and transit authorities. Other stakeholders should be consulted as appropriate, such as transportation research centers, the vehicle manufacturing industry, and groups representing pedestrians, bicyclists, consumers, and other interested parties. Lead agency should inform the automated safety technology committee of requests from manufacturers to test in their jurisdiction and the status of the designated agency's response to the manufacturers. Lead agency should use or establish statutory authority to implement a framework and regulations for (1) licensing and registration; (2) driver education and training; (3) insurance and liability; (4) enforcement of traffic laws and regulations; and (5) administration of motor vehicle inspections in order to address unnecessary barriers to safe testing, deployment, and operation of HAVs. Each state should develop an internal process to develop an application for manufacturers to test in the jurisdiction, issue test vehicle permits, and identify any legal issues that need to be addressed prior to the deployment and operation of automated vehicles.
Application by manufacturers and others to test HAVs on public roadways	Each manufacturer or other entity should submit an application to the designated lead agency in each jurisdiction in which they plan to test their HAVs. The application should include the following types of information: • Evidence that each vehicle used for testing by manufacturers or other entities follows the performance guidance established by NHTSA and meets applicable Federal Motor Vehicle Safety Standards. • Name of the manufacturer or other entity, corporate physical and mailing addresses of the manufacturer or other entity, in-state physical and mailing addresses of manufacturer, if different from corporate address, name of the program administrator or director, and contact information for the program administrator or director. • Identification of each vehicle that will be used on roadways for testing purposes by VIN, vehicle type, and other unique identifiers such as the year, make, and model. • Identification of each test operator, their driver's license number, and the jurisdiction or country in which the operator is licensed. • The manufacturer's or other entity's safety and compliance plan for testing vehicles, which should include a self-certification of testing and compliance with NHTSA's vehicle performance guidance for the technology in the test vehicles under controlled conditions that simulate real-world conditions (e.g., various weather, types of roads, times of day and night, and seasonal variations in traffic flow density and routes) to which the applicant intends to subject the vehicle on public roadways. • Evidence of the manufacturer's or other entity's ability to satisfy a judgment or judgments for damages for personal injury, death, or property damage caused by a vehicle in testing in the form of an instrument of insurance, a surety bond, or proof of self-insurance, for no less than five million U.S. dollars. Depending on the circumstances, states may wish to establish a higher minimum insurance requirement. • Summary of the training provided to the employees, contractors, or other persons designated by the manufacturer or other entity as operators of the test vehicles. Approval should be granted by the designated lead agency if evidence of insurance, operator training, self-certification, and fulfillment of other conditions discussed below are demonstrated.

(Continued)

Table 11.10 (Continued) NHTSA 2016 model state policy framework for test, deployment, and operation of HAVs

Function	State and manufacturer (or other entity) responsibilities
Jurisdictional permission to test	Each jurisdiction's lead agency should involve the jurisdictional law enforcement agency before responding to the request from the manufacturer or other entity.
	The lead agency may request additional information or require the manufacturer or other entity to modify its application before granting authorization.
	The lead agency may choose to grant authorization to test in a jurisdiction with restrictions, and/or may prohibit manufacturers or other entities from testing in certain areas or locations, such as school zones, construction zones, or other safety-sensitive areas.
	The authorization may be suspended if the manufacturer or other entity fails to comply with state insurance or driver requirements, or fails to comply with its self-certification compliance plan.
	The lead agency should issue a letter of authorization to the manufacturer or other entity to allow testing in the state, and the state's motor vehicle agency should issue a permit to each test vehicle. The authorization and permits may be renewed periodically. The jurisdiction may determine that it is appropriate to charge fees for the application and for each vehicle-specific permit.
	The vehicle-specific permit must be carried in the test vehicle at all times.
	Each test vehicle should be properly registered and titled in accordance with state laws.
Testing by manufacturer or other entity	Manufacturers or other entities must comply with federal law and applicable NHTSA regulations before operating vehicles on public roadways, whether or not they are in testing or in "normal" operation.
	The test vehicle must be operated solely by persons designated by the manufacturer or other entity, who have received training and instruction concerning the capabilities and limitations of the vehicle.
	The operators testing the vehicles must hold a valid state driver's license from the state in which testing takes place or from any other state (in accordance with the testing state's laws).
	Before being allowed to operate a test vehicle, the persons designated by the manufacturer or other entity as operators of the test vehicles may be subjected to a background check including, but not limited to, a driver history review and a criminal history check.
	The test operators are responsible for following all traffic rules and will be responsible for all traffic violations.
	All crashes involving test vehicles must be reported in accordance with the laws in the state in which the crash occurred.
Defining drivers of deployed vehicles and their responsibilities	States regulate human drivers. Licensed drivers are necessary to perform the driving functions for motor vehicles equipped with automated safety technologies that are less than fully automated (SAE Levels 3 and lower). A licensed driver has responsibility to operate the vehicle, monitor the operation, or be immediately available to perform the driving task when requested or the lower level automated system disengages.
	Fully automated vehicles are driven entirely by the vehicle itself and require no licensed human driver (SAE Levels 4 and 5), at least in certain environments or under certain conditions, i.e., the entire driving operation is performed by a motor vehicle automated system from origin to destination. For dual-capable vehicles defined as vehicles entirely driven either by the vehicle itself or by a human driver, the states would have jurisdiction to regulate (license, etc.) the human driver.
	In order to make the transition from human-driven motor vehicles equipped with automated safety technologies to fully automated vehicles, gaps in current regulations should be identified and addressed by the states (with the assistance of NHTSA). Some examples of instances where gaps may exist are as follows:

(Continued)

Table 11.10 (Continued) NHTSA 2016 model state policy framework for test, deployment, and operation of HAVs

Function	State and manufacturer (or other entity) responsibilities
	• Law enforcement and emergency response • Occupant safety • Motor vehicle insurance • Crash investigations and crash reporting • Liability (tort, criminal, etc.) • Motor vehicle safety inspections • Education and training • Vehicle modifications and maintenance • Environmental impacts
Registration and titling of deployed vehicles	HAV technologies that allow the vehicle to be operated without a human driver either at all times or under limited circumstances should be identified on title and registration documentation by states with the code HAV in a new data field. When HAV technologies that allow the vehicle to be operated without a human driver either at all times or under limited circumstances are installed on a vehicle after the initial purchase of the vehicle, the motor vehicle agency should be notified by the installer. The vehicle registration and title should be marked with the code HAV in a new data field. Regulations governing labeling and identification for HAVs should be issued by NHTSA.
First-responder training and HAV driving regulations	First responders and law enforcement should be aware and understand how HAVs may affect their duties. Law enforcement personnel will require training and education regarding their interaction with drivers or operators in both the testing and deployment of these technologies. Responders to crashes of HAVs should understand the unique hazards they may encounter. These include silent operation, self-initiated or remote ignition, high voltage, and unexpected vehicle movement. For vehicles that offer less than full automation capabilities, there is potential for increased distracted driving. Contributors to distracted driving, such as using an electronic device, eating, drinking, and conversing with passengers, could significantly increase in HAVs. Regulations to limit these activities, especially in vehicles providing less than full self-driving capabilities, should be developed and be consistent across jurisdictions.
Liability and insurance	States are responsible for determining liability rules for HAVs and should consider how to allocate liability among HAV owners, operators, passengers, manufacturers, and others when a crash occurs. For example, if an HAV is determined to be at fault in a crash then who should be held liable? Resolution of who or what is the "driver" of an HAV in a given circumstance does not necessarily determine liability for crashes involving that HAV. For example, states may determine that in some circumstances liability for a crash involving a human driver of an HAV should be assigned to the manufacturer of the HAV. States need to determine who (owner, operator, passenger, manufacturer, etc.) must carry motor vehicle insurance. Rules and laws allocating tort liability could have a significant effect on both consumer acceptance of HAVs and their rate of deployment. Such rules also could have a substantial effect on the level and incidence of automobile liability insurance costs in jurisdictions in which HAVs operate. It may be desirable to create a commission to study liability and insurance issues and make recommendations to the states.

Source: Federal Automated Vehicles Policy, U.S. Department of Transportation, National Highway Traffic Safety Administration, Washington, DC, Sep. 2016. https://www.transportation.gov/sites/dot.gov/files/docs/AV%20policy%20guidance%20PDF.pdf.

conducts the driving task and monitors the driving environment (typically SAE Levels 3–5) to be the driver of the vehicle. For vehicles and circumstances in which a human is primarily responsible for monitoring the driving environment (typically SAE Levels 1–2), NHTSA recommends the state consider that human to be the driver for purposes of traffic laws and enforcement. States may still wish to experiment with different policies and approaches to formulating consistent standards. The goal of state policies in this realm need not be uniformity or identical laws and regulations across all states. Rather, the aim should be sufficient consistency of laws and policies to avoid a patchwork of inconsistent state laws that could impede innovation and the expeditious and widespread distribution of safety-enhancing automated vehicle technologies. In such an approach, NHTSA generally would regulate motor vehicles and motor vehicle equipment (including computer hardware and software that perform functions formerly performed by a human driver) and the states would continue to regulate human drivers, vehicle registration, traffic laws, regulations and enforcement, insurance, and liability.

11.4.7 Next steps

NHTSA anticipates continuation of its collaboration with state, international, and other stakeholders in developing subsequent steps and future Model State Policy updates. These actions include the following:

- Gathering public comments concerning the Model State Policy and the entire policy.
- Holding public workshops to provide interactive discussions of the Model State Policy and gather additional input for future considerations.
- Participating in discussions with stakeholders at the state level who implement the Model State Policy to understand more about what states learned through their regulation of HAVs.
- Exploring a mechanism with vehicle manufacturers to help state officials gain a better understanding of available vehicle technologies and NHTSA's roles and activities.
- Discussing with relevant stakeholders (e.g., environmental groups and disability advocacy groups) the development of a work plan that facilitates policy refinements, or convening a commission to study a particular issue (e.g., insurance and liability) and make recommendations.
- Engaging with Canadian and Mexican authorities to leverage this policy to promote North American cross-border coordination.
- Coordinating with state partners and other safety stakeholders on a continuous basis to ensure that the Vehicle Performance Guidance and the Model State Policy continue to complement each other.

11.5 SYSTEMS REQUIRED BY AUTOMATED VEHICLES

Enabling the various levels of automated vehicles to operate on public roads requires interaction among many types of technologies. Two classification schemes for the hardware and software systems needed by these vehicles and the associated longitudinal and lateral vehicle control system requirements are described below.

11.5.1 The four-system taxonomy

The first method of examining the systems required by automated and autonomous vehicles is based on four types of systems [26]. The first is a satellite navigation system containing digital maps to establish location. These have been available for over two decades either from vehicle manufacturers or aftermarket devices. Today most smart phones make navigation mapping systems available to their owners.

The second system fulfills the requirement to see 360° around the vehicle under a wide variety of conditions including inclement weather and darkness. A typical system with this capability utilizes a central computer to collect information from a variety of sensors, including long-, medium-, and short-range microwave and millimeter-wave radars, lidars, video cameras (including visible spectrum and infrared models), and ultrasonic sensors as illustrated in the top portion of Figure 11.6. The computer activates alerts or brakes automatically based on its analysis of inputs from these sensors. Each sensor type has its strengths and limitations. Radar and lidar do not need natural or artificial light to function and, unlike cameras, can determine relative velocity and estimate distance. Radar sensors can be designed to detect cars and pedestrians out to specific distances as indicated in the bottom section of Figure 11.6. However, radar sensors may be more limited than visible spectrum cameras in determining shapes around them. Lidar provides higher resolution information than radar over a narrow field of view (unless the lidar beam is scanned), whereas radar provides wider area coverage. Lidar is also capable of providing detailed information about object shapes by using algorithms to analyze the object profiles or other features generated by the sensor. Visible spectrum cameras require light for operation, but they can readily identify objects. Ultrasonic sensors can warn a vehicle that it is approaching another object such as a parked vehicle, curb, sign post, or tree. Chapter 5 and Table 16.1 contain additional information concerning the operation, strengths, and limitations of sensor technologies suited for automated vehicle applications.

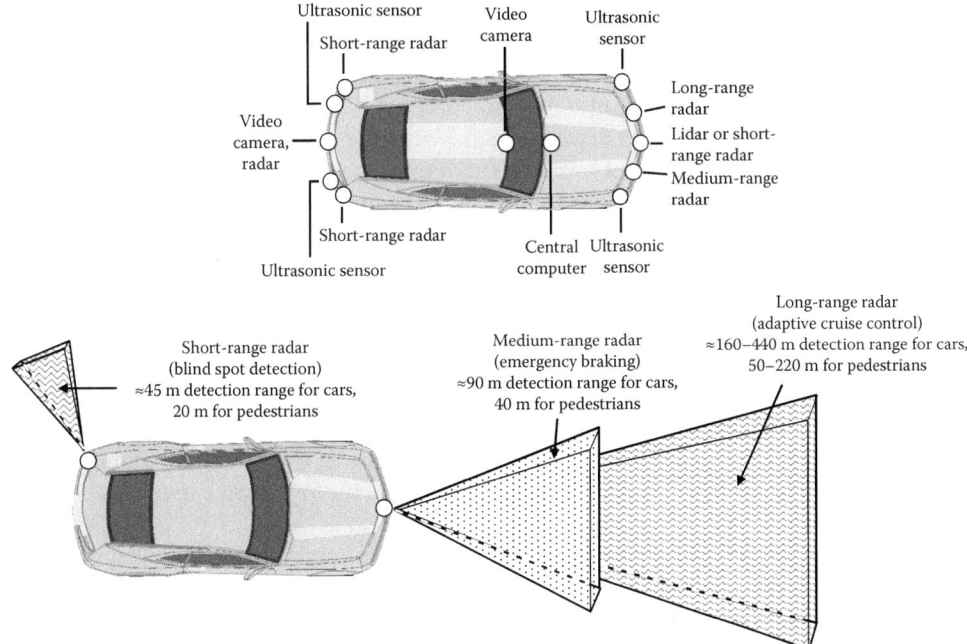

Figure 11.6 Sensor concepts to aid a vehicle's ability to see 360° around itself. The depiction on the top is notional and is not meant to represent a specific vehicle, manufacturer, or sensor grouping. The bottom picture shows typical detection ranges of short-, medium-, and long-range automotive radar sensors that detect cars and pedestrians.

The third system provides the capacity to communicate with other vehicles and parts of the infrastructure such as traffic signals, school zone warning devices, and parking and tolling facilities. This is accomplished by V2X communications discussed earlier in this chapter and again in Chapter 12. This system may not be required by autonomous vehicles, depending on their operational capabilities [27].

The fourth item is intelligent software that determines and initiates actions a vehicle must take in a given situation. This includes the attendant hardware to control longitudinal and lateral positions of the vehicle. Currently, there is no available method for efficiently developing, verifying, and validating software that can be certified as being dependable enough to make safety-of-life critical decisions. This issue is explored further in Section 12.9.

11.5.2 Homocentric description of autonomous vehicle systems

In this scheme, autonomous vehicle operation is enabled by a homocentric grouping of five systems [28]. These provide an interface between a human user and an autonomous vehicle, several types of data input sensors, automated controls for the vehicle functions, and the artificial intelligence that integrates input data and determines when and how to activate the automated vehicle controls. The systems are as follows:

1. Human–vehicle interface. This interface may be biometric such as a fingerprint reader or speaker recognition, or may take the form of a fob, a push-button, password entry, or other on–off control. For security purposes, at least two means of verification (access factors) are likely to be required.

2. Sensors that provide data about internal operation of the vehicle and its components. These include sensors for brakes, transmission, steering, throttle, and tires that are already embedded in many modern vehicles.

3. Sensors that provide location and real-time external roadway environment data. Precise, real-time mapping, tracking, and other technologies embedded in autonomous vehicles to make them aware of environmental conditions are essential to safe vehicle operation. As a result, most driverless cars will routinely receive and generate mapping updates at frequent intervals. This dynamic mapping data could potentially be provided as cloud-sourced autonomous vehicle roadway data.

4. Automated control of vehicle functions and operation. In an autonomous vehicle, control over vehicle operation is automated through networks of actuator microprocessors (sometimes called electronic control units [ECUs]) triggered by the vehicle's artificial intelligence. Automated controls in conventional vehicles appear remarkably reliable in accomplishing specific vehicle operations from antilock brakes to electronic stability control. However, some automated vehicle controls seem more reliable than others. For example, automatic lane-keeping controls appear less reliable than electronic stability control. Automated controls have proved to be the most vulnerable aspect of vehicle automation to car hacking and present legal as well as technical challenges for autonomous vehicles.

5. Artificial intelligence that integrates in-vehicle operational data with external roadway data and activates automated vehicle controls. This machine ability to control all vehicle operations distinguishes autonomous vehicles from other automated technologies that either assist or warn human drivers. It is likely that autonomous vehicle artificial intelligence will be functionally distributed across multiple parts of a vehicle's decision and control systems, rather than being located in a single central processing unit. It also will be self-learning in the sense that the algorithms utilized in operating a vehicle modify themselves over time in response to previous operations, new information, and feedback.

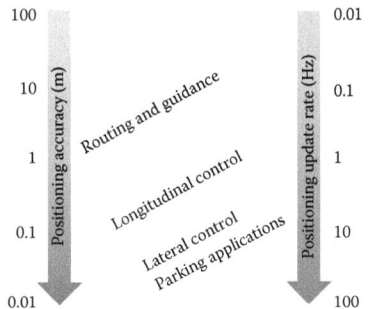

Figure 11.7 Summary of accuracy and update rates for vehicle positioning system applications. (From M. van Schijndel-de Nooij et al. *Definition of Necessary Vehicle and Infrastructure Systems for Automated Driving*, European Commission SMART 2010/0064 Study Report, Brussels, BE, June 2011.)

Application	Accuracy	Update rate
Basic routing and guidance	1 m to tens of meters	0.1 to 1 Hz
Longitudinal control	0.1 m to 1 m	1 to 10 Hz
Lateral control	0.01 to 0.1 m	1 to 10 Hz and greater
Parking	0.01 to <0.1 m	10 Hz or greater

11.5.3 Longitudinal and lateral vehicle control system requirements

Positioning systems that can automate longitudinal or lateral vehicle control are required for even Level 1 (SAE and NHTSA automation levels) automation applications. Positioning system requirements are directly related to the particular application. For example, Figure 11.7 summarizes the accuracy and update rate requirements for four different generic applications: routing and guidance (including lane matching) based on map information, longitudinal control in urban or highway settings, lateral control in urban or highway settings, and parallel parking along a street curb [5].

Basic routing and guidance requires a positioning system accuracy of decameter level order, in combination with map matching. If lane matching is included, the accuracy increases to the meter level. The update rate for this application is between 0.1 and 1 Hz. Longitudinal control (e.g., platooning, collision warning and avoidance) requires an accuracy between decimeter and meter level along with an increase in update rate to 1–10 Hz. Lateral control (e.g., lane keeping) accuracy increases even more to between centimeter and decimeter level. The update rate has the same range as that for longitudinal control. Parallel parking application accuracy requirements are in the lateral control category. However, the update rate is of the order of 10 Hz or greater. Parking applications are also affected by time delays in the positioning system, which are directly related to positioning accuracy and positioning update rate.

11.6 DRIVER ASSISTANCE AND CRASH AVOIDANCE FEATURES

There are a variety of driver assistance and crash avoidance features offered in vehicles. Several of them can be categorized as SAE Level 1 automation or above. The information in this section is intended as a sampling of these features and is not meant to be a complete listing of all offerings from all car manufacturers. The pictures in Figure 11.8 display several of these aids.

11.6.1 Blind spot warnings

A number of manufacturers offer blind spot warnings and cross-traffic alerts that employ radar to detect a vehicle in the driver's blind spot. Upon activating the turn signal to show

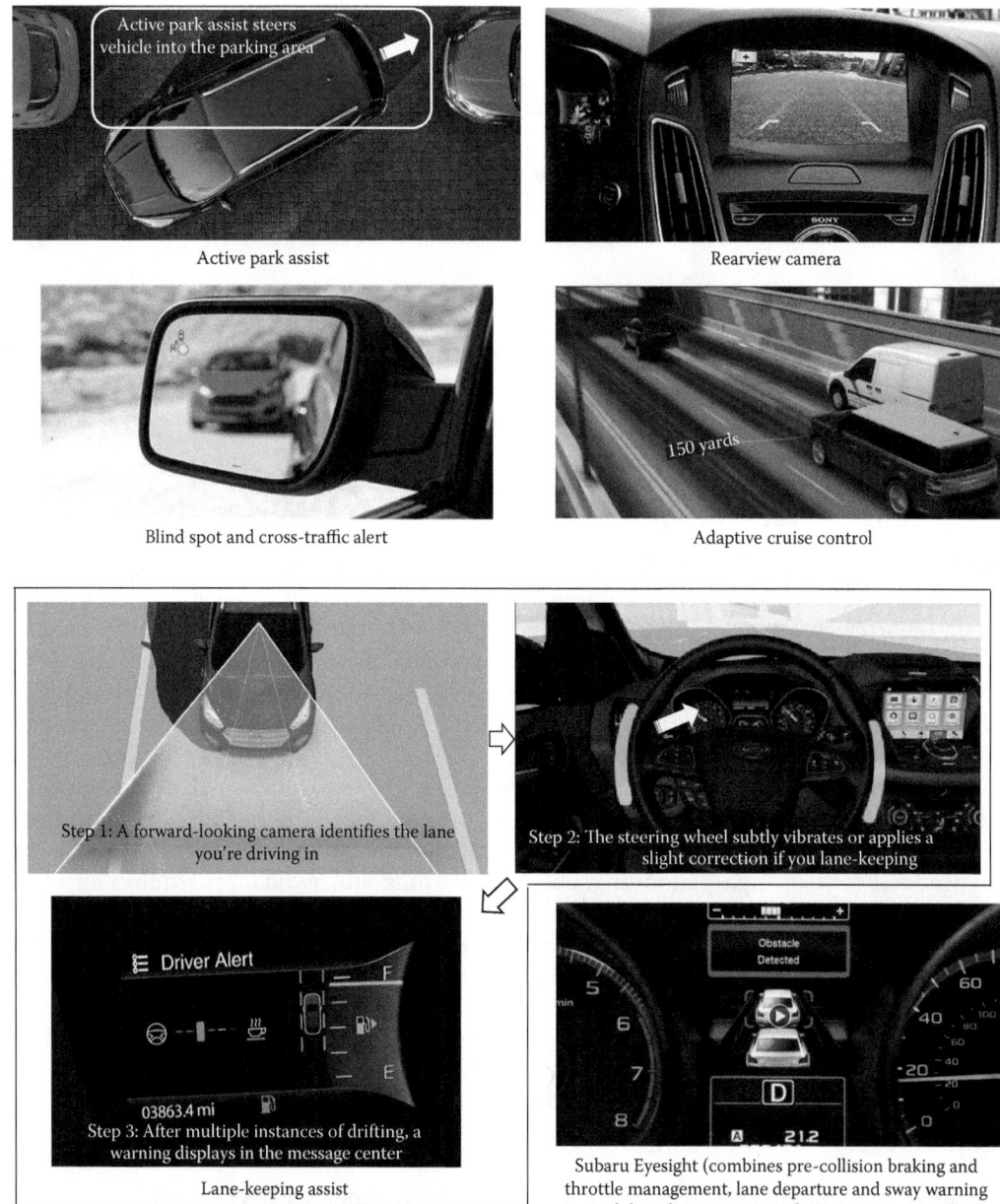

Figure 11.8 Driver assist and crash avoidance features found in vehicles. (Photographs reprinted with permission of Ford Motor Company and Subaru of America.)

the intention to change lanes, the blind spot warning system recognizes and alerts the driver through a visible signal in the exterior mirror casing if a vehicle is in the driver's blind spot or a vehicle is approaching at high speed in the overtaking lane. Cross-traffic alert also uses radar to watch for traffic behind a vehicle as the driver backs out of a parking spot or driveway. Audible and visual alerts are given if cross-traffic sensors detect a vehicle approaching up to 15 yards (14 m) away.

11.6.2 Adaptive cruise control and forward collision warning

Adaptive cruise control (a Level 1 driver assistance function) operates like traditional cruise control, with one difference. When the sensors detect traffic slowing ahead, the vehicle also slows down. When traffic clears, the vehicle resumes the set speed. A further refinement of this concept is automatic emergency braking (AEB), which applies the brakes for the driver. The systems use on-vehicle sensors such as radar, cameras, or lidars to detect an imminent crash, warn the driver, and apply the brakes if the driver does not take sufficient action quickly enough. In March of 2016, NHTSA and the Insurance Institute for Highway Safety (IIHS) announced that 20 automakers entered into a voluntary agreement to make AEB standard by September 1, 2022 on more than 99% of the U.S. auto market. Trucks with gross weights between 8501 and 10,000 pounds will mostly be equipped with AEB by September 1, 2025, 3 years after the first agreement begins [29].

Forward collision warning with brake support alerts the driver if it senses a potential collision with the car in front. A heads-up display, which simulates brake lights, flashes on the windshield and provides an audible warning. If the driver does not react in time, the brakes precharge and increase brake-assist sensitivity to provide full responsiveness when the driver does brake.

11.6.3 Lane-keeping system

A camera mounted behind the windshield monitors road lane markings that are not obscured by rain, snow, or ice to determine vehicle position and detect a lane departure. The lane-keeping alert consists of a series of steering wheel vibrations that mimic a rumble strip. The aid actively applies steering torque, which alerts the driver to direct the vehicle back into the target lane should the system detect an unintended lane departure.

11.6.4 Rearview camera

With the gear selector in reverse, the rearview camera automatically transmits the image of what is behind a vehicle to a screen on the vehicle's console. Interactive lane lines may appear to indicate whether a parking space is large enough for the vehicle. When maneuvering with a trailer, the rearview camera may also provide support via an extended zoom function.

11.6.5 Active park assist

When activated and while driving slowly near parking spots, this feature scans for available parking spots. Ultrasonic sensors measure the distance to the curb and other parked cars. Once a big enough spot is identified, the driver is signaled to stop and accept the system's assistance. In some vehicles, the driver still controls the shifting, accelerating, and braking. In others, the park assist function selects the gear on its own, guides the steering, and automatically accelerates or brakes, thus fulfilling the requirements for a Level 2 automation system.

One manufacturer offers a vehicle that can park itself in a parking structure (Level 3+ automation). The feature is activated via a smart phone once the driver exits the vehicle. The vehicle then drives itself through the parking structure until it finds a vacant space, upon which it executes the steering, acceleration, and braking maneuvers necessary to safely park the vehicle. When the driver wants to retrieve the vehicle, he once again activates the application on the smart phone. The car will start up, pull out of the parking spot, and maneuver its way through the parking structure to the waiting driver.

11.6.6 360° sensing

Some manufacturers increase the visual range of their vehicles to up to 50 m in front of the vehicle and environment recognition to 500 m. Vehicles driving ahead, oncoming and crossing traffic, pedestrians, and a variety of traffic signs and road markings are detected and assigned a spatial classification. A thermal imaging camera is sometimes present to warn drivers of the potential danger of pedestrians or animals in unlighted areas. In-vehicle sensors may be installed to warn drivers of inattentiveness and drowsiness over an extended speed range.

11.6.7 Level 3 automated truck functionality

Daimler Trucks North America has unveiled trucks that operate on highways at Level 3 automated vehicle capability, enabling the driver to cede full control of all safety-critical functions to the vehicle under certain traffic or environmental conditions. The Inspiration Truck by Daimler underwent extensive testing before the Nevada Department of Motor Vehicles granted it a license to operate on public roads in the state using Level 3 automation [30]. The Inspiration is equipped with highway pilot sensors and computer hardware based on what is produced for another Freightliner model, and is fully certified to meet all U.S. FMVSS.

The pilot vehicle links together a set of camera technology and radar systems with lane stability, collision avoidance, speed control, braking, steering, and other monitoring systems. This combination creates a Level 3 automated vehicle operating system that can perform safely under a range of highway driving conditions. The automated vehicle system maintains legal speed, stays in the selected lane, keeps a safe braking distance from other vehicles, and slows or stops the vehicle based on traffic and road conditions. The vehicle monitors changes in conditions that require transition back to driver control when necessary in highway settings. The driver is in control of the vehicle for exiting the highway, on local roads, and in docking for making deliveries.

A 2016 American Transportation Research Institute report estimates the cost of Level 3 truck hardware and software in the range of $13,000 (U.S.), but expects prices to decrease as the technologies become more widely adopted [31]. The report also lists a number of impediments to autonomous truck deployment and proposed trucking industry solutions as enumerated in Table 11.11.

11.6.8 Monitoring driver fatigue

The EU's HARKEN (Heart and Respiration In-Car Embedded Nonintrusive Sensors) Research Project monitors cardiac and respiratory rhythms in a nonintrusive manner in an environment where vehicle vibrations and driver movements may make the desired signals difficult to detect. The car is able to determine if its driver is suffering from fatigue and issue warnings when driving ability becomes impaired. The sensors are composed of smart materials embedded in the car's seat cover and safety belt [32,33].

11.7 FULLY AUTOMATED VEHICLES

While fully automated vehicles are expected to make appearances on the roadways of many countries in the next several years notably for research and evaluation purposes, their availability, acceptance, and purchase by the general public may be limited for a decade or more.

Table 11.11 Potential impediments to autonomous truck deployment

Issue	Impediment	Proposed solution
Autonomous truck operational environment	Autonomous truck operations require high-quality roadways. Deficient infrastructure, such as potholes and poor lane markings, can impede autonomous technology.	Increase infrastructure funding to improve and maintain infrastructure.
Liability for autonomous truck-involved accidents	Liability across a variety of state laws needs to be addressed.	Legal system will, over time, set legal precedents causing state liability laws related to vehicle crashes to likely change significantly.
State and Federal trucking regulations	State law and the Federal Motor Carrier Safety Regulations (FMCSRs) do not sufficiently address the autonomous environment. Many rules within the FMCSRs currently conflict with or do not address autonomous trucks. For the trucking industry, federal leadership and possibly federal preemption is critical in providing a seamless national transportation system that benefits from autonomous technology.	Major overhaul of state laws pertaining to commercial vehicles and FMCSRs.
Traffic laws	Following too close is a moving violation. The congestion mitigation aspect of autonomous vehicle technology requires close vehicle proximity during movement. For truck platooning, close proximity is also required to realize fuel savings.	Changes in state law.

Source: J. Short and D. Murray, *Identifying Autonomous Vehicle Technology Impacts on the Trucking Industry*, American Transportation Research Institute, Arlington, VA, 12–14.

Google already utilizes a self-driving fleet of Lexus and research-prototype vehicles that it developed on its corporate campus [27]. Volvo and Nissan expect autonomous vehicles to make an appearance in 2017 and 2018, respectively. Google, Nissan, Ford, and Audi, among others, expect true driverless operations within the 2020–2025 time period [34,35]. BestMile, a spin-off of École Polytechnique Fédérale de Lausanne (EPFL) and CarPostal, a subsidiary of the Swiss Post, will operate autonomous shuttle vehicles in Sion, Valais for locals and visitors [36]. This 2-year trial, scheduled to begin in 2016, will feature the vehicles shown in Figure 11.9 that can carry nine passengers and operate on public roads. Finland deployed EasyMile EZ10 10-passenger, driverless shuttles onto the public roads of Helsinki, as part of a pilot project called SOHJOA [37]. The project aims to solve the challenges of urban mobility in the face of city traffic and safety concerns, while providing a satisfactory user experience. The vehicles run on virtual tracks in a defined area that can be reconfigured to accommodate sudden changes in demand. The shuttles will be operating in real traffic sequentially at three different locations (Helsinki, Espoo, and Tampere) for 1 year. The service began in Helsinki and will stop at first snow to focus on vehicle testing under extreme weather conditions (e.g., snow and ice) and is scheduled to resume service in spring 2017.

In 2018, carmakers Tesla and Mercedes plan to introduce autonomous driving to cities in Britain with vehicles that drive without your hands or your feet touching the controls [38]. However, other forecasts are more conservative in their outlooks for the widespread use of autonomous vehicles in this time frame. Typical of these are comments reported by Tom Simonite, San Francisco bureau chief of *MIT Technology Review* [39]. "Probably what Ford would do to meet their 2021 milestone is have something that provides low-speed taxi service limited to certain roads—and don't expect it to come in the rain," according to

Figure 11.9 Nine-passenger autonomous shuttle vehicle planned for Sion, Valais CH. (Reprinted with permission of BestMile, Lausanne, Switzerland.)

Steven Shladover of the University of California, Berkeley. Alain Kornhauser, professor and director of Princeton University's transportation program, also expects 2021's vehicles to be very restricted as, for example, operating in a defined and limited region where self-driving vehicles can move about. "The challenge will be making that fenced-in area large enough so that it provides a valuable service." Chris Urmson, former director of the Self-Driving Cars project at Google's parent Alphabet, said at *MIT Technology Review*'s EmTech Digital Conference in May 2016 that he expected self-driving vehicles to be offered in certain urban pockets first. He did not elaborate on how limited they would be, or how quickly it would be possible to expand their availability.

11.8 POLICY IMPLICATIONS

Potential applications of autonomous vehicles that may impact city planning and policy decisions include individually owned personal and family transportation, on-demand personal mobility services in urban areas, rental vehicles for short-term mobility and transport needs, long-haul movement of goods and commodities, commercial local delivery services, paratransit driverless vehicles (services for persons with disabilities), fleets owned by corporations or other entities, fleet ownership by groups of users for cooperative use, and urban low-speed vehicles on restricted roadways [27]. Another effect may be to complement transit by addressing the first- and last-mile conundrum. On the other hand, automated vehicles could have a negative influence on public transit by stealing market share or completely redefining transit [40].

Autonomous vehicles promise many advantages. They may be easier and less expensive to operate than conventional automobiles. They may be fleet-owned rather than individually owned if the fleets can guarantee arrival at the requested time and destination. This will free up land currently designated for parking and will enable bicycle lanes to be added to many streets or widened on existing ones. In fact, ITS Finland's CEO predicts that there will be no roadside parking in Helsinki, Finland 15 years from now because self-driving cars will pick up clients within 10 min of their request for the vehicle. The vehicle would continue on its journey once the clients reached their destination [41]. Fleet ownership will also reduce the number of vehicles on the roadways and thus alleviate much of the congestion that is currently experienced by travelers. Assuming that automated vehicles will travel at speeds within the speed limit and eliminate or vastly reduce the numbers of drunk drivers on the roads, their use can diminish the numbers of accidents and deaths and bring about savings in healthcare and auto repair. One of the negative impacts of this scenario is that a significant number of driving jobs may be eliminated, for example, truck, transit, and taxi drivers that total about 2% of the U.S. workforce. The loss of these jobs will have a ripple effect on other businesses [42].

If the entry of large numbers of autonomous vehicles is a certainty in some future decade, then policy and transportation infrastructure planners should be factoring this new paradigm into their discussions. For example, is building a multi-billion dollar bus terminal in the best interests of a municipality if much of the current transit ridership will be converted to using autonomous vehicles? Perhaps, the PRT concept in Figure 11.3 or the mini-bus concept in Figure 11.9 or self-driving taxis, such as the pilot program in Singapore [43] and elsewhere, will replace traditional buses. What will be the impact on the light and heavy rail modes of transportation? The reduction in demand for bus and truck drivers may require policy leaders to consider winding down vocational schools that offer courses in bus and truck driving as a career. A related issue is how to raise money to keep highway trust funds solvent since the majority of their funding comes from gasoline taxes. If autonomous and other future vehicles are predominantly hybrid, electric, or hydrogen fuel-cell powered, where will these funds come from?

The increasing popularity of car sharing, alternative taxi services, and alternative travel modes such as bicycle sharing, along with the likely presence of autonomous vehicles in the not too distant future, lead to contemplation of a broader concept, Mobility as a Service (MaaS). This notion is based on transporting people and goods from one location to another without being concerned about how it is achieved. MaaS links capacity and demand by making available all the options for traveling between the trip origin and destination. The price reflects the time of day, cost of the travel mode, road capacity, whether the conveyance is shared, and the congestion level of the selected route [44–46]. Figure 11.10 illustrates the framework for the model. The alternative mobility services include car and bicycle sharing, crowd-sourced logistics, person-to-person car rental, fleet and ride sharing, autonomous transport, multimodal transport, and the growing Uber and Lyft fleets. Users will request the various travel options through mobile cell phone apps and pay for all the transport services in a single transaction. The bottom portion of the figure indicates the entities that share or assist in sharing data.

In a workshop that addressed multimodal mobility and best practices sponsored by the USDOT, participants discussed use cases for mobility on demand in urban, suburban, and rural settings [47]. The benefits and challenges that were identified are summarized in Table 11.12.

Automobile manufacturers are developing platforms for car sharing and other mobility services. To enhance platform-based car sharing, Toyota has developed the Smart Key Box (SKB) that can be placed in a vehicle without modification. Car-sharing users can lock and unlock doors and start the engine with their smartphone, providing a safer and more secure way of lending and renting cars. A smartphone application will receive codes to access the key-box device, which the assigned vehicle owner has placed in the vehicle. The time and period when the user can access the SKB is set and managed by Toyota based on the vehicle reservation [48].

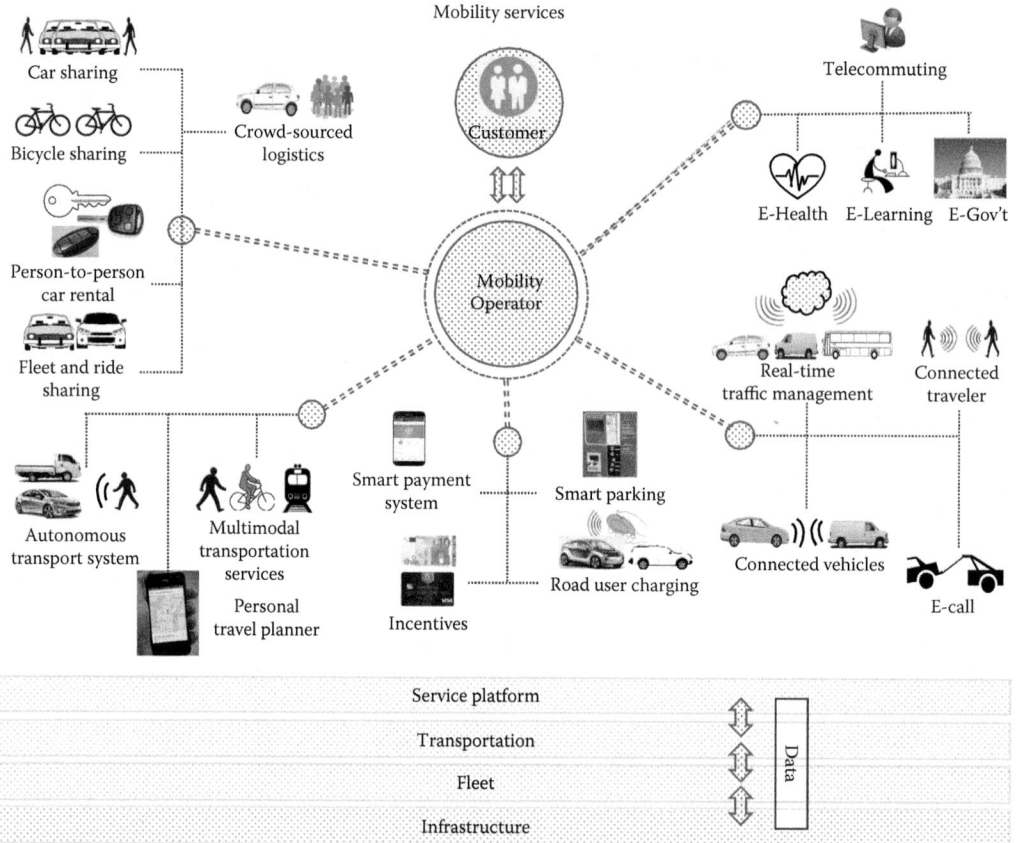

Figure 11.10 MaaS framework showing travel and accessibility options. (Adapted from P.E. Ross, Delphi to test self-driving taxi service in Singapore, *IEEE Spectrum*, August 2016.)

11.9 SUMMARY

Several methods of classifying vehicle automation levels were presented. The most widely utilized are the six-level SAE and five-level BASt models. Fully autonomous vehicles must be designed to operate under a variety of conditions, for example, large variations in road surface materials, road maintenance and lane demarcation quality, traffic flow levels, weather, unplanned event types and durations, and vehicle models and automation levels. While vehicles containing Level 1 and 2 driver assist or automation features are common and those with Level 3 automation (SAE automation taxonomy) are making limited appearances, vehicles with higher levels of automation may be anywhere from several years to decades removed from use by large swatches of the general public. Even as autonomous vehicles promise many benefits to the driver and society in general, they also raise many policy questions. They require planners, regulators at the national and state level, and the public to take a tough and perhaps intricate gaze into the future and prepare accordingly. In the United States, NHTSA's 2016 automated vehicles policy is among the latest approaches to address these concerns. Another tactic to address mobility concerns is MaaS, which makes a variety of transportation modes and services for getting from Point A to Point B available to travelers and allows them to choose a mode based on travel time and cost. Residents of rural

Table 11.12 Mobility on demand benefits and challenges for urban, suburban, and rural residents

Region served	Use cases	Benefits	Challenges
Urban	First- and last-mile connections to public transportation, urban goods movement, daily commuting and other business trips, school trips, special events, disaster response, and trips for people with special needs, such as disabled users, caregivers, those with medical appointments	Reduced number of single-occupant vehicles, congestion, and emissions; increased affordability of transportation and housing (the latter through less parking infrastructure); improved access to jobs and health care, especially for people with disabilities	Perceived negative effects on retail by reducing parking or vehicle access; resistance from political stakeholders and lobby groups (e.g., taxi industry, trucking industry); and perceived and real lack of leverage by local governments to enforce standards and performance targets on private-sector operations
Suburban	Similar to urban use cases that include first- and last-mile connections to public transportation, daily commuting and other business trips, school trips, trips for low-income and carless households, and trips for people with special needs	Enhanced accessibility, improved convenience, reduced travel times and parking demand, increased densification by eliminating obsolete land uses (e.g., park-and-ride lots), increased social and economic inclusion, leveraging that allows users to make spontaneous transportation decisions and the potential to facilitate goods movement (e.g., courier network services and drones)	Competition with personal and single-occupant vehicles; need for incentives to ensure affordability due to fewer drivers and longer trip lengths in the suburbs; improving real-time data services, partnering with the private sector, and creating open platforms (e.g., data sharing and data commons); DOT funding incentives to encourage public and private partnerships; more flexible parking and zoning codes; and technological integration and interoperability between the public and private sector
Rural	Improved access to resource-based jobs (e.g., farms and mining), special needs populations (e.g., older adults, low-income and carless households, and people with disabilities), nearby airports and medical care, commercial goods delivery, and transporting goods to market (e.g., produce and poultry from rural areas to more urbanized areas)	Leveraging of mobility on demand to overcome physical and social isolation or improving high-speed Internet connectivity as an alternative to transportation accessibility, e.g., medical diagnosis services	Sharing of vehicles in areas with low population densities; use cases not requiring technological access, e.g., casual carpooling, to overcome digital poverty and poor cellular data access in rural areas; opportunities to partner with faith-based organizations and other types of associations and gathering places common in rural communities; multijurisdictional issues, e.g., subsidizing ride sourcing trips with origins and destinations in different counties or across state lines; American Disability Act (ADA) requirements that could prevent services from existing in rural communities when there are fewer ADA users because of lower population densities

areas may face different issues from those in urban and suburban regions in obtaining access to alternative mobility choices. These include more limited opportunities to share vehicles due to lower population densities and restricted access to Internet services in some areas.

REFERENCES

1. M. Hallenbeck and M. Wachs, *Influence of Technical Advances on Transportation Behavior*, White Paper prepared for the Transportation Futures Task Force, March 2015.
2. California Vehicle Code, Order to Adopt, Title 13, Division 1, Chapter 1, Article 3.7—Autonomous Vehicles § 227.00.
3. R. Parasuraman, T.B. Sheridan, and C.D. Wickens, A model for types and levels of human interaction with automation, *IEEE Transactions on Systems, Man and Cybernetics—Part A: Systems and Humans*, 30(3):286–297, May 2000.
4. T.B. Sheridan and W.L. Verplank, *Human and Computer Control of Undersea Teleoperators*, MIT Man-Machine Systems Laboratory Tech. Report, Cambridge, MA, 1978.
5. M. van Schijndel-de Nooij, B. Krosse, T. van den Broek, S. Maas, E. van Nunen, H. Zwijnenberg, A. Schieben et al. *Definition of Necessary Vehicle and Infrastructure Systems for Automated Driving*, European Commission SMART 2010/0064 Study Report, Brussels, BE, June 2011.
6. S.E. Shladover, Preparing California for automated vehicles, *Berkeley Transportation Letter*, Winter 2014. http://its.berkeley.edu/btl/2014/winter/automatedcars. Accessed November 21, 2015.
7. S.E. Shladover, Automated vehicles: Dreaming with deadlines, *Thinking Highways*, 7(3):20–22, October 2012.
8. SAE International, *Taxonomy and Definitions for Terms Related to Driving Automation Systems for On-Road Motor Vehicles*, SAE International Standard J3016™, Revised 2016-09. http://standards.sae.org/j3016_201609/. Accessed January 17, 2017.
9. R. Bishop and S.E. Shaldover, Opening plenary session, In *Towards Road Transport Automation—Opportunities in Public–Private Collaboration: Summary of the Third EU-U.S. Transportation Research Symposium, Conference Proceedings 52*, K.F. Turnbull, Rapporteur, Transportation Research Board, Washington, DC, 7–10, April 14–15, 2015.
10. K. Dopart, U.S. Department of Transportation connected vehicle and automated vehicle research update, In *Automated and Connected Vehicles, Transportation Research Board Conference Proceedings on the Web 19*, K.F. Turnbull, Rapporteur, Washington, DC, 6–10, November 4–5, 2015.
11. T.M. Gasser and D. Westhoff, *BASt-study: Definitions of Automation and Legal Issues in Germany*, German Federal Highway Research Institute, Bergisch Gladbach, DE, July 25, 2012.
12. S.E. Shladover and R. Bishop, Appendix A: Commissioned white paper 1, road transport automation as a public–private enterprise, In *Towards Road Transport Automation—Opportunities in Public–Private Collaboration: Summary of the Third EU-U.S. Transportation Research Symposium, Conference Proceedings 52*, K.F. Turnbull, Rapporteur, Transportation Research Board, Washington, DC, 40–64, April 14–15, 2015.
13. *U.S. Department of Transportation Releases Policy on Automated Vehicle Development*, May 30, 2013. http://www.nhtsa.gov/About+NHTSA/Press+Releases/U.S.+Department+of+Transportation+Releases+Policy+on+Automated+Vehicle+Development. Accessed February 23, 2016.
14. *Preliminary Statement of Policy Concerning Automated Vehicles*, National Highway Traffic Safety Administration, U.S. Department of Transportation, National Highway Traffic Safety Administration, Washington, DC, May 30, 2013. http://www.nhtsa.gov/staticfiles/rulemaking/pdf/Automated_Vehicles_Policy.pdf. Accessed September 24, 2016.
15. Vectus PRT Test Track Video, Uppsala Sweden. http://faculty.washington.edu/jbs/itrans/prtquick.htm. Accessed August 21, 2015.
16. S.E. Shladover, *Automated Vehicles: Terminology and Taxonomy*, Taxonomy Working Group, University of California PATH Program, Richmond, CA.

17. B.W. Smith, *Automated Vehicles Are Probably Legal in the United States*, 1 Tex. A&M L. Rev. 411, 2014.

18. H. Bradshaw-Martin and C. Easton, Autonomous or 'driverless' cars and disability: A legal and ethical analysis, *European Journal of Current Legal Issues*, 20(3):Web JCLI, 2014. http://webjcli.org/article/view/344/471. Accessed November 23, 2015.

19. *Prioritising the Safety Potential of Automated Driving in Europe—Briefing*, European Transport Safety Council, April 2016. http://etsc.eu/automated-driving-report/. Accessed July 18, 2016.

20. D.J. Glancy, Autonomous and automated and connected cars—Oh my! First generation autonomous cars in the legal ecosystem, *Minnesota Journal of Law, Science & Technology*, 16(2):619–692, 2015. http://scholarship.law.umn.edu/mjlst/vol16/iss2/3. Accessed October 20, 2016.

21. Automated vehicles need "driving tests" says ETSC, *ITS International*, 6, May/June 2016.

22. *Federal Automated Vehicles Policy*, U.S. Department of Transportation, National Highway Traffic Safety Administration, Washington, DC, September 2016. https://www.transportation.gov/sites/dot.gov/files/docs/AV%20policy%20guidance%20PDF.pdf. Accessed September 20, 2016.

23. A. Christensen, A. Cunningham, J. Engelman, C. Green, C. Kawashima, S. Kiger, D. Prokhorov, L. Tellis, B. Wendling, and F. Barickman, Crash avoidance metrics partnership (CAMP) automated vehicle research (AVR) consortium, Key considerations in the development of driving automation dystems, *Proceedings of the 24th Enhanced Safety of Vehicles Conferences*, 2015. http://www-esv.nhtsa.dot.gov/Proceedings/24/files/24ESV-000451.pdf.

24. *DOT/NHTSA Policy Statement Concerning Automated Vehicles, 2016 Update to Preliminary Statement of Policy Concerning Automated Vehicles*. http://www.nhtsa.gov/staticfiles/rulemaking/pdf/Autonomous-Vehicles-Policy-Update-2016.pdf. Accessed September 26, 2016.

25. Highway Safety Act, 23 U.S.C. § 401 et seq.

26. J. Capp and B. Litkouhl, The rise of the crash-proof car, *IEEE Spectrum*, 51(5):32–37, May 2014.

27. R. Medford, Progress to fully driverless cars, In *Automated and Connected Vehicles, Transportation Research Board Conference Proceedings on the Web 19*, K.F. Turnbull, Rapporteur, Washington, DC, 10–12, November 4–5, 2015.

28. D.J. Glancy, R.W. Peterson, and K.F. Graham, A Look at the Legal Environment for Driverless Vehicles, *NCHRP Legal Research Digest 69 Pre-Publication Draft*, Transportation Research Board, Washington, DC, October 2015.

29. R. Rader, *U.S. DOT and IIHS Announce Historic Commitment of 20 Automakers to Make Automatic Emergency Braking Standard on New Vehicles*, U.S. Department of Transportation, National Highway Traffic Safety Administration, Mar. 17, 2016. http://www.nhtsa.gov/About+NHTSA/Press+Releases/nhtsa-iihs-commitment-on-aeb-03172016. Accessed March 22, 2016.

30. USA's first fully-licensed autonomous truck unveiled, http://www.traffictechnologytoday.com/news.php?NewsID=68831. Accessed May 8, 2015.

31. J. Short and D. Murray, *Identifying Autonomous Vehicle Technology Impacts on the Trucking Industry*, American Transportation Research Institute, Arlington, VA, 12–14, November 2016.

32. *The HARKEN Concept*, http://harken.ibv.org/index.php/about. Accessed November 23, 2015.

33. *D7.3. Laboratory Evaluation Report of the Integrated System*, HARKEN Project, June 2014. http://harken.ibv.org/index.php/documents/doc_download/108-d7-3-laboratory-evaluation-report-of-the-integrated-system. Accessed November 23, 2015.

34. N.J. Goodall, Can you program ethics into a self-driving car?, *IEEE Spectrum*, 53(6):26–31, 57–58, North American Edition, June 2016.

35. P.E. Ross, CES 2017: Nvidia and Audi say they'll field a level 4 autonomous car in three years, *IEEE Spectrum*, January 5, 2017. http://spectrum.ieee.org/cars-that-think/transportation/self-driving/nvidia-ceo-announces/?utm_source=CarsThatThink&utm_medium=Newsletter&utm_campaign=CTT01182017. Accessed January 17, 2017.

36. N. Jaynes, Autonomous buses will hit Swiss streets this spring, http://mashable.com/2015/11/12/swiss-autonomous-bus/#IraQ88y4buqk. Accessed November 30, 2015.

37. http://easymile.com/portfolio/sohjoa-project-finland/. Accessed November 29, 2016.

38. K. Palmer, Why bad driving will be eliminated by 2020—and car insurance costs will plummet, *The Telegraph*, UK, March 22, 2016. http://www.telegraph.co.uk/finance/personalfinance/insurance/motorinsurance/11614723/Why-bad-driving-will-be-eliminated-by-2020-and-car-insurance-costs-will-plummet.html. Accessed March 22, 2016.

39. T. Simonite, Prepare to be underwhelmed by 2021's autonomous cars, *MIT Technology Review*, August 23, 2016. https://www.technologyreview.com/s/602210/prepare-to-be-underwhelmed-by-2021s-autonomous-cars/. Accessed September 13, 2016.

40. R. Pendyala, Understanding the potential impacts of connected and automated vehicles on activity-travel behavior: Implications for transit modeling, In *Automated and Connected Vehicles, Transportation Research Board Conference Proceedings on the Web 19*, K.F. Turnbull, Rapporteur, Washington, DC, 42–46, November 4–5, 2015.

41. Helsinki ponders smart city solutions, *ITS International*, 21, September/October 2015.

42. S. Strauss, What cars and horses have in common, *Los Angeles Times*, A20, November 8, 2015. http://www.huffingtonpost.com/steven-strauss/as-the-age-of-autonomous_b_8504540.html. Accessed November 30, 2015.

43. P.E. Ross, Delphi to test self-driving taxi service in Singapore, *IEEE Spectrum*, August 2016.

44. S. Hietanen, MaaS—a vision of the future, *ITS International*, 14–17, May/June 2016.

45. A.B. Williams, Xerox serves up mobility as a service, *ITS International*, NAFTA 1–2, May/June 2016.

46. J. Czako, Focusing on ITS innovations for future mobility, *ITS International*, 16–18, July/August 2016.

47. S. Shaheen, A. Cohen, and E. Martin, *The U.S. Department of Transportation's Smart City Challenge and the Federal Transit Administration's Mobility on Demand Sandbox: Advancing Multimodal Mobility and Best Practices Workshop*, January 8, 2017, Transportation Research Circular E-C219, DOI: 10.17226/24718, Transportation Research Board, Washington, DC, March 2017. http://www.trb.org/main/blurbs/175826.aspx. Accessed March 21, 2017.

48. Toyota to establish platform for car-sharing and other mobility services, *Traffic Technology Today Digital Edition*, November 2, 2016. http://www.traffictechnologytoday.com/news.php?NewsID=82338. Accessed November 3, 2016.

Chapter 12

Connected vehicles

Benefits promised by the connected vehicle environment lie in the power of wireless connectivity among vehicles, infrastructure, and mobile devices to bring about transformative changes in highway safety, mobility, and environmental impacts of the transportation system. In the United States, the Federal Highway Administration (FHWA) has implemented the Connected Vehicle Program to achieve this transformation. In the European Union (EU), the European Commission has created the Platform for the Deployment of Cooperative Intelligent Transport Systems.

The overriding purpose of the Connected Vehicle Program is to improve driver and pedestrian safety. This is achieved through two-way communication of data and information from vehicle-to-vehicle (V2V), vehicle-to-pedestrian, and vehicle- and pedestrian-to-infrastructure using wireless devices as indicated in Figure 12.1. Vehicles include motorcycles, passenger cars, commercial trucks of all types, emergency response vehicles, and transit vehicles. The concept also extends to compatible aftermarket devices brought into vehicles and to pedestrians, motorcyclists, bicyclists, and transit users carrying compatible devices, often referred to as nomadic devices. The connectivity portion of the infrastructure incorporates traffic and transportation management centers, dynamic message sign displays, highway advisory radio broadcasts to drivers, traffic signal controllers, Bluetooth® and dedicated short-range communication (DSRC) receivers, cellular telephone devices, and roadway sensors. Collectively, these components form the connected vehicle environment. Safety-critical data are currently envisioned to be transmitted using the DSRC protocol. Non-safety–critical data and information can be transmitted via cellular networks, Bluetooth if the range allows, or other transmission media.

The first part of this chapter describes the Connected Vehicle Program and the major results of tests that studied its feasibility. The second portion further explores technology, security, policy (technical, legal, and implementation), and institutional issues being addressed by vehicle manufacturers, governmental agencies, and professional transportation organizations as they develop the vehicle functions, systems, regulations, and infrastructure that promote vehicle connectivity and autonomous or self-driving vehicles. Of particular interest are the December 2016 U.S. National Highway Traffic Safety Administration's (NHTSA's) Notice of Proposed Rulemaking (NPRM) for V2V Communications and the FHWA's Vehicle-to-Infrastructure (V2I) Deployment Guidance. The NHTSA rulemaking would mandate V2V communications for new light vehicles and standardize the message and format of the transmissions. Unlike the NHTSA rules, the FHWA V2I guidance would not be mandatory, but is intended to inform transportation system owners and operators of Federal-aid highway program requirements and practices to help ensure interoperability and effective planning, procurement, and operations throughout the full life cycle of a deployment. The final sections of the chapter discuss connected vehicle pilot deployment concept demonstrations in the United States, the EU approach for promoting a shared vision for the

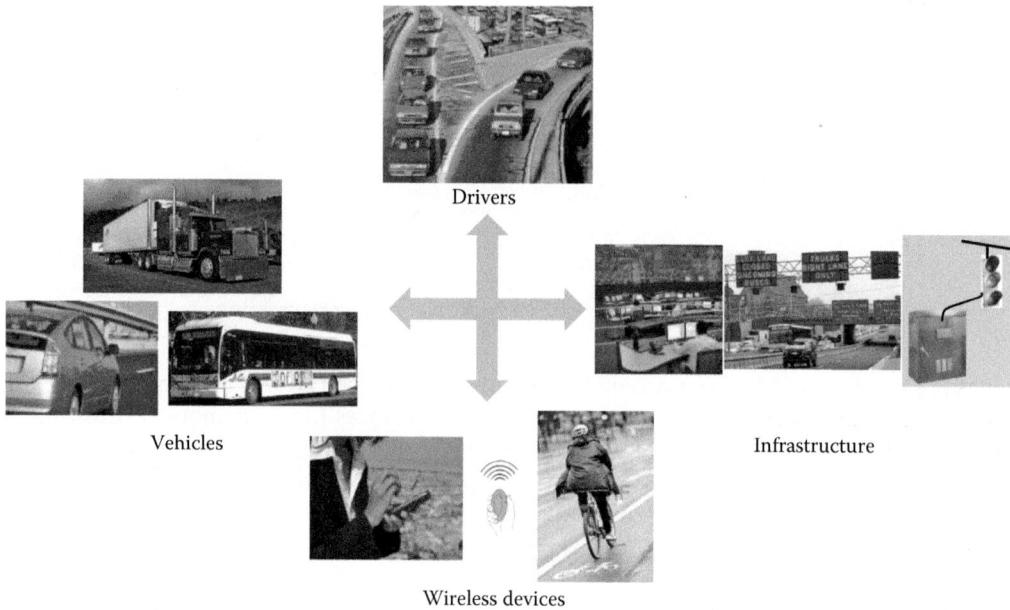

Figure 12.1 Wireless connectivity among drivers, vehicles, infrastructure, and pedestrians carrying wireless devices.

interoperable deployment of cooperative vehicles, and several cooperative vehicle pilot and operational projects in Europe.

12.1 CONNECTED VEHICLE BENEFITS

The Connected Vehicle Program strives to increase safety and mobility, reduce environmental impact, and increase efficiencies of public agency transportation system management and operations through the following mechanisms:

- Reducing highway crashes that currently contribute up to 81% of unimpaired crashes in the United States.
- Improving mobility by providing timely and accurate information about travel conditions and travel-mode options to drivers, transit riders, freight managers, and system operators. System operators include roadway agencies, public transportation providers, public safety agencies, and port and terminal operators that need actionable information and tools to influence the real-time performance of the transportation system.
- Mitigating the environmental impact of travelers by enabling them to make informed decisions about travel modes and routes, and of vehicles through communication with the infrastructure to avoid or reduce unnecessary stops and enhance fuel efficiency.
- Benefiting public agency and private freight transportation system management and operation such that system operators can continuously monitor the status and direct the various assets under their control to improve their efficiency of operations.

Before the connected vehicle concept becomes operational, a number of issues have to be addressed. These include technical concerns such as hardware, software, and standards selection; testing to determine the extent of the benefits to safety, mobility, and the environment and if the magnitudes of the benefits are sufficient to warrant the implementation costs; public acceptance with respect to security and privacy of personal data and information; liability assignment; and finally identification of funding sources to build out the system.

12.2 CONNECTED VEHICLE SAFETY PILOT TEST

In 2012 and 2013, a connected vehicle safety pilot test was conducted in Ann Arbor, Michigan to assist in making informed decisions about the effectiveness of V2V core technologies to reduce crashes [1]. The pilot program was a scientific research initiative that explored a real-world implementation of connected vehicle safety technologies, applications, and systems using everyday drivers. The effort evaluated performance, human factors, and usability; observed policies and processes; and collected empirical data to gather a more accurate, detailed understanding of the potential safety benefits of these technologies. The empirical data supported the 2013 NHTSA decision on using DSRC for vehicle communications that involve safety.

This test involved 2800 vehicles (cars, buses, and trucks) equipped with V2V communications devices, including vehicles with embedded equipment and others that incorporated aftermarket devices or a simple communications beacon. These vehicles emitted a basic safety message (BSM) 10 times per second, which forms the basic data stream that other vehicles analyzed to determine when a potential conflict exists. When these data are further combined with a vehicle's internally generated data, it creates a highly accurate data set that is the foundation for cooperative, crash-avoidance safety applications. The safety pilot test also evaluated the effectiveness of V2I communications to communicate safety messages containing driver alerts and warnings to vehicles and their drivers.

12.2.1 V2V safety pilot test objective

The primary objective of V2V connected vehicle communications is to increase safety [2]. On the left of Figure 12.2, a vehicle broadcasts a BSM that lets other vehicles know its location. More specific information generated by onboard vehicle sensors can also be communicated as shown in the right of the figure.

Only a relatively small number of connected vehicles are needed to improve the flow of traffic. For example, appropriate V2V applications can reduce the potential for accidents

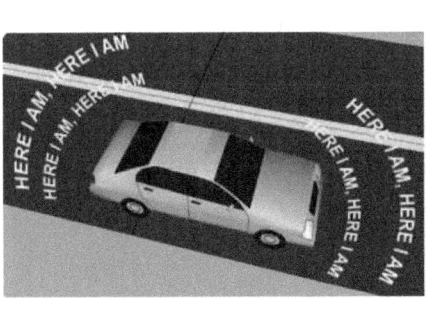
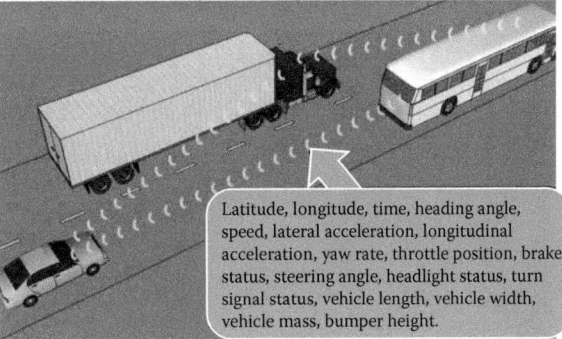

Latitude, longitude, time, heading angle, speed, lateral acceleration, longitudinal acceleration, yaw rate, throttle position, brake status, steering angle, headlight status, turn signal status, vehicle length, vehicle width, vehicle mass, bumper height.

Figure 12.2 V2V data communications. (From A.L. Svenson, IntelliDrive vehicle to vehicle safety applications research plan, In *IntelliDrive Webinar Safety Applications for Commercial Vehicles*, National Highway Traffic Safety Administration, U.S. Department of Transportation, Washington, DC, January 20, 2010.)

and thus improve safety through transmission of a signal that warns drivers that a vehicle in front of them has applied its brakes. Without the V2V communications, the sudden braking action by a leading car forces vehicles further back to brake a few seconds later, and so on down the line of cars. The resulting shock wave may even gain in amplitude and finally form a standing wave. The result is a long-lived traffic jam at some random section of roadway. The back-propagating shock wave can be stopped, however, by notifying people in vehicles a kilometer or more upstream to reduce their speed, for example, from 120 km/h (75 mi/h) to 110 km/h (68 mi/h). Such action completely dampens the shock wave as demonstrated in the Netherlands. Unequipped vehicles, that is, those that are not connected, can be notified of the speed reduction by dynamic message signs or dynamic speed limit signs [3].

12.2.2 V2V applications

The safety pilot test utilized V2V communications to exchange position, speed, and location data among vehicles for the six applications shown in Figure 12.3. These are as follows:

- Forward collision warning (FCW) that notifies drivers of stopped, slowing, or slower vehicles ahead.
- Electronic emergency brake light (EEBL) that warns drivers of heavy braking ahead in the traffic queue.
- Blind spot warning/lane change warning that alerts drivers to the presence of vehicles approaching or in their blind spot in the adjacent lane.
- Intersection movement assist (IMA) that warns drivers of vehicles approaching from a lateral direction at an intersection.
- Do not pass warning that advises a driver of an oncoming, opposite-direction vehicle when attempting to pass a slower vehicle on an undivided two-lane roadway.
- Left turn assist (LTA) that warns drivers of the presence of oncoming, opposite-direction traffic when attempting a left turn.

Each vehicle analyzed the data to determine if a threat or hazard was posed by the relative positions of the other vehicles, calculated risk, issued driver advisories or warnings, and initiated other actions to avoid or mitigate crashes. Additional development was found to be needed to address more complex crash scenarios, such as head-on collision avoidance, intersection

Safety pilot V2V test applications included:
- Forward collision warning
- Electronic emergency brake light
- Blind spot warning/lane change warning
- Intersection movement assist
- Do not pass warning
- Left turn assist

Figure 12.3 V2V communications environment. (From *ITS ePrimer—Module 13: Connected Vehicles*, Intelligent Transportation Systems Joint Program Office, Research and Innovative Technology Administration, U.S. Department of Transportation, Washington, DC, 2013.)

collision avoidance, pedestrian crash warning, and extending the capabilities to prevent motor-cycle crashes.

12.2.3 V2I safety pilot test objective

The safety pilot test also evaluated the ability of infrastructure-based devices to identify high-risk situations and communicate safety messages containing driver alerts and warnings to vehicles. Conventional roadway infrastructure equipment was transformed by incorporating algorithms that analyzed vehicle data to provide early recognition of high-risk driving situations. The resulting driver alerts, indicated in the upper left portion of Figure 12.4, were designed to avoid or mitigate crashes [4]. Other types of warnings that were weather and work-zone related are illustrated on the right of Figure 12.4. Traffic signals communicated signal phase and timing (SPaT) data to vehicles that were converted into active safety messages and warnings to drivers as displayed in the lower left of Figure 12.4. The SPaT data are intended to reduce red-light running violations and assist a driver in negotiating turns and other maneuvers that occur at an intersection.

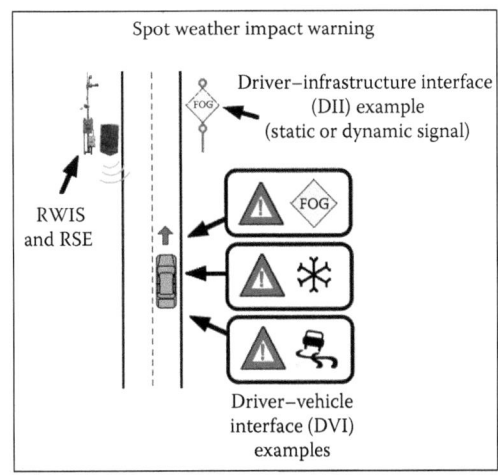

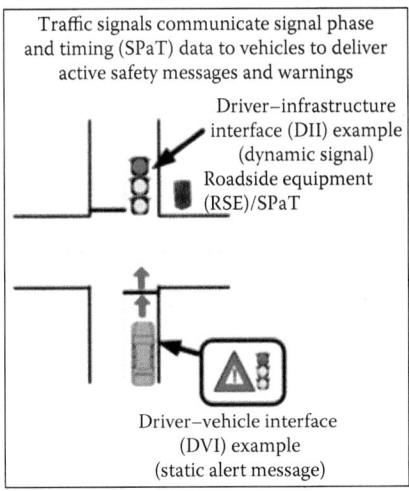

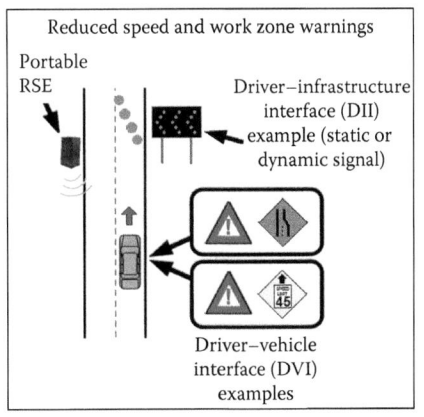

Figure 12.4 Candidate V2I safety applications. (From *ITS ePrimer—Module 13: Connected Vehicles*, Intelligent Transportation Systems Joint Program Office, Research and Innovative Technology Administration, U.S. Department of Transportation, Washington, DC, 2013.)

12.2.4 V2I applications

Four categories of V2I applications that impact safety and mobility are included in the Connected Vehicle Program, namely, intersection applications to prevent crashes, speed-reduction applications, vulnerable road user (VRU) applications, and other miscellaneous applications [5].

- Intersection applications are intended to prevent crashes at intersections from events such as drivers running red lights and stop signs, and by facilitating driver gaps at signalized intersections and stop-controlled intersections.
- Speed applications are meant to target crashes involving one or more vehicles when speeding contributed to the crash and include curve speed warning (CSW), school zone speed warning, work zone warning for reduced speed in work zones, spot treatments for inclement weather conditions, and speed zone warning.
- VRU applications are aimed at targeting crashes involving pedestrians or vehicles in vulnerable situations and include work zone alerts, infrastructure pedestrian detection, priority assignment for emergency vehicle preemption, at-grade rail crossing, and bridge clearance warning.
- Other applications that do not fit into the aforementioned categories include secondary accident warning, lane departure warning, and advanced traffic management, for example, variable lane-by-lane speed limits, reversible traffic flow lanes, and hard shoulder running.

12.2.4.1 SPaT V2I applications

SPaT applications benefit transportation network stakeholders, including pedestrians, bicyclists, and drivers of privately owned vehicles and drivers and operators of long-haul and short-haul commercial fleets, public transit, jurisdictional fleets (i.e., police, fire, ambulance, road maintenance vehicles, and tow trucks), and taxi vehicles. V2I safety applications that rely on SPaT data are listed in Table 12.1. The major systems needed for SPaT communications are the following:

- Roadside equipment (RSE) associated with the intersection, including the traffic signal controller, traffic sensors, safety applications processor(s), communications devices, and GPS time-reference equipment.
- Onboard equipment (OBE) associated with mobile entities using the intersection, including mobile communications transceivers, positioning sensors, time references, and applications processors. In some applications, the OBE includes interfaces to other vehicle systems via a network interface, typically the Controller Area Network (CAN) bus.

Table 12.1 V2I safety applications requiring SPaT messages

Application	Application
Red-light warning to reduce red-light running	Curve speed warning
Left turn assist	Right turn assist
Stop gap assist	Railroad crossing red-light violation warning
Oversize vehicle warning	Transit signal priority
Freight signal priority	Emergency vehicle preemption
Pedestrian signal assist	Spot weather impact warning (see upper right of Figure 12.4)
Reduced speed and work zone warning (see lower right of Figure 12.4)	

An important benefit of SPaT messaging is prevention of red-light running accidents. In this instance, the SPaT data extend the red phase for opposing traffic so that a vehicle traveling too fast to stop at the intersection can proceed safely through it. Sensors convey vehicle speed and location data to the intersection controller, which then adjusts the signal timing as required. The current SPaT information is transmitted via the RSE communications devices from the traffic signal controller to OBE communications devices within communications range of the RSE. SPaT messages themselves are generally time critical, especially since the SPaT status may change from one transmission to the next (such as associated with traffic signal emergency preemption). However, some SPaT-related communications are not time critical. For example, the geometric intersection description (GID) can be transmitted infrequently and may be carried using communications media other than DSRC. Communications options for non-safety–critical V2I applications are described in Section 12.7.2.

12.2.4.2 SPaT communications requirements

Figure 12.5 contains the basic geometry and timing considerations related to SPaT for the red-light running application [6]. The SPaT message representing the current SPaT must be received by an approaching vehicle prior to it reaching the stopping sight distance as defined by the American Association of State Highway and Transportation Officials (AASHTO) [7,8]. For example, a vehicle approaching an intersection at 50 km/h (31 mi/h) requires 4.9 s to react to a red signal and stop before entering the intersection [8]. The tasks performed by a driver during this period include maintaining a safe lane position, checking surroundings for unsafe situations, decelerating, observing vehicle stopping trajectory, maintaining a safe distance from the decelerating lead vehicle, maintaining a safe distance from decelerating following vehicles, observing status of the traffic signal, and finally stopping.

Requirements for SPaT communications equipment are listed in Table 12.2 [6]. Table 12.3 shows the impact of the proposed V2I intersection, speed-reduction, VRU, and

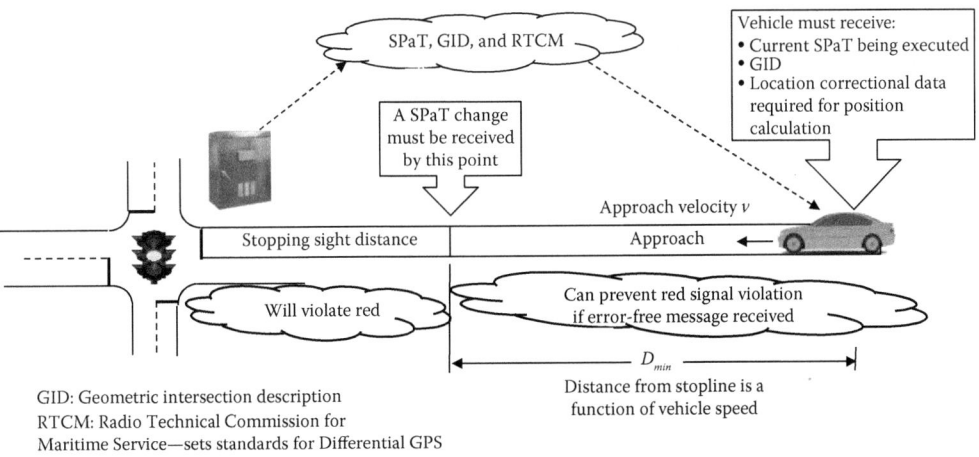

Figure 12.5 SPaT geometry and timing considerations for red-light running application. (From *Signal Phase and Timing (SPaT) Applications, Communications Requirements, Communications Technology Potential Solutions, Issues and Recommendations*, Draft Final Report FHWA-JPO-13-002, April 3, 2012. Prepared by Bruce Abernethy, ARINC Incorporated; Scott Andrews, Cogenia Partners; and Gary Pruitt, ARINC Incorporated. www.its.dot.gov/index.htm.)

Table 12.2 SPaT communications equipment requirements

SPaT communications specification	Requirement
Communications range for high probability of message receipt	High-end range: 331 m (1087 ft) High-end range assumes signal preemption and time to clear the intersection based on posted speed and stopping sight distance at 0.2 G deceleration from 45 mi/h (72 km/h) Nominal range: 176 m (579 ft) Nominal range is based on stopping sight distance at 0.3 G deceleration from 75 mi/h (121 km/h)
Maximum bit error rate (BER) and confidence factor	10^{-4} Achieved by 4 message transmissions based on safety integrity level (SIL)[a] = 1 (message reliability = Packet error rate = 10^{-2})
Data throughput, SPaT messages	40 kbps Includes single intersection GID associated with the application. Does not include other message traffic in channel
Background BSM/here I am (BSM/HIA) data load on DSRC in which SPaT messages must compete	4.77 Mbps considering J2735 part 1 29.44 Mbps considering J2735 parts 1 and 2 Based on 176 vehicles within communication range
Data rate required for GIDs using wide-area broadcast	Function of population and number of intersections 200 K population = 54 kbps 500 K population = 135 kbps 1 M population = 270 kbps
Weather	Meet SPaT communications requirements in all weather conditions (rain, sleet, snow, and fog)
Radio-frequency environment	Must operate in an RF environment consisting of licensed and unlicensed emitters both in the intersection and near the intersection (details in the full report)
Size and weight	Compatible with small car (approximate size 500 in.^{3} [8195 cm^3]) Approximate weight 2 lbs (0.91 kg)
Cost	Affordable to purchaser of a private vehicle. Generally considered to be <$300

Source: Signal Phase and Timing (SPaT) Applications, Communications Requirements, Communications Technology Potential Solutions, Issues and Recommendations, Draft Final Report FHWA-JPO-13-002, 2012, April 3, 2012. Prepared by Bruce Abernethy, ARINC Incorporated; Scott Andrews, Cogenia Partners; and Gary Pruitt, ARINC Incorporated. www. its.dot.gov/index.htm.

[a] SIL is defined as a relative level of risk reduction provided by a safety function.

miscellaneous safety applications on the number and cost of crashes [5]. These applications have the potential to affect over 2 million crashes and save over 200 billion dollars (U.S.) annually.

12.2.4.3 SPaT deployments

A commercial implementation of SPaT is already available in the marketplace. Audi of America introduced a system in 2016 that enables the car to communicate with the infrastructure in select U.S. cities and metropolitan areas. The car receives real-time signal information from an advanced traffic management system that monitors the traffic signals. The link between the vehicle and the infrastructure is routed through the onboard 4G long-term evolution (LTE) data connection and the service provider. When approaching a connected traffic light, the system displays the time remaining until the signal changes to green in the driver instrument cluster and the head-up display (if equipped) [9].

Table 12.3 Impact of proposed V2I safety applications on annual number and cost of crashes

Application area		Estimated annual no. of crashes targeted	Annual cost of crashes targeted (millions of dollars)
Intersection applications	Running red signal	234,881	13,152
	Running stop sign	44,424	2034
	Driver gap assist at signalized intersections	200,212	10,252
	Driver gap assist at stop-controlled intersections	278,886	18,273
Speed applications	Curve speed warning	168,993	29,080
	Work zone warning for reduced speed	16,364	1335
	Spot treatment/weather conditions	211,304	13,019
	Speed zone warning	360,695	28,500
VRU applications	Work zone alerts	86,611	4563
	Infrastructure pedestrian detection	17,812	3333
	At-grade rail crossing	1314	653
Other applications	Lane departure warning	1,236,647	145,347
Total (accounting for overlaps)		2,288,021	202,344

Source: *Crash Data Analyses for Vehicle-to-Infrastructure Communications for Safety Applications*, Publication No. FHWA-HRT-11-040, FHWA Research, Development, and Technology, Turner Fairbank Highway Research Center, McLean, VA 22101-2296, November 2012.

State and local public-sector transportation infrastructure owners and operators in each of the 50 states are being encouraged to participate in the SPaT Challenge to equip at least one coordinated corridor or network (about 20 signalized intersections) with DSRC infrastructure to broadcast SPaT information by January 2020 and maintain the operations for at least 10 years [10]. The Challenge is intended to provide the transportation agencies experience with procurement, licensing, installation, and operation of DSRC-based V2I deployments. Sponsors are the V2I Deployment Coalition Technical Working Group (TWG) 1 and the AASHTO Connected-Automated Vehicle Working Group.

12.3 CONNECTED VEHICLE MOBILITY APPLICATIONS

We next examine the motivation for developing connected vehicle applications to improve mobility. Figure 12.6 shows a USDOT forecast for the widespread occurrence of congested and highly congested highways in the United States by the year 2035 [11]. The need for congestion relief and increased mobility is apparent.

A report prepared by Inrix in the United States and the Center for Economics and Business Research (Cebr) in the United Kingdom examines economic and environmental costs of congestion in London, Paris, Stuttgart, and Los Angeles and extends the forecast to the year 2030 [12]. The forecast includes additional driving time that drivers will allow for uncertainty over the level of congestion they will encounter during their journey.

Across all four national economies, the costs imposed by congestion are predicted to rise 46% between 2013 and 2030, from $200.7 billion to $293 billion with a total cumulative cost of $4.4 trillion. The United Kingdom has the greatest cost increase (up 63% from $20.5 billion in 2013 to $33.4 billion by 2030). Of the cities studied, London is expected to have the largest congestion problem with economic costs increasing by 71%. Los Angeles is close

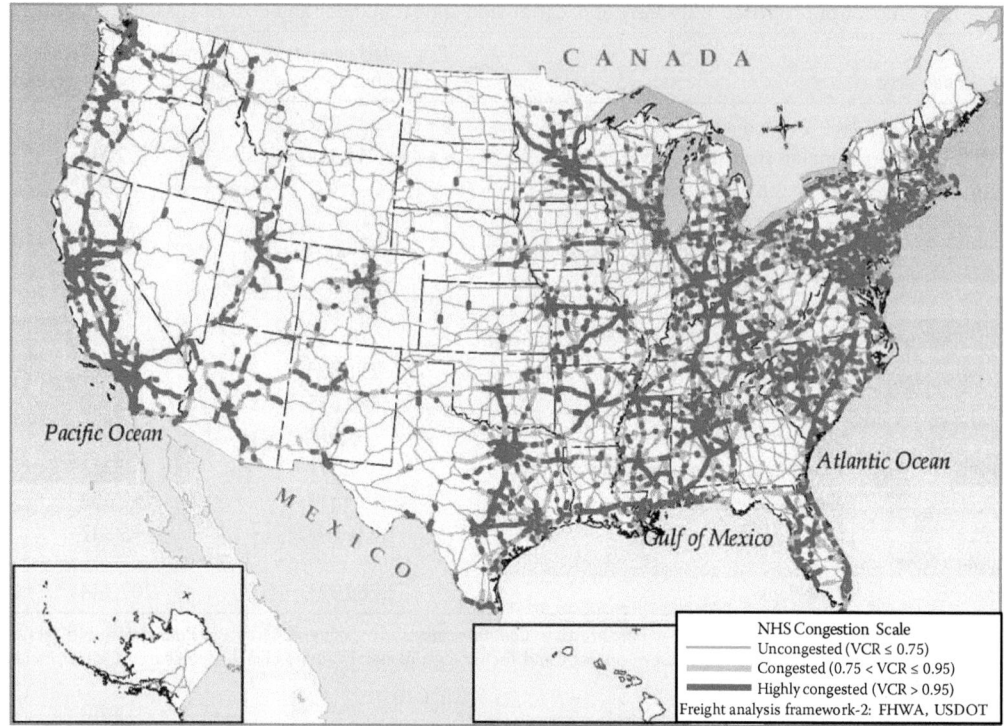

Figure 12.6 U.S. congestion forecast for 2035. VCR represents the volume/capacity ratio. (From T. Kearney, *IntelliDrive Webinar Safety Applications for Commercial Vehicles*, FHWA, USDOT, January 20, 2010.)

behind with a 65% rise in impact giving a cumulative cost of congestion of $559 billion by 2030 according to Cebr.

12.3.1 How mobility applications function

Connected vehicle mobility applications enable ease of travel by providing a data-rich journey environment to travelers with the appropriate telematics. The network captures real-time data from equipment located onboard vehicles (automobiles, trucks, and public transit vehicles) and within the infrastructure. The data are transmitted wirelessly or on wired networks in support of a wide range of dynamic, multimodal applications that assist in managing the transportation system (including arterials, freeways, toll facilities, transportation corridors, and regional facilities) to minimize delays and congestion.

12.3.2 FHWA mobility enhancement programs

The six FHWA programs designed to enhance mobility are the following:

1. *Enable Advanced Traveler Information Systems (Enable ATIS)* serves a region by providing the traveler network-focused information concerning multimodal integration, data sharing, end-to-end trip perspectives, and predictive information specific to users.
2. *Integrated Dynamic Transit Operations (IDTO)* protects transfers between transit and non-transit modes, requests a trip and generates itineraries containing multiple

transportation services, or requests carpooling where drivers and riders arrange trips within a relatively short departure time.

3. *Multimodal Intelligent Traffic Signal Systems (MMITSS)* supports an overarching system optimization that accommodates transit and freight signal priority, preemption for emergency vehicles, and pedestrian movements while maximizing overall arterial network performance. This application is explored further in the next chapter to illustrate systems engineering principles.

4. *Intelligent Network Flow Optimization (INFLO)* consists of applications related to queue warning, speed harmonization, and cooperative adaptive cruise control.

5. *Response, Emergency Staging and Communications, Uniform Management, and Evacuation (R.E.S.C.U.M.E.)* utilizes data from freeways to solve issues faced by emergency management agencies, emergency medical services (EMS), and persons requiring assistance during traffic incidents and mass evacuations.

6. *Freight Advanced Traveler Information Systems (FRATIS)* provides freight-specific dynamic travel planning and performance information, or optimizes drayage operations so that load movements are coordinated between freight facilities to reduce empty-load trips in the region served.

12.4 CONNECTED VEHICLE TRANSIT APPLICATIONS

Connected vehicle transit applications are designed to increase the safety of transit vehicle operations and to provide additional services to users of these facilities. Two programs with these objectives are described below: IDTO and the Transit Safety Retrofit Package Pilot Deployment.

12.4.1 Integrated Dynamic Transit Operations

The Integrated Dynamic Transit Operations developed within the Dynamic Mobility Applications (DMA) Program for connected vehicles include three additional mobility applications:

1. T-DISP enables a traveler to access real-time information about available travel options, including costs and predicted travel time, in order to best manage their commute. The application integrates information from multiple modes and providers, and combines schedule and vehicle location information with the position-locating and connectivity capabilities of smart phones.

2. T-CONNECT improves traveling by transit by increasing the likelihood of making successful transfers, particularly when these transfers are multimodal or multiagency. The system determines, through a series of decisions, whether the request can be fulfilled and communicates the result to the traveler. If granted, the traveler will continue to receive status updates, particularly if subsequent conditions prevent the connection from being met.

3. D-RIDE takes the concept of traditional preplanned ride-sharing (i.e., carpooling) and expands it by leveraging the positioning, messaging, and computing capabilities of smart phones, and advancing a near real-time application that lets drivers and travelers exchange information about needs or, in the case of a nondriver, available space in a particular vehicle.

12.4.2 Transit Safety Retrofit Package

Connected vehicle high-priority concerns identified by transit agencies were also part of the USDOT's Safety Pilot Model Deployment in Ann Arbor, Michigan [13,14]. Here, a team led by Battelle implemented a Transit Safety Retrofit Package (TRP) on University of Michigan transit buses.

TRP project objectives were to design and develop safety applications for transit buses using V2V and V2I technologies to enhance transit bus and pedestrian safety. The project also investigated if DSRC technologies could be combined with onboard safety applications to provide bus drivers real-time alerts of potential and imminent crashes.

The TRP contained three V2V and two V2I collision avoidance applications. These were as follows:

- V2V applications—EEBL, FCW, and vehicle turning right in front of bus warning (VTRW).
- V2I applications—CSW and pedestrian in crosswalk warning (PCW).

The project leveraged components and approaches proven on other Safety Pilot Model Deployment vehicles. It included the following system elements that were used in the testing scenario of Figure 12.7:

- *Transit vehicle OBE*: The OBE included a DENSO mini wireless safety unit (WSU) to receive and transmit BSMs via 5.9 GHz DSRC. The mini WSU interoperated with other model deployment vehicles and RSE according to IEEE 802.11p and 1609.2 standards and the SAE International J2735 message standard. A Samsung Galaxy Tablet computer provided the driver-vehicle interface (DVI) and additional processing. The mini WSU interfaced with the DVI and vehicle CAN bus.
- *Safety applications*: Battelle developed two new transit-specific safety applications, namely, PCW and VTRW, hosted on the tablet computer. Three basic safety applications, FCW, EEBL, and CSW, common with other model deployment vehicles, were preloaded on the mini WSU.
- *Crosswalk motion sensors*: The MS SEDCO SmartWalk XP was deployed to detect pedestrians in intersection crosswalks in support of the PCW safety application. These units were mounted to existing poles at the recommended height of 10–12 ft (3.0–3.7 m), and employed microprocessor-analyzed Doppler microwave detection technology.
- *Data acquisition system (DAS)*: The University of Michigan Transportation Research Institute (UMTRI) DAS was employed to record data for TRP evaluation, including data from the vehicle CAN bus, four video cameras, a range and position sensor, and the basic safety applications.

Quantitative requirements were also imposed on the TRP such as the three latency specifications below for the transit vehicle display [14]:

- [SYSREQ_020] The TRP latency from safety application event detection to aural and visual display shall be less than 250 ms.
- [SYSREQ_021] The TRP system latency for pedestrian detection shall be no more than 2 s from detecting pedestrian to warning.
- [SYSREQ_022] The TRP system latency for right turning vehicle shall be no more than 2 s from receipt of path prediction of right turn conflict data to warning.

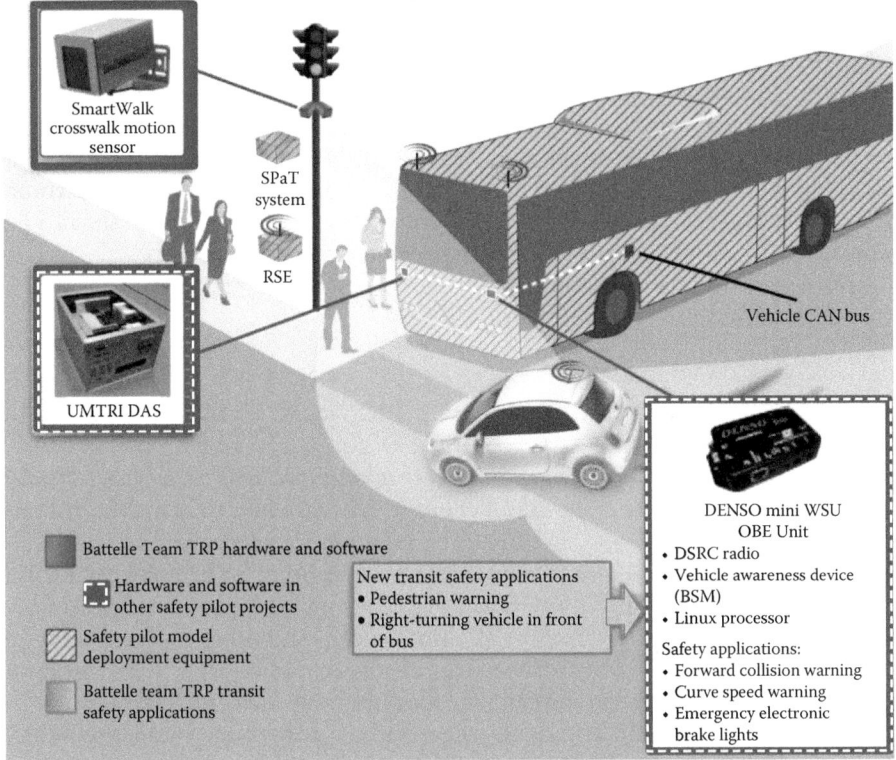

Figure 12.7 Transit safety retrofit package testing scenario. (From D. Valentine et al., *Transit Safety Retrofit Package (TRP): Leveraging DSRC for Transit Safety—Fielding Results and Lessons Learned*, Intelligent Transportation Systems Joint Program Office, Federal Highway Administration, Federal Transit Administration, U.S. Department of Transportation, Washington, DC, November 2014.)

The major conclusions and lessons learned from the TRP are the following [13,14]:

• The TRP on-bus software was effective at providing alerts to transit drivers.
• Any TRP acceptance issues by the transit drivers were due to TRP inaccuracies and in-vehicle display weaknesses.
• There was a high rate of false alerts for the PCW application due primarily to a combination of GPS limitations and pedestrian detector limitations. The PCW application should not assume specific bus routes and should be suppressed after the bus enters the crosswalk.
• There was a high rate of false alerts for the VTRW application due to GPS limitations.
• Wide-Area Augmentation System (WAAS)-enabled GPS accuracy is insufficient for the PCW and VTRW applications. Typical lane width is 3.35 m (11 ft), which requires location accuracy within 1.675 m (5.5 ft). This cannot reliably be achieved with WAAS-enabled GPS. A more precise technology, such as differential GPS (DGPS), is needed on future systems to achieve the required performance levels.
• The Doppler microwave crosswalk detectors are insufficient for the PCW application. A more discerning technology, such as high-speed imaging, should be employed on future systems to achieve expected performance levels.
• DSRC radio technology performed well as no TRP problems were traced to DSRC radio communications.

12.5 CONNECTED VEHICLE FREIGHT APPLICATIONS

Figure 12.8 shows the USDOT connected vehicle freight applications [11]. The Coordinated Federal Lands Highway Technology Implementation Program (C-TIP) is a cooperative technology deployment and sharing program between the FHWA Federal Lands Highway office and the Federal Land Management agencies. It provides a forum for identifying, studying, documenting, and transferring new technology to the transportation community. Commercial Vehicle Information Systems and Networks (CVISN) refers to the ITS information system elements that support commercial vehicle operations (CVOs). Clarus is an integrated surface transportation weather observing, forecasting, and data management system. The other applications are self-explanatory.

12.6 CONNECTED VEHICLE ENVIRONMENTAL APPLICATIONS

Environmental applications are designed to capture relevant, real-time transportation data to support environmentally friendly travel choices. The data and information they generate can be used in three general ways. The first is by assisting travelers avoid congestion by notifying them of less congested alternate routes or transit options, or by rescheduling their trip and thus become more fuel and time efficient. The second is by providing system operators real-time information concerning vehicle location, speed, and other operating conditions to improve system operation. Finally, drivers can choose to use relevant information to optimize the vehicle's operation and maintenance for maximum fuel efficiency.

The Applications for the Environment—Real-Time Information Synthesis (AERIS) program focuses on the capture, synthesis, and delivery of real-time, vehicle- and infrastructure-based, ecologically relevant information to support system management that advances environmentally friendly choices within the transportation system [15]. The program goal is to find and promote transformational applications and strategies that can affect a significant decrease in emissions and fuel consumption and also bring incremental improvements to existing capabilities. To achieve this goal, AERIS first estimates the amount and percentage of emissions (pollutants and greenhouse gases) and fossil-fuel consumption that can

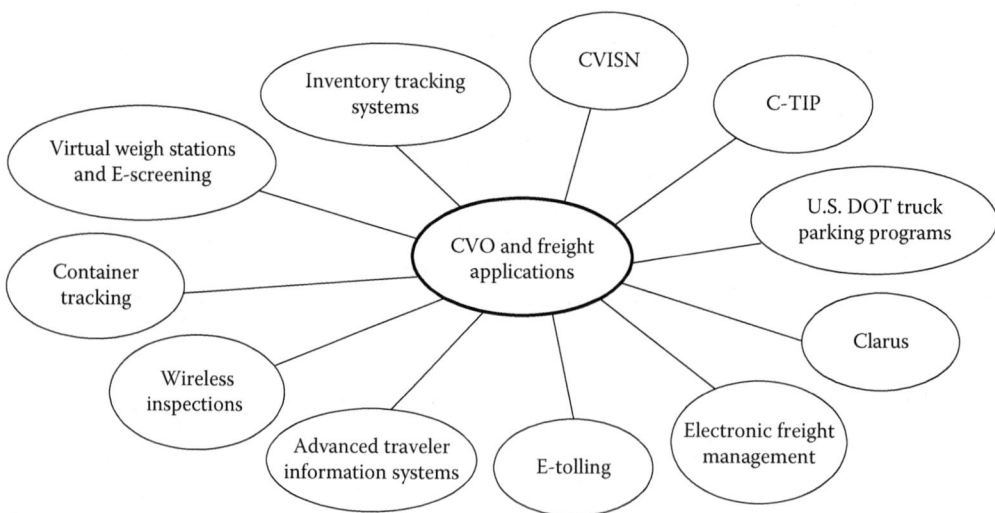

Figure 12.8 Connected vehicle freight applications.

be reduced or avoided through the use of real-time data and strategies. Then it applies the selected strategies to various travel scenarios.

AERIS contains three proposed connected vehicle concepts for reducing the environmental impact of vehicles. These are the following:

- Eco-signal operations that will optimize roadside and traffic signal equipment in their collection and sharing of relevant positional and emissions data to lessen transportation environmental impact. This will be accomplished by decreasing fuel consumption, greenhouse gas, and criteria air pollutant emissions by reducing idling, number of stops, and unnecessary accelerations and decelerations, and improving traffic flow at signalized intersections.
- Dynamic eco-lanes, similar to HOT and HOV lanes, but optimized to support freight, transit, alternative fuel, or regular vehicles operating in eco-friendly ways. These dedicated eco-lanes target low-emission, high-occupancy, freight, transit, and alternative-fuel vehicles (AFVs).
- Dynamic low-emissions zones, similar to cordon areas with fixed infrastructure, but designed to provide incentives for eco-friendly driving. They include a geographically defined area that seeks to restrict or deter access by specific categories of high-polluting vehicles to improve the air quality within the area. Low-emissions zones can be dynamic, allowing the operating entity to change the location, boundaries, fees, or time of operation of the zone.

12.7 CONNECTED VEHICLE TECHNOLOGY

Various types of equipment and technologies are needed in vehicles, infrastructure, and back-office systems to implement the connected vehicle concept. These include OBE and RSE, communication systems, and security credentials management systems. The first two were discussed briefly in conjunction with SPaT applications in Section 12.2.4.1.

- Onboard or mobile equipment located in the vehicle are the systems or devices through which most end users will interact with the connected vehicle environment. Data communicated through these systems, including location, speed, and heading from GPS or other sensors, contribute to the basic information used in connected vehicle applications. Additional sensor data, such as inter-vehicle spacing, windshield wiper status, turn-signal activation, or antilock braking or traction control activation, may be beneficial in certain applications.
- RSE provides connectivity between vehicles and roadside systems, such as integration with traffic signal controllers or dynamic lane speed-limit devices.
- Communications systems are the infrastructure that supports network connectivity from RSE to other system components, whether on vehicles or pedestrians. Different types of communications systems may be needed to send data to vehicles and pedestrians for safety-related applications as compared to other data needed by traffic and transportation management centers.
- Support systems that include security credentials management allow devices and systems in the connected vehicle environment to establish trust relationships. These systems facilitate interactions among vehicles, field infrastructure, and back-office users.

The following sections describe communication options for safety-critical and non-safety–critical communications. Security credentials management is treated in Section 12.8.

12.7.1 V2I communications for safety-critical applications

The current communication choice for V2V and V2I safety-critical communication is DSRC, primarily because of its low latency. Safety applications designed with this technology reduce collisions and other types of accidents by providing real-time advisories such as forward collision and cross-traffic warnings; requests to traffic signal controllers for green-time extensions; spot weather road condition warnings to alert drivers to slippery patches of roadway ahead; warnings concerning veering close to the edge of the road, vehicles suddenly stopped ahead, collision paths during merging, sharp curves, and work zones; and alerts to the presence of nearby communications devices on pedestrians, bicyclists, motorcyclists, and vehicles.

DSRC utilizes a two-way, short-to-medium range wireless communications protocol to enable high data rate transmission (3–27 Mbps). In Report and Order FCC-03-324, the U.S. Federal Communications Commission (FCC) allocated 75 MHz of spectrum in the 5.9-GHz band for use by ITS vehicle safety and mobility applications. The 75-MHz DSRC bandwidth has key functional and technical attributes that make it suitable for safety-related applications. These are listed in Table 12.4. Figure 12.9 shows the latency of DSRC compared to alternative communications technologies. The USDOT committed to DSRC for active safety, but will explore alternative wireless technologies for other connected vehicle

Table 12.4 DSRC functional and technical attributes suited for safety applications

Functional attributes	Technical attributes
Priority for safety applications over non-safety applications.	DSRC utilizes a communications protocol similar to Wi-Fi that addresses the technical issues associated with sending and receiving data among vehicles and between moving vehicles and fixed roadside access points.
Low latency characterized by very short delays in opening and closing connections between vehicles or a vehicle and the infrastructure. Active safety applications must recognize and transmit messages to the participating devices within milliseconds without delay. The most stringent latency requirement for safety applications is 0.02 s. DSRC satisfies this requirement since its latency is approximately 0.0002 s, that is, a factor of 100 lower than required [6]. The latency of DSRC and other communications technologies is plotted in Figure 12.9.	DSRC includes the Wireless Access in Vehicular Environments (WAVE) Short Message protocol defined in the IEEE 1609 standard.
High-reliability communications link with fast network acquisition. Its high immunity to interference provides robust performance in the face of other radio interference. Its short communications range makes it largely unaffected by distant radio sources.	Typical range of a DSRC access point is 300–450 m (ranges up to 1 km are possible). This range is typical of installations at intersections and other roadside locations.
Reliably connects with high-speed vehicles.	Derived from the IEEE 802.11p standard.
Ability to prioritize safety messages.	
Tolerance to multipath transmissions typical of roadway environments.	
Protection of security and privacy of messages by providing safety message authentication.	
Maintains performance levels during inclement weather conditions (e.g., rain, fog, snow, dust).	
Interoperability through use of widely accepted communications standards that support V2V and V2I communications.	

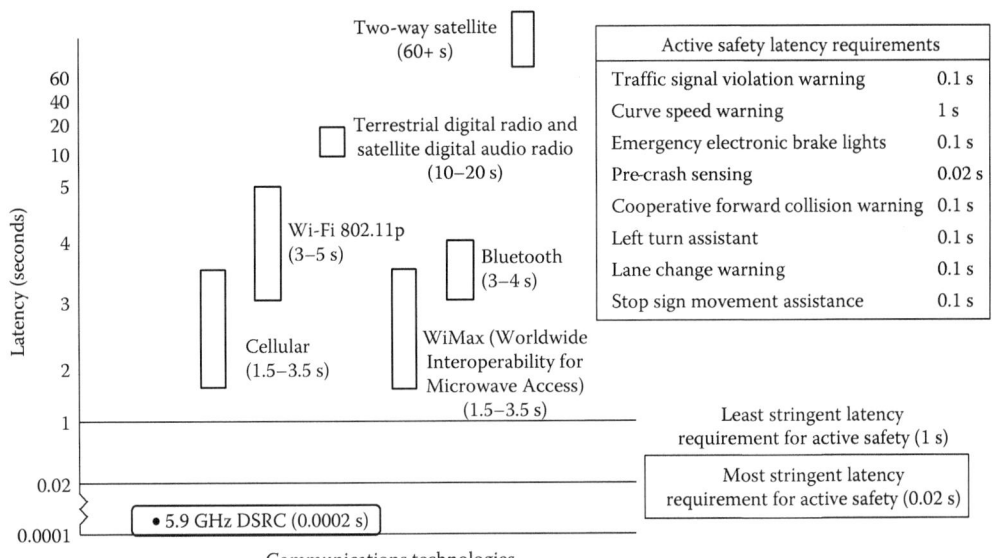

Figure 12.9 Latency of potential V2V and V2I communications technologies. (From *ITS ePrimer— Module 13: Connected Vehicles*, Intelligent Transportation Systems Joint Program Office, Research and Innovative Technology Administration, U.S. Department of Transportation, Washington, DC, 2013.)

applications. ITS America, USDOT, and FHWA have collaborated to produce a report that describes recommended practices for DSRC licensing and spectrum management [16]. The U.S. FCC was expected to announce a DSRC-use rule by 2017.

In addition, convenience V2I services such as e-parking and toll payment are able to communicate using DSRC. Information concerning the number of passengers in a vehicle could be used to avoid toll payments for HOV and HOT lane travel. Anonymous information from electronic sensors and other devices in vehicles can also be transmitted over DSRC to provide better traffic, travel time, and road closure information to travelers and transportation managers.

In Europe, the European Telecommunications Standards Institute (ETSI) is developing the ITS-G5 communications standard for safety-critical applications, for which the 5.875–5.905 GHz band has been set aside [17,18]. The protocol will transmit the BSM (in Europe referred to as a Cooperative Awareness Message [CAM]) at a 1–10 Hz rate in support of Cooperative-ITS (C-ITS) projects, such as the C-ITS Corridor described in Section 12.15. The first identified applications are infrastructure-to-vehicle road works warnings and V2I transmission of vehicle data such as vehicle position, speed, direction, and dimensions, along with event-driven environmental messages that contain rain, slippery road surface, congestion, and other hazard information.

12.7.2 V2I communications options for non-safety–critical applications

Communications technologies other than DSRC may be suitable for non-safety–related applications. These include DGPS, agency-owned ITS networks, cellular non-fee for service networks, ITS wide-area networks (WANs), computer-aided dispatch (CAD) networks, and licensed wireless networks. The choice depends on the types of facilities being connected (e.g., TMCs, cellular service providers, network operations centers, fleet dispatch centers, transit operations centers, and emergency response providers), bandwidth required to transmit the data of interest, types of communications infrastructures already available in the region, and cost considerations.

Mobile broadband communications in the 315–2690 MHz portion of the RF spectrum could be used for many infrastructure- and vehicle-oriented non-safety–critical services. These frequencies already support terrestrial HD digital wireless radio; digital audio and media broadcasting; remote keyless entry and tire pressure monitoring; 2G, 3G, and 4G cellular; tolling via radio-frequency identification (RFID); GPS navigation; satellite radio; and Bluetooth connectivity. Additional applications include locating unoccupied parking bays, idle taxis or rental vehicles, and vacant passenger seats on public transit vehicles and private ride-sharing vehicles. Other vehicle services, such as diagnostics, fleet management, and pay-as-you-go insurance, could also be provided by mobile broadband communications. Arterial and highway infrastructure-mounted radar frequencies at 10 and 24 GHz find use in vehicle sensing for signal control and incident detection. Higher frequency radars operating from 76 to 77 GHz are located in vehicles for vehicle and obstacle detection and location. In-vehicle–mounted lidar sensors operating in the near-infrared frequency band also provide vehicle and obstacle detection and location, but at shorter ranges [19]. The knowledge gathered by the latter two categories of sensors could be transmitted using DSRC to other vehicles and the infrastructure as part of the connected vehicle information stream.

Figure 12.10 depicts several of the V2I alternate communications networks that could be utilized to transmit non–time-critical safety-support data. These are DGPS, jurisdictionally owned ITS networks, cellular networks, ITS WANs, CAD, and mobile wireless networks.

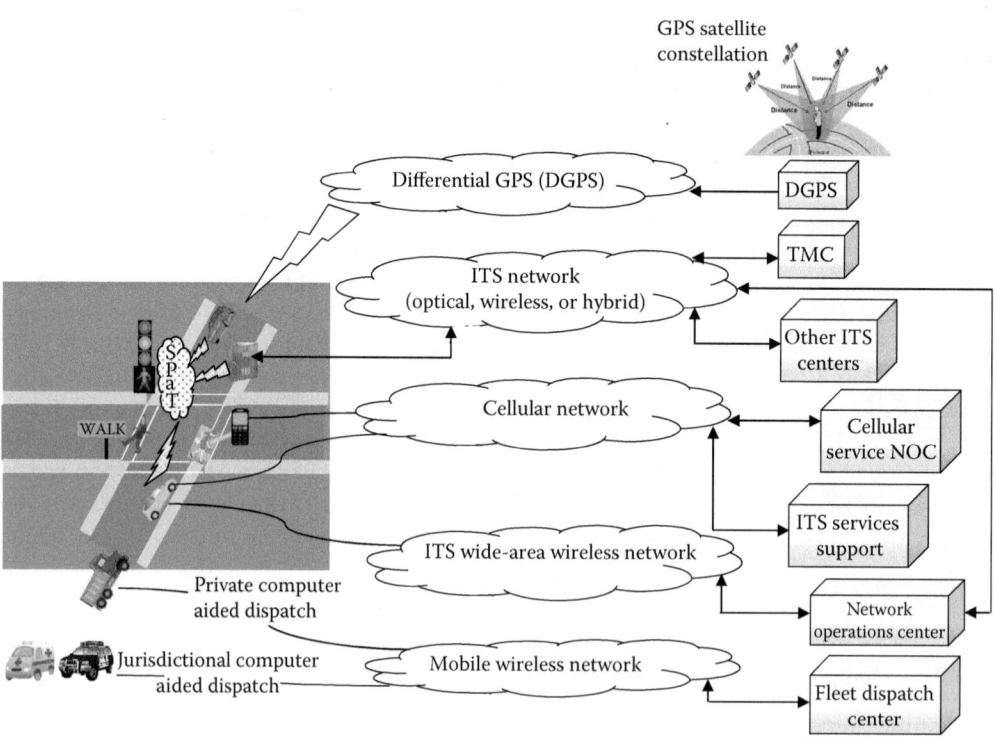

Figure 12.10 Infrastructure-to-vehicle communications options. (From *Signal Phase and Timing (SPaT) Applications, Communications Requirements, Communications Technology Potential Solutions, Issues and Recommendations*, Draft Final Report FHWA-JPO-13-002, April 3, 2012. Prepared by Bruce Abernethy, ARINC Incorporated; Scott Andrews, Cogenia Partners; and Gary Pruitt, ARINC Incorporated.)

- Differential GPS. DGPS helps correct satellite location errors caused by delays in the propagation velocity of their signals due to atmospheric effects or from multipath caused by signals bouncing off buildings, trees, or other structures. The concept calculates the GPS inaccuracy at a stationary receiver station having a known location. Using the known receiver location, the DGPS hardware can easily calculate the receiver's inaccuracy as determined from the satellite signals alone. The station then broadcasts a radio signal to all DGPS-equipped receivers in the area, providing signal correction information for that area. Second-generation DGPS receivers have typical errors of 80 cm, while more advanced DGPS techniques can achieve errors as low as 4 cm. In contrast, GPS receivers have errors of about 1.6×10^3 cm.
- Jurisdictionally owned ITS networks (optical, wireless, or hybrid). The traffic management center utilizes this class of network to communicate to intersection traffic controllers and associated sensors when monitoring traffic control devices and managing traffic flow. ITS networks may also be utilized to communicate from one TMC to another or to freight, transit, and emergency response provider dispatch centers. Non–time-critical safety data such as GID data can be transmitted from the infrastructure to a vehicle using this network. One technology option for an ITS network is a digital wireless radio system. Four types of digital wireless systems are recognized by the International Telecommunication Union worldwide for V2I or pedestrian communications. In the United States, it is HD radio also known as the digital terrestrial radio system.
- Cellular networks. The fee-for-service cellular networks serve mobile customers with good coverage in all urban areas and most major highways in the United States. The increased deployment of LTE technologies offers high-speed data rates to a large number of users simultaneously, but security may be an issue. No-fee for service LTE cellular networks are designated by the FCC to support emergency, interoperable communications using the 700-MHz emergency frequency band. These networks may be adequate for several connected vehicle applications including some safety, mobility, and environmental applications. Many regions have sufficient bandwidth in their LTE infrastructure to support safety information broadcasts, although it is yet to be determined if the bandwidth will be allocated for these transmissions, or if those jurisdictions implementing it will allow the bandwidth to be used for this purpose [6]. Germany is evaluating a dedicated 5G-network in the 700 MHz band for V2V and V2I communications along a 30-km (18.6-mi) test route [20].
- ITS wide-area networks. A WAN is a telecommunications network or computer network that extends over a large geographical distance. WANs are often established with leased telecommunication circuits. WANs are used to connect local area networks (LANs) and other types of networks together, so that users and computers in one location can communicate with users and computers in other locations. Many WANs are built for one particular organization and are private. Others, built by Internet service providers, provide connections from an organization's LAN to the Internet.
- Computer-aided dispatch. An ITS center can use a jurisdictional ITS network to communicate to fleet dispatching centers, which then distribute messages to fleet vehicles via the associated CAD link. However, many of the CAD wireless communications links have narrow bandwidths and would require transition to broadband services such as planned for emergency vehicle fleets. Furthermore, use of fleet CAD does not provide a universal solution for GID and other safety-related data to be transmitted to vehicles. In addition, centralized quality oversight is lost. Reliance on the CAD system reduces probability of safety data delivery due to additional communications systems added to the communications path to vehicles.

- Mobile wireless networks. Fleet dispatch centers communicate through mobile wireless networks, either private or jurisdictional. Wireless network configurations can be broadband or narrowband. Broadband options include conventional microwave, spread spectrum that provides CCTV and data transmission capabilities if there are a limited number of cameras on a channel, and short- and long-range wireless Ethernet. Narrowband options are conventional low data rate microwave or spread spectrum radio and area-wide radio networks with data transmission capability.

Table 12.5 summarizes the capabilities of a number of wireless technologies as they relate to V2X communications [21].

12.8 SECURITY AND CREDENTIALS MANAGEMENT

Security and credentials management (SCM) is a set of support applications that ensure trusted communications between one mobile device and another or between a mobile device and a roadside device, and the protection of data they control from unauthorized access. The applications allow credentials to be requested and revoked, and secure the exchange of trust credentials between parties so that no other party can intercept and use those credentials illegitimately. Thus, they provide security to the transmissions between connected devices, hopefully ensuring authenticity and integrity of the messages. Additional security features include privacy protection, authorization and privilege class definition, and non-repudiation of origin. Not to be overlooked is the need to provide security to back-office systems and the data they store in the cloud.

12.8.1 PKI systems

Research indicates a public key infrastructure (PKI) security system, involving the exchange of digital certificates among trusted users, can support both the need for message security and for appropriate anonymity to users. Digital certificates are used to sign the messages that pass between vehicles in the connected vehicle environment and, therefore, allow the receiver of a message to verify that the message came from a legitimate source. The receiver of the message also needs to check if the sender has the correct certificate to send not only the message, but to give the commands it contains.

Most readers will be familiar with one form of a PKI system, namely, the secure Hypertext Transfer Protocol Secure (HTTPS), which utilizes secure socket layer (SSL) connections. PKI systems rely on encryption for security. In a symmetric encryption system, there is one key

Table 12.5 Wireless technology capabilities

Capabilities	5.9 GHz DSRC	Bluetooth Class II	Nationwide DGPS	IEEE 802.11p wireless LAN	2.5–3 G PCS and digital cellular	Remote keyless entry (RKE)
Range	1000 m	10 m	300–400 m	1000 m	≈4–6 km	30 m
One-way to vehicle	Yes	–	Yes	–	–	Yes
One-way from vehicle	Yes	–	–	–	–	–
Two-way	Yes	–	–	–	–	–
Point-to-point	Yes	Yes	–	Yes	Yes	Yes
Point-to-multipoint	Yes	Yes	Yes	Yes	Yes	–
Latency	200 μs	3–4 s	n/a	3–5 s	1.5–3.5 s	n/a

that is used both to encrypt and to decrypt the message. In an asymmetric encryption system such as PKI, keys come in pairs—the sent message contains one-half of the key pair, while the receiving device has the other half—potentially enabling a PKI system to be more secure.

12.8.1.1 Basic elements of a PKI system

Before beginning the discussion of PKI systems, it is worthwhile to reference the abbreviations encountered when describing the operation of PKI systems. These are shown in Table 12.6. PKI systems contain, at a minimum, the following basic elements and functions [22]:

- Securities and Credentials Management System (SCMS)—contains personnel and procedures whose role is to manage the overall system, protect and maintain the computer hardware and facilities, update software and hardware, remove or revoke entities that do not comply with standards or misbehave, and address unanticipated issues.
- Certificate Authority (CA)—an entity that acts as the trusted third party to provide the action to authenticate the entities within a network. It typically does so by signing and distributing digital certificates. To protect privacy, these short-term certificates contain no information about users, but serve as credentials that permit them to participate in the V2V system. The CA also typically revokes certificates and publishes a Certificate Revocation List (CRL) so that valid users know to ignore certificates of users that have been revoked. A CA is considered the root of trust in a PKI.
- Registration Authority (RA)—the entity certified to register users and issue certificates. This function is performed by the CA in the simplest PKI systems.
- Root Certificate Authority (sometimes the CA and sometimes a separate entity)—the highest trusted entity within a PKI security system. The Root CA typically has a self-signed and issued certificate. A certificate that is issued by a CA to itself is referred to as a trusted root certificate as it establishes a point of ultimate trust for a CA hierarchy. Once the trusted root has been created, it can be used to authorize subordinate CAs to issue certificates on its behalf [23].
- Digital Certificates (also known as public key certificates)—electronic documents that use a digital signature to bind a public key with an identity. Digital certificates are verified using a chain of trust. The trust anchor for the digital certificate is the Root CA. Many software applications assume these root certificates are trustworthy on the user's behalf. For example, a web browser employs them to verify identities within SSL/transport layer security (TLS) connections and encrypt confidential data sent over the insecure Internet network [24]. However, the utilization of this protocol implies that the user trusts their browser's publisher, the certificate authorities, and any intermediates authorized by the certificate authority that may have issued a certificate, to

Table 12.6 PKI system abbreviations

Abbreviation	Element	Abbreviation	Element
CA	Certificate Authority	PCA	Pseudonym Certificate Authority
CME	Certificate Management Entity	PKI	Public Key Infrastructure
CRL	Certificate Revocation List	RA	Registration Authority
DCM	Device Configuration Manager	SCM	Securities and Credential Management
ECA	Enrollment Certificate Authority	SCMS	Securities and Credential Management System
LA	Linkage Authority		
LOP	Location Obscurer Proxy	SSL	Secure Socket Layer
MA	Misbehavior Authority	TLS	Transport Layer Security

faithfully verify the identity and intentions of all parties that own the certificates. This (transitive) trust in a root certificate is the usual case. The most common commercial variety is based on the International Telecommunication Union Telecommunication Standardization Sector Standard X.509 [25].

- Secure hardware and software (servers, stores, repositories; also known as a central directory)—hardware and software to support the processing of certificate requests, save issued certificates before they are distributed, or save revoked certificates. This hardware and software may generate certificates, validate received certificates, and also be used in back-up systems.
- Communications—wire line, wireless, or Internet services that provide the communications capacity over which management capabilities are enacted to receive requests, distribute certificates, collect misbehavior reports, revoke certificates, and distribute the CRL. Average sizes of PKI objects are
  - Private/public key pair = 1 KB (typical).
  - Local certificate = 2 KB.
  - CA certificate = 2 KB.
  - CA authority configuration = 500 bytes.
  - CRL (average size is variable, depending on how many certificates have been revoked by a particular CA) = 300 bytes to 2 MB and more (typical sizes).

12.8.1.2 Limitations of existing PKI systems

Off-the-shelf PKI systems existing today are not broad enough in their functions to serve as a key safety-critical element for V2V communications. Most PKI systems are concerned with data exchange among parties that are either known to each other as trusted sources (e.g., the military knows each of its communication points) or are identifiable (e.g., air traffic controllers can identify each of the planes involved in safety-critical data exchange). Also, the majority of other safety-critical systems employ highly secure networks (e.g., the military) or private networks (e.g., the military) and cannot leverage either existing communications systems or the Internet (to keep capital investment costs to a minimum and to achieve widespread access) in a manner that does not introduce additional vulnerabilities and risks [25].

Nearly all of the existing commercial systems, by comparison, do leverage the Internet and wireless systems. These systems enable online purchasing or online financial transactions in a way that allows for easy accessibility to millions of users. They do not, however, meet the level of privacy protection required for V2V data exchanges, as these organizations have preexisting agreements with the CA and thus user identity resides within databases and is typically used as part of the authentication process.

12.8.2 V2V security system

Figure 12.11 shows a simplified view of a V2V security system that has enhanced security features as compared to a basic PKI system. The interactions between the components in the figure are executed by automatic machine-to-machine operations using processors in the various V2V components, including the OBE in the vehicle. No human judgment is involved in creation, granting, or revocation of the digital certificates [22]. The enhanced security addresses the following limitations of basic PKI systems:

- Protects privacy with a system that divides and separates some of the functionality to ensure that no one entity has the ability to match records that would lead to identification of a specific driver or specific vehicle.

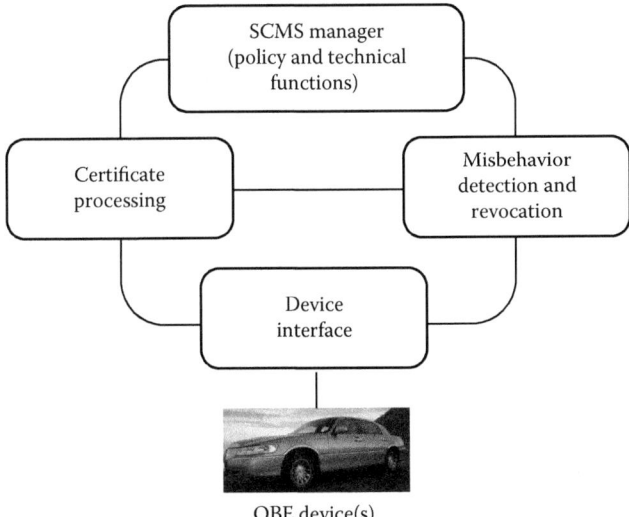

OBE device(s)

Figure 12.11 V2V security system (simplified). (Adapted from J. Harding et al., *Vehicle-to-Vehicle Communications: Readiness of V2V Technology for Application*, Chapter IX. V2V Communications Security, DOT HS 812 014, National Highway Traffic Safety Administration, U.S. Department of Transportation, Washington, DC 20590, August 2014. http://www.nhtsa.gov/staticfiles/rulemaking/pdf/V2V/Readiness-of-V2V-Technology-for-Application-812014.pdf. Accessed August 5, 2016.)

- Utilizes two linkage authorities (LAs) to create linkage values that allow one entry on the CRL to revoke an entire batch of certificates, instead of having to list each certificate. An LA has enough information such that an inside attacker can track a user. Therefore, if the linkage value comes from the output of two separate LAs, neither has enough information to track anyone.
- Allows for a greater number of digital certificates to be issued. Digital certificates employ random identifiers that change frequently to lower the risk of identifying any one vehicle or driver with a particular certificate.
- Addresses privacy considerations by adding elements that obscure location coordinates when a vehicle or device communicates with the system (e.g., requesting additional digital certificates or reporting misbehavior detected locally near the vehicle).
- Enhances the misbehavior authority (MA) in the V2V PKI so that it can detect and take actions to mitigate or remove malicious behavior.
- Improves the trust requirement through a direct interface with a Certification Lab entity to verify that each type of device meet standards proving their capabilities to be trusted, secure, and interoperable.
- Adds request coordination to ensure that an OBE cannot obtain multiple batches of certificates by sending requests to several RAs at the same time.

To implement these features, practical V2V security systems add the following entities or enhance the functions of those found in the basic system.

12.8.2.1 SCMS manager

The SCMS manager provides the security policy and technical standards for the entire connected vehicle industry. Personnel within the SCMS operate the overall system, select and adhere to standards, protect and maintain the computer hardware and facilities, update software and hardware, and address unanticipated issues. Just as any large-scale industry ensures consistency and standardization of technical specifications, standard operating procedures (SOPs), and other industry-wide practices such as auditing, the SCMS manager would perform and monitor these types of activities. This can happen in a number of ways. Often in commercial industries, volunteer industry consortiums take on this role (see reference [26] for example). In other industries, or in public or quasi-public industries, this role may be assumed by a regulatory or other legal or policy body. Regardless of the final choice of how to implement a central administrative body, it is expected that one would be established for the SCMS. The hope is that the SOPs, audit standards, and other practices set by this body would then be executed and complied with by each Certificate Management Entity (CME) individually. It is also assumed that any guidance, practices, SOPs, auditing standards, or additional industry-wide procedures would be established based on any Federal guidance or regulation. Section 12.9.3.2 describes the implementation of the SCMS manager as envisioned in a pilot program sponsored by NHTSA.

Generally, SCMS operating functions fall into two categories: pseudonym functions and initialization or bootstrap functions [22].

12.8.2.2 Pseudonym functions and certificates

The V2V security design utilizes short-term digital certificates used by a vehicle's OBE to authenticate and validate sent and received BSMs that form the foundation for V2V safety technologies. These short-term certificates contain no information about users to protect privacy, but serve as credentials that permit users to participate in the V2V system. Pseudonym functions create, manage, distribute, monitor, and revoke short-term certificates for vehicles. They include the following:

- Intermediate Certificate Authority (Intermediate CA) is an extension of the Root CA shielding it from direct access to the Internet. It can authorize other CMEs or possibly an Enrollment Certificate Authority (ECA) using authority from the Root CA. It does not hold the same authority as the Root CA in that it cannot self-sign a certificate. The Intermediate CA provides system flexibility because it obviates the need for the highly protected Root CA to establish contact with every SCMS entity as they are added to the system over time. Additionally, the use of Intermediate CAs lessens the impact of an attack by maintaining protection of the Root CA.
- Location Obscurer Proxy (LOP) obscures the location of OBE seeking to communicate with the SCMS functions, so that the functions are not aware of the geographic position of a specific vehicle. All communications from the OBE to the SCMS components must pass through the LOP. Additionally, the LOP may shuffle misbehavior reports that are sent by OBEs to the MA (see below) during full deployment. This function increases participant privacy but does not increase or reduce security.
- LA is the entity that generates linkage values. The LA has been designed to come in pairs of two, referred to as LA1 and LA2. The LAs for most operations communicate only with the RA (see below) and provide linkage values in response to a request by the RA and Pseudonym Certificate Authority (PCA) (see below). The linkage values provide the PCA with a means to calculate a certificate ID and a mechanism to connect all short-term certificates from a specific device for ease of revocation in the event of misbehavior.
- MA is the central function to process misbehavior reports and produce and publish the CRL. It works with the PCA, RA, and LAs to acquire necessary information about a certificate to create entries to the CRL through the CRL Generator. The MA eventually

may perform global misbehavior detection, involving investigations or other processes to identify levels of misbehavior in the system. The MA is not an external law enforcement function, but rather an internal SCMS function intended to detect when messages are not plausible or when there is potential malfunction or malfeasance within the system. The extent to which the CMEs share externally information generated by the MA about devices sending inaccurate or false messages—either with individuals whose credentials the system has revoked or with law enforcement—will depend on law, organizational policy, and/or contractual obligations applicable to the CMEs and their component functions.

- PCA issues the short-term certificates used to ensure trust in the system. In earlier designs, their lifetime was fixed at 5 min. The validity period of certificates is still on the order of "minutes" but is now a variable length of time, making them less predictable and thus harder to track. Certificates are the security credentials that authenticate messages from a device. In addition to certificate issuance, the PCA collaborates with the MA, RA, and LAs to identify linkage values to place on the CRL if misbehavior has been detected.
- RA performs the necessary key expansions before the PCA performs the final key expansion functions. It receives certificate requests from the OBE (by way of the LOP), requests and receives linkage values from the LAs, and sends certificate requests to the PCA. It shuffles requests from multiple OBEs to prevent the PCA from correlating certificate IDs with users. It also acts as the final conduit to batching short-term certificates for distribution to the OBE. Lastly, it creates and maintains a blacklist of enrollment certificates so it will know to reject certificate renewal requests from revoked OBEs.
- Request Coordination prevents an OBE from receiving multiple batches of certificates from different RAs by synchronizing activities with the RAs such that certificate requests during a given time period are responded to without duplication. This function is necessary only if there is more than one RA in the SCMS.
- Root CA is the master root for all other CAs; it is the "center of trust" of the system. It issues certificates to subordinate CAs in a hierarchical fashion, providing their authentication within the system so all other users and functions know they can be trusted. The Root CA produces a self-signed certificate (verifying its own trustworthiness) using out-of-band communications. This enables trust that can be verified between ad hoc or disparate devices because they share a common trust point. It is likely that the Root CA will operate in a separate, offline environment because compromise of this function is a catastrophic event for the security system.

12.8.2.3 Initialization functions and enrollment certificates

The security design also includes functions that perform the bootstrapping process, which establishes the initial connection between a motor vehicle's OBE and the SCMS. The principal element in this process is the ECA that assigns a long-term enrollment certificate to each OBE. To the extent required by NHTSA or other stakeholders, the bootstrap process creates a link between the SCMS and specific OBEs or production lots of OBEs and enrollment certificates that later may be used by OEMs and NHTSA to identify defective V2V equipment. The design does not indicate when bootstrapping should take place, but NHTSA has suggested that it might need to occur at the time of OBE manufacture to facilitate the level of linkage between long-term enrollment certificates and equipment production lots that NHTSA requires for enforcement purposes (e.g., to identify defective equipment).

At the time the Harding et al.'s report [22] was prepared in 2014, bootstrap functions had been fairly well defined for OBEs. The process for establishing the connection between aftermarket safety devices and the SCMS had not been defined; nor will it be by Crash Avoidance Metrics Partners (CAMP) [26]. It will need to be defined by aftermarket safety-device manufacturers working with the final structure of the SCMS.

Initialization functions are performed by the following security elements:

- Certification Lab, which does not take part in the particular use cases of the SCMS, instructs the Enrollment CA on polices and rules for issuing enrollment certificates. This is usually done when a new device is released to the market or if the SCMS manager releases new rules and guidelines. The Enrollment CA uses information from the Certification Lab to confirm that devices of the given type are entitled to an enrollment certificate.
- ECA verifies the validity of the device type with the Certification Lab. Once verified, the ECA produces the enrollment certificate and sends it to the OBE. After the OBE has a valid enrollment certificate, it is able to request and receive certificates from the SCMS.
- Device Configuration Manager (DCM) is responsible for giving devices access to new trust information, such as updates to the certificates of one or more authorities, and relaying policy decisions or technical guidelines issued by the SCMS manager. It also sends software updates to the OBEs. The DCM coordinates initial trust distribution with OBE by passing on credentials for other SCMS entities, and provides the OBE with information it needs to request short-term certificates from an RA. The DCM also participates in the bootstrap process by ensuring that a device is cleared to receive its enrollment certificate from the ECA and provides a secure channel to the ECA. There are two types of connections used from devices to the DCM: in-band and out-of-band communications. In-band communication uses the LOP, while out-of-band communication is sent directly from the OBE to the ECA by way of the DCM.

12.8.2.4 Unique support technologies

1. Butterfly keys. Butterfly keys are a novel cryptographic construction that allows a device to request an arbitrary number of certificates, each with different signing keys and each encrypted with a different encryption key. The request protocol contains only one verification public key seed and one encryption public key seed, but two "expansion functions" that allow the second party to calculate an arbitrarily long sequence of statistically uncorrelated (as far as an outside observer is concerned) public keys such that only the original device knows the corresponding private keys.

   Without butterfly keys, the device would have to send a unique verification key and a unique encryption key for each certificate. Thus, butterfly keys reduce the upload size of certificate requests, and allow requests to be made when there is only spotty connectivity (although they also increase the size of the certificate upload). They also reduce the work done by the requester to calculate the keys, thus reducing computational burden.

2. Linkage values. To support efficient revocation, end-entity certificates contain a linkage value that is derived from cryptographic seed material. Publication of the seed is sufficient to revoke all certificates belonging to the revoked device, but without the seed an eavesdropper cannot tell which certificates belong to a particular device. The revocation process is designed such that it does not give up backward privacy. For protection against insider attacks, the seed is the combination of two seed values produced by two LAs. This ensures that no single organizational entity possesses enough

information to identify a single device. An extension to the linkage values approach allows for group revocation. This permits all devices of a particular type having a flaw to be revoked with a single entry on the revocation list, while keeping group membership secret until the relevant group seed is revealed. Group revocation is considered an option along with revocation of single devices.

Linkage values and LAs enable the SCMS to support seven requirements:

a. There should exist an efficient way of revoking all the certificates within a device.
b. There should exist an efficient way of revoking all the certificates within a group of devices.
c. Certificates should not be linkable by an eavesdropper unless the owner has been revoked.
d. Membership in a group should not be disclosed unless that group has been revoked.
e. If a vehicle's security credentials are revoked, the vehicle should be identifiable going forward, but its movements before it was revoked should not be trackable.
f. Similarly, if a group of vehicles' security credentials are revoked, a device belonging to that group should be identifiable as a member. However, it should not be possible to determine the membership to a group before the group revocation took place.
g. No single entity within the system should be able to determine that two certificates belong to the same device or to the same group. An exception to this rule is the MA.

If there is a requirement that no single entity within the SCMS should be able to identify a vehicle once an LA is established, this requirement is no longer fulfilled. For that reason, two LAs are introduced and the information that allows for identification is split between them.

3. MA and CRL generation. Most SCMS functions listed above are fairly well developed. One critical function still under development is the MA, whose misbehavior detection policies critically influence system integrity and system costs [22]. The MA is the central security element responsible for processing misbehavior reports generated by the OBE and producing and publishing the CRL. This list, once distributed, identifies digital certificates that are no longer valid and that the OBE should no longer rely on for messages. The size of the CRL depends on the frequency of list distribution and rate of misbehavior across the vehicle fleet. Onboard storage and distribution costs for the CRL are two major cost generators in the technical design.

The MA is also responsible for performing global misbehavior detection, involving the collection of a sampling of misbehavior reports from the OBE for detecting system-wide misbehavior and revoking misbehaving entities. A NHTSA decision to move forward with regulatory action will require maturation of the misbehavior detection processes through the NHTSA-CAMP collaboration and perhaps the involvement of other consultants [22].

12.8.3 Hacking of connected vehicle communications

Connected vehicle applications rely on V2V and V2I transmissions to send and receive messages and warnings, monitor congestion, alter traffic signal timing, and optimize transit. However, any person within a few hundred meters of a V2V or V2I transmitter can also receive these messages using a wireless "sniffing station."

Data in BSM vehicle transmissions are unencrypted to allow other vehicles to use the broadcast speed and position information. In addition to there being no personally identifiable data (such as a license plate) within the message itself, each wireless bulletin is digitally signed to ensure that fake messages cannot be introduced to disrupt traffic or possibly even cause accidents.

It is these digital signatures that sniffing stations track. A solution proposed by NHTSA and European authorities is for vehicles to sign their messages using pseudonyms that automatically change every 5 min. But researchers found that even this was not enough to outfox the sniffing system. "Changing pseudonyms every five minutes just leads to a 50 percent increase of the cost for the attacker, meaning they'll have to install 50 percent more sniffing stations" as reported by University of Twente in the Netherlands [27].

12.9 POLICY AND INSTITUTIONAL ISSUES

Connected vehicle policy, implementation, and institutional issues are often solved through research and analysis of the challenges that may limit successful deployment of connected vehicle technologies. The vision for connected vehicle policy research is to create a collaborative effort among the USDOT, key industry stakeholders, vehicle manufacturers, state and local governments, representative associations, citizens, and others. Input from all stakeholders allows the structuring and implementation of a research agenda that weighs the benefits and risks of all viable options and produces a strong policy foundation for successful deployment of connected vehicle technologies and applications.

Specific technical policy, legal policy, implementation policy, implementation strategy, and institutional issues affecting connected vehicle deployment are highlighted below. Due to the rapid evolution of automated, connected, and autonomous vehicles, some of these may have been resolved since the referenced studies were performed. However, the value in discussing them here is to bring awareness to future planners and policymakers of pertinent issues that could affect other strategic programs and initiatives.

12.9.1 Technical policy issues

Technical policy issues involve the following:

- Analysis of technical choices for V2V and V2I technologies and applications to identify if options require new institutional models or can leverage existing assets and personnel. These analyses extend to policies related to the core system, system interfaces, and device certification and standards. For example, the radio-frequency environment near a signalized intersection may require analysis and modeling to determine if there is any interference or minimal interference with signals transmitted for safety-critical V2V and V2I applications. Figure 12.12 illustrates the mix of communications technologies that may cause such concerns. In addition to the communications media available to a TMC to communicate with vehicles, controllers, and other infrastructure devices (see, e.g., Figure 12.10), there may be high-power emitters in operation by other entities that can saturate the RF front end of DSRC receivers and devices. Any harmonics in the 5.8–5.92 GHz frequency band produced by these emitters can negatively impact safety-related communications [6].
- Certification standards for issuing enrollment certificates. Once NHTSA establishes a Federal Motor Vehicle Safety Standard (FMVSS), vehicle and device manufacturers are required to certify that they comply with it in order to sell vehicles and devices. Noncompliance could result in enforcement action by NHTSA (e.g., a requirement to recall affected vehicles and devices, an injunction from selling affected vehicles and devices until remedied, and civil penalties). Additionally, if V2V devices develop a safety defect, manufacturers (both of vehicles and V2V devices) may also be ordered to recall the devices. It is possible that manufacturers may choose to rely on third-party

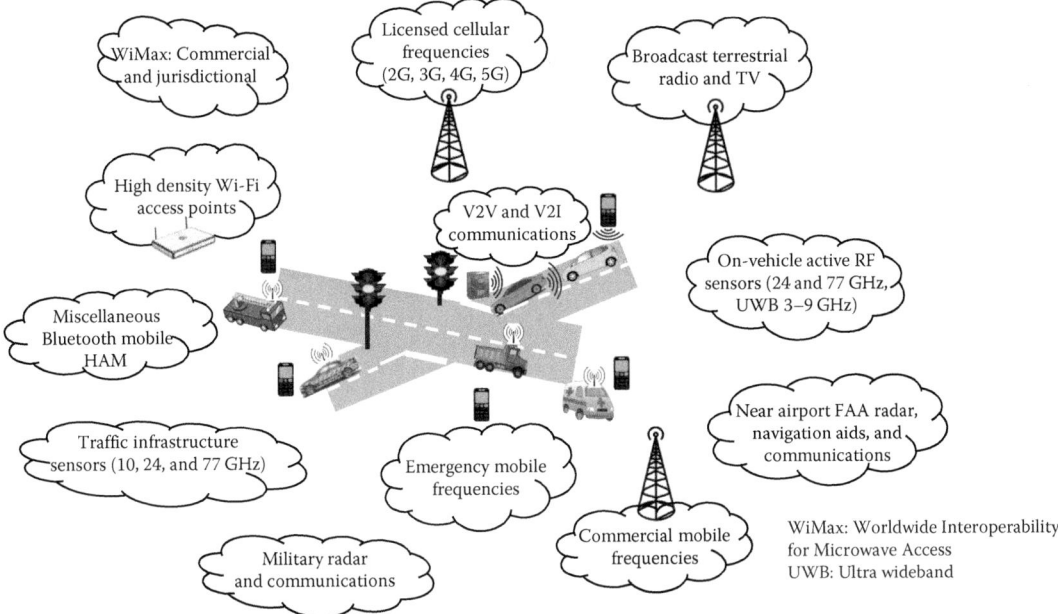

Figure 12.12 Radio-frequency environment near a signalized intersection may include a complex mix of frequencies. (From *Signal Phase and Timing (SPaT) Applications, Communications Requirements, Communications Technology Potential Solutions, Issues and Recommendations*, Draft Final Report FHWA-JPO-13-002, April 3, 2012. Prepared by Bruce Abernethy, ARINC Incorporated; Scott Andrews, Cogenia Partners; and Gary Pruitt, ARINC Incorporated.)

certification for V2V devices to ensure uniform adherence to NHTSA's requirements, but NHTSA would not expect to participate in that certification [22].

- Selection of analytic methods and performance measures for analyzing data, modeling, and decision support system effectiveness [28].
- A common linear reference system to integrate multiple sources of data.
- Software safety verification and validation methods. Currently, no method exists for efficiently developing, verifying, and validating software to ensure its reliability is sufficient to make safety-of-life critical decisions. The complex nature of software for an application as complicated as automated driving creates a situation where it is not possible to prove its completeness or correctness analytically. Exhaustive enumeration or testing is also impossible because the number of possible combinations of paths through the software logic, given the diversity of the input data that the software will encounter in driving, is too vast to be manageable.
- Analytical methods have been applied to verification and validation on simple example problems, and even those have been found to become extremely complicated. The existing analytical methods are not scalable to a problem of the complexity of automated driving. In practice, software verification and validation are currently executed with costly and time-consuming brute-force methods.

Shladover and Bishop [29] comment that if a fully sufficient solution is not available, the question of what constitutes "good enough" is raised. In the domain of product development in the competitive automotive industry, each system developer is answering this question individually.

Google (through Waymo–its self-driving technology company), for example, reports that as of May 2017 its self-driving cars drove 3 million actual road miles (mainly on city streets). In addition, it logged more than one billion computer-simulated miles [30], while Tesla says its Autopilot feature has been activated for more than 130 million miles as of July 2016.

Techniques for addressing software safety that build upon extensive techniques already developed for active safety systems are under active development at this time. The specifics of these approaches are proprietary and not published. At the same time, public agencies with responsibility for protecting the public safety have to exercise their own due diligence regarding the safety claims of system developers rather than simply accept those claims at face value. Because of the technical complexity of automation software and the absence of specific reference standards, it is difficult for an external entity (such as an impartial test lab or government agency) to independently verify the safety of automation software, which thus remains one of the primary unresolved technological challenges. The situation is further complicated by aftermarket systems offered by new market entrants that may not have a long legacy of developing robust and safe complex vehicle control systems.

12.9.2 Legal policy issues

Legal policy issues deal with the following:

- Analysis of the federal role and authority in connected vehicle system development and deployment.
- Analysis of liability and limitations to risk.
- Policy and practices regarding privacy.
- Policies on intellectual property and data ownership.
- A Rand Corporation study recommends the following policy considerations [31]:
    - Policymakers should avoid passing regulations prematurely while the technology is still evolving.
    - Distracted-driving laws will need updating to incorporate autonomous vehicle technology.
    - Policymakers should clarify who will own the data generated by this technology and how it will be used, and address privacy concerns.
    - Regulations and liability rules should be designed by comparing the performance of automated and self-driving vehicles to that of average human drivers and the long-term benefits of the technology should be incorporated into determinations of liability.

The following sections of 12.9.2 amplify upon the legal concerns facing autonomous or self-driving vehicles. Many of these also share commonalities with connected vehicles.

12.9.2.1 2013 NHTSA recommendations for licensing drivers for self-driving vehicle testing

According to NHTSA, very few Level 3 (NHTSA taxonomy from Section 11.2.4 implied) automated systems exist and the systems that do exist are at early stages of testing and development. Because Level 4 automated systems are not yet a reality and the technical specifications for Level 3 automated systems are still in flux, the agency believes that regulation of the technical performance of automated vehicles is premature at this time.

In 2013, NHTSA offered the recommendations in Table 12.7 for *testing* self-driving vehicles [32]. These are based on observations of self-driving vehicle technology development, including in-depth discussions with developers of those technologies, direct experience with several of the vehicles under development, and knowing that some states are anxious for guidance on how to proceed with regard to self-driving vehicles. The recommendations are divided into four categories, those for licensing drivers to operate self-driving vehicles for testing, those for states developing regulations for self-driving vehicles, basic principles for testing of self-driving vehicles, and guidance to defer development of regulations for the operation of self-driving vehicles for purposes other than testing. Further guidance was provided in 2016 as was summarized in Table 11.10.

12.9.2.2 American Association of Motor Vehicle Administrator's best practices working group

The American Association of Motor Vehicle Administrators (AAMVA) Autonomous Vehicle Best Practices Working Group in partnership with NHTSA embarked on developing a best practices guide for utilization by NHTSA and the states [33]. The guide will not be a mandate for states, but rather a first step in addressing some of the challenges associated with autonomous vehicles and innovative technologies. The two-year project involves jurisdictions, law enforcement, federal agencies, and the automobile, automation, insurance, and legal communities in the gathering, organizing, and sharing of information on testing and public use of autonomous vehicles with the AAMVA community. Concerns relate to liability, insurance, testing standards, and safety. (Refer to Section 11.4 for a description of the 2016 NHTSA automated vehicles policy.)

The working group is divided into three subgroups whose functions are as follows. The driver subgroup is examining driver licensing requirements, driver training and testing for SAE Level 3 and Level 4 operation, training for state examiners, defining operators versus drivers, possible license restrictions and endorsements, and license suspensions and revocations. The vehicle subgroup is focusing on vehicle testing requirements, insurance requirements, consumer registration and title requirements, state reciprocal agreements for testing, and safety requirements for testing vehicles. The law enforcement subgroup is considering traffic laws associated with Level 3 and 4 automation, violation codes, crash investigations with Levels 3 and 4, accessing black box autonomous information, road restrictions, and criminal activity. Another area worthy of attention by the law enforcement subgroup is hacking of data and information residing in or being sent to and from patrol cars that are part of the connected vehicle environment. These vehicles may need special hacking detection and protection mechanisms.

12.9.2.3 State-by-state self-driving car regulations

Many states are developing standards and passing regulations to govern the operation of self-driving vehicles to ensure the safety of the traveling public [34]. As of May 2017, sixteen states—Alabama, Arkansas, California, Florida, Georgia, Louisiana, Michigan, New York, Nevada, North Dakota, Pennsylvania, South Carolina, Tennessee, Utah, Virginia, and Vermont—and Washington DC passed legislation related to autonomous vehicles. Governors in Arizona, Massachusetts, and Wisconsin issued executive orders related to autonomous vehicles. It is likely that testing is legal on public highways in the absence of legislation, so the effect of legislation is often to narrow the circumstances under which testing can take place [35]. The Uniform Law Commission is studying whether to adopt a proposed model law for the states, including insurance requirements for drivers.

Table 12.7 2013 NHTSA recommendations to states for testing self-driving vehicles

Category	Recommendations
Licensing drivers to operate self-driving vehicles for testing	Ensure that the driver understands how to operate a self-driving vehicle safely. • Issuance of a driver's license endorsement (or separate driver's license) to a person should be conditioned upon prerequisites, such as passing a test concerning the safe operation of a self-driving vehicle and presentation of certification by a manufacturer of self-driving vehicles (or the manufacturer's designated representative) that the person has successfully completed a training course provided by that manufacturer (or representative), or certification by that manufacturer (or representative) that the person has operated a self-driving vehicle for a certain minimum number of hours.
State regulations governing testing of self-driving vehicles	Ensure that on-road testing of self-driving vehicles minimizes risks to other road users by • Requiring manufacturers to certify that the vehicle has already operated for a certain number of miles in self-driving mode without incident before seeking the license to test the vehicle on public roads. • Requiring these manufacturers to submit data from previous testing involving the technology. • Requiring manufacturers to submit a plan to the state regulatory body describing how the manufacturer plans to minimize safety risks to other road users. • Strongly recommending that states require that a properly licensed driver be seated in the driver's seat and ready to take control of the vehicle while the vehicle is operating in self-driving mode on public roads. Limit testing operations to roadway, traffic, and environmental conditions suitable for the capabilities of the tested self-driving vehicles. • States should require manufacturers of self-driving vehicles to submit a testing plan that includes the operating conditions of the test. Manufacturers should supply states with test data or other information to demonstrate that their self-driving vehicles are capable of operating under the test conditions with limited driver intervention. • States are encouraged to limit the conditions in which a vehicle may be operated in self-driving mode to ensure their safe operation. • Regulations governing self-driving vehicle testing could limit testing to the operating conditions for which the self-driving system is specifically designed, such as driving on a limited-access highway. Likewise, depending on the self-driving vehicle, regulations could limit testing of the self-driving vehicle to roads in only certain geographical locations, e.g., those known for having light traffic or for having heavy traffic at low travel speeds. Establish reporting requirements to monitor the performance of self-driving technology during testing and, by so doing, expand the body of data and support research concerning self-driving vehicles. • Require businesses testing self-driving vehicles to submit to the state certain information, including — Instances in which a self-driving vehicle, while operating in or transitioning out of self-driving mode, is involved in a crash or near crash. — Incidents in which the driver of a self-driving vehicle is prompted by the vehicle to take control of the vehicle while it is operating in the self-driving mode because of a failure of the automated system or the inability of the automated system to function in certain conditions.

(Continued)

Table 12.7 (Continued) 2013 NHTSA recommendations to states for testing self-driving vehicles

Category	Recommendations

Basic principles for testing of self-driving vehicles	NHTSA does not recommend that states attempt to establish safety standards for self-driving vehicle technologies as they are in the early stages of development and a number of technological and human performance issues must still be addressed. Until such time as NHTSA develops vehicle safety standards pertinent to self-driving technologies, states may want to require that self-driving test vehicles adhere to the following guidelines:
	• Ensure that the process for transitioning from self-driving mode to driver control is safe, simple, and timely.
	  – During the testing phase of the development of self-driving vehicles, a driver familiar with the particular vehicle's automated systems is necessary to guarantee that a failure of the automated system or the occurrence of conditions in which the automated system is not intended to operate does not put other road users at risk.
	  – A regulation may require that the driver be able to retake control of the test vehicle by an immediately overriding, relatively simple, and non-distracting method such as pressing a button located within the driver's reach.
	  – The automated functions of a test vehicle should defer to the driver's input by allowing the driver to retake control by using the brakes, the accelerator pedal, or the steering wheel.
	  – The self-driving vehicle should alert the driver when the driver must take control of the vehicle because the automated system cannot operate due to road conditions, environmental conditions, a malfunction, or any other condition or circumstance that would require manual driving for safe operation.
	• Self-driving test vehicles should be capable of detecting and recording that the system of automated technologies has malfunctioned or is operating in a degraded state, and informing the driver in a way that enables the driver to regain proper control of the vehicle. The recording should enable personnel to establish the cause of any such malfunction, degradation, or failure.
	• Ensure that self-driving test vehicles record information about the status of the automated control technologies in the event of a crash or loss of vehicle control.
	  – Self-driving test vehicles should record data from the vehicle's sensors, including sensors monitoring and diagnosing the performance of the automated vehicle technologies, in the event of a crash or other significant loss of vehicle control. The recording should note whether the automated technology system was in control of the vehicle at the time of the crash.
	  – Data recorded by the vehicle's event data recorder in the event of a crash should be made available to the state.
	• Ensure that installation and operation of any self-driving vehicle technologies does not disable any federally required safety features or systems.

Operation of self-driving vehicles for purposes other than testing	NHTSA does not recommend that states authorize the operation of self-driving vehicles for purposes other than testing at this time. The agency believes there are a number of technological and human performance issues that affect safety to be addressed before self-driving vehicles can be made widely available.

Source: *Preliminary Statement of Policy Concerning Automated Vehicles*, U.S. Department of Transportation, National Highway Traffic Safety Administration, Washington, DC, 2013. http://www.nhtsa.gov/staticfiles/rulemaking/pdf/Automated_Vehicles_Policy.pdf. Accessed September 29, 2016.

California Department of Motor Vehicles (DMV) testing regulations first went into effect on September 16, 2014. They were modified in October 2016 to allow pilot projects to test autonomous vehicles that have no driver and are not equipped with a steering wheel or pedals, but only if the testing is conducted at specified locations and the driverless vehicle operates at speeds of less than 35 mi/h (56 km/h) [36,37]. In March of 2017, the California DMV issued a notice of proposed regulatory action that seeks to clarify rules for testing of autonomous vehicles without a driver in the vehicle and to adopt new rules that would allow deployment of autonomous vehicles for use by the public on California roads [38–40]. These proposed regulations for autonomous vehicle testing and deployment are summarized in Table 12.8.

Weiner and Smith [43] provide a summary of state legislative actions concerning automated driving in the United States. The state-by-state compilations outline legislation that has been or is under consideration or has been enacted. The synopsis also includes state executive orders and regulations. In February 2016, NHTSA responded to Google's request to interpret a number of provisions in the FMVSS and to designate its artificial intelligence Self-Driving System (SDS) as the legal driver of its Level 4 full self-driving automated vehicle [44]. Most of Google's requests were granted, but several require additional submittals of information or rulemaking.

12.9.2.4 Insurance for autonomous vehicles

Insurance for autonomous vehicle operation on public roads is predicted to shift from drivers to original equipment manufacturers (OEMs), road operators, and local authorities. The factors leading to this change are the following [34,45]:

- Autonomous vehicles are likely to increase insurance claims related to product specifications rather than driver liability. Analysis from Frost & Sullivan finds that motor insurers will move away from the driver-centric strategy to follow one or a combination of three models as autonomous vehicles become common, namely, product-centric evaluation, brand-centric evaluation, and system-centric evaluation.
- The current system of calculating motor insurance premiums is based on driver-related factors such as age, gender, and driving record, and insured vehicle characteristics. However, the introduction of autonomous vehicles will focus importance on vehicle-related parameters. This will result in higher product liability, causing the responsibility of insuring the vehicle to shift from vehicle owners to manufacturers, road operators, and local transport authorities. Further, all excess insurance coverage currently carried by the insured will be shared among several stakeholders, such as road operators and local transport authorities.
- Assuming the risk of accidents will fall drastically with the advent of autonomous vehicles, the insurance premium to cover that risk too will drop significantly. Nevertheless, OEMs and suppliers will increase insurance spending to cover their share of product liability risk, thereby offsetting the shrinkage in consumer-driven insurance revenues. In the wake of plummeting premiums, motor insurance will become part of other insurance policies and value-added packages as stakeholders look to new avenues of profit generation in a changing environment. The traditional method of underwriting that uses historic accident and repair data will take a back-seat, paving the way for a new breed of underwriters capable of evaluating driving algorithms and assigning relevant risks. Insurance for cyber protection against cyber-attacks and hacks will be part of the new coverages.

Table 12.8 Summary of California proposed regulations for testing and deploying autonomous vehicles

Application	Proposed regulations
Testing with a licensed driver in the vehicle	Manufacturers must provide proof the vehicle being tested was successfully tested under controlled conditions. Manufacturers must obtain a Manufacturers Testing Permit. Anyone who gets behind the wheel of a self-driving vehicle must first complete a training program. While the vehicle is moving, the driver must be in the driver's seat and be able to take over, if needed. Requires the manufacturer to have a $5 million (U.S.) insurance or surety bond. Any incident involving an accident or an incident where the driverless technology disengages has to be immediately reported to the DMV. Prohibits operation of any test vehicle when members of the public pay a fee or the manufacturer receives compensation for providing a ride to members of the public. This regulation is intended to ensure that vehicles are operated only for testing purposes, and not for generating revenue from providing transportation services.
Testing without a licensed driver in the vehicle	Applies to driverless vehicles for which the manufacturers have obtained a Manufacturers Testing Permit—Driverless Vehicles. Clarifies autonomous mode by defining it as vehicle operation using a combination of hardware and software, both remote and onboard, that performs the dynamic driving task with or without a natural person actively monitoring the driving environment. Defines an autonomous test vehicle as one equipped with technology that allows it to operate at Levels 3, 4, or 5 of the SAE *Taxonomy and Definitions for Terms Related to Driving Automation Systems for On-Road Motor Vehicles* [41]. Clarifies that an autonomous test vehicle does not include vehicles equipped with one or more systems that provide driver assistance or enhance safety benefits, but are not capable singularly or in combination of performing the dynamic driving task on a sustained basis without the constant control or active participation of a natural person. Requires submission of a copy of the manufacturer's 15-point safety assessment letter provided to NHTSA pursuant to the "Vehicle Performance Guidance for Automated Vehicles" in NHTSA's *Federal Automated Vehicles Policy* [42]. Requires the manufacturer to provide written support from the jurisdiction in which the vehicles will be tested, certify there is a communications link in the vehicles, provide information related to the intended operational design domain, maintain a training program for remote operators, provide certain disclosures to any passengers, and submit a copy of the law enforcement interaction plan that instructs police officers, fire fighters, and paramedics how to deal with the vehicle in the event of a breakdown or accident. Requires the manufacturer to have a $5 million (U.S.) insurance or surety bond. Requires the autonomous vehicle to have a mechanism to engage and disengage the autonomous technology that is easily accessible to the operator.
Deployment on public roadways	Submission of the specified application and NHTSA exemption if the vehicle is not equipped with manual controls but complies with all other FMVSS. Certification that the vehicles have a communication link that can transfer vehicle owner information and certification that the vehicle has been registered with NHTSA. Submission of a consumer education plan; copies of law enforcement interaction plan, specified written disclosures, and safety assessment letter provided to NHTSA; and test data demonstrating the vehicle has been tested in its intended operational design domain. Requires a manufacturer that has identified a safety-related defect in its autonomous technology to submit to the department a copy of the report prepared in accordance with Part 573 of Title 49, Code of Federal Regulations.

Source: Department of Motor Vehicles Notice to Amend Sections in Article 3.7 and Adopt Sections in Article 3.8 of Chapter 1, Division 1, Title 13 of the California Code of Regulations, relating to Autonomous Vehicles, March 10, 2017. https://www.dmv.ca.gov/portal/dmv/detail/vr/autonomous/auto. Accessed March 13, 2017; Initial Statement of Reasons, Title 13, Division 1, Chapter 1, Article 3.7—Testing of Autonomous Vehicles, Article 3.8—Deployment of Autonomous Vehicles. March 10, 2017. https://www.dmv.ca.gov/portal/dmv/detail/vr/autonomous/auto. Accessed March 13, 2017; M. Harris, California gives the green light to self-driving cars, *IEEE Spectrum*, March 10, 2017. http://spectrum.ieee.org/cars-that-think/transportation/self-driving/california-gives-the-green-light-to-selfdriving-cars/?utm_source=CarsThat Think&utm_medium=Newsletter&utm_campaign=CTT03152017. Accessed March 16, 2017.

12.9.3 Implementation policy issues

Implementation policy issues concern the following:

- Agreeing upon model structures for governance with identified roles and responsibilities for the various participants.
- Identification of viable options for financial and investment strategies.
- Analysis and comparisons of communications systems for data delivery and data sharing among agencies.
- Analyses that support NHTSA decisions concerning cost–benefit analyses, value proposition analyses, and market penetration analyses.
- A requirement from NHTSA to mandate V2V communications for new light vehicles and to standardize the message and format of V2V transmissions. The communications assume DSRC to transmit the BSM containing a vehicle's speed, heading, brake status, and other vehicle information to surrounding vehicles, and receive the same information from them [46,47]. Section 12.10 discusses the proposed 2016 V2V communications rule further.

12.9.3.1 Reservations concerning use of DSRC for safety data transmission

As recently as 2014, four reservations were clouding the potential for near-term requirements regarding DSRC V2V safety data transmissions [31]. The first was Congressional pressure to reallocate parts of the dedicated 5.9 GHz DSRC spectrum to other types of wireless users. It was argued that non-vehicle uses of this spectrum could cause interference and make DSRC V2V communications unreliable, particularly in congested urban areas. The second issue involved objections about the absence of adequate measures to protect both privacy and security and to prevent the use of V2V for surveillance. To address these concerns, the NHTSA V2V communications rule proposes that V2V equipment be hardened against intrusion by entities attempting to steal its security credentials. This requirement would be met by requiring Federal Information Processing Standard (FIPS)-140 Level 3 validation. Thirdly, legal objections surfaced based on lack of express statutory authorization for such an agency requirement. In its 2016 Notice of Proposed V2V Communications Rulemaking [47], NHTSA addresses this issue by stating

> Under the Vehicle Safety Act, 49 U.S.C. 30101 et seq., the agency has the legal authority to require new vehicles to be equipped with V2V technology and to use it, as discussed in Section VI [... of the notice of proposed rulemaking]. NHTSA has broad statutory authority to regulate motor vehicles and items of motor vehicle equipment, and to establish FMVSSs to address vehicle safety needs.

Finally, some transportation technology experts view DSRC as 1990s technology that needs reassessment in light of newer and better communications technologies. As yet, alternative communication technologies, such as those used in commercial mobile wireless applications, have not attained the speed and low latency that make DSRC essential for vehicle safety communications. However, NHTSA acknowledges that its mandate could also be satisfied using non-DSRC technologies that meet certain performance and interoperability standards.

12.9.3.2 SCM system manager

The SCM system manager performs and monitors activities that ensure consistency and standardization of technical specifications, SOPs, and other industry-wide security practices such as auditing as described in Section 12.8.2.1 [22]. The system also removes or revokes entities that do not comply with standards or misbehave.

In the United States, a Security Credential Management System (SCMS) Proof-of-Concept (POC) Implementation Project (SCMS POC Project) is being conducted by the CAMP LLC Vehicle Safety Communications 5 (VSC5) Consortium. Members of the consortium are Ford Motor Company; General Motors LLC.; Honda R&D Americas, Inc.; Hyundai-Kia America Technical Center, Inc.; Mazda; Nissan Technical Center North America, Inc.; and Volkswagen Group of America. One goal of the SCMS POC design is to provide security services to support V2V and V2I communications at current production levels of passenger vehicles (up to 17 million vehicles annually) for the first year of deployment. Another goal is to provide a flexible architecture that is capable of scaling to support larger numbers of V2V and V2I devices in the years following initial deployment. It is also anticipated that the SCMS POC design will provide a stable platform and a research platform to support USDOT and industry research needs prior to deployment. The project is sponsored by NHTSA [26].

12.9.4 Implementation strategies

Several implementation issues will influence the wide-scale deployment of connected vehicle technologies in the United States. Among these are identifying sources of infrastructure funding and when connected vehicle OBE and RSE will be present in sufficient numbers to realize the anticipated benefits, and developing a set of automotive cybersecurity best practices. The cybersecurity matter was addressed in 2016 by the Automotive Information Sharing and Analysis Center (Auto-ISAC) in collaboration with the Alliance of Automobile Manufacturers (representing 77% of all car and light truck sales in the United States) and the Association of Global Automakers in a report that outlined automotive cybersecurity best practices [48]. The practices apply primarily to U.S. light-duty, on-road vehicles, but are applicable to other automotive markets including heavy-duty and commercial vehicles. The practices acknowledge that participating automakers share a common commitment to vehicle cybersecurity, although their electrical architectures, connected services, and organizational compositions vary. Accordingly, they do not prescribe specific technical or organizational solutions, but provide considerations for organizational design to align functional roles and responsibilities.

The recommended practices, enumerated in Table 12.9, emphasize risk management, including the identification of risks and implementation of reasonable risk-reduction measures since cybersecurity experts agree that a future vehicle with zero risk is unobtainable and unrealistic. As shown in the table, the practices include seven functions, namely, governance, risk assessment and management, security by design, threat detection and protection, incident response and recovery, training and awareness, and collaboration and engagement with appropriate third parties.

Key issues that affected transportation agency plans to move forward with their connected vehicle programs were identified by AASHTO from survey responses and agency personnel discussions [49]. Those dealing with DSRC appear close to resolution. Since implementation options are affected by technological advances and governmental policy and funding decisions, it is possible that the issues highlighted by AASHTO may become moot in the future.

AASHTO's Connected Vehicle Field Infrastructure Deployment Analysis of 2011 found the following [49]:

- Infrastructure deployment decisions by state and local transportation agencies depend on the nature and timing of benefits.
- Benefits depend on availability of connected vehicle equipment installed in vehicles as original equipment and aftermarket equipment.

Table 12.9 Automotive cybersecurity best practices proposed by Auto-ISAC

Function	Description	Specific best practices
Governance	Aligns a vehicle cybersecurity program with an organization's broader mission and objective.	Define executive oversight for product security. Functionally align the organization to address vehicle cybersecurity, with defined roles and responsibilities across the organization. Communicate oversight responsibility to all appropriate internal stakeholders. Dedicate appropriate resources to cybersecurity activities across the enterprise. Establish governance processes to ensure compliance with regulations, internal policies, and external commitments.
Risk assessment and management	Mitigates the potential impact of cybersecurity vulnerabilities by identifying, categorizing, prioritizing, and treating cybersecurity risks that could lead to safety and data security issues. Assists automakers identify and protect critical assets and develop protective measures.	Establish standardized processes to identify, measure, and prioritize sources of cybersecurity risk. Establish a decision process to manage identified risks. Document a process for reporting and communicating risks to appropriate stakeholders. Monitor and evaluate changes in identified risks as part of a risk assessment feedback loop. Include the supply chain in risk assessments. Establish a process to confirm compliance by critical suppliers to verify security requirements, guidelines, and trainings. Include a risk assessment in the initial vehicle development stage, and reevaluate at each stage of the vehicle life cycle.
Security by design	Integrates hardware and software cybersecurity features during the product development process.	Consider commensurate security risks early on and at key stages in the design process. Identify and address potential threats and attack targets in the design process. Consider and understand appropriate methods of attack surface reduction. Layer cybersecurity defenses to achieve defense-in-depth. Identify trust boundaries and protect them using security controls. Include security design reviews in the development process. Emphasize secure connections to, from, and within the vehicle. Limit network interactions and help ensure appropriate separation of environments. Test hardware and software to evaluate product integrity and security as part of component testing. Perform software-level vulnerability testing, including software unit and integration testing. Test and validate security systems at the vehicle level. Authenticate and validate all software updates, regardless of the update method. Consider data privacy risks and requirements in accordance with the Consumer Privacy Protection Principles for Vehicle Technologies and Services.

(Continued)

Table 12.9 (Continued) Automotive cybersecurity best practices proposed by Auto-ISAC

Function	Description	Specific best practices
Threat detection and protection	Proactive detection of threats, vulnerabilities, and incidents to raise awareness of suspicious activity and hence enable proactive remediation and recovery activities.	Assess risk and disposition of identified threats and vulnerabilities using a defined process consistent with overall risk management procedures. Inform risk-based decisions with threat monitoring to reduce enterprise risk by understanding and anticipating current and emerging threats. Identify threats and vulnerabilities through various means, including routine scanning and testing of the highest risk areas. Support anomaly detection for vehicle operations systems, vehicle services, and other connected functions, with considerations for privacy. Outline how the organization manages vulnerability disclosure from external parties. Report threats and vulnerabilities to appropriate third parties based on internal processes.
Incident response and recovery	Documents processes to respond to cybersecurity incidents affecting the motor vehicle ecosystem through reliable and efficient protocols and methods that ensure continuous process improvement.	Document the incident response life cycle, from identification and containment through remediation and recovery. Ensure an incident response team is in place to coordinate an enterprise-wide response to a vehicle cyber incident. Perform periodic testing and incident simulations to promote incident response team preparation. Identify and validate where in the vehicle an incident originated. Determine actual and potential fleet wide impact of a vehicle cyber incident. Contain an incident to eliminate or lessen its severity. Promote timely and appropriate action to remediate a vehicle cyber incident. Restore standard vehicle functionality and enterprise operations; address long-term implications of a vehicle cyber incident. Notify appropriate internal and external stakeholders of a vehicle cyber incident. Improve incident response plans over time based on lessons learned.
Training and awareness	Cultivates a culture of security, enforces vehicle cybersecurity responsibilities, and strengthens stakeholders' understanding of cybersecurity risks.	Establish training programs for internal stakeholders across the motor vehicle ecosystem. Include IT, mobile, and vehicle-specific cybersecurity awareness. Educate employees on security awareness, roles, and responsibilities. Tailor training and awareness programs to roles.
Collaboration and engagement with appropriate third parties	Commits industry to engage with third parties to enhance cyber threat awareness and cyber-attack response.	Review information and data using a standardized classification process before release to third parties. Engage with industry bodies, such as the Auto-ISAC, Alliance of Automobile Manufacturers, Association of Global Automakers, and others. Engage with governmental bodies, including NHTSA, National Institute of Standards and Technology (NIST), Department of Homeland Security, United States Computer Emergency Readiness Team, Federal Bureau of Investigation, and others. Engage with academic institutions and cybersecurity researchers, who serve as an additional resource on threat identification and mitigation. Form partnerships and collaborative agreements to enhance vehicle cybersecurity.

- It may be difficult for public agencies to justify the investment in roadside infrastructure if there are few equipped vehicles that will interact with it and use the agency-deployed V2I applications.
- The situation is even more significant for V2V applications, which require two vehicles (of the small number equipped) to interact in, most likely, a crash-imminent situation. Since the appearance of the AASHTO report in 2011, vehicles with autonomous, driver-assist, and driver-warning features have been emerging on roadways. They have the potential to supplement V2V communications as originally envisioned using DSRC. The automated features are enabled with OEM sensors such as radar, lidar, visual- and infrared-spectrum cameras, and others. These devices detect and provide alerts to the driver when cross-traffic, slow-moving vehicles, vehicles in a driver's blind zone, pedestrians, and obstacles on the roadway are identified. Some assist with parking and lane change maneuvers. NHTSA believes, however, that V2V information can be fused with existing radar- and camera-based systems to provide even greater crash avoidance capability than either approach alone by conveying safety information about a particular vehicle to other vehicles. V2V communication can thus detect threat vehicles that are not in the sensors' field of view, and can use V2V information to validate a return signal from a vehicle-based sensor [47].

Additional details concerning AASHTO's conclusions are discussed below. Many of these are similar to the conclusion of the EU's C-ITS Platform, which is discussed in Section 12.15.

12.9.4.1 Communications

Communications infrastructure design and deployment present challenges that arise from the need to satisfy the demands of different applications and the rapidly evolving technology landscape. Fiber-optic cable, in which many states have invested heavily, and radio systems including 800 MHz were frequently mentioned as either needing expansion or requiring new infrastructure to support backhaul communications. Technical issues remain where signalized intersections will be equipped with DSRC RSE for safety applications, such as line-of-sight and interference. These could potentially impact a safety system's ability to function properly. Respondents also mentioned that many of the connected vehicle system applications of interest to the agencies have less restrictive communications requirements than the safety systems, and could use existing cellular technologies.

12.9.4.2 Power for infrastructure devices

Providing power to equipment in the field is a concern, particularly in rural areas. The question of AC or DC power supplies, solar power, and the ability of batteries to meet specifications are real. Supplying power to a DSRC unit for a curve warning system in a remote area will be a practical concern to the people who have to design and install connected vehicle technology applications.

12.9.4.3 Back-office systems

Most applications will require back-office systems for data processing, storage, retrieval, and end-user presentation. Defining these systems often presents opportunities for agencies to better understand their needs and streamline procedures by utilizing the systems engineering process. However, challenges may arise when integrating systems across multiple agencies. For example, automating the commercial trucking credentialing, permitting, and taxing back-office systems within a state involves the DOT, State Police, Department of Revenue, Departments of Motor Vehicles and Licensing, Federal Motor Carrier Safety Administration, and usually others. Additional factors, such as assessing the data needs and interfaces between all system stakeholders, may appear when integrating new data elements from mobile sources and other added systems into the existing network. These include 511, Highway Performance Monitoring System (HPMS), Maintenance Decision Support System (MDSS), and emergency response CAD systems. Not to be overlooked is the potential threat posed by hackers to data security that must be addressed before these systems become operational and trusted by all users.

12.9.4.4 Standards

Communications standards, interoperability standards, data dictionaries and message sets, and open systems where appropriate have been and continue to be a major issue facing the transportation industry. Existing and evolving consumer electronic devices such as personal navigation devices, smart phones, and tablets highlight the need for interoperability and coordination within the market place. Agencies typically want to use standards when procuring systems because it simplifies their jobs and allows lower purchase costs. Some agency personnel actively participate in standards-creation organizations and are familiar with emerging standards, but most agencies wait until standards have matured before they implement them. Agencies also desire guidance and training concerning standards as they are adopted. The AASHTO analysis determined that the USDOT will have to continue to take the lead for the thoughtful and timely development of the required standards.

12.9.4.5 Funding, staging, and USDOT leadership

Since the population of vehicles equipped with the necessary DSRC and other vehicle technology will start very small, the state DOTs must weigh the benefits and costs of deploying RSE. Until there is a national DSRC RSE deployment strategy, pending NHTSA decisions are made, and vehicle penetration rates increase dramatically, most agencies feel they have limited ability to define long-range programs. Many of the survey respondents noted that making the political and financial commitment to connected vehicles is more immediately important than fulfilling the infrastructure needs. That commitment also needs to recognize the maintenance and operations costs and expertise needed to sustain a successful connected vehicle program in the long term. Many of the respondents were optimistic that the eventual NHTSA decisions will be the primary catalyst for moving ahead with infrastructure deployment. However, they were also quick to note that there are some serious technical and policy challenges that must be overcome. This would be the first time that the auto manufacturers and public agencies would be operating a truly cooperative vehicle and infrastructure system. While the respondents were optimistic, they were also realistic and understand the amount of effort that will be needed, along with leadership from USDOT. If NHTSA actions do not result in requirements for onboard DSRC equipment for light and heavy vehicles, the agencies will likely still carry on with many of their plans for applications that will benefit their management and operations. Additional funding guidance from FHWA is found in Section 12.11.

12.9.4.6 Integration across existing infrastructure

States and other agencies have invested heavily in ITS and connected vehicle technologies over the years in such areas as hardware, software, training, and importantly in enhancing planning and programming processes to accommodate technology deployments. With that investment in place, it will be important to continue to leverage and effectively use current systems so that connected vehicle applications contribute to them, rather than render them obsolete.

12.9.4.7 DSRC certificate authority

AASHTO noted that DSRC certificate authority was also an issue that needed resolution. This security function controls the process by which a vehicle's onboard system is authenticated and deemed trustworthy on a regular basis. Who issues the certificates, on what communications networks are they transmitted, on whose servers are they hosted, and who is responsible if there is a system failure? Some of these questions are technical, some are policy, and others are legal issues. The USDOT through NTHSA appears to have taken the lead in addressing them [26], but state agencies and auto manufacturers will have to be active participants as discussed in Section 12.9.3.2.

12.9.5 Institutional issues

Institutional issues are often of concern when the stakeholder community consists of multiple agencies. The issues include defining an interagency concept of operations that is supported by all agencies, planning for integrated multiagency and multimode transportation management center operations, harmonizing agency contracting and procurement practices and policies, settling on ownership of development products, managing operations and maintenance, and identifying funding for operations and maintenance [28,50]. Other institutional challenges are lack of staff with the necessary technical skills, lack of benefit and cost information to support deployment decisions, lack of information to build a business case for deployment, not knowing the plans of vehicle manufacturers and technology companies, and data access, ownership, and support issues [51]. Several of these are discussed in the next chapter in conjunction with systems engineering and the need to get all stakeholders to buy-in to the proposed system concept of operations, architecture, and benefits.

12.9.6 Technical, implementation, and policy crosscutting issues

One concern that appears to encompass both technical and implementation issues is the lack of standard vehicle data formats that can be integrated with off-the-shelf CAN-bus technology to transmit inclement weather information to other vehicles and to traffic and transportation management centers. This is of particular concern in rural areas where dangerous situations caused by inclement weather could otherwise be ameliorated by effective connected vehicle technology [52].

Some advocates involved with connected vehicle policy development espouse conservative views with respect to infrastructure deployment of V2I technologies. They propose that whatever functionality is needed to drive safely should be onboard each individual vehicle rather than in the roadside infrastructure. Their outlook comes from linking the scarcity of funds often available to maintain even current equipment and highway road surfaces with inefficient traffic management systems and poor road quality [53,54]. However, these arguments are often driven by nothing more than opposition to any spending by federal, state, and local governments and the myopia that causes them to lose sight of the benefits a modern infrastructure brings to commerce, especially in a global economy, and quality of life. An added argument for infrastructure funding for improvements and maintenance of existing facilities is that, one way or another, users of these facilities pay the costs of driving on them. For example, the American Society of Civil Engineers estimates that the annual cost of substandard infrastructure to the U.S. economy will be $210 billion by 2020 and $520 billion by 2040. In California, drivers pay an average of $762 a year in vehicle repairs and operating costs due to poorly maintained roads [55].

12.10 NHTSA 2016 PROPOSED V2V COMMUNICATIONS RULE

In December of 2016, NHTSA issued a NPRM to establish a new FMVSS, No. 150, which mandates V2V communications for new light vehicles and standardizes the message and format of V2V transmissions [47]. Light vehicles, in the context of this rulemaking, refers to passenger cars, multipurpose passenger vehicles, trucks, and buses with a gross vehicle weight rating of 10,000 pounds (4536 kg) or less. Without such a mandate, the agency believes that V2V will not achieve sufficient coverage. The currently envisioned V2V system would consist of a combination of a radio technology for the transmission and reception of messages, a common specification for BSMs that is independent of the potential communications technology, authentication of incoming messages by receivers, and, depending on a vehicle's behavior, triggering of one or more safety warnings to drivers. NHTSA is also proposing that vehicles be capable of receiving over-the-air (OTA) security and software updates (and to seek consumer consent for such updates where appropriate), and contain firewalls between V2V modules and other vehicle modules connected to the data bus as a security measure.

The agency considers V2V communications as a source of information that can be fused with existing in-vehicle radar, camera, and other sensor systems to provide even greater crash avoidance capability than either approach alone. Vehicles equipped with onboard sensors reap added benefits from the complementary information provided by V2V systems. Furthermore, instead of relying on each vehicle to sense its surroundings on its own, V2V communications enable nearby vehicles to assist each other by conveying safety and operational information about themselves to other vehicles, for example, brake pedal status, transmission state, stability control status, and vehicle at rest versus moving. Similarly, vehicle-based sensor systems can augment V2V systems by providing information concerning crash scenarios not reported by V2V communications, such as lane and road departure. These complementary capabilities can potentially lead to more timely warnings and a reduction in the number of false warnings, thereby adding confidence to the overall safety system and increasing consumer satisfaction and acceptance.

In the longer term, NHTSA believes that the fusion of V2V and vehicle-resident technologies will advance the development of vehicle automation systems, including the potential for self-driving vehicles. Although most existing automated vehicle systems currently rely on data obtained from vehicle-resident technologies, data acquired from GPS and V2V communications could significantly augment the automated systems by improving the performance of onboard crash warning systems, and by supporting further development and deployment of safe and reliable automated vehicles.

Highlights from the NPRM for V2V Communications for light vehicles are found in Table 12.10 [47]. A decision on guidance for V2V in new medium- and heavy-duty vehicles and nomadic devices will likely follow at some point.

12.10.1 Costs and benefits

Although the NHTSA 2016 proposed rulemaking mandates that all light vehicles be equipped with V2V communications, it has decided not to mandate any specific safety

Table 12.10 Highlights from NHTSA's proposed rulemaking for V2V communications for light vehicles

Technology area	Mandate	Description
Communication	DSRC technology	DSRC units in a vehicle are to transmit and receive BSMs in the 5.850–5.925 MHz frequency band. The governing regulations are FCC 47 CFR Parts 0, 1, 2 and 95 for OBE and Part 90 for roadside units. The OSI model physical and data link layers (Layers 1 and 2) are addressed by IEEE 802.11p and P1609.4; network, transport, and session layers (Layers 3, 4, and 5) by P1609.3; security communications by P1609.2; and additional session and prioritization related protocols by P1609.12. This mandate could also be satisfied using non-DSRC technologies that meet certain performance and interoperability standards.
	Channel and data rate	All vehicles are to transmit the BSM on Channel 172 via a dedicated radio at a data rate of 6 Mbps.
	Transmission frequency	10 times per second under non-congested conditions.
	Time between transmissions	Staggered transmission of BSMs every 100 ± 0–5 ms
	Transmission envelope	BSM broadcast range is 300 m from the vehicle and at elevation angles extending from $+10°$ to $-6°$.
	Reliability	A packet error rate of less than 10% is required to ensure that BSMs are received. This requirement aligns with ASTM standard E2213-03 (2003) 4.1.1.2.
Message format and information	Standardizes content, initialization time, and transmission characteristics of BSMs regardless of the V2V communication technology used	Proposed content requirements for BSMs are largely consistent with voluntary consensus standards SAE 2735 and SAE 2945 that contain data elements such as speed, heading, acceleration, yaw, trajectory, exterior lights status, transmission gear (forward, reverse, neutral), steering wheel angle, and other information, although NHTSA purposely does not require some elements to alleviate potential privacy concerns. Standardizing the message will facilitate the use of a common language among V2V devices to ensure interoperability and the ability to inform drivers of potential crashes.
Message authentication	PKI	V2V devices are to sign and verify their BSMs using a PKI digital signature algorithm in accordance with performance requirements and test procedures for BSM transmission and the signing of BSMs. This will establish a level of confidence in the messages exchanged between vehicles and ensure that BSM information is received from devices that have been certified to operate properly, are enrolled in the security network, and are in good working condition. Use of PKI can ensure that safety applications are able to distinguish valid messages from those that have been modified or changed while in transit, originated by bad actors such as hackers, or defective devices.
		Comments concerning two alternative approaches are also solicited. The first scheme is a performance-only method requiring a receiver of a BSM to be able to validate the contents of the message such that it can reasonably confirm that the message originated from a single valid V2V communications device and the message was not altered during transmission. The second alternative does not require specific message authentication. BSMs would still be validated with a checksum or other integrity check, and be passed through a misbehavior detection system to attempt to filter malicious or misconfigured messages. Implementers would be able to include message authentication as an optional function.

(Continued)

Table 12.10 (Continued) Highlights from NHTSA's proposed rulemaking for V2V communications for light vehicles

Technology area	Mandate	Description
Misbehavior detection and reporting	Security credential management system	The SCMS contains procedures to report misbehavior and learn of misbehavior by other participants, and methods to detect if a device's hardware and software have been altered or tampered with to prevent their intended behavior. This approach enhances the ability of V2V devices to identify and block messages from other misbehaving or malfunctioning V2V devices. Comments concerning an alternative misbehavior detection approach are also sought. This method imposes no requirement to report misbehavior or implement device blocking to an authority. However, implementers would need to identify methods that verify a device's hardware and software functionality to ensure its intended behavior has not been altered or tampered with. Implementers would be free to include misbehavior detection and reporting as optional functions.
Hardware security		V2V equipment is to be hardened against intrusion to FIPS-140 Level 3 to protect against entities attempting to steal its security credentials.
Privacy and security	Excludes information that directly identifies a specific vehicle or individual regularly associated with a vehicle	Minimizes risks to consumer privacy by excluding from V2V communications information such as owner's or driver's name, address, and vehicle identification numbers, and data reasonably linkable to an individual (i.e., data that are under the control of a covered entity, not otherwise generally available to the public through lawful means, and are linked, or as a practical matter linkable by the covered entity, to a specific individual, or linked to a device that is associated with or routinely used by an individual). Additionally, the proposal contains specific privacy and security requirements with which manufacturers would be required to comply.
Safety applications	None specifically proposed	NHTSA believes the proposed V2V communications will create the standardized information environment that will allow innovation and market competition to develop improved safety and other applications. Additionally, the agency judges that more research is likely needed in order to create regulations for safety applications. Therefore, the agency is seeking comment on information that could inform a future decision to mandate specific safety applications.

applications, instead allowing them to be developed and adopted as determined by the market. This market-based approach makes estimating the potential costs and benefits of V2V communications difficult because the technology would improve safety only indirectly, by facilitating the deployment of previously developed OEM safety applications. However, the agency is confident that these technologies will be developed and deployed once V2V communications are mandated and interoperable. Considerable research has already been done on various potential applications, and the agency believes that functioning systems are likely to become available within a few years if their manufacturers can be confident that V2V communications will be mandated and interoperable.

In order to provide estimates of the rule's costs and benefits, NHTSA considered a scenario where two V2V-enabled safety applications, IMA, and LTA are voluntarily adopted on hypothetical schedules similar to those observed in the actual deployment of other advanced communications technologies. The agency believes that IMA and LTA will reduce the frequency of crashes and concurrent loss of life that cannot be avoided by in-vehicle systems, and will thus generate significant safety benefits that would not be realized in the absence of universal V2V communications capabilities. In addition, the marginal costs of including the IMA and LTA applications are extremely low once the V2V system is in place. NHTSA has not quantified any benefits attributable to the wide range of other potential uses of the technology. Recognizing its experience with other technologies, NHTSA believes that focusing on the implementation of these two inexpensive applications provides a reasonable approach to estimating potential benefits of the proposed rule, and is likely to understate the breadth of potential benefits of V2V communications.

NHTSA's evaluation of the total annual costs to comply with this proposed mandate in the 30th year after it takes effect would range from $2.2 billion to $5.0 billion (U.S.), corresponding to a cost per new vehicle of roughly $135–$301. This estimate includes costs for equipment installed on vehicles and the annualized equivalent value of initial investments necessary to establish the overarching security manager and the communications system, but due to uncertainty, does not include opportunity costs associated with use of the frequency spectrum, which will be included in the final cost–benefit analysis. The primary source of the wide range between the lower and upper cost estimates is based on the assumption that manufacturers could comply with the rule using either one or two DSRC radios. Table 12.11 summarizes the costs and benefits of requiring light vehicle manufacturers to include V2V technology in support of IMA and LTA applications [47].

12.10.2 Effective date

NHTSA is proposing that the effective date for manufacturers to begin implementing the V2V requirements be two model years after the final rule is adopted, with a 3-year phase-in period at rates of 50%, 75%, and 100%, respectively, to accommodate vehicle manufacturers' product cycles. Assuming a final rule is issued in 2019, this implies that the phase-in

Table 12.11 Costs (excluding spectrum opportunity costs) and benefits (based on IMA and LTA applications only) in year 30 of deployment (2051)

Total annual costs	Per vehicle costs	Crashes prevented and lives saved	Monetary benefits
$2.2 billion–$5.0 billion	$135–$301	Crashes: 424,901–594,569 Lives: 955–1321	$53 billion–$71 billion (in 2016 dollars)

period would begin in 2021, and all vehicles subject to that final rule would be required to comply in 2023.

12.11 FHWA V2I DEPLOYMENT GUIDANCE

FHWA indicates that the final guidance for V2I communications will be similar to the draft guidance issued in their September 2014 and December 2016 reports [56,57]. Unlike the NHTSA NPRM, the FHWA guidance is not mandatory and is intended to provide assistance to transportation system owners and operators in meeting Federal-aid highway program requirements and ensuring interoperability of V2I operations among state and local transportation agencies.

The proposed FHWA policy is designed to build upon emerging V2V technologies and services that enhance safety, mobility, and environmental benefits. The deployment guidance addresses planning, Federal-aid eligibility of V2I equipment, relation of V2I deployments to the National Environmental Policy Act (NEPA) and National Historic Preservation Act (NHPA), interoperability, evaluation, ITS equipment capability and compatibility, hardware device certification, reliability, right-of-way use, private-sector use, facility design, use of existing structures and infrastructure, use by public-sector fleets, procurement process, legacy systems and devices, communication technology and licensing, data connection and latency, connected vehicle privacy and security, data access, consistency with the Manual on Uniform Traffic Control Devices, and use of public–private partnerships. Table 12.12 summarizes the key aspects of the 2016 guidance document.

12.12 INFRASTRUCTURE FUNDING MECHANISMS

Connected vehicle technology will have a large impact on local DOT funding needs. New infrastructure will require capital and ongoing operations and maintenance funds. Day-to-day operations costs may increase as a result of staffing needs, purchases of energy to run the infrastructure equipment, and upkeep of backhaul communications from connected vehicle field sites. Maintenance cost budgets must ensure that funds are made available for scheduled and unscheduled costs, and replacement costs for field and back-office equipment at the end of their life. Furthermore, TMC systems may be integrated with the payment systems and variable toll displays, increasing requirements and costs to provide system reliability and security.

Funding issues are still being addressed by local, state, and national agencies. New funding sources will be necessary to replace shrinking gas tax revenues. Not only are vehicles becoming more efficient, but alternatively fueled vehicles reduce the correlation between vehicle miles traveled and gas consumption. Raising gas taxes may be problematical as it is often politically unpalatable. Alternative methods of generating revenue are road-use pricing including both traditional fixed-rate tolling and congestion pricing zones, mileage-based user fees, tolling by the responsible agency for an interstate highway in need of repair, tolling by a private entity (company or consortium) granted a concession through a public–private partnership to repair and improve a highway in return for the tolling revenue for some number of years, or general taxation sources allocated to the maintenance of roadways and systems that are vital to the country's transportation network and commerce.

Agencies can consider other various funding categories, such as those listed in Table 12.12, to support deployment of connected vehicle equipment. These include ITS budget or federal and state funds with ITS eligibility, safety improvement program funding, and funds set aside for congestion mitigation or air quality improvement projects.

Table 12.12 FHWA preliminary guidance for V2I project deployment

Guidance area	Guidance
Planning	Metropolitan planning organizations, local public agencies, transit operators, and state transportation agencies should begin considering V2I strategies in their long-range planning. Potential topics include understanding system benefits and operation, application options that are effective under a variety of conditions, advantages and limitations of each option, capital and maintenance costs, funding sources, integration with existing systems such as traffic management and communications networks, long-term impacts, coordination across metropolitan planning organizations and state boundaries, staff needs, and integration of V2I into existing state and regional ITS architectures.
Federal-aid eligibility of V2I equipment	Several Federal programs are available to support funding of V2I equipment. However, specific program requirements must be satisfied and eligibility determined on a case-by-case basis. Equipment procurement, installation, preventive maintenance, and operational costs that support V2I applications compatible with connected vehicle standards for interoperability and security qualify for Federal-aid funding where eligibility for ITS investments has been previously established.
	V2I safety applications are eligible for Highway Safety Improvement Program (HSIP) funds if they address a state's Strategic Highway Safety Plan priority, are identified through a data-driven process, and contribute to reduction in fatalities and serious injuries.
	Costs related to mobility and safety applications compatible with basic connected vehicle standards for interoperability and security may be eligible for National Highway Performance Program (NHPP) and Surface Transportation Program (STP) funds. NHPP-funded projects shall be used for a facility located on the National Highway System (NHS) (per 23 U.S.C. 119[c]) and should lead to advancement of the mobility, freight, condition, and safety goals established at 23 U.S.C. 150(b) (per 23 U.S.C. 119[e][2]).
	Congestion mitigation and air quality (CMAQ) projects must meet three criteria for funding: it should be a transportation project, generate an emissions reduction, and located in or benefit a nonattainment or maintenance area. All phases of eligible projects, such as equipment procurement, installation, construction, and studies that are part of project development, qualify for CMAQ funding.
	ITS projects that support V2I applications, including equipment and installation costs, may be suitable for CMAQ funding if they meet the basic eligibility requirements. Examples of V2I applications that satisfy CMAQ eligibility are SPaT, eco-drive, congested intersection adjustment, and traveler information systems. Transit project eligibility is determined by the Federal Transit Administration. Transit applications that support V2I applications may also be eligible for CMAQ funds.
Interoperability	V2I deployments will need to be interoperable and coordinated with other modes of transportation (e.g., light- and heavy-duty vehicles, transit systems, and railroad crossings) to operate on a national level. To the maximum extent possible, all devices and applications deployed by jurisdictions and modes should leverage the applicable standards and should not be standalone deployments. V2I deployments should be compatible with connected vehicle security policies that may be developed; support the distribution, receipt, and use of security certificates to the maximum extent possible; and protect privacy at the highest level appropriate to the connected vehicle environment. Since vehicles and V2I devices will interact with equipment throughout the United States, BSM transmission and receipt requirements should be standardized across the United States and vehicle manufacturers and models, device types, and applications. Similarly, V2I deployments should conform to the same requirements across the United States to ensure successful exchange of information between vehicles, devices, and applications. To the greatest extent possible, the equipment, functionality and utility of V2I deployments should be consistent between transportation modes and regions to be eligible for Federal-aid funding, and should comply with existing ITS and connected vehicle guidance provided by the USDOT.
Evaluation	Evaluations should be performed to determine the effectiveness of V2I technology and applications in satisfying stakeholder needs, benefits and cost goals, and in assessing user satisfaction.

(Continued)

Table 12.12 (Continued) FHWA preliminary guidance for V2I project deployment

Guidance area	Guidance
ITS equipment capability and compatibility	Early deployments of connected vehicle field infrastructure are likely to be installed alongside or as part of existing ITS equipment (e.g., dynamic message signs, CCTV cameras, and detection stations) and traffic signal controllers. In many cases, connected vehicle infrastructure will be integrated with existing equipment to enable direct data communications between the infrastructure and vehicles. Integration will require using standard interfaces and message sets developed by FHWA. Furthermore, FHWA recommends that the systems engineering process be used to identify any equipment purchased for V2I applications and deploy the equipment in an environment that is V2I ready. The guidance defines V2I ready as a roadside installation having the following characteristics: • Reliable power supply. • At least one secure backhaul communication link and two secure backhaul communication links if required by the implementing agency (one for ITS or traffic signal data and one for connected vehicle data). • ITS equipment or controllers that are National Transportation Communications for Intelligent Transportation System Protocol (NTCIP) standards compliant. • Electronic map or geometric description of the surrounding area available in SAE J2735 compliant format. • Roadside cabinet space sufficient to house an external processor (size depends on the intended application) that may be installed in the future. • Planning for mounting locations for a DSRC roadside unit to enable V2I communications.
Hardware device certification	Certification research will primarily be focused on understanding the interoperability needs for device compliance, systems security, and privacy. The USDOT will conduct the following research activities in support of certification [57,58]: • Track 1: Policy Research Related to Certification: The USDOT will establish a forum for solving policy-related issues, including a determination of what is to be certified, the entity that will be responsible for certification, and the parties that will need to obtain certification. The Policy and Institutional Issues research program will involve industry and federal, state, and local government stakeholders to provide input. • Track 2: Technical Requirements for Certification: The level of components within devices, or which interfaces need to be certified will be defined. Additionally, how this certification is to be accomplished will be determined. It is envisioned that the responsibility for work in this area will be shared by Government and industry. However, Government will have a primary role in funding development prior to the emergence of a consumer market for certified products. In that sense, the Government will serve as an enabler and coordinator of this function. • Track 3: Implementation Support and Oversight: A third-party entity is expected to conduct implementation of the planned certification process. The implementation process will include development of test tools and methods. The Federal Government may have a role in assisting with startup, and in overseeing operation and adherence to standards. This implies an ongoing operational role for the Federal Government beyond the scope of this research. The USDOT has proposed a four-layer approach to certification to ensure that a device and software meet environmental and communications protocol requirements. The four certification layers are as follows [56]:

(Continued)

Table 12.12 (Continued) FHWA preliminary guidance for V2I project deployment

Guidance area	Guidance
	1. Environmental abilities, e.g., temperature, vibration, weather. 2. Communication protocol abilities, e.g., radio service interoperability for DSRC. 3. Interface abilities to ensure that message syntax and contents are formatted properly. 4. Overall application abilities to verify the system-level function. A basic device should be certified at layers one and two, whereas an application should be certified at all four layers if it resides on a basic device.
Reliability	Determining the reliability of deployed equipment should be incorporated into the systems engineering process. To ensure availability of the V2I application, an equipment maintenance and replacement plan should be established for all connected vehicle deployments. Equipment and applications deployed on the NHS or using Federal-aid highway funds will be purchased from a qualified provider and certified through an industry-approved process when available.
Right-of-way (ROW) use	Use of ROW for V2I RSE follows current regulations and funding eligibility. Installation of connected vehicle infrastructure within the ROW will be allowed if its use has a public benefit and does not impair the safety of the roadway. Private-sector secondary use may be approved by the FHWA Administrator under 23 CFR 1.23(c) as long as (1) it is determined to be in the public interest; (2) it does not interfere with or degrade free and safe flow of traffic or the current and future primary safety and mobility applications; and (3) the private application will be opt-in and the highway user will be able to disable it at any time, at no cost to the user. Private use arrangements are subject to 23 U.S.C. 156 and require the state to charge the fair market value for non-highway use or to obtain an exception from FHWA based on a social, environmental, or economic purpose. For Interstate highway ROW use, an airspace agreement is necessary. The federal non income from revenues obtained by the state for Title 23 (Highways) projects, preferably for deployment and operation of the V2I network. For federal purposes, secondary use messages are not considered advertising when displayed in a vehicle or on a personal device under the Highway Beautification Act. However, state law or policies could supersede this distinction.
Private-sector use	FHWA supports policies that maximize the possibility of private investment to leverage costs for deployment and operations. Public owners and operators of infrastructure communication systems can allow off-public ROW private-sector use of the system for V2I communication if the following conditions are met: • There is a public-sector benefit of the V2I RSE. • Private-sector use does not interfere with or degrade the primary safety and mobility applications. • The private-sector application will be an opt-in by highway users and they will have the ability to disable the application at any time at no cost to the user. The roadway facility owner or public owners and operators of the infrastructure system are not charged for any private-sector use. The cost associated with installation and operation of private-sector use or components may be eligible for Federal-aid beyond the operation of the communication backhaul.
Facility design	V2I applications may be utilized to mitigate safety and operational impacts due to substandard geometric features of highways on a network level or as part of a design exception. Deployed equipment needs to be interoperable and coordinated with other transportation modes such as light- and heavy-duty vehicles, transit systems, and railroad crossings. Projects to construct or reconstruct highways should be designed to accommodate the concurrent or potential future installation of V2I RSE.

(Continued)

Table 12.12 (Continued) FHWA preliminary guidance for V2I project deployment

Guidance area	Guidance
Use of existing structures and infrastructure	Installation of connected vehicle equipment on existing structures and infrastructure will be allowed if its use has a public benefit and does not create potential safety issues. Private-sector secondary use is permissible if it complies with the conditions in the ROW use section. Designers should consider how connected vehicle technologies will affect pedestrians, bicyclists, and other nonmotorized users within the highway ROW and how connected vehicle technologies may affect access to transit services and enhance livability.
Use by public-sector fleets	Depending on specific program requirements, Federal-aid highway funds can be used to procure components (including those for the collection and dissemination of data and information) that enable V2I applications in public-sector vehicles. The same types of funds can be used to procure infrastructure-based components that enable V2I applications, e.g., RSE and components installed in maintenance garages and traffic and emergency operations centers. Private contractor equipment providing maintenance and response or work zone safety capabilities can be federally funded if it provides a public benefit and the public sector retains ownership and control of the equipment. V2I messages should comply with the standards in the NHTSA V2V Communication rulemaking. Messages should be prioritized in a manner that considers the urgency of the message and its consequence, e.g., safety messages should have priority over mobility and environment-related messages. The applications deployed on public-sector fleets should be consistent and integrated with other connected vehicle deployments on private-sector vehicles. Although the ability of first responders to preempt traffic signals is an important safety application, its unwarranted use could have serious mobility implications. Therefore, if this application is deployed, guidance should be provided to the operators of such vehicles concerning the negative impacts of misuse, when it should be and not be activated, and disciplinary consequences of misuse.
Procurement process	An effective procurement process should incorporate the following actions: • Consult with stakeholders affected through the V2I deployment to ensure their needs are satisfied. • Identify existing assets that are candidates for modification in order to reduce the costs for delivery of power and communications to the deployment location. • Consult the ITS Costs Database for comparisons and ranges of unit and system costs. • Consult with agencies that manage statewide acquisition of information technology products to streamline acquisition of connected vehicle assets. Since most people outside the transportation community are not aware of connected vehicles, deployers will be expected to provide basic education and training concerning the technology and its benefits to public agencies. • Apply the specifications produced by USDOT as the basis for amending existing state and local agency qualified product lists. This enables vendors and application developers to become more effective as there is less variation in the specifications they are required to meet. An updated qualified products list also introduces rigor that better secures interoperability among agencies across the United States. • Modify existing guides for project cost estimation to include connected vehicle assets. Allow for contingencies consistent with other ITS roadside assets such as dynamic message signs. • Apply the Connected Vehicle Footprint Analysis produced by AASHTO to develop typical profile and plan view drawings for implementation of the project. These plans should also reflect previous assessments of existing assets and supporting infrastructure. • Establish test procedures and systems acceptance methods that minimize the risk to the agency. During systems acceptance, ensure that data produced from the locations are not in the sole possession of the design-build contractor; if one is used, but that the data can be validated by the agency working with an independent party.

(Continued)

Table 12.12 (Continued) FHWA preliminary guidance for V2I project deployment

Guidance area	Guidance
	• Coordinate with the agency's asset management unit to maintain accurate records of when items were deployed to establish a baseline for the life cycle of the units. This activity supports the development of funding plans to sustain the operation of the system over time. • Establish operations and maintenance plans for the system taking into consideration functions such as providing staff appropriate training and technical expertise, having responsive and timely repairs once the project is operational, and exhibiting resourcefulness in obtaining support services through a number of providers.
Legacy systems and devices	Legacy systems and devices may be owned and operated by public agencies, private companies, or the general public. This guidance applies to systems and devices under the jurisdiction of public agencies, although they can apply to the others through agreements, coalitions, and similar cooperative efforts. Legacy systems and devices critical to the functioning of active V2I safety applications should be retrofitted or replaced. Other legacy systems and devices may be augmented by V2I safety applications that improve safety or mitigate a design exception. Upgrading and replacement of legacy systems and devices should incorporate support for V2I and other connected vehicle applications. The systems engineering process should be used to establish equipment maintenance and replacement plans for legacy systems. Issues to be considered are the functionality needed to support V2I application requirements, number of devices affected, useful life of equipment and life-cycle costs, and processes for replacing equipment.
Communication technology and licensing	Selection of the V2I communications technology will be based on a systems engineering analysis consistent with application interoperability across the United States and noninterference with licensed users. Although DSRC is the anticipated communication technology for V2V safety applications, other technologies may emerge for V2I mobility applications. Alternative options would be evaluated based on capability of V2V onboard units, national interoperability, certified application support, and its attributes that support the needs of the application and installation environment. DSRC-enabled V2V and V2I safety applications always have primary status over non-safety applications. Roadside infrastructure in the 5.9 GHz band is licensed to both public safety and nonpublic safety entities pursuant to 47 CFR Part 90, while onboard units are licensed by rule under 47 CFR Part 95 (no individual license is required). Thus, the National Telecommunications and Information Administration authorizes the use of the DSRC frequency spectrum for governmental entities while the FCC issues licenses for the private sector based on each applicant's area-of-operation, e.g., county state, multi-state, or nationwide. Other entities, such as those engaged in the operation of a commercial activity (e.g., toll collection and parking guidance), can also be authorized to operate roadside units in the DSRC band. Further, the licensing requirement only applies to equipment used for transmission; equipment designed and operated only to receive does not require siting licenses. Although safety communications have precedence, identification and removal of interfering non-safety signals could be problematic, particularly if the interfering equipment was licensed and sited first. Therefore, FHWA recommends that site licensing should be undertaken once an application deployment is identified in the planning process.
Data connection and latency	DSRC can communicate directly between vehicles and between vehicles and infrastructure with low latency. However, its range is limited to 300 m and it is a line-of-sight technology. Therefore, key infrastructure and deployment features may differ for the same application when it is utilized at different locations or with different communications technologies. (The communications options were discussed in Section 12.7.) A reliable data connection and low latency should be maintained with any choice of communications technology. Technology reliability should be based on three principles: (1) elimination of a single point of failure, (2) reliable crossover, and (3) detection of failures in real time. Latency should be measured from end-to-end or between the originating and the responding application.

(Continued)

Table 12.12 (Continued) FHWA preliminary guidance for V2I project deployment

Guidance area	Guidance
Connected vehicle privacy	The USDOT is committed to supporting deployment of a connected vehicle environment that both protects personal privacy and promotes connected vehicle safety, mobility, and environmental benefits. FHWA will endeavor to identify and disseminate data privacy best practices and new technical or policy controls as applications and the V2I system develop. The privacy aspects are guided by the Fair Information Privacy Principles based on the tenets of the Federal Privacy Act of 1974 and mirrored in the laws of many states. These include the following: • Transparency that ensures that customers have pertinent information concerning the data being collected and transmitted by the V2I system and the use of the data. • Individual participation and redress that ensures that consumers have a reasonable opportunity to make informed decisions about the collection, use, and disclosure of their personally identifiable information (PII) or other data that may be used to identify them directly or indirectly, reasonable access to their PII, and the opportunity to have it corrected, amended, or deleted as appropriate. • Purpose disclosure to make clear the rationale for which the V2I system collects, uses, maintains, or disseminates specific data elements, such as basic safety functions and potential mobility, environmental, and commercial applications. • Data minimization to explain why the data collection is not excessive and how long data will be retained. • Use limitation to assure consumers that collected data will not be used for purposes incompatible with the specified purposes. • Data quality and integrity to explain how the V2I system will assure data quality and integrity throughout the data life cycle and in all associated uses. • Security to identify the physical, technical, and procedural measures system administrators will take to protect collected data. • Accountability and auditing that explain how the V2I system will ensure that the privacy controls are executed.
Connected vehicle security	This area appears in the 2014 document [56], but not the 2016 document [57]. Connected vehicle equipment in motor vehicles transmits generic information used by V2V and V2I applications in a limited geographical range. The generic BSMs do not identify specific drivers or vehicles. There are multiple mechanisms incorporated into the connected vehicle environment to minimize risks to individual privacy, including a security certificate management system. As currently envisioned, this system will be a complex PKI that creates, manages, stores, distributes, and revokes digital certificates that accompany and validate each BSM. The PKI allows users to securely and privately exchange BSMs through a public and private cryptographic key pair obtained and shared through a trusted authority. No single entity within the PKI holds enough information about a participant to link a BSM transmitted by a vehicle to a specific driver or vehicle. Additional information about PKI is found in Section 12.8.
Data access	In general, federal law does not assign ownership, access, and use limitations to broadcast data. As a result, USDOT and FHWA do not currently have a specific policy assigning data ownership or limiting access to BSM data. However, there is a need to balance access to data used in mobility, environmental, and other applications with the consumer's privacy concerns. To address consumer concerns about privacy and enhance consumer acceptance, V2I applications should contain sufficient controls to mitigate potential privacy and security risks such as transparency and consent. These controls and the Privacy Principles will assist consumers in understanding the application and the data it will gather and, for opt-in applications, in granting consent for the collection of their data.

(Continued)

Table 12.12 (Continued) FHWA preliminary guidance for V2I project deployment

Guidance area	Guidance
Consistency with Manual on Uniform Traffic Control Devices (MUTCD)	V2I applications that provide traffic control information should be consistent with the MUTCD. It is not expected that every roadside sign, signal, or pavement marking will be mirrored within the vehicle. If they are, the displays will most likely be implemented by the OEMs. Information received by the onboard unit should be sufficient for it to generate the appropriate symbol or convey that information in a manner consistent with the MUTCD. While the traffic control devices are governed by the provisions of the MUTCD, the communication medium and message composition within the vehicle are not addressed in the MUTCD. All information conveyed to the driver should comply with and cannot contradict information conveyed by the signs, signals, and markings along the road. In-vehicle systems should also express the priority of the signs, signals, and markings, e.g., regulatory signs take priority over warning signs.
Use of public–private partnerships (P3s)	To the extent possible, P3s and other commercial relationships should be considered for deployment. Such arrangements should ensure that any commercial applications protect the public interest, do not compromise the safety and mobility objectives or their precedence, provide for safe maintenance practices, and hold the jurisdiction harmless due to lack of both public and private services. The P3 agreement should guarantee the precedence of the V2V or V2I environment. The public owner or operator of the V2I infrastructure should ensure that the arrangement produce a net life-cycle benefit to the public and value to the transportation agency. Such deployments should be compatible with the connected vehicle security system; support the distribution, receipt, and use of certificates in support of message authentication approaches as needed; and implement the privacy protections and controls of the connected vehicle environment. Current FHWA policy allows state DOTs to use Federal-aid funds in innovative long-term contracts with private developers under certain conditions. In such agreements, the state grants exclusive rights (a concession) to a developer (concessionaire) who assumes responsibility for the highway's construction, operations, and upkeep. In the case described here, the developer would take responsibility for the deployment and operation of the V2I system. P3 concession contracts often allow the concessionaire to collect and retain revenue from tolls, but in this instance tolls would not be applicable. The business model for this type of P3 is therefore different in that the private entity would most likely receive revenue from use of the communications channel to generate fee-for-service opportunities or advertising revenue.

12.13 CONNECTED VEHICLE PROJECTED MARKET GROWTH

Figure 12.13 displays projected connected vehicle market growth under various growth rate assumptions [4]. It illustrates the population ratio of connected vehicle equipment (i.e., the percentage of vehicles in the U.S. light vehicle fleet with OBE) based on both a step-function introduction (i.e., all new vehicles are built with the feature) and a more typical "S-curve" application rate in which the feature is introduced into the fleet over time.

One key conclusion is that it will take 13 years for 90% of the U.S. light vehicle fleet to be equipped if a step-function growth curve is assumed (such as with a NHTSA mandate for installation of DSRC in all new light vehicles). In contrast, a phased introduction takes an additional 20 years before 90% of the U.S. light vehicle fleet is equipped. The phased introduction reaches 50% of the U.S. light vehicle fleet in about 12–13 years.

A phased introduction may be especially problematic for V2V safety applications. While individual equipped vehicles would receive immediate benefits from V2I applications (assuming the infrastructure has been upgraded to support these applications), V2V benefits would only occur when both interacting vehicles are equipped. The figure indicates that the probability of obtaining benefits from V2V communications is less than 50% for more than 17 years after an initial phased introduction of the feature. For a step-function introduction of the feature, this point is reached at about 9 years.

Figure 12.14 plots projected sales of autonomous vehicles from 2020 through 2070 [59–61]. According to these forecasts, fully self-driving cars probably will not see large-quantity purchases by the general public in the United States until the 2040s, although the graph indicates that self-driving cars will be available sometime after 2020. This forecast appears to be given credence by a Ford announcement in mid-August 2016 of their intent to have fully autonomous vehicles in commercial operation for a ride-hailing or ride-sharing service beginning in 2021 [62]. Furthermore, other car manufacturers (e.g., Volvo, Audi, Mercedes Benz, Tesla, General Motors, Fiat-Chrysler, BMW, and Renault-Nissan) are currently testing and evaluating their

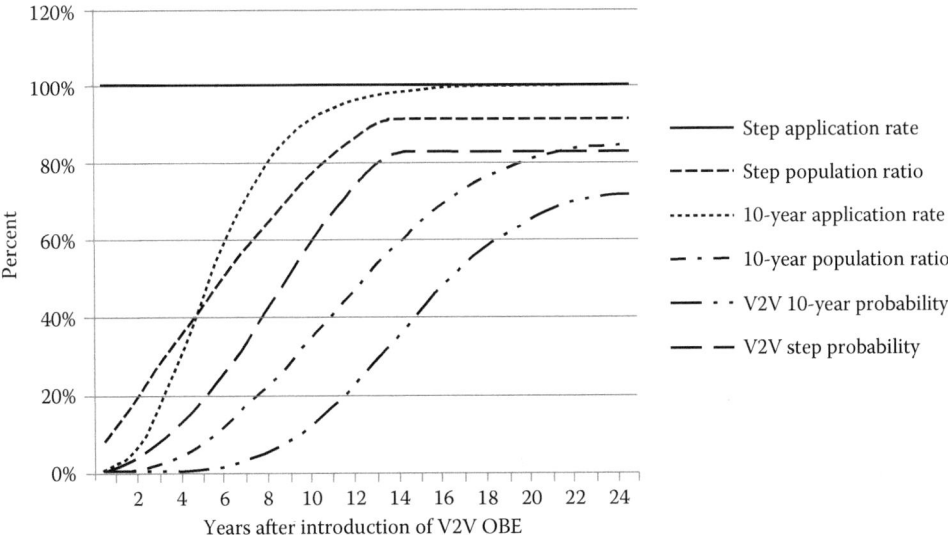

Figure 12.13 Projected connected vehicle market growth for various growth rates. (From *ITS ePrimer— Module 13: Connected Vehicles*, Intelligent Transportation Systems Joint Program Office, Research and Innovative Technology Administration, U.S. Department of Transportation, Washington, DC, 2013.)

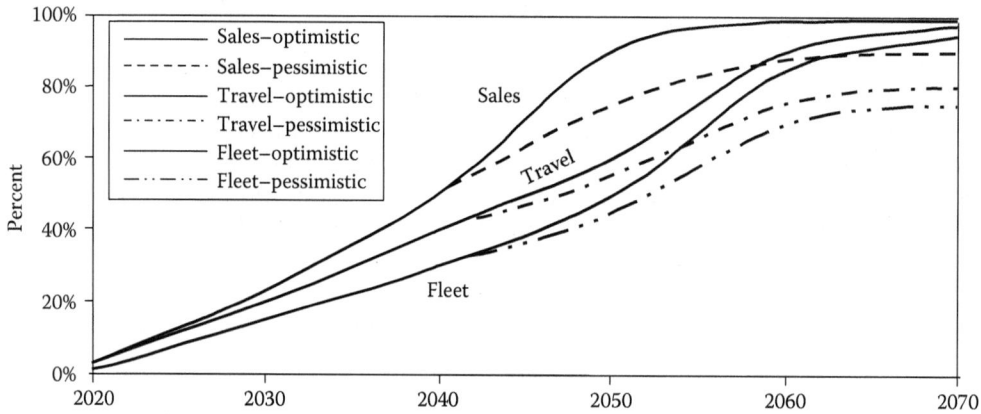

Figure 12.14 Projected autonomous vehicle sales. The curves are based on historical data from implementation rates of other vehicle technologies to predict the uptake of autonomous vehicles. (Adapted from T.A. Litman, Victoria Transport Policy Institute, Victoria, BC, 2016. http://www.vtpi.org/avip.pdf.)

autonomous vehicle concepts, some through partnerships with ride-hailing services such as Uber and Lyft [63]. ABI Research reports that semiautonomous systems will continue to dominate the market over the next decade, with SAE Level 2 and 3 systems accounting for 86% of autonomous vehicles shipping in 2026. Higher levels of autonomy will gain traction quickly, representing just under one-third of autonomous vehicles shipping in 2030 [64].

Predictions from Figure 12.14 indicate that in the 2040s autonomous vehicles are likely to represent approximately 50% of new vehicle sales, 30% of the total vehicle fleet, and 40% of total vehicle travel. Only in the 2050s would more than 50% of the vehicles be capable of self-driving. The dashed lines designate pessimistic projections of saturation levels that include the possibility that at saturation, a portion of motorists will choose to continue to drive their present vehicles, that is, those without full automation. Even these curves may be too optimistic in their prediction of the year at which the indicated percentage levels will be attained. New technologies normally require about three decades to be implemented in 90% of operating vehicles. Furthermore, a large percentage of motorists may not want to pay the additional costs that will be commanded by fully autonomous vehicles.

Factors likely to discourage interest in driverless vehicles include cost; psychological queasiness about loss of control; roadway risks from other vehicles, the infrastructure, and non-vehicle road users such as pedestrians, bicycles, and skaters; concerns about surveillance and tracking of individuals; and insecurity about potential defects in and hacker attacks on driverless vehicle technical systems. It is unclear whether driverless vehicles will be preferred for long or short journeys. Some espouse that initial adoption of driverless vehicles may be limited until controlled operating environments, such as segregated roadways, are established [34].

12.14 ONGOING CONNECTED VEHICLE CONCEPT DEMONSTRATIONS AND TESTS

In March 2014, U.S. FHWA announced a procurement action for one or more pilot deployment concept demonstrations to begin the following year. The purpose of this request for information was to refine plans for initial deployments of combinations of connected vehicle and mobile device technologies to improve traveler mobility and system productivity, while

Table 12.13 Procurement timetable for additional connected vehicle concept demonstrations

Item	Date[a]
Regional pre-deployment workshops and webinars	Summer–Fall 2014
Solicitation for Wave 1 pilot deployment concepts	Early 2015
Wave 1 pilot deployment awards	September 2015
Solicitation for Wave 2 plot deployment concepts	Early 2017
Wave 2 pilot deployment awards	September 2017
Pilot deployments complete	September 2020

a   Schedule is from Katherine Hartman, ITS Joint Program Office. http://www.its.dot.gov/pilots/.

reducing environmental impacts and enhancing safety. As part of these tests, USDOT will provide a prototype national-level SCMS [65]. Table 12.13 contains the FHWA schedule for Wave 1 and Wave 2 deployment concept demonstrations [66].

12.14.1 Wave 1 pilot deployment program

Three Wave 1 awards were made in Fall 2015 to New York City; Tampa, Florida; and the state of Wyoming to address safety, congestion, and freight movement issues. All projects involve DSRC and contain four phases: concept development; design, deploy, and test; maintain and operate; and ongoing post-pilot operations. All projects were in Phase 2 as of early 2017. The three selected pilot deployment sites focus on combinations of applications that result in improved and measurable system performance in one or more of these areas [67]:

- System productivity.
- Mobility, including impact on freight movements.
- Livability and accessibility, where accessibility is defined as the ability to reach goods, services, and activities.
- Environment and fuel use.
- Traveler and system safety, including advising of potentially unsafe conditions and mitigating the impact of events that may cause vehicle crashes.

The New York City sites exploit V2V and V2I communications to improve vehicle flow and pedestrian safety in high-priority corridors. The intersections are closely spaced and typical of those in a dense urban transportation environment. Approximately 400 intersections in midtown Manhattan and central Brooklyn will be instrumented with DSRC RSE to communicate with up to 8500 vehicles (taxis, buses, commercial fleet delivery trucks, city-owned vehicles, and some private vehicles) equipped with aftermarket safety devices [68]. These devices will provide alerts and warnings to vehicle drivers. The V2V applications include forward collision alerts, blind spot warnings, intersection cross-traffic warnings, and emergency brake light alerts. V2I audio warnings include red light and speed-compliance broadcasts and information cautioning motorists that their vehicle is too large or too tall to negotiate the city's bridges, tunnels, and underpasses. The aftermarket devices will also transmit data to the city's TMC using the citywide wireless network as the communication backbone. At the TMC, real-time information from traffic signal controllers equipped with roadside units will be analyzed and used to optimize traffic signal operation.

In Tampa, the pilot demonstration deploys safety and mobility applications that rely on communications between vehicles, the infrastructure, and personal devices on and in proximity to reversible freeway lanes on the Lee Roy Selmon Expressway in downtown Tampa.

In addition to the Expressway, the deployment area contains bus and trolley services, high pedestrian densities, special-event trip generators, and highly variable traffic demand over the course of a typical day. The primary objective is to improve safety and alleviate congestion on the roadway during morning commuting hours for motorists, pedestrians, and transit operation. The pilot will employ DSRC to enable transmissions among approximately 1500 cars, 10 buses, 10 trolleys, 500 pedestrians with smartphone applications, and approximately 40 roadside units along city streets. V2V safety applications include EEBL warning, forward collision warning, and IMA. V2I safety applications include curve speed warning, pedestrian in signalized crosswalk warning, and red-light violation warning. In addition, there are mobility (mobile accessible pedestrian signal system, intelligent traffic signal system, and transit signal priority) and agency data (probe-enabled data monitoring) connected vehicle applications [67].

The Wyoming project uses V2I and V2V technologies to increase the safety and efficiency of truck transportation and reduce the impact of inclement weather in the I-80 corridor [69–71]. Plans are to equip 400 vehicles (200 large trucks, 100 small-to-medium sized trucks, and a combination of 100 highway patrol cars and DOT snowplows) with onboard DSRC communications and install 75 roadside units along I-80. The applications support a flexible range of services from advisories, roadside alerts, parking notifications, and dynamic travel guidance. Information is made available directly to the equipped fleets or through data connections to fleet management centers who then communicate it to their trucks using their own systems. The operational readiness test is planned for Spring 2018 with full operation later in the year.

Other testing programs in the State of Michigan are also ongoing. These projects are exploring V2V, V2I, bicycle, and pedestrian interactions.

12.14.2 University of Michigan and Michigan Department of Transportation tests

The University of Michigan (U of M) Mobility Transformation Center is continuing testing and evaluation of connected vehicle concepts in the United States. This public–private research and development partnership focuses on prototyping an entire system of connected and automated transportation on the streets of southeast Michigan through 2021 [72,73]. The center's operations rely on three programs conducted in collaboration with the Michigan Department of Transportation (DOT), namely, the Ann Arbor Connected Vehicle Test Environment, the Southeast Michigan Connected Vehicle Deployment, and the Ann Arbor Automated Vehicle Field Operational Test (FOT).

The emphasis of the Ann Arbor Connected Vehicle Test Environment is testing V2I and vehicle-to-pedestrian functions. The test includes up to 9000 vehicles, 12 freeway sites, 60 intersections, OTA security, a backhaul communication network, and back-end data storage.

Up to 20,000 vehicles will be included in the Southeast Michigan Connected Vehicle Deployment by 2019–2020. This program builds on the Michigan DOT smart corridors and includes the Michigan Connected Vehicle Pilot project and, with OEM participation, product development, and deployment.

Elements of the Ann Arbor Automated Vehicle FOT are 2000 connected and automated vehicles, including Level 4 automated vehicles. The project consists of personal vehicles, public transit buses, trucks, bicycles, and pedestrians. It covers 27 mi^2 of densely instrumented infrastructure in Ann Arbor. In addition to the University campus, the area contains two major hospitals approximately 0.5 mi (0.8 km) apart and an assisted living facility.

Mcity, which is part of the Ann Arbor Automated Vehicle FOT, opened in July 2015 providing a 32-acre off-roadway test environment for connected and automated vehicles [74]. It contains a four-lane 1000-ft (305-m) straight asphalt highway, merge lanes, a network of asphalt and concrete urban streets, a traffic circle, a crushed-gravel road segment, a concrete calibration pad, traffic signs and signals, and street lights to simulate everyday driving conditions. Cars will merge into a series of lanes and travel down a 5-mi (8-km) road with twists and turns. The test environment will eventually include mechanical pedestrians designed to do the unexpected things we humans do, like stepping out into traffic when we should not or crossing in the middle of a block to determine just what level of complexity an automated car's systems can handle. Its $6.5 million cost was split equally between Michigan DOT and the university.

Another Michigan DOT public–private project is the smart corridor that will deploy V2I on over 120 mi of roadways in the Detroit metropolitan area. Initially, 17 roadside units will be installed to supplement the existing USDOT testbed in Oakland County, Michigan. Other elements include upgrading 80 roadside DSRC devices and nine intersections with controllers capable of transmitting SPaT messages [75].

12.15 COOPERATIVE VEHICLE PROGRAMS IN EUROPE

The Platform for the Deployment of Cooperative Intelligent Transport Systems in the EU and several operational and pilot projects are discussed in the following sections.

12.15.1 Platform for the Deployment of Cooperative Intelligent Transport Systems in the EU

The Platform for the Deployment of Cooperative Intelligent Transport Systems in the EU (C-ITS Platform) was created by the European Commission services (Directorate-General for Mobility and Transport [DG MOVE]) in November 2014 to address issues concerning how to foster business cases, promote interoperability and support the emergence of a common vision across all actors involved in the value chain, determine the basis for cooperation among public and private stakeholders, and establish where investments should start first. The C-ITS Platform enabled dialogue, exchange of technical knowledge, and facilitated cooperation among the Commission, public stakeholders from member states, local and regional authorities, and private stakeholders (such as vehicle manufacturers, service providers, road operators, telecommunication companies, and Tier 1 suppliers) on technical, legal, organizational, administrative, and governing aspects concerning cooperative vehicles. Its first phase (November 2014–January 2016) contribution was a report that depicted its shared vision of the interoperable deployment of Cooperative Intelligent Transport Systems in the EU [76].

The conclusions and recommendations of the C-ITS Platform in the areas of common technical framework; legal questions; legitimacy of the deployment, that is, justifying and fostering the deployment of C-ITS at all levels; and international cooperation necessary for the deployment of C-ITS are discussed below.

12.15.1.1 A common technical framework

The common technical framework addressed issues that included Day 1 services, security and certification, radio-frequency and hybrid communication, standardization, decentralized congestion control, and access to in-vehicle data and resources.

Day 1 services and beyond. The C-ITS Platform Day 1 services are those with societal benefits and technology maturity that are expected to be available in the short term. Personal benefits, users' willingness to pay, business cases, and market-driven deployment strategies were not taken into account at this stage. The platform also identified Day 1.5 services considered as mature and highly desired by the market, but for which specifications or standards might not be completely ready. The Day 1 and 1.5 services are listed in Table 12.14 [76].

Security and certification. To address security and its importance to the deployment of C-ITS in the EU, the C-ITS Platform recommended the following:

- One common standardized C-ITS trust model and certificate policy all over the EU, based on a PKI and defined in an appropriate regulatory framework, to support full secure interoperability of C-ITS Day 1 services at the European level.
- C-ITS may be extended beyond the Day 1 phase with multiple interoperable trust domains if deemed necessary to take into account the variety of stakeholders and the responsibilities of the private and public entities involved.
- International cooperation beyond the EU to discuss how interoperability of other domains (outside Europe) with the single EU trust domain can be realized by identifying areas where harmonization is needed. This topic is even more relevant for the future where the emergence of multiple trust domains in Europe may occur.
- Participation of all disciplines concerned with security (e.g., standardization, revocation of trust, compliance assessment, identification, and involvement of actors regarding the governance of the PKI) and adherence to a well-defined time plan that will ensure the secure deployment of C-ITS.

Radio-frequency and hybrid communication. The C-ITS Platform concluded that as of January 2016, neither ETSI ITS-G5 nor cellular systems can provide the full range of necessary services for C-ITS. Consequently, a hybrid communication concept was advocated to incorporate complementary technologies. Their suggestion was to transmit C-ITS messages independently of the underlying communications technology (access-layer agnostic) wherever possible. The pertinent C-ITS recommendations are as follows:

- Initial use of the IEEE802.11p/ETSI ITS-G5 communications protocol for short-range communications in the 5.9 GHz band, with further study of whether geographical coverage obligations can be introduced to increase coverage of C-ITS services through the existing cellular communications infrastructure to foster uptake of C-ITS services.

Table 12.14 C-ITS Day 1 and Day 1.5 services

Day 1 services	Day 1.5 services

• Hazardous location notifications:
  – Slow or stationary vehicle(s) and traffic ahead warning
  – Road works warning
  – Weather conditions
  – Emergency brake light
  – Emergency vehicle approaching
  – Other hazardous notifications
• Signage applications:
  – In-vehicle signage
  – In-vehicle speed limits
• Signal violation and Intersection safety:
  – Traffic signal priority request by designated vehicles
  – Green light optimal speed advisory (GLOSA)
• Other:
  – Probe vehicle data
  – Shockwave damping (falls under the local hazard warning of the ETSI)
	Information concerning fueling and charging stations for AFVs
	VRU protection
	On-street parking management and information
	Off-street parking information
	Park and ride information
	Connected and cooperative navigation into and out of the city (first and last mile, parking, route advice, coordinated traffic lights)
	Traffic information and smart routing

- Development of mitigation techniques to ensure coexistence between 5.8 GHz tolling DSRC and 5.9 GHz ITS applications. Other coexistence issues (e.g., with urban rail) need to be studied and alleviated.
- Designation of the 5855–5875 MHz, 5905–5925 MHz, and 63–64 GHz bands for C-ITS services to cope with future capacity demand and mitigate risks related to possible wireless access system and radio local area network (WAS/RLAN) expansion in the 5 GHz band.
- International cooperation, for example, via joint studies and positions, to protect the 5.9 GHz band and the allocation of additional spectrum in the 63 GHz frequency band.

Standardization. Standardization and specific standards needed to support the interoperability of near-future C-ITS deployments were addressed by each working group.

Decentralized congestion control (DCC). DCC assists in avoiding interference and degradation of C-ITS applications by addressing network stability issues in the absence of an access point or base station and when encountering an increasing number of emitted C-ITS messages. DCC is standardized and defined in ETSI Technical Specification 102 687 V1.1.1 in sufficient detail for the Day 1 applications. DCC will also be needed when C-ITS applications are introduced for pedestrians and other VRUs.

Access to in-vehicle data and resources. This topic is of importance to existing C-ITS applications and future in-vehicle applications and services. In-vehicle data and resource access are influenced by existing legislation, in particular the eCall type-approval regulation, that requests the European Commission to "assess the need of requirements for an interoperable, standardized, secure and open-access platform" (Article 12[2] of Regulation 2015/758). eCall is the EU's standardized automatic road emergency alert service that is to be installed in all new cars and pickup trucks sold in member states by 2018.

A set of five guiding principles that apply when granting access to in-vehicle data and resources was approved as a basis for all agreements and discussions:

1. *Data provision conditions and consent*: The data subject (owner or user of the vehicle or nomadic device) decides if data can be provided and to whom, including the specific purpose for the use of the data (and hence for the identified service). An opt-out option is always available for end customers and data subjects. This is without prejudice to requirements of regulatory applications.
2. *Fair and undistorted competition*: Subject to prior consent of the data subject, all service providers should be in an equal, fair, reasonable, and nondiscriminatory position to offer services to the data subject.
3. *Data privacy and data protection*: Data subjects need to have their vehicles and movement data protected for privacy reasons, and in the case of companies, for competition and security reasons.
4. *Tamper-proof access and liability*: Services utilizing in-vehicle data and resources should not endanger the proper safe and secure functioning of the vehicles. In addition, the access to vehicle data and resources shall not impact the liability of vehicle manufacturers regarding the use of the vehicle.
5. *Data economy*: With the caveat that data protection provisions or specific technologic prescriptions are respected, standardized access favors interoperability between different applications, notably regulatory key applications, and facilitates the common use of same vehicle data and resources.

Three technical solutions were identified for access to in-vehicle data and resources: the onboard application platform, the in-vehicle interface, and the data server platform. Also agreed upon were standardization needs inputs to the 2015 Rolling Plan for Information and Communication Technology (ICT) Standardization, a technical solution for the in-vehicle interface, and steps toward the identification of possible use cases and related data needs.

There remain, however, disagreements between vehicle manufacturers and the independent operators and service providers in their views of how data can be accessed, onboard application platforms, governance of the data server platform, concrete implementation strategies, and possible legislation. As many of these were not simply technical issues, but included concerns linked to the lack of trust between direct competitors, exploring new ways to improve cooperation was recommended. Further progress would benefit from a scenario-based analysis of legal, liability, technical, and cost–benefit aspects.

12.15.1.2 Legal questions

Legal issues address liability and data protection and privacy.

Liability. Many types of stakeholders may be involved in providing C-ITS information and services. Since Day 1 applications concern information only, the driver always remains in control of the vehicle, and no changes are required concerning liability as compared to the current situation. Hence, the C-ITS Platform concluded that the current amendment to the Vienna Convention (Amendment Article 8, paragraph 5) was sufficient.

However, there are two mitigating factors. First is the tendency of consumers to trust technology, this effect being even stronger with information provided by public authorities. Therefore, a recommendation was made that vehicle manufacturers, service providers, and public authorities use the appropriate level of information (e.g., disclaimers) to raise the user's awareness of the limitation of the information provided, in particular regarding safety-critical messages and information provided in the absence of physical traffic signage. The second aspect is related to the trends toward higher levels of connectivity and automation, where information provided via C-ITS may trigger subsequent action from the vehicle. A review of liability in these cases was recommended for the second phase of the C-ITS Platform.

Data protection and privacy. The continual broadcasting by C-ITS equipped vehicles of CAMs and Decentralized Environmental Notification Messages (DENMs), which contain data such as vehicle speed and location, raises potential concerns of how to guarantee privacy and data protection. Various consultations, in particular with the European Data Protection Supervisor (EDPS) and privacy experts, encouraged the C-ITS Platform to treat these messages as personal data because of their potential ability to indirectly identify users. Therefore, the EU legislation (Directive 95/46/EC) on data privacy and data protection applies. Accordingly, their recommendation is to implement the principle of Informed Consent by providing the vehicles with ad hoc technologies allowing the attachment of consent markers to personal data. The Platform recommends that an opt-out possibility be offered to the drivers, authorizing them to shut down the broadcast, while fully informing the drivers about possible adverse consequences.

Other identified potential legal bases are "vital interests of data subject" and "public interest" (articles 7[d] and 7[e] of the Directive, respectively), which could allow the processing of data without a driver's explicit consent. For C-ITS road safety and traffic management applications, where a vital or public interest is at stake and is demonstrated, a limited number of applications could process the data without a driver's explicit consent, provided that the legal basis to process the data and these applications is strictly defined, and the data collected under these conditions are not further processed or re-purposed beyond these applications. In any case, it is recommended to foster the principle of Privacy by Design and develop systems flexible enough to guarantee full control of personal data by the data subject.

12.15.1.3 Legitimacy of C-ITS deployment

The legitimacy of C-ITS deployment is concerned with road safety issues, acceptance and readiness to invest, and costs and benefits.

Road safety issues. The deployment of C-ITS poses some obvious road safety issues, linked in particular to the driver's lack of knowledge of C-ITS functionalities, false perception, and overreliance on the system. Likewise, the simultaneous presence on the same road networks of C-ITS equipped and non-equipped vehicles may create some safety challenges. Therefore, the C-ITS Platform proposed several recommendations related to the following:

1. Revision of the European Statement of Principles on Human Machine Interface, namely, that the Statement adapt its content to current scientific knowledge and the technology found in new vehicles.
2. Coexistence of equipped and non-equipped vehicles including promoting research to better understand the safety risks posed by the interaction of equipped and non-equipped vehicles, adapting driving regulations to reflect the presence on the roadways of equipped and automated vehicles, sharing of safety-relevant information between equipped and non-equipped users, and exploitation of infrastructure-based C-ITS to compensate for non-equipped users.
3. Training and awareness including campaigns to inform road users about the existence, functionalities, and limitations of the new technologies; adapting driving license education programs to update the public about the technologies that new drivers are likely to experience; encouraging post-license training, possibly linked to the acquisition of an equipped vehicle, to update drivers concerning new safety-related technologies; and encouraging vehicle manufacturers to offer complete information on the new technologies incorporated into a vehicle, for example, through a demonstration or training session as part of the sales package.

Acceptance and readiness to invest. A major obstacle for C-ITS deployment is the significant upfront investment required to develop cooperative vehicles and the corresponding infrastructure. Hence, enhanced collaboration between stakeholders, synchronization of actions, and acknowledgment of existing interdependencies are needed to ensure that promised benefits and program success are realized.

Therefore, the recommendations are for the European Commission to continue its financial support of C-ITS deployment projects in the context of the Connecting Europe Facility (CEF), and for all existing and upcoming projects to exchange results and experiences through stable mechanisms. The second phase of the C-ITS Platform should also consider how to consolidate the engagement of key stakeholders in the future. A further recommendation is that the Commission support public investment through harmonized C-ITS pre-commercial procurement methods and practical tools such as investment guidelines for infrastructure managers.

C-ITS benefits and problem-solving approaches must be clearly communicated to private and professional end users and infrastructure owners or operators to ensure a persuasive and rapid uptake of C-ITS deployment and to secure investments in vehicle and infrastructure equipment. Difficulties in developing business models in urban environments were specifically highlighted and, hence, the importance of having quick-win cases and ambassadors for C-ITS projects was recommended. The second phase of the C-ITS Platform could help better define the measures and messages that address the legal and technical certainty for infrastructure owners, reduction of operating costs for fleets, societal benefits (e.g., safety, reduction of congestion and emissions), tracking fears, and knowledge sharing between stakeholders.

Costs and benefits. A cost–benefit analysis was performed centered on the list of Day 1 services. In addition, several additive scenarios based on multiple combinations of services were analyzed, taking into account the likely type of communication, different geographical environments, and the purpose of the services, for example, road safety, traffic information, and freight services.

Using the 2018–2030 timeframe to assess the impact of C-ITS deployment, the Platform found that significant benefits only started to accumulate between 5 and 10 years after initial investments, depending on the scenario and uptake rates. Ultimately, benefits significantly outweighed costs on an annual basis and, depending on the scenario, by a ratio of up to 3:1 when evaluated over the entire 2018–2030 period.

An overall conclusion is that a strong uptake is an essential prerequisite for achieving meaningful benefits, and that services will most probably always be bundled. Benefits of deploying C-ITS services are very large, but they will not necessarily appear in the short term.

To ensure interoperability and maximize benefits, the list of Day 1 applications and common standards must be the basis for deployment throughout the EU. In parallel, as investments will not be dependent on the number of services, it is necessary to deploy a maximum number of services as quickly as possible in order to ensure the quickest possible positive return on investment. The need to have low entry barriers for access to in-vehicle data to encourage the deployment of new C-ITS-enabled services and applications was also highlighted.

12.15.1.4 International cooperation

International cooperation is essential for the acceptance of cooperative systems as worldwide markets have global players requiring global strategies. Subjects such as C-ITS security policy and standard harmonization have benefitted from cooperation with the United States and Japan since 2009 and 2011, respectively. Well-established dialogue among nations has also brought substantial progress in other areas such as communication and spectrum issues, and data protection. Continued progress of C-ITS requires a change in activities, moving from research and pilot projects to the stages of early deployment. This was recognized as an important milestone to determine when to revisit aspects and (possibly new) priorities for future international cooperation.

Learning from collaborations with partners within the same geographical region or at the international level is also a key requirement for future progress. The C-ITS Platform recommends the Commission encourage the exchange of technical, organizational, and political learning from pilots in different regions, while the private sector, in parallel, addresses other aspects more closely linked to commercial issues. The C-ITS Platform also suggests the Commission enlarge cooperation concerning deployment practices at the government level with Canada, Australia, South Korea, and others, and closely follow international developments in this field and in the area of automation.

12.15.1.5 Conclusions of the C-ITS Platform

The first and paramount general conclusion of the C-ITS Platform is the need for coordinated action to deploy C-ITS in the EU. This involves establishing a unique legal and technical framework and harmonized efforts to ensure quick acceptance of C-ITS.

The second general conclusion is urgency. It was noted that the technology is ready, the industry is already deploying C-ITS equipped vehicles in other parts of the world, and is intending to deploy in the EU by 2019, provided the legal and technical framework is

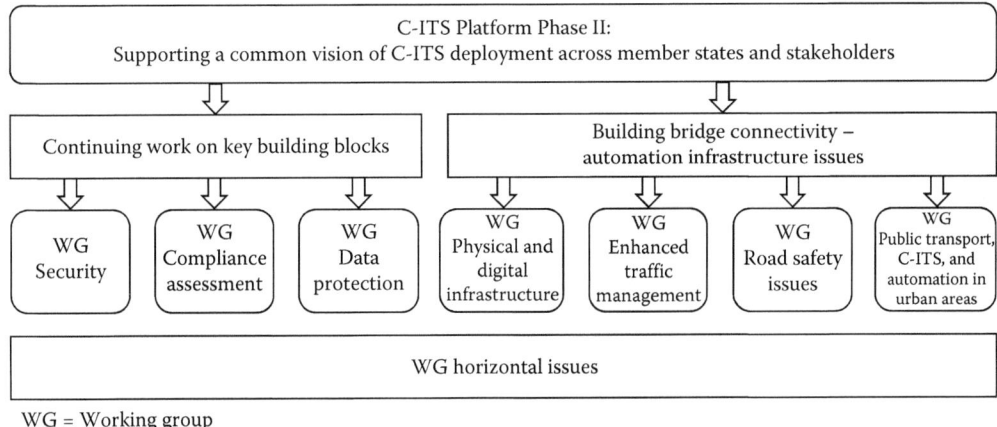

WG = Working group

Figure 12.15 Vision and organization of C-ITS Platform Phase II.

sufficiently in place by that time. Access to in-vehicle data and resources requires a scenario-based analysis of legal, liability, technical, and cost–benefit concerns for further progress and to help answer legislators' requests regarding an open-access platform.

12.15.1.6 Phase II of the C-ITS Platform

Figure 12.15 displays the intention and organization of the second phase of the C-ITS Platform [77]. Some of its objectives are as follows:

- C-ITS deployment in 2019.
- European Commission publishing of guidance on European C-ITS security and certificate policy in 2017.
- European Commission publishing of first guidance regarding data protection by design and by default in 2018.
- Using the C-Roads Platform [78] as the coordination mechanism for C-ITS deployment at the operational level, including testing and validation, to ensure interoperability of Day 1 services across the EU.
- Establishing a compliance assessment process for Day 1 services by 2018.
- Engaging with the Horizon 2020 Projects [79].

12.15.2 Cooperative vehicle pilot and operational projects in Europe

Looking to another country's cooperative vehicle designs and infrastructure for concepts and funding mechanisms that lead to interagency and intercountry cooperation may produce benefits for the developer and other stakeholders. For example, the Netherlands, Germany, and Austria are jointly developing a C-ITS Corridor shown in Figure 12.16 that converses with a vehicle as it travels from Rotterdam through Munich, Frankfurt, and on to Vienna without a single interruption in the initial, basic service: warning drivers of upcoming roadwork and other obstacles [80–83]. To make the project a reality, the countries must ensure interoperability of V2V and V2I communications across all jurisdictions. The lead organization for this project is the Dutch Ministry of Infrastructure and the Environment. Many countries have said they will eventually connect their local smart-road projects to the Corridor. France,

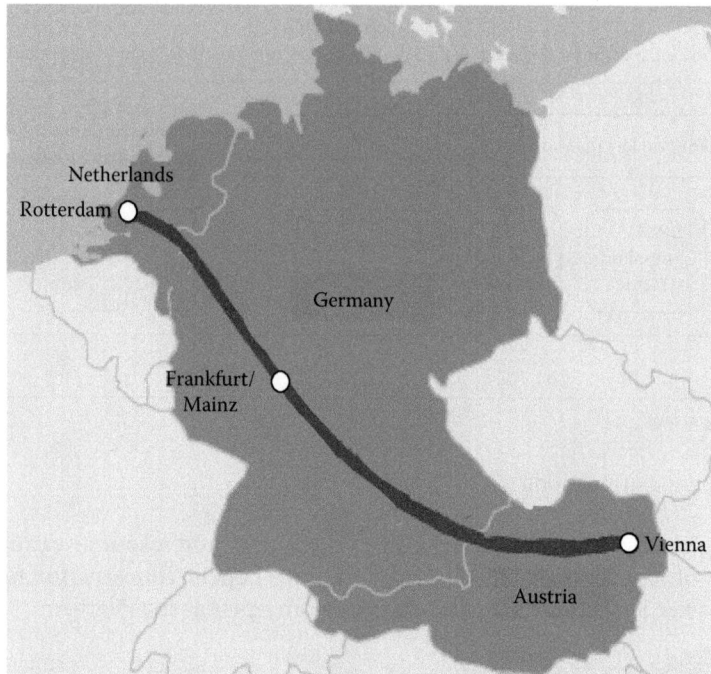

Figure 12.16 C-ITS Corridor connecting Rotterdam, Frankfurt/Mainz, and Vienna. (Adapted from Cooperative ITS Corridor Joint Deployment, Ministry of Infrastructure and the Environment of the Netherlands, Bundesministerium für Verkehr, Bau und Stadtentwicklung of Germany, Bundesminiterium für Verkehr, Innovation und Technologie of Austria. http://www.bmvi. de/SharedDocs/EN/Anlagen/VerkehrUndMobilitaet/Strasse/cooperative-its-corridor. pdf?__blob=publicationFile.)

Poland, and the Czech Republic will likely be among the first. The true leader in smart-road systems is Japan, where drivers are already informed about traffic conditions and speed limits by collecting and disseminating data through radio and infrared transceivers.

In the EU, the European Commission launched a connected and automated vehicles project in the Autumn of 2016 that aims to manage automated vehicles in urban environments with signalized intersections and mixed traffic. The Managing Automated Vehicles Enhances Network (MAVEN) project is a 3-year effort, with €3.15 million (≈$3.5 million [U.S.]) of funding under the EC's Horizon 2020 Research and Innovation Framework Program. Since highly automated vehicles and C-ITS technology with V2V and V2I communications will be more prevalent in the near future, combining both in connected autonomous vehicles (CAVs) could considerably improve traffic flow, particularly in urban areas. The project's three demonstration cities are Helmond (Netherlands), Braunschweig (Germany), and the London Borough of Greenwich (United Kingdom). The project is coordinated by the DLR German Aerospace Center and led by Dynniq Netherlands [84].

MAVEN will develop infrastructure-assisted platoon organization and negotiation algorithms for the management of automated vehicles at signalized intersections and corridors. The algorithms will extend and connect in-vehicle systems for trajectory and maneuver planning, which will interface with adaptive traffic signal optimization systems. These will optimize signal timing to facilitate movement of organized platoons, providing better use of existing infrastructure capacity, increasing traffic efficiency by reducing vehicle delays, and lowering vehicle emissions. MAVEN will also contribute to the development of communication standards and high-precision maps, and develop advanced driver assistance systems that include provisions for VRUs such as pedestrians and cyclists.

In the United Kingdom, £20 million (≈$26 million [U.S.]) has been awarded for eight projects to develop the next generation of CAVs [85]. The Jaguar Land Rover (JLR) consortium is investing in a 41-mi (66-km) project to develop new connected and autonomous vehicle technologies. This particular CAV test corridor will evaluate different communication technologies that could share information at very high speeds between cars, and between cars and roadside infrastructure, including traffic signals and overhead gantries. These connected and autonomous vehicle features are designed to improve road safety, enhance the driving experience, reduce the potential for traffic jams, and improve traffic flow.

A 3-year project, the £5.5 million (≈$7 million [U.S.]) UK-CITE (UK Connected Intelligent Transport Environment) project, will create the first test route capable of testing both V2V and V2I systems on public roads in the country. The CAV corridor, which includes public roads around Coventry and Solihull, will evaluate new systems in real-world driving conditions. Roadside communications will enable testing of a fleet of up to 100 connected and highly automated cars. Four connectivity technologies will be evaluated: 4G-based LTE, DSRC, LTE-V (a more advanced version of LTE), and local Wi-Fi hotspots. V2X technologies to be explored are cooperative adaptive cruise control and over-the-horizon warning systems that inform drivers or future autonomous vehicles of hazards, changing traffic conditions, and approaching emergency vehicles.

In another U.K. trial, Highways England is investing £150 million (≈$195 million [U.S.]) in driverless car technology on the A2/M2 motorway to collect real-world data on performance and potential impacts on capacity and operations. Journey information will be sent wirelessly to specially adapted vehicles on the motorway between London and Kent. The trial will also deploy radar technology to improve breakdown detection [86].

12.16 SUMMARY

The connected vehicle environment is composed of wireless connectivity among vehicles, infrastructure, and mobile devices carried by travelers, whether in vehicles, on bicycles or on other transport and conveyances, or walking. Connected vehicles and devices are designed to bring about transformative changes in highway safety, mobility, and environmental impact by enlisting a broad stakeholder base encompassing government, industry, researchers, drivers, and other travelers. A connected vehicle safety pilot was conducted in 2013 and 2014 to support NHTSA and FHWA decisions concerning the effectiveness of V2V and V2I technologies to reduce crashes. Additional deployment of pilot concepts began in 2015.

In December 2016, NHTSA issued a NPRM that would mandate V2V communications in new light vehicles and solicited comments concerning the rulemaking. The effective date for manufacturers to begin implementing the V2V requirements is proposed as two model years after the final rule is adopted, with a 3-year phase-in period to accommodate vehicle manufacturers' product cycles. FHWA also issued new non-mandatory guidance in December 2016 concerning Federal-aid requirements for highway programs and interoperability of V2I operations among state and local transportation agencies.

In brief, the U.S. Connected Vehicle Program has the following benefits, communications options, and challenges:

- Connected vehicle systems and technologies deliver services and benefits to users in three broad categories: safety (including those based on V2V or V2I communications), dynamic mobility, and environmental.

- V2V and V2I communications may address 81% of unimpaired crashes in all vehicle types and reduce congestion and vehicle emissions.
- DSRC technologies have been developed specifically for vehicular communications and are currently reserved for transportation safety by the U.S. FCC.
- DSRC will be used for V2V and V2I safety-critical applications, pending NHTSA decisions. Cellular communications can be explored for other safety, mobility, and environmental applications. An example is the alerting of drivers to pedestrians and cyclists using a peer-to-peer wireless standard such as Wi-Fi Direct that allows smart phones to communicate directly with one another and to a receiver in a vehicle that alerts the driver to the presence of pedestrians.
- A PKI security system, involving the exchange of digital certificates among trusted users, can support both the need for message security and appropriate anonymity to users.
- Current strategic challenges exist in many areas, including technical and policy issues, benefit analysis and prediction, deployment costs, public acceptance, and security. They include changes to driver, vehicle, and law enforcement regulations in areas that affect the testing or operations procedures that account for the unique attributes of connected and automated vehicles.
- An AASHTO connected vehicle field infrastructure deployment analysis indicates infrastructure deployment decisions of state and local transportation agencies will be based on the nature and timing of benefits and that benefits will depend on the availability of connected vehicle equipment installed in vehicles, either as original equipment or as aftermarket devices.
- An automobile industry advocacy group, Auto-ISAC, recommended risk management practices for car manufacturers that include seven functions: alignment of the cybersecurity program with the organization's broader mission and objective, risk assessment and management, security by design, threat detection and protection, incident response and recovery, training and awareness, and collaboration and engagement with appropriate third parties.

The first phase of the C-ITS Platform established in 2014 addressed issues and delivered recommendations that affect the common technical framework, international cooperation, legal questions, and the legitimacy of C-ITS deployment in Europe and elsewhere. Its contributions include endorsement of the following:

- Specific Day 1 services, applications, and common standards as the basis for deployment throughout the EU.
- A PKI trust model and certificate policy.
- A hybrid communications concept that utilizes complementary technologies to provide all necessary services.
- Five guiding principles encompassing data provision conditions and consent, fair and undistorted competition, data privacy and data protection, tamper-proof access and liability, and data economy that apply when granting access to in-vehicle data and resources.
- Three technical solutions for access to in-vehicle data and resources: the onboard application platform, the in-vehicle interface, and the data server platform.
- Reevaluating liability concerns in the second phase of the C-ITS Platform.
- The principle of Informed Consent by providing vehicles with ad hoc technologies allowing the attachment of consent markers to personal data and offering an opt-out possibility to drivers concerning sharing of their personnel data.
- Several recommendations related to human–machine interface issues, coexistence of equipped and non-equipped vehicles, and training and awareness.

- Continued European Commission financial support of C-ITS deployment projects.
- Exchanges of results and experiences from existing and upcoming projects.
- Finding mechanisms to consolidate the engagement of key stakeholders in the future.
- European Commission support of public investment through harmonized C-ITS pre-commercial procurement methods and tools for infrastructure managers.
- Deploying a maximum of services as quickly as possible in order to ensure the quickest possible positive return on investment.
- Need for low entry barriers for access to in-vehicle data to encourage deployment of new C-ITS-enabled services and applications.
- European Commission exchange of technical, organizational, and political learning from pilots in different regions and countries.

Among the objectives of Phase II of the C-ITS Platform are C-ITS deployment in 2019 and use of the C-Roads Platform as the coordination mechanism for C-ITS at the test, validation, and operational level to ensure interoperability of Day 1 services across the EU.

Pilot deployment projects and other initiatives that test and evaluate critical portions of concepts and applications needed for the success of connected vehicle and cooperative vehicle programs are ongoing in the United States, Europe, and other parts of the world. In many instances, these trials gather information that is also indispensable to the refinement and safety enhancement of autonomous or self-driving vehicles.

REFERENCES

1. *DOT Launches Largest-Ever Road Test of Connected Vehicle Crash Avoidance Technology.* http://www.nhtsa.gov/About+NHTSA/Press+Releases/2012/DOT+Launches+Largest-Ever+R oad+Test+of+Connected+Vehicle+Crash+Avoidance+Technology. Accessed August 16, 2016.
2. A.L. Svenson, IntelliDrive vehicle to vehicle safety applications research plan, In *IntelliDrive Webinar Safety Applications for Commercial Vehicles*, National Highway Traffic Safety Administration, U.S. Department of Transportation, Washington, DC, January 20, 2010.
3. P.E. Ross, Thus spoke the Autobahn, *IEEE Spectrum*, 52(1):52–55, January 2015.
4. *ITS ePrimer—Module 13: Connected Vehicles*, Intelligent Transportation Systems Joint Program Office, Research and Innovative Technology Administration, U.S. Department of Transportation, Washington, DC, 2013.
5. *Crash Data Analyses for Vehicle-to-Infrastructure Communications for Safety Applications*, Publication No. FHWA-HRT-11-040, FHWA Research, Development, and Technology, Turner Fairbank Highway Research Center, McLean, VA 22101-2296, November 2012.
6. *Signal Phase and Timing (SPaT) Applications, Communications Requirements, Communications Technology Potential Solutions, Issues and Recommendations*, Draft Final Report FHWA-JPO-13-002, April 3, 2012. Prepared by Bruce Abernethy, ARINC Incorporated; Scott Andrews, Cogenia Partners; and Gary Pruitt, ARINC Incorporated. www.its.dot.gov/index.htm.
7. *A Policy on Geometric Design of Highways and Streets ("Green Book")*, Sixth Edition, American Association of State Highway and Transportation Officials Publications Order Department, Atlanta, GA, 2011.
8. C.M. Richard, J.L. Campbell, and J.L. Brown, *Task Analysis of Intersection Driving Scenarios: Information Processing Bottlenecks*, FHWA-HRT-06-033, Federal Highway Administration, U.S. Department of Transportation, McLean, VA, August 2006. https://www.fhwa.dot.gov/publications/research/safety/06033/06033.pdf. Accessed March 1, 2016.
9. Audi cars to tell when lights turn green, *ITS International*, 68, September/October 2016.
10. AASHTO SPaT Challenge Webinar, September 15, 2016. http://stsmo.transportation.org/Documents/AASHTO%20SPaT%20Challenge%20Webinar%20-%20Slide%20deck%20ver%202%20-%20FINAL%2009152016.pdf. Accessed January 17, 2017.

11. T. Kearney, IntelliDrive CVO/freight vehicle to infrastructure overview, In *IntelliDrive Webinar Safety Applications for Commercial Vehicles, Federal Highway Administration*, U.S. Department of Transportation, Washington, DC, January 20, 2010.

12. J. Masters, Seeking solutions to congestion's costs, *ITS International*, 36–37, March/April 2015.

13. D. Valentine, R. Zimmer, S. Mortensen, and R. Sheehan, *Transit Safety Retrofit Package (TRP): Leveraging DSRC for Transit Safety—Fielding Results and Lessons Learned*, Intelligent Transportation Systems Joint Program Office, Federal Highway Administration, Federal Transit Administration, U.S. Department of Transportation, Washington, DC, November 2014.

14. R.E. Zimmer, M. Burt, G.J. Zink, D.A. Valentine, and W.J. Knox Jr, *Transit Safety Retrofit Package Development Final Report*, FHWA-JPO-14-142, Federal Highway Administration, U.S. Department of Transportation, Washington, DC, July 30, 2014. http://ntl.bts.gov/lib/54000/54500/54592/FHWA-JPO-14-142_v1.pdf. Accessed November 13, 2015.

15. R. Glassco, M. Barba, E. Escalera, S. Jung, C. Kain, Y. Li, M. Mercer, and M. Vasudevan, *State of the Practice of Techniques for Evaluating the Environmental Impacts of ITS Deployment*, Report FHWA-JPO-11-142, Intelligent Transportation Systems Joint Program Office, Research and Innovative Technology Administration, U.S. Department of Transportation, Washington, DC, August 2011.

16. S. Bayless, A. Guan, A. Shaw, M. Johnson, G. Pruitt, and B. Abernathy, *Recommended Practices for DSRC Licensing and Spectrum Management*, FHWA-JPO-16-267, Intelligent Transportation Society of America, Washington, DC, December 2015. ntl.bts.gov/lib/56000/56900/56950/FHWA-JPO-16-267.pdf. Accessed January 17, 2017.

17. *ETSI EN 302 663 V1.2.0 (2012-11), Intelligent Transport Systems (ITS); Access Layer Specification for Intelligent Transport Systems Operating in the 5 GHz Frequency Band—Draft*, Reference REN/ITS-0040028, ETSI, 650 Route des Lucioles F-06921 Sophia Antipolis Cedex—France.

18. L. Lin, ETSI G5 technology: The European approach, DRIVE C2X @ TSS, Gothenburg, June 13, 2013.

19. Vehicle and infrastructure communications, navigation, and active sensing technologies, *ITS International*, NA1, March/April 2015.

20. *Ericsson Initiates 5G Motorway Project with Cross-Industry Consortium in Germany*, https://www.ericsson.com/news/161117-ericsson-initiates-5g-motorway-project-with-cross-industry-consortium-in-germany_244039853_c. Accessed January 30, 2017.

21. M. Weigle, *Standards, WAVE/DSRC/802.11p*, Old Dominion University, Spring, 2008.

22. J. Harding, G. Powell, R. Yoon, J. Fikentscher, C. Doyle, D. Sade, M. Lukuc, J. Simons, and J. Wang, *Vehicle-to-Vehicle Communications: Readiness of V2V Technology for Application*, Chapter IX. V2V Communications Security, DOT HS 812 014, National Highway Traffic Safety Administration, U.S. Department of Transportation, Washington, DC 20590, August 2014. http://www.nhtsa.gov/staticfiles/rulemaking/pdf/V2V/Readiness-of-V2V-Technology-for-Application-812014.pdf. Accessed August 5, 2016.

23. *Certification Authority Certificates*, https://technet.microsoft.com/en-us/library/cc778623. Accessed February 16, 2016.

24. *Overview of SSL/TLS Encryption*, https://technet.microsoft.com/en-us/library/cc781476%28v=ws.10%29.aspx. Accessed February 16, 2016.

25. *Information Technology—Open Systems Interconnection—The Directory: Public-Key and Attribute Certificate Frameworks*, International Standard ISO/IEC 9594-8, Recommendation ITU-T X.509, International Telecommunication Union, October 2012. http://www.itu.int/rec/T-REC-X.509-201210-I/en. Accessed August 14, 2016.

26. Crash Avoidance Metrics Partners Vehicle Safety Communications 5 Consortium, *Security Credential Management System Proof-of-Concept Implementation, End Entity (EE) Requirements and Specifications Supporting SCMS Software Release 1.1*, U.S. Department of Transportation, National Highway Traffic Safety Administration (NHTSA), May 4, 2016. http://www.its.dot.gov/pilots/pdf/SCMS_POC_EE_Requirements.pdf. Accessed August 5, 2016.

27. M. Harris, *Researchers Prove Connected Cars Can Be Tracked*, http://spectrum.ieee.org/cars-that-think/transportation/advanced-cars/researchers-prove-connected-cars-can-be-tracked/, posted October 21, 2015.
28. J.N. Spiller, N. Compin, A. Reshadi, B. Umfleet, T. Westhuis, K. Miller, and A. Sadegh, *Advances in Strategies For Implementing Integrated Corridor Management (ICM)*, Scan Team Report, NCHRP Project 20-68A, Scan 12-02, October 2014. http://www.domesticscan.org/wp-content/uploads/NCHRP20-68A_12-02.pdf. Accessed November 17, 2015.
29. S.E. Shladover and R. Bishop, Appendix A: Commissioned White Paper 1, Road transport automation as a public–private enterprise, In *Towards Road Transport Automation—Opportunities in Public–Private Collaboration: Summary of the Third EU-U.S. Transportation Research Symposium, Conference Proceedings 52*, K.F. Turnbull, Rapporteur, Transportation Research Board, Washington, DC, 40–64, April 14–15, 2015.
30. A. Ohnsman, Google spins off self-driving car unit as "Waymo", *Forbes*, December 13, 2016. http://www.forbes.com/sites/alanohnsman/2016/12/13/googles-spins-off-self-driving-car-unit-as-waymo/#7470551d2af4. Accessed December 13, 2016. Also https://waymo.com/ontheroad/. Accessed June 1, 2017.
31. J.M. Anderson, N. Kalra, K.D. Stanley, P. Sorensen, C. Samaras, and O.A. Oluwatola, *Autonomous Vehicle Technology: A Guide for Policymakers*, 2014. http://www.rand.org/content/dam/rand/pubs/research_reports/RR400/RR443-1/RAND_RR443-1.pdf. Accessed November 15, 2015.
32. *Preliminary Statement of Policy Concerning Automated Vehicles*, U.S. Department of Transportation, National Highway Traffic Safety Administration, Washington, DC, May 30, 2013. http://www.nhtsa.gov/staticfiles/rulemaking/pdf/Automated_Vehicles_Policy.pdf. Accessed September 29, 2016.
33. J. Hurin, American Association of Motor Vehicle Administrator's Autonomous Vehicle Best Practices Working Group, *Automated and Connected Vehicles, Transportation Research Board Conference Proceedings on the Web 19*, K.F. Turnbull, Rapporteur, Washington, DC, 15–16, November 4–5, 2015.
34. D.J. Glancy, R.W. Peterson, and K.F. Graham, *A Look at the Legal Environment for Driverless Vehicles, NCHRP Legal Research Digest 69 Pre-Publication Draft*, Transportation Research Board, Washington, DC, October 2015.
35. D.J. Glancy, Autonomous and automated and connected cars—Oh my! First generation autonomous cars in the legal ecosystem, *Minnesota Journal of Law, Science & Technology*, 16(2):619–692, 2015. http://scholarship.law.umn.edu/mjlst/vol16/iss2/3. Accessed October 20, 2016.
36. http://sacramento.cbslocal.com/2014/05/20/dmv-announces-manufacturers-rules-for-testing-self-driving-cars-on-california-roads/.
37. New California law allows testing of autonomous vehicles without a driver, *Traffic Technology Today.com*, October 6, 2016. http://www.traffictechnologytoday.com/news.php?NewsID=81838. Accessed October 6, 2016.
38. Department of Motor Vehicles Notice to Amend Sections in Article 3.7 and Adopt Sections in Article 3.8 of Chapter 1, Division 1, Title 13 of the California Code of Regulations, relating to Autonomous Vehicles, March 10, 2017. https://www.dmv.ca.gov/portal/dmv/detail/vr/autonomous/auto. Accessed March 13, 2017.
39. Initial Statement of Reasons, Title 13, Division 1, Chapter 1, Article 3.7—Testing of Autonomous Vehicles, Article 3.8—Deployment of Autonomous Vehicles. March 10, 2017. https://www.dmv.ca.gov/portal/dmv/detail/vr/autonomous/auto. Accessed March 13, 2017.
40. M. Harris, California gives the green light to self-driving cars, *IEEE Spectrum*, March 10, 2017. http://spectrum.ieee.org/cars-that-think/transportation/self-driving/california-gives-the-green-light-to-selfdriving-cars/?utm_source=CarsThatThink&utm_medium=Newsletter&utm_campaign=CTT03152017. Accessed March 16, 2017.
41. SAE International, *Taxonomy and Definitions for Terms Related to Driving Automation Systems for On-Road Motor Vehicles*, SAE International Standard J3016™, Revised 2016-09. http://standards.sae.org/j3016_201609/. Accessed January 17, 2017.

42. *Federal Automated Vehicles Policy*, U.S. Department of Transportation, National Highway Traffic Safety Administration, Washington, DC, September 2016. https://www.transportation.gov/sites/dot.gov/files/docs/AV%20policy%20guidance%20PDF.pdf. Accessed September 20, 2016.

43. G. Weiner and B.W. Smith, *Automated Driving: Legislative and Regulatory Action*, http://cyberlaw.stanford.edu/wiki/index.php/Automated_Driving:_Legislative_and_Regulatory_Action#cite_note-4. Accessed February 23, 2016.

44. P.A. Hemmersbaugh, *Interpretation of a Number of Provisions in the Federal Motor Vehicle Safety Standards as They Apply to Google's Described Design for a Motor Vehicle*, U.S. Department of Transportation, National Highway Traffic Safety Administration, Washington, DC, 2016. http://isearch.nhtsa.gov/files/Google%20--%20compiled%20response%20to%2012%20Nov%20%2015%20interp%20request%20--%204%20Feb%2016%20final.htm. Accessed February 23, 2016.

45. Motor insurance for autonomous vehicles 'will shift from drivers to OEMs,' *ITS International*, Electronic Edition, October 19, 2015.

46. *U.S. DOT Advances Deployment of Connected Vehicle Technology to Prevent Hundreds of Thousands of Crashes*, National Highway Traffic Safety Administration, U.S. Department of Transportation, Washington, DC, December 13, 2016. https://www.nhtsa.gov/About-NHTSA/Press-Releases/nhtsa_v2v_proposed_rule_12132016. Accessed December 15, 2016.

47. Notice of Proposed Rulemaking (NPRM), V2V Communications, 49 CFR Part 571 [Docket No. NHTSA-2016-0126], RIN 2127-AL55, Federal Motor Vehicle Safety Standards, National Highway Traffic Safety Administration, U.S. Department of Transportation, Washington, DC, December 13, 2016. http://www.safercar.gov/v2v/pdf/V2V%20NPRM_Web_Version.pdf. Accessed December 15, 2016.

48. *Automotive Cybersecurity Best Practices—Executive Summary*, Automotive Information Sharing and Analysis Center, July 21, 2016.

49. C.J. Hill and J. Kyle Garrett, *AASHTO Connected Vehicle Infrastructure Deployment Analysis—Final Report*, performed by Mixon/Hill, Inc., ITS Joint Program Office, Research and Innovative Technology Administration, U.S. Department of Transportation, Washington, DC, June 17, 2011.

50. *Systems Engineering Guidebook for Intelligent Transportation Systems*, Ver. 3.0. U.S. Department of Transportation, Federal Highway Administration—California Division, and California Department of Transportation, November 2009. http://www.fhwa.dot.gov/cadiv/segb/. Accessed November 9, 2015.

51. J. Barbaresso, Infrastructure deployment considerations for connected vehicles, *Automated and Connected Vehicles, Transportation Research Board Conference Proceedings on the Web 19*, K.F. Turnbull, Rapporteur, Washington, DC, 26–28, November 4–5, 2015.

52. B. Hammit and R. Young, *Connected Vehicle Weather Data for Operation of Rural Variable Speed Limit Corridors*, MPC 15-299, Mountain-Plains Consortium, December 2015. http://www.ugpti.org/resources/reports/details.php?id=835.

53. R. O'Toole, Policy implications of autonomous vehicles, *Policy Analysis*, 758, Cato Institute, September 2014.

54. R. Bailey, Is the connected vehicle mandate needed for autonomous vehicles? *Reason Foundation, Surface Transportation Innovations*, 153, July 2016.

55. M. Click, Funding's future is mileage-based, *ITS International*, NAFTA 1–4, July/August 2016.

56. *2015 FHWA Vehicle to Infrastructure Deployment Guidance and Products*, Draft v9a, Federal Highway Administration, U.S. Department of Transportation, Washington, DC, September 29, 2014. www.its.dot.gov/meetings/pdf/V2I_DeploymentGuidanceDraftv9.pdf. Accessed January 13, 2017.

57. *FHWA Vehicle-to-Infrastructure (V2I) Deployment Guidance and Products*, Draft Report, V2I Guidance Rev., FHWA-HOP-15-015, U.S. Department of Transportation, Federal Highway Administration, Washington, DC, December 30, 2016. http://www.its.dot.gov/v2i/. Accessed January 19, 2017.

58. *Connected Vehicle Technology: Certification–Research Plan,* ITS Joint Program Office, http://www.its.dot.gov/research_archives/connected_vehicle/connected_vehicle_cert_plan.htm. Accessed January 19, 2017.

59. T. Litman, Ready or waiting? *Traffic Technology International,* 36–42, January 2014.

60. T.A. Litman, Autonomous Vehicle Implementation Predictions: Implications for Transport Planning, Paper 15-3326, *Presented at the Transportation Research Board 94th Annual Meeting,* Washington, DC, 2015. Also available at Victoria Transport Policy Institute, Victoria, BC, September 1, 2016. http://www.vtpi.org/avip.pdf. Accessed October 19, 2016.

61. T. Litman and A. Hars, Will most new cars be self-driving by 2030? *Traffic Technology International,* 14–19, February/March 2016.

62. M. Felds, *Ford's Road to Full Autonomy,* https://shift.newco.co/fords-road-to-full-autonomy-36cb9cca330#.207d5k3pk. Accessed August 19, 2016.

63. S. Wilson, Beyond Uber, Volvo and Ford: Other automakers' plans for self-driving vehicles, *Los Angeles Times Business/Auto Section,* August 24, 2016. http://latimes.us10.list-manage.com/track/click?u=f089ecc9238c5ee13b8e5f471&id=69c31929d6&e=d6fa5dfe66. Accessed August 24, 2016.

64. Car OEMs target 2021 for rollout of SAE Levels 4 and 5 of autonomous driving, *The Market Potential for Semi-Autonomous Driving,* ABI Research, 2016. https://www.abiresearch.com/press/car-oems-target-2021-rollout-sae-levels-4-and-5-au/. Accessed December 1, 2016.

65. W. Fehr, *Connected Vehicle Pilot Deployment Program,* Intelligent Transportation Systems Joint Program Office, Research and Innovative Technology Administration, U.S. Department of Transportation, Washington, DC, September 2014. www.its.dot.gov/pilots/pdf/CVPilot_Webinar4_SCMSv2.pdf. Accessed August 6, 2016.

66. *Federal Register,* 79(48), Wednesday, March 12, 2014/Notices, p. 14105.

67. *CV Pilot Deployment Program—Technical Assistance Events for Concept Development Phase,* http://www.its.dot.gov/pilots/technical_assistance_events.htm#phase1. Accessed August 25, 2016.

68. The future of New York, *Traffic Technology International,* 46–50, October/November 2016. http://viewer.zmags.com/publication/55821023#/55821023/48. Accessed January 31, 2017.

69. K. Dopart, U.S. Department of Transportation connected vehicle and automated vehicle research update, *Automated and Connected Vehicles, Transportation Research Board Conference Proceedings on the Web 19,* K.F. Turnbull, Rapporteur, Washington, DC, 6–10, November 4–5, 2015.

70. K.K. Hartman, *Connected Vehicle Pilot Deployment Program—Ready for Deployment,* Intelligent Transportation Systems Joint Program Office. http://www.its.dot.gov/pilots/index.htm. Accessed April 28, 2016.

71. Wyoming CV Pilot, *Traffic Technology International,* 52–55, January 2017. http://viewer.zmags.com/publication/9492f4ae#/9492f4ae/54. Accessed January 30, 2017.

72. M. Hall, Engineering the future, *Traffic Technology International,* 7, August/September 2015.

73. J. Maddox, University of Michigan Mobility Transformation Center, *Automated and Connected Vehicles, Transportation Research Board Conference Proceedings on the Web 19,* K.F. Turnbull, Rapporteur, Washington, DC, 32–35, November 4–5, 2015.

74. *Mcity Test Facility,* http://www.mtc.umich.edu/test-facility. Accessed August 6, 2016.

75. M. Smith, Michigan Department of transportation connected vehicle initiative, *Automated and Connected Vehicles, Transportation Research Board Conference Proceedings on the Web 19,* K.F. Turnbull, Rapporteur, Washington, DC, 39–42, November 4–5, 2015.

76. *C-ITS Platform Final Report,* January 2016. http://ec.europa.eu/transport/themes/its/doc/c-its-platform-final-report-january-2016.pdf. Accessed August 26, 2016.

77. C. DePre, European Community's Cooperative ITS Program, *Presented at the Intelligent Transportation Committee Meeting, 96th Annual Meeting of the Transportation Research Board,* Washington, DC, January 11, 2017.

78. *C-Roads Platform,* https://www.c-roads.eu/platform.html. Accessed January 17, 2017.

79. *Horizon 2020—The EU Framework Programme for Research and Innovation,* https://ec.europa.eu/programmes/horizon2020/en/what-horizon-2020. Accessed January 17, 2017.

80. P.E. Ross, Europe's smart highway will shepherd cars from Rotterdam to Vienna, *IEEE Spectrum*, posted 30 Dec 2014. http://spectrum.ieee.org/transportation/advanced-cars/europes-smart-highway-will-shepherd-cars-from-rotterdam-to-vienna. Accessed November 12, 2015. Also published as "Thus Spoke the Autobahn," *IEEE Spectrum*, 52–55, January 2015.
81. J. Stojaspal, Misconnected vehicles? *Traffic Technology International*, 28–34, April/May 2016.
82. *Cooperative ITS Corridor Joint Deployment Web Site*, http://www.c-its-korridor.de/?menuId= 1&sp=en. *Amsterdam Group Web Site*, https://amsterdamgroup.mett.nl/Road+Map/Road+Map+Activities/default.aspx. Accessed October 16, 2016.
83. *Cooperative ITS Corridor Joint Deployment*, Ministry of Infrastructure and the Environment of the Netherlands, Bundesministerium für Verkehr, Bau und Stadtentwicklung of Germany, Bundesminiterium für Verkehr, Innovation und Technologie of Austria, http://www.bmvi. de/SharedDocs/EN/Anlagen/VerkehrUndMobilitaet/Strasse/cooperative-its-corridor.pdf?__ blob=publicationFile. Accessed October 30, 2016.
84. European Commission launches new connected and automated vehicles project, *Traffic Technology Today Digital Edition*, November 2, 2016. http://www.traffictechnologytoday. com/news.php?NewsID=82340. Accessed November 3, 2016.
85. Consortium to create UK's first "connected roads, *Traffic Technology Today*, February 4, 2016. http://www.traffictechnologytoday.com/news.php?NewsID= 77223. Accessed February 5, 2016.
86. UK to trial wireless CVs and driverless cars, *ITS International*, 8, May/June 2016.

Chapter 13

Systems engineering process

Systems engineering is an interdisciplinary approach utilized to develop and build complex systems such as intelligent transportation and traffic management systems. It focuses on understanding and defining customer needs and functionality early in the development cycle, documenting requirements and system design options, selecting a system design option as part of a review process, and only then proceeding with the subsystem and component-level design and build phases followed by testing and system validation. Systems engineering takes into consideration the points of view of all the stakeholders, that is, the owners, operators, managers, maintainers, and users of the system. A concept of operations (ConOps) is a vital part of the systems engineering process as it is the narrative that conveys the functioning and benefits of the system to each of the stakeholder classes. System design and operation are evaluated through performance measures that compare performance with expected benefits.

Risks are addressed as early as feasible in the design process to keep their cost impacts as small as possible. Technology choices are made at the last possible moment so that they do not drive design decisions and to allow the option of incorporating the latest technological advances into the system build. Interfaces are an important part of the system design since their proper specification and configuration will help ensure a smooth integration of individual components, subsystems, and data flows. The systems engineering discipline is especially critical when developing systems to accommodate new concepts such as connected and autonomous vehicles, which will operate together with older vehicles and legacy traffic management systems that do not incorporate the newer technologies and features.

13.1 SYSTEMS ENGINEERING

The systems engineering process illustrated in the Vee diagram of Figure 13.1 is applicable to ITS projects that may be part of a larger regional or national architecture or simply stand-alone projects. The diagram shows the life cycle of a project and how its early phases directly affect end-of-project tasks [1]. This model is widely applied in the systems engineering community and is included in systems engineering process standards such as International Organization for Standardization/International Electrotechnical Commission (ISO/IEC) TR 19760, ISO 15288, and Electronic Industries Alliance (EIA) 632.

Development of transportation management systems encompasses concept exploration, planning, and design phases, the latter including benefits analysis, high-level system design, and lower-level subsystems and component designs. The identification of the standards and data transfer protocols that specify the interfaces between components, subsystems, system operators, and users of the information cannot be overemphasized as their use will ensure a smooth flow of data across these boundaries.

Interfacing to the regional architecture	Concept exploration and benefits analysis	Project planning, SEMP, and concept of operations	System and subsystem requirements, detailed design	Development, testing, integration, verification	System validation, operations and maintenance	Changes and upgrades	System retirement and replacement
Phase–1	Phase 0	Phase 1	Phase 2	Phase 3	Phase 4		Phase 5

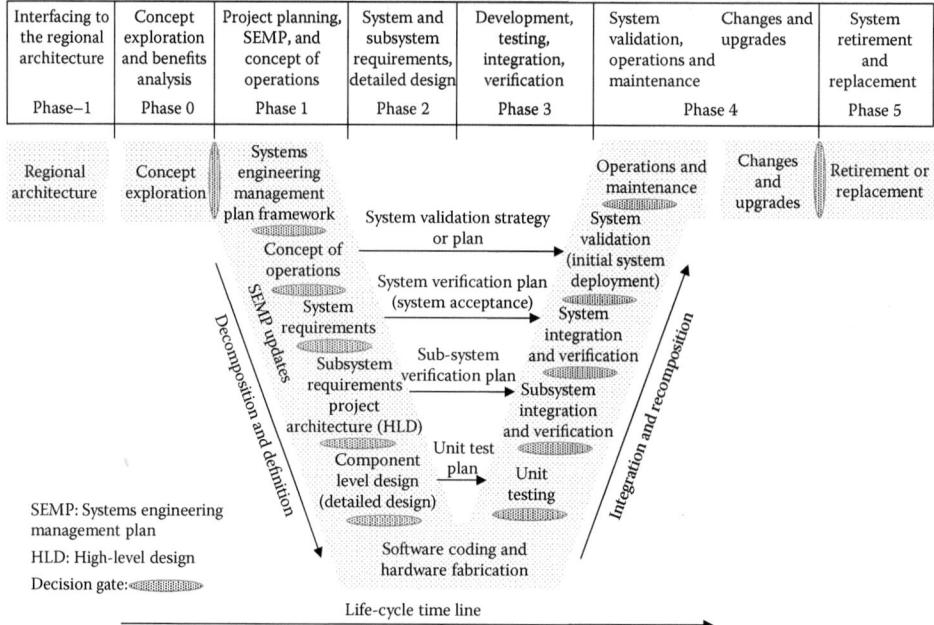

Figure 13.1 Vee model of systems engineering process. (Adapted from *Systems Engineering Guidebook for Intelligent Transportation Systems*, Ver. 3.0, U.S. Department of Transportation, Federal Highway Administration, California Division and California Department of Transportation, November 2009. http://www.fhwa.dot.gov/cadiv/segb/.)

When applying systems engineering to ITS development, the process is sometimes divided into the seven phases displayed at the top of Figure 13.1. These correlate with the more detailed steps that appear in the Vee model below it. The first phase is interfacing with a regional architecture that is developed in accordance with National ITS Architectures described in Chapter 14. Other ITS personnel identify systems engineering with the 13 steps in Table 13.1. The correlation with the seven phases in Figure 13.1 is indicated by the phase number in the last column of the table.

Table 13.1 Thirteen-step system engineering process

Step	Task	Vee phase
1	Identify the stakeholders, that is, the users and operators of the system	−1
2	Interview them and ask what benefits they want from the system	0
3	Prepare a ConOps that addresses the needs of all user groups	1
4	Develop the system and subsystem requirements	2
5	Develop an architecture for the system	2
6	Design the subsystems and system, including the interfaces between subsystems	3
7	Build the subsystems and system	3
8	Test the subsystems and system	4
9	Redesign as needed	4
10	Retest as needed	4
11	Operate the system for its users	4
12	Maintain the system	4
13	Design and build a new system to satisfy new user demands	5

Some highlights of the Vee model are as follows:

- Emphasis on developing and using a systems engineering management plan (SEMP) and ConOps, stakeholder involvement, and validation of requirements and output products.
- The importance of the ConOps to clearly define the system for the system's owner and stakeholders.
- Depicting the relationship between the system requirements and other activities on the left side of the Vee with the attainment of the end products that appear on the right side. The link between the requirements and the end product is a document, usually a verification or testing plan.
- Recognizing the importance of beginning verification planning when requirements are first defined at every level.
- Mandating definition and control of the evolving baseline at each phase of the project through the use of a configuration management plan, risk management plan, and other requirements and validation test documents.
- Including decision gates to ensure that the system's owner and other stakeholders have a say in whether to proceed with the project as it is currently designed.
- Applying interface standards to ensure the smooth flow of data and information from the hardware and software subsystems through the communications media and finally to the operators and users of the system. These standards are typical of those developed through a consensus process by the following organizations in the United States and similar groups elsewhere:
  - AASHTO (American Association of State Highway and Transportation Officials).
  - ANSI (American National Standards Institute).
  - APTA (American Public Transportation Association).
  - ASTM (American Society for Testing and Materials).
  - IEEE (Institute of Electrical and Electronics Engineers).
  - ITE (Institute of Transportation Engineers).
  - NEMA (National Electrical Manufacturers Association).
  - SAE (Society of Automotive Engineers).

A list of National Transportation Communications for Intelligent Transportation System Protocol (NTCIP) standards is found in [2]. Brief descriptions of the Vee model system development phases are presented below [1].

13.1.1 Interfacing with planning agencies and the regional ITS architecture

Key actions that occur during this phase are the identification of the regional stakeholders and the building of consensus for information sharing and long-term operations and maintenance. These are typically accomplished by coordinating the architecture with the region's long-range transportation plan and programming candidate ITS projects with national, statewide, and agency capital plans.

Institutional integration is usually a major component of this system development phase, especially if multiple agencies are involved. Using integrated corridor management (ICM) as an example, institutional partnerships are needed among the operating agencies in the following areas [3]:

- Agreements and memorandums of understanding to establish policies concerning joint operations and information sharing.
- Funding source identification for initial and sustained operations.

- Champion identification.
- Executive buy-in and commitment.
- Organizational structure definition.
- Roles and responsibilities definition.
- Involvement of all transportation modes and stakeholders in the corridor.
- External and internal marketing, outreach, and education.

13.1.2 Concept exploration and benefits analysis

Concept exploration examines the initial feasibility and incorporates a benefits analysis and needs assessment for the candidate projects identified with the regional ITS architecture. Included is the business case and benefit-cost analysis for alternative project concepts. The output of this stage is a definition of the problem space, key technical metrics, and refinements to the needs, goals, objectives, and vision. The highest benefit-to-cost concept is selected and moved forward into development. The purpose of the decision gate shown in the Vee diagram is to gain management support and approval for the project to progress into the planning and definition phases of the project.

13.1.3 Systems engineering planning

Planning takes place in two parts. In part one, the system's owner develops a set of master plans and schedules that identifies what additional plans are needed and, at a high level, the schedule for implementation of the project. This becomes the framework for part two, where the ConOps and the high-level design plans are completed. These documents, once approved by the system's owner, become the control documents for completion of project development and implementation.

13.1.4 Concept of operations

The ConOps provides the initial definition of the system. It describes the problem space, goals and operation of the envisioned system, and how the system will meet the needs and expectations of the stakeholders. System operation is defined from multiple viewpoints consisting of the owner, operators, users, maintenance personnel, and managers. Users consist of anyone who will come into contact with the system whether they are walking, driving, riding, cycling, maintaining, overseeing, or monitoring its operation. Included are the general public (e.g., drivers, transit riders, cyclists, and pedestrians), public workers and administrators, system operating agencies, transit providers, commercial freight operators, and emergency response providers (e.g., city police, state highway police, sheriff, fire departments, and ambulance and paramedic services). The ConOps also specifies how the system will be validated, that is, through procedures or measures of effectiveness (MOEs) that demonstrate how well it meets project objectives. Additionally, it contains the updated, distilled summary of work done at the concept exploration phase.

13.1.5 System-level requirements

System-level requirements include definitions of what the system is to do, how well it is to do it, and under what conditions it will do it. System requirements are based on the user needs from the ConOps. Requirements do not state how the system will be implemented, unless it is intended to constrain the development team to a specific solution. The specific implementation is defined in the design statements that follow later.

13.1.6 High-level design and subsystem requirements

High-level design defines the project-level architecture for the system. System-level requirements are further refined and allocated to the subsystems of hardware, software, databases, and people. Requirements for each subsystem element are documented in the same manner as the system-level requirements. This process is repeated until the system is fully defined and decomposed. Each subsystem has its own set of interfaces defined (these may include hardware, software, and data flow specifications, standards, and protocols) and each requires an integration plan for incorporation into the eventual system. The control gate used for the final review of this stage is referred to as the preliminary design review (PDR).

13.1.7 Component-level detailed design

During component-level detailed design, the development team defines how the system will be built. Each subsystem is decomposed into hardware, software, database elements, firmware, or process components. Design specialists in each of these fields create documentation ("build-to" specifications) that is used to build or procure the individual components. A final check on the build-to specifications allows the design to move into the actual coding and hardware fabrication. At this level, the specific commercial off-the-shelf (COTS) hardware and software products are specified. They are not purchased until the review is completed and approved by the system's owner and stakeholders. The control gate for this final review is the critical design review (CDR).

13.1.8 Hardware and software procurement or development and unit testing

In this stage, the development team is involved with hardware fabrication, software coding, database implementation, and the procurement and configuration of COTS products. The system's owner and stakeholders monitor this process with planned periodic reviews, for example, code walkthroughs and technical review meetings. Concurrent with this effort, unit test procedures are developed that will be used to demonstrate how the products will meet the detailed design. At the completion of this stage, the developed products are ready for unit test.

13.1.9 Unit testing

The hardware and software components are verified in accordance with the unit verification plan during this stage. Unit testing establishes that the delivered components match the documented component-level detailed design. The decision gate serves as a review point for the system's owner and stakeholders.

13.1.10 Subsystem integration and verification

This step integrates and verifies performance of the components at the lowest level of the subsystems. Verification proceeds according to the procedures in the verification plan developed for this stage. Prior to the actual verification, a test readiness review is held to determine if the subsystems are ready for this stage of testing. When the integration and verification are completed, the next level of subsystem is integrated and verified in a similar manner. This process continues until all subsystems are integrated and verified.

13.1.11 System verification

System verification occurs in two parts. The first is performed under a controlled environment, sometimes called a factory test. The second occurs within the environment in which the system is intended to operate after initial system deployment. This part of the verification test is sometimes called on-site testing and verification. At this stage, the system is verified in accordance with the system verification plan. A control gate appears for conditional system acceptance by the owner and other stakeholders.

13.1.12 Initial system deployment

At initial system deployment, the system is finally integrated into its intended operational environment. This step, sometimes called system burn-in, may take several weeks or longer to complete to ensure that the system operates satisfactorily in the long term. Many system issues surface when the system is operating in the real-world environment for an extended period of time. This is due to the uncontrollable nature of inputs to the system, such as long-term memory leaks in software coding (a condition where memory allocations are not managed in a way that releases memory that is no longer needed) and race conditions (those where an output is dependent on the sequence or timing of other uncontrollable events that affect the intended sequence of execution of software actions or introduce a delay into them).

13.1.13 System validation

This key activity for the system's owner and stakeholders assesses the system's performance against the intended needs, goals, and expectations documented in the ConOps and the validation plan. System validation takes place as early as possible after the acceptance of the system in order to assess its strengths, weaknesses, and the opportunities it offers. The activity does not check on the work of the system integrator or the component suppliers, which is the role of system verification. System validation is performed after the system has been accepted and paid for. As a result of validation, new needs and requirements may be identified, which lead to the next evolution of the system.

13.1.14 Operations and maintenance

After initial deployment and system acceptance, the system moves into the operations and maintenance phase. Here, the system performs the intended operations for which it was designed. Routine maintenance and staff training are implemented during this phase. It is the longest phase, usually lasting for decades and extending through the evolution of the system. It ends when the system is retired or replaced. When adequate resources are not allocated to carry out needed operations and maintenance activities, the life of the system is significantly shortened due to neglect.

13.1.15 Changes and upgrades

Changes and upgrades should be implemented in accordance with the technical process found in the *Systems Engineering Guidebook for Intelligent Transportation Systems* [1] in order to maintain system integrity, that is, synchronization between the system components and supporting documentation. When the existing system is not well documented, it is necessary to reverse engineer the affected area of the system to develop the needed documentation for the forward-looking engineering process.

13.1.16 Retirement and replacement

Eventually, every ITS system will be retired or replaced for one or more of the following reasons:

- The system may no longer be needed.
- It may not be cost effective to operate.
- It may no longer be maintainable because of obsolescence of key system elements.
- It might be an interim system that is being replaced by a more permanent system.
- Technical advances render modernization and replacement with a more capable system a necessity.

Therefore, the system owner and operator must know how to monitor system performance, assess needed changes, and formulate change and upgrade decisions.

13.1.17 Crosscutting activities

A number of crosscutting activities are needed to support the development of intelligent transportation systems. The following ones enable one or more of the life-cycle process steps described above.

13.1.17.1 Stakeholder involvement

Stakeholder involvement is one of the most critical enablers within the development and life cycle of the project and system. Without effective stakeholder involvement, the systems engineering and development team will not gain the insight needed to understand the key issues and needs of the system's owner and other stakeholders. This increases the risk of not acquiring a valid set of requirements to build the system or to obtain buy-in on changes and upgrades.

13.1.17.2 Eliciting stakeholder needs

The process of eliciting stakeholder expectations, needs, and requirements (sometimes referred to as elicitation) employs a collection of techniques to extract, clarify, and document stakeholder preferences concerning the objectives and operation of the system. Some information may be in written form or stated clearly by the stakeholders, but much of it may be implied or assumed. The elicitation process helps draw out and refine the information, resolve conflicting information, build consensus, and validate the information. Typical activities during this process are identifying the stakeholders; conducting a literature search; performing day-in-the-life studies, surveys of stakeholder desires, and interviews; conducting workshops; and documenting the results of these activities.

13.1.17.3 Project management

Project management practices promote the various development activities. For example, they support obtaining needed resources, monitoring and controlling costs and schedules, and communicating status between and across the development team members, the system's owner, and stakeholders.

13.1.17.4 Risk management

The risk management process consists of formalizing project risk management, identifying and analyzing potential risks, prioritizing risk items, assigning responsibility to a team member to plan for each risk, and obtaining tools that aid in monitoring and mitigating the risks. Risk management practices result in risk mitigation, avoidance, transference, or acceptance.

13.1.17.5 Project metrics

Project metrics are measures that allow the project manager and systems engineer to track and monitor the project schedule and the technical milestones and performance of the system development effort.

13.1.17.6 Configuration management

Configuration management is the process that administers and documents changes that occur throughout the life cycle of the system. It assists in establishing system integrity by ensuring that the documentation matches the functional and physical attributes of the system throughout its life. Lapses in change and configuration management may shorten the life of the system and prevent it from being implemented and deployed in the first place.

13.1.17.7 Process improvement

Continuous process improvement through learning from previous efforts provides insight into procedures and other items that worked successfully and those that require change and modification.

13.1.17.8 Decision gates

Decision gates are formal decision points along the life cycle that allow the system's owner and stakeholders to determine if the current phase of work satisfies its requirements and is completed and if the team is ready to move on to the next phase of the life cycle. This is accomplished by setting entrance and exit criteria for each gate.

13.1.17.9 Decision support and trade studies

The presentation and evaluation of alternative solutions are needed to optimize the concepts and designs that are part of all phases of system development.

13.1.17.10 Technical reviews

Technical reviews assess the completeness of a product, identify defects in work, and align team members in a common technical direction.

13.1.17.11 Traceability

Traceability is a crosscutting process that supports verification and validation of requirements by ensuring that all needs are traced to requirements and that all requirements are implemented, verified, and validated. Traceability also supports impact analysis for changes, upgrades, and replacement.

13.2 CONOPS AND ARCHITECTURE CREATION USING THE MULTIMODAL INTELLIGENT TRAFFIC SIGNAL SYSTEM AS AN EXEMPLAR

The USDOT has identified 10 high-priority mobility applications under the Dynamic Mobility Applications (DMA) program for the connected vehicle environment. These mobility applications share position, velocity, acceleration, and other pertinent data from vehicles, infrastructure, pedestrians, transit vehicle riders, and so on through wireless communications. Three of the ten DMA applications (Intelligent Traffic Signal System, Transit Signal Priority, and Mobile Accessible Pedestrian Signal System) derive benefits from the transformative Multimodal Intelligent Traffic Signal System (MMITSS). Therefore, it is used to exemplify the process involved in creating a ConOps and system architecture. The MMITSS is a comprehensive traffic signal system that exploits the connected vehicle environment for a variety of transportation modes, including general passenger vehicles, transit, pedestrians, freight vehicles, and emergency vehicles.

MMITSS is part of the USDOT's Cooperative Transportation System Pooled Fund Study (CTS PFS) titled "Program to Support the Development and Deployment of Cooperative Transportation System Applications." CTS PFS members are the actual owners and operators of transportation infrastructure. The users that are part of the stakeholder community for the MMITSS include the following:

- Drivers of passenger vehicles and bicyclists.
- Transit operating agencies and drivers.
- Pedestrians.
- Freight company operators and drivers.
- Emergency response agencies and drivers.

13.3 MMITSS DEVELOPMENT PLAN

The MMITSS has five important functions, namely, intelligent traffic signal system operations for all stakeholders, transit signal priority, pedestrian mobility, freight signal priority, and emergency vehicle priority (EVP) [4]. The MMITSS project is divided into four technical segments that align with the systems engineering process, namely, the following:

1. The first technical stage where the solicitation of stakeholder inputs followed by development of the ConOps, including stakeholder feedback, occurs.
2. The second technical stage where the reviewed stakeholder inputs and ConOps are utilized to develop, define, and populate the MMITSS system requirements.
3. The third stage, which applies the system requirements and prior research to define the MMITSS architecture design. The design and test efforts use the California Test Bed and the Maricopa County Test Bed as the target implementation networks.
4. The final stage that defines implementation, integration, deployment, and test plans based on the design.

A user-oriented operational description is prepared for each user group or stakeholder group so that each may understand the benefits they will receive from the system. The different user groups often have unique needs or desire different services from an intelligent traffic signal control system. Hence, each group will have their own requirements, which may be in conflict with those of other user groups. With that in mind, the system developers must prepare operational descriptions of the traffic signal system as seen by each set of users so that they can understand how the system operates to benefit them.

13.4 MMITSS STRUCTURE

The MMITSS project provides the foundational analysis (stakeholder input solicitation, ConOps, and system requirements) and design (conceptual design, implementation plan, integration plan, and test plans) necessary to develop and field test or demonstrate an MMITSS. Brief descriptions of the arterial traffic signal control system structure that supports the applications identified through the DMA template are provided below [4].

13.4.1 Intelligent traffic signal system structure

Data collected from vehicles through wireless communications will facilitate accurate measurements and predictions of lane-specific platoon flow, platoon size, and other driving characteristics. Real-time data availability has the potential to transform the design, implementation, operation, and monitoring of traffic signal systems. Furthermore, systems that collect data via V2V and V2I wireless communications to control signals to maximize flows in real time can improve traffic conditions significantly.

When integrated into an overarching system optimization application, such a system can accommodate transit or freight signal priority, preemption, and pedestrian movements to maximize overall arterial network performance. In addition, the system design should incorporate the effects of traffic flow between arterial signals and ramp meters (i.e., traffic signals installed on freeway on-ramps).

13.4.2 Transit signal priority structure

Providing reliable transit service is an important transportation system goal and is one that makes transit an attractive travel mode. Traffic signal timing can impede service by contributing to the delay of buses and light rail. Transit signal priority (TSP) strategies adjust signal timing at intersections to better accommodate transit vehicles that are behind schedule or, if running behind the planned headway, resume their schedules.

Connected vehicle technologies offer additional opportunities to enhance current TSP systems by (1) providing more accurate estimates of prevailing traffic conditions at signalized intersections by integrating conventional loop detector data and wireless data, (2) allowing earlier detection and continuous monitoring of transit vehicles as they approach and progress through intersections, and (3) supporting additional intelligent priority strategies based on trade-off analyses between traffic and transit delay at network intersections. In a connected vehicle environment, transit vehicles can transmit data characterizing the need for priority (i.e., the level of priority) to the roadside infrastructure. This facilitates the provision of differential priority that grants varying levels of priority to multiple transit vehicles. The priority level depends on a number of factors, including prevailing traffic conditions, current status of the traffic signal controller, and the status of the transit vehicle.

13.4.3 Pedestrian mobility structure

MMITSS will facilitate pedestrian mobility at intersections by meeting a pedestrian's special needs or by balancing utilization of the intersection by vehicles and pedestrians. This application integrates traffic and pedestrian information from roadside or intersection sensors and new forms of data from wirelessly connected, pedestrian-carried mobile devices (nomadic devices) to request dynamic pedestrian signal timing or to inform pedestrians when to cross and how to remain aligned with the crosswalk based on real-time signal phase and timing (SPaT) and message access profile (MAP) information [5]. In some cases, priority is given to pedestrians, such as persons with disabilities that need additional crossing time, or under special conditions (e.g., weather or special events) when pedestrians may warrant priority or additional crossing time.

Pedestrian calls can be routed to the traffic signal controller from the nomadic device of a registered person with disabilities after confirming the direction and orientation of the roadway this person desires to cross. Pedestrian crosswalks are managed by the intelligent traffic control system to accommodate certain predetermined conditions that improve the efficiency of intersection utilization, or avoid overcrowding pedestrians at curbs in large downtown areas or at special events, such as sports or concerts.

13.4.4 Freight signal priority structure

The use of public roadways by freight vehicles imposes greater interactions with and requirements on transportation operating agencies, such as street maintenance due to increased pavement wear caused by transporting heavy loads and congestion mitigation due to double-parked trucks when parking is not available for scheduled deliveries. In a connected vehicle environment, signal priority techniques for transit can be applied to freight vehicles to grant right-of-way over general traffic. Priority strategies for freight can consider the special operating characteristics associated with freight vehicles. For example, freight vehicles require greater stopping distance than passenger cars and the severity of accidents is greater when they are unable to stop at red signals. After stopping, additional time and fuel are required to resume nominal travel speeds due to vehicle dynamics, which can impose delays to surrounding vehicles. The goals of freight signal priority include reduced stops, reduced delays, and increased travel-time reliability for freight vehicles, which can reduce negative environmental impacts and pavement damage, and enhance safety at intersections. An integrated framework can be utilized to respond to priority requests from freight vehicles to better accommodate the collective needs of multimodal travelers.

13.4.5 EVP structure

EVP provides a high level of precedence for emergency first-responder vehicles (fire, ambulance, police, sheriff, highway patrol, paramedics, react personnel, and National Guard and Border Patrol agents). Historically, priority for emergency vehicles was granted by special traffic signal timing strategies called preemption. The goal of EVP is to facilitate safe and efficient movement through intersections. As such, clearing queues and holding conflicting phases can facilitate emergency vehicle movement. For congested conditions, it may take additional time to clear a standing queue, so the ability to provide information in a timely fashion is important. In addition, a plan for transitioning back to normal traffic signal operations after providing EVP is required since the control objectives are significantly different.

13.5 OPERATIONAL EXPECTATIONS

The first technical stage, as described in Section 13.3, begins by identifying the MMITSS users and culminates in the creation of the ConOps. Users include the general public, public workers and administrators, traffic signal system operating agencies, transit providers, emergency response agencies, and commercial freight operators. These transportation users, whether in an urban or rural area, often provide information to the system through cell phones, smart phones, mobile GPS devices, OnStar, and related devices or services. It seems reasonable for the public to anticipate that traffic signal control has benefitted from these and other technological advancements and they will demand that the system respond to situations such as those described below.

13.5.1 Driving public operational expectations

The driving public wants traffic signals that are aware of the following:

- Cars waiting for the traffic signal to turn green, while there is not another vehicle or pedestrian in sight.
- Icy roadways making it harder to stop at the yellow light.
- Dilemma zone protection.
- Majority of traffic leaving a sports venue traveling on specific arterial routes in order to reach a controlled access highway.
- Other opportunities for improved performance.

13.5.2 Transit rider operational expectations

The transit rider interacts with the signalized intersection in ways that offer opportunities for enhancement that include

- Information on the expected time of arrival of the next bus to prevent stepping out into traffic to view if the bus is on the way.
- Updated information on a connection while sitting on a moving transit vehicle either through a display in the vehicle or a smart phone application.
- Bus scheduling based on realistic, observable, and seasonal data.
- Other desirable improvements.

13.5.3 Walking or cycling public operational expectations

The walking or cycling public wants traffic signals that are aware of the following:

- Utilization pattern differences for crosswalks near major universities or in downtown metropolitan areas from those in rural areas.
- Decreasing or changing mobility and eyesight in an aging population that requires additional crossing time and assistance.
- The increasing price of fuel causing more people to rely on alternative transportation such as walking, bicycles, and transit.
- Other opportunities for assisting the nonmotorized traveler.

13.5.4 Freight operator operational expectations

From a direct user perspective, freight operators and supporting fleet management system operators desire traffic signals that are attentive to

- Consequences of idling at a traffic signal or series of traffic signals that influences the cost of goods, diminishes local air quality, impacts pavement lifespan due to acceleration and deceleration of loaded vehicles, and contributes to increased engine and exhaust heat and noise.

- Technological advances that allow the status of the signalized intersections and roadway sections to be reported to the freight dispatch center or fleet management system for routing and rerouting decisions.
- Minimizing delay of large freight vehicles at intersections that potentially form queues of other types of vehicles, thereby compounding delay and decreasing freight vehicle maneuverability due to obstacles.
- Cargo that is heterogeneous (e.g., perishable, express, hazmat), which affects the transportation objective beyond getting from an origin to a destination.
- Other desirable improvements.

13.5.5 First-responder operational expectations

First responders interact with the traffic signal system in more complex ways than freight, transit, or passenger vehicles. They want an intelligent traffic signal system to be aware that

- Costs associated with traffic delays are greater than loss of productivity or inconvenience—the costs can be measured in loss of life or limb if arrival at an incident is too late to render help.
- Since most emergencies are multifaceted, diverse emergency vehicles will approach the incident site from nearly all directions, resulting in safety concerns when transitioning through nearby intersections even after priority has been granted.
- As the demands on the transportation infrastructure exceed design capacity, first responders need to rely on an intelligent system to alleviate queues and congestion, thus permitting the maneuvering of emergency vehicles around traffic.
- While traveling to an incident, first responders need to depend on effective dispatch operations that could benefit from real-time status information from nearby roadside equipment (RSE).
- Other enhanced interactions with signalized intersections.

13.6 THE CONOPS NARRATIVE

Once the needs of the stakeholders are understood by the system developers, the ConOps, in this case for a transportation network with significant commuter traffic, can be prepared in its narrative form to describe how the system is designed to satisfy the desires of each user group. The ConOps descriptions given below are representative of the full ConOps that is found in the *MMITSS Final Concept of Operations* [4].

The first part of the ConOps establishes that the MMITSS may operate differently in different areas, depending on the mix of users in the area. It states

> Traffic signal priorities may differ in different sections of the network. For example, one section may be operated as a coordinated system with a long cycle length to accommodate the heavy flow of trucks, giving them priority treatment when needed. Transit vehicles will receive priority, but their priority level will be lower than trucks.
>
> In another section, transit receives higher priority than other vehicles, while pedestrians are provided priority when not conflicting with higher priority demands. Within the class of transit vehicles, school buses are determined to have the highest level of priority (e.g., 4 on a scale of 1 to 10, 1 being highest, where railroad crossings are assigned a 1, emergency vehicles are assigned 2–3, transit vehicles are assigned levels 4–6, with trucks next). Transit vehicles are required to determine their eligibility for priority based on their schedule lateness.

This portion of the ConOps describes how the MMITSS operates to the benefit of transit vehicles.

To provide a high quality of service, the transportation management agency has established a policy whereby the network is pedestrian and transit friendly, but the agency is aware of the volume of commuter vehicles in the mornings and the evenings. In addition to regular schedule-based transit service, there is a large volume of school buses that ferry children to and from school each morning and afternoon.

Since there may be multiple transit vehicles in the network section at any time, the transportation management agency has established a policy of reducing the total delay to the collection of transit vehicles at any intersection. Transit vehicles are required to determine their eligibility for priority based on their schedule lateness. Transit vehicles that are more than 3 min behind schedule (20% late assuming a 15-min scheduled headway) can receive priority (e.g., a request at a level of 6). Transit vehicles that are more than 50% full of passengers are allowed to request a higher level of priority than just late vehicles (e.g., a request at a level of 5). Due to random boarding and alighting times, it is difficult to provide route-based priority for transit vehicles, but downstream signals can be "priority aware" of vehicles that may be one or two cycles away and prepare by serving nonpriority phases to ensure vehicles that are waiting receive minimal delay.

This piece of the ConOps explains how the MMITSS operates to provide additional benefits for pedestrians, transit vehicles, and passenger vehicles. The system is also designed to accommodate the extra traffic volume produced by special events.

Since this network is transit and pedestrian friendly, the signals are coordinated but at a relatively short cycle length based primarily on vehicle volumes and pedestrian crossing times. A time-of-day plan control strategy is used to provide different cycle lengths depending on the different mix of the different modes of travelers. Longer cycle times may be required before and after school due to the volume of pedestrians near the school and school bus stops. Shorter cycles may by feasible during off-peak times when passenger vehicles are the predominant mode. The morning and evening commuter traffic may cause congestion and require special consideration and require limiting the level of priority control for transit while oversaturated conditions are managed.

In the off-peak operational periods or during lower volume periods, the signals may operate in free mode (non-coordinated) with intelligent phase actuation (phase call and gap out logic). This will provide a high quality of service to the vehicles that are present in the network. The network and individual signals may self-determine when to coordinate and when to operate in a free mode based on the observed vehicle tracks at the approach to the intersections and performance measures compiled by the traffic signal system. If the signals are operating in a free mode and there is a significant volume of vehicles arriving randomly over time (at an interior signal), then it may be beneficial to change to a coordinated plan and platoon vehicles so they can progress together through the network. Similarly, if the signals are operating in a coordinated plan and there are very few vehicles arriving during the coordinated phase green time, it may be beneficial to drop to a free mode.

Special events are likely in a network section where there are schools, for example, sporting events, open school parent–teacher events, concerts, and plays. Extra transit service may be provided for these occasions and traffic volumes may vary significantly.

Special transit vehicles may be provided a high level of priority (e.g., 5 of 10). Vehicle volumes may result in long queues that could block intersections and short-term congestion may occur. The traffic signal system will provide mitigation for these special congestion circumstances.

This portion of the ConOps portrays how the MMITSS operates in the event of an emergency in the network and states that performance measures will be evaluated to ensure the network is functioning to meet the needs of all its users.

In the event of an emergency in the network, emergency vehicles will receive the highest priority consideration and will override trucks and transit vehicles. When the number of transit vehicles and pedestrians in the network is significant, route-based EVP will be provided to assist in the clearance of pedestrians (e.g., long clearance times). Since the route of the emergency vehicle may not be known, the route assumptions may extend only one or two signals downstream, but this will help reduce queuing in the emergency vehicle's path.

The presence of multiple emergency vehicles will be accounted for by reducing the total delay to all active emergency vehicles at an intersection. Each intersection will recover from EVP by considering the maximum delay of any single vehicle at the intersection.

Performance measures will be continuously collected to characterize the operational effectiveness of the signals in the network. The goals of being transit and pedestrian friendly will be supported by measures that include transit vehicle delay for those vehicles that request priority (e.g., late vehicles) and pedestrian delay (time from when the pedestrian requests service—either by pedestrian detection or nomadic device).

13.7 OPERATIONAL SCENARIO FOR A TRANSPORTATION NETWORK WITH SIGNIFICANT COMMUTER TRAFFIC

In conformance with the ConOps and the systems engineering process, the system designer next develops operational scenarios for different network conditions and demands that are served by the MMITSS. This process is part of the second technical stage of the system development. The scenario below is for a road network with significant commuter traffic. It includes several transit routes and pedestrian activity, and serves morning and evening commuters and daily travelers going to and from school, shopping, and other events. The scenarios addressed by the intelligent traffic signal system are the following:

- Basic signal actuation.
- Coordinated section of signals.
- Congestion control.
- Dilemma zone protection.
- Utilization and performance measures that adjust and adapt signal timing to improve operations. These are collected and updated as part of the equipped vehicle scenarios. Equipped vehicles are those that have some type of onboard equipment (OBE) or nomadic device that is connected-vehicle or MMITSS aware and can operate as part of the traffic signal control system. Equipped nonmotorized travelers include pedestrians, bicyclists, and other modes such as equestrians that are not required to be licensed to operate on the public roadway and are in possession of a nomadic device.

13.7.1 Basic signal actuation

Basic signal actuation provides services for single unequipped, single equipped, and multiple unequipped and equipped vehicles. This scenario utilizes features from other scenarios that support the accommodation of multiple vehicles of unequipped, equipped, or mixed configurations.

13.7.2 Coordinated section of signals

A coordinated section of signals provides green bands or progression bands for safe and efficient movement of a group (platoon) of vehicles through a section of traffic signals. A coordinated section is five to seven signals, but this depends on the geometry (e.g., signal spacing). A good offset between signals will allow the platoon to progress without delay, while a poor offset causes delay and stops. Queue size is random; however, the time to clear the queue should be sufficient, but not too long, for minimal impact to the platoon.

Adjustment of coordination parameters, primarily offset, is based on performance measures related to platoon progression in the desired direction of the green band. Adjustment of cycle length and phase splits is related to phase failure (e.g., the failure to completely discharge a queue during a green service interval) or excessive phase time that results in early return to green for the coordinated phase and inefficient use of green time. Accurate estimation of these performance measures can be significantly improved using probe data from equipped vehicles. Vehicle trajectories allow accurate assessment of offset and true phase failure estimation.

13.7.3 Congestion control

From a stakeholder or user perspective, congestion is perceived to exist when large groupings of vehicles sit idling through multiple signal cycles with little progress or relief. This situation is repeated at an adjacent signalized intersection or several neighboring intersections.

Congestion can be characterized by its duration (amount of time that one or more intersections have persistent phase failures) and its extent (distance in space where the intersections are congested). Continued failure of a phase over many cycles can cause queues to spillback to upstream intersections resulting in network-wide congestion. Traditional traffic control systems can estimate phase failures by considering stop bar sensor occupancy at the beginning and end of the green service phase, but these systems cannot distinguish between situations where newly arriving vehicles are stopped or when vehicles are stopped for two or more cycles. Connected vehicle data provide the opportunity to accurately estimate phase failures and the persistence of congested conditions.

13.7.4 Dilemma zone protection

A dilemma zone occurs when a vehicle on a high-speed approach cannot stop safely when the traffic signal changes from green to yellow. Dilemma zones can complicate traffic control when coupled with inclement weather, such as rain, sleet, ice, snow, and dust, or when they involve heavy or large vehicles, such as loaded freight vehicles, tanker vehicles, and wide loads.

In a connected vehicle environment, basic signal actuation as used for single and multiple vehicles usually manages the dilemma zone situation. However, the same condition exists when the phase reaches the maximum green time, except the controller can decide to terminate the phase early (rather than start the first extension timer) since it can track the approaching vehicle over a sufficiently long distance.

13.8 MMITSS ARCHITECTURE

The MMITSS architecture, defined as part of the third technical development stage, depicts the infrastructure, vehicle, and pedestrian-carried components and their interconnections that are utilized to provide the functionality of the system. Several of the National ITS Architectures described in Chapter 14 also include the communications media that transmit the data.

Two types of travelers, namely, motorized vehicles and nonmotorized travelers, are accommodated by the basic MMITSS system architecture displayed in Figure 13.2. Motorized vehicles consist of passenger vehicles, trucks, transit vehicles, emergency vehicles, and motorcycles. This type of traveler includes any vehicle that must be licensed to operate on the public roadway. Nonmotorized travelers, those not requiring a license, are either unequipped or equipped.

Nodes or connection points depicted in 3D boxes and ovals in the top portion of the figure are part of the connected vehicle, traffic management, and fleet management systems (or nodes that can be modified or assigned MMITSS responsibilities). Nodes toward the bottom of the figure in framed boxes represent the vehicles, travelers, and field detection system. MAP, in the upper right portion of the figure, contains the digital description of the intersection geometry and associated traffic control definitions used by devices that exchange message objects. It is based on a client–server interaction model where the client initiates the

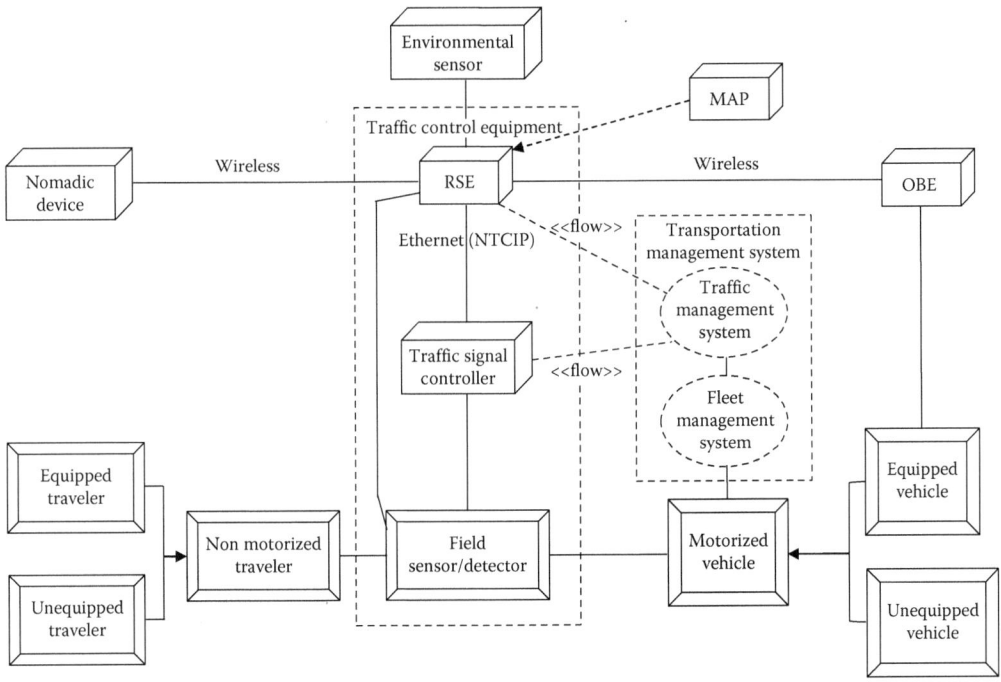

Figure 13.2 MMITSS basic conceptual system architecture. Nodes depicted in 3D boxes and ovals in the top portion of the figure are part of connected vehicle, traffic management, and fleet management systems, while nodes toward the bottom of the figure in framed boxes represent vehicles and travelers. (Adapted from *MMITSS Final Concept of Operations*, University of Arizona (Lead), University of California PATH Program, Savari Networks, Inc., SCSC, Econolite, Volvo Technology, Version 3.1, University of Arizona, Tucson, AZ, December 4, 2012. http://www.cts.virginia.edu/wp-content/uploads/2014/05/Task2.3._CONOPS_6_Final_Revised.pdf.)

transactions such as in Bluetooth® communications. Systems that are active participants in the MMITSS (e.g., connected vehicle, traffic management, and fleet management) can have different responsibilities, and in alternative system designs some of these responsibilities can be assigned to different components.

Both motorized and nonmotorized travelers can be detected by the Field Sensor/Detector node at the intersections using a variety of detection technologies, including inductive loop detectors, magnetometers, video detection systems, Doppler microwave sensors, presence-detecting radar sensors, passive infrared sensors, and pedestrian push button. The detection system at an intersection provides information to the traffic signal controller that stimulates the control algorithms. For example, a vehicle that triggers a traffic flow sensor will call a signal control phase for service or extension. A pedestrian may activate a pedestrian push button to request the traffic signal pedestrian interval associated with a crosswalk movement.

Figure 13.3 illustrates a simple two-intersection section of a signalized transportation network with both unequipped and equipped travelers using the different travel modes listed at the bottom, center of the figure. The operational environment is constrained by the physical, technical, and institutional policies that govern the control of systems that span multiple travel modes. The MMITSS requires a great many sensors at each intersection and elsewhere to be effective. While the need for these sensors affects the cost of the system, they do provide a data-rich and spatially and temporally dynamic data environment.

The three MMITSS operational scenarios discussed in the following sections provide additional insight into how the system functions for equipped travelers [4]. These scenarios are single and multiple-equipped vehicles service, basic transit signal priority, and dilemma zone protection.

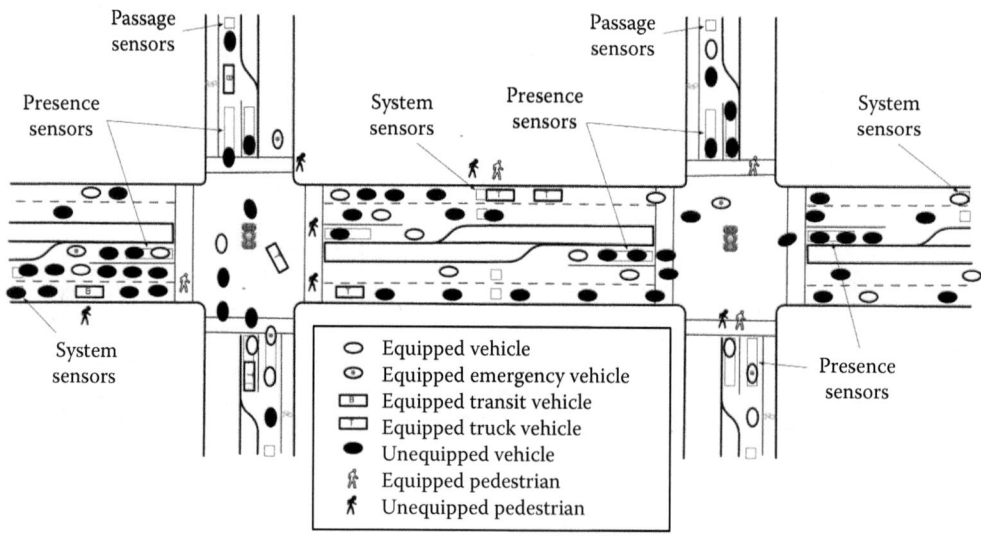

Figure 13.3 Two-intersection section of a signalized transportation network with both unequipped and equipped travelers from different travel modes. (Adapted from *MMITSS Final Concept of Operations*, University of Arizona (Lead), University of California PATH Program, Savari Networks, Inc., SCSC, Econolite, Volvo Technology, Version 3.1, University of Arizona, Tucson, AZ, December 4, 2012. http://www.cts.virginia.edu/wp-content/uploads/2014/05/Task2.3._CONOPS_6_Final_Revised.pdf.)

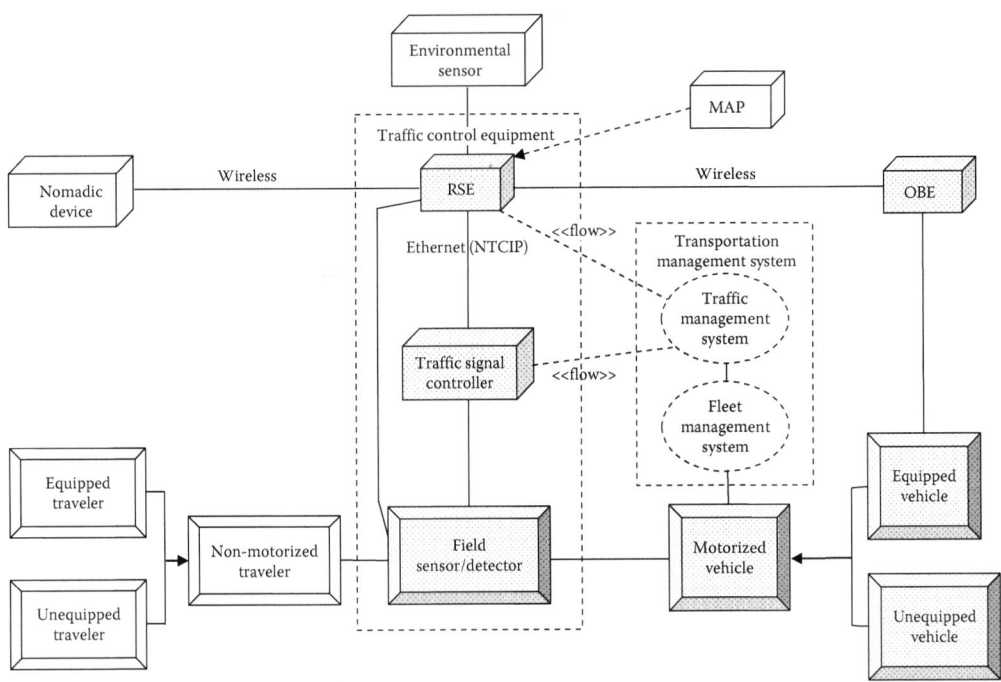

Figure 13.4 Active architecture nodes (darker color) for single and multiple-equipped vehicle signal actuation. (Adapted from *MMITSS Final Concept of Operations*, University of Arizona (Lead), University of California PATH Program, Savari Networks, Inc., SCSC, Econolite, Volvo Technology, Version 3.1, University of Arizona, Tucson, AZ, December 4, 2012. http://www.cts.virginia.edu/wp-content/uploads/2014/05/Task2.3._CONOPS_6_Final_Revised.pdf.)

13.8.1 Single and multiple-equipped vehicles service

Figure 13.4 depicts how single and multiple-equipped vehicles are accommodated by the MMITSS. The active architecture nodes for this application appear in a darker color. To understand how the system serves these vehicles, assume a single vehicle trajectory where the vehicle is traveling 25 mi/h (36.7 ft/s). When this vehicle reaches communications range (e.g., DSRC at approximately 300 m or 984 ft), the RSE begins to receive basic safety messages (BSMs) from the vehicle. The RSE calls the desired service phase, but the signal continues to serve other phases that have active calls. As the vehicle reaches the extension sensor, the RSE notes that the detection event was generated by an equipped vehicle, and instead of resetting the gap timer, it places a hold on the phase that has recently changed to green. The hold is maintained until the vehicle crosses the stop bar 6.8 s later. This is 1.8 s after the signal would have gapped out under normal extension operations. Comparing the scenarios of the equipped and unequipped vehicles, the equipped vehicle would have been served by the green, whereas an unequipped vehicle would have been stopped, or could have entered the intersection during the clearance interval. This scenario is relevant to connected vehicles configured by original equipment manufacturers (OEMs) or retrofitted with compatible OBE.

Consider now several equipped vehicles approaching an intersection from conflicting directions. This scenario includes the single signal actuation scenarios where each equipped vehicle or unequipped vehicle will call the signal control phases so that they are served in the order determined by the programming of the traffic signal controller. The equipped vehicles are treated in a similar fashion, except that the controller logic will assess the likelihood that a vehicle will arrive at the stop bar before the phase will terminate due to the phase max-out time or a coordinator force-off point. If the vehicle will not arrive in time, then the signal should be allowed to terminate early to maintain efficiency of the intersection.

13.8.2 Basic transit signal priority

Basic transit signal priority scenarios address transit vehicles approaching an equipped intersection. In Figure 13.5, the active nodes are again shown in a darker color. The operation of the system assumes that transit vehicles communicate TSP requests with the immediate downstream intersection and the TSP decisions are granted locally by the intersection. Each vehicle continuously monitors its schedule, headway adherence, and passenger loads to determine whether there is a need to request signal priority. When conditions are met, the vehicle sends a priority request in the form of a signal request message (SRM) to the roadside. The priority request includes the level of priority assigned to the vehicle according to the established priority policy. While approaching the intersection, the vehicle will periodically send location updated messages (i.e., BSMs) to the roadside until it clears the intersection.

The RSE processes SRMs and BSMs sent from the transit vehicles and determines the most appropriate priority control strategy based on a number of factors, including the prevailing traffic condition and the requested level of priority. In urban areas, requests for priority may occur on conflicting approaches of an intersection. The priority timing routine

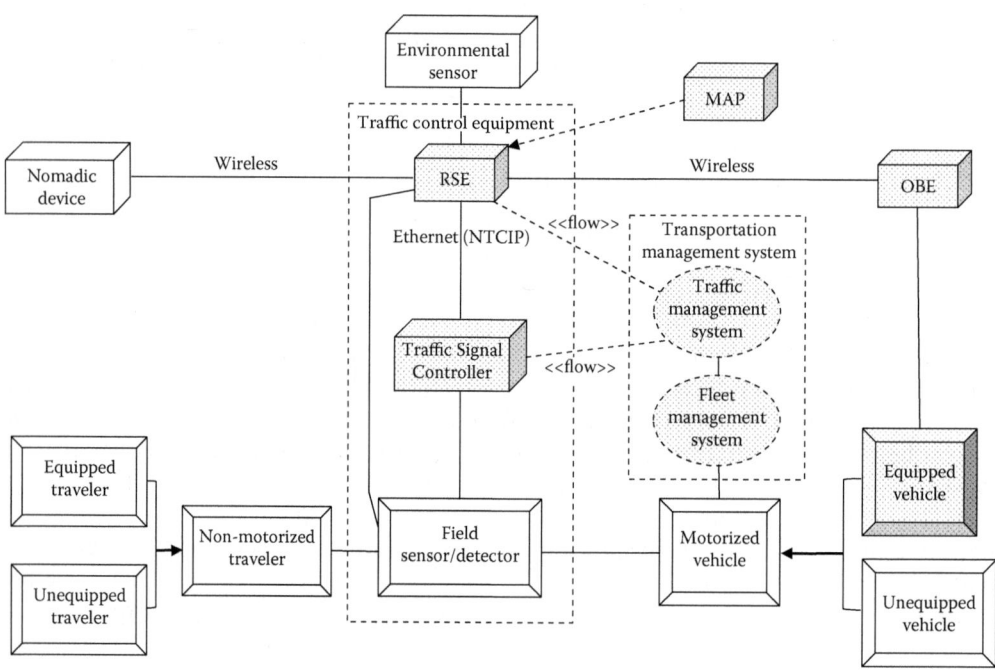

Figure 13.5 Active architecture nodes (darker color) for transit signal priority, freight signal priority, and emergency vehicle priority. (Adapted from *MMITSS Final Concept of Operations*, University of Arizona (Lead), University of California PATH Program, Savari Networks, Inc., SCSC, Econolite, Volvo Technology, Version 3.1, University of Arizona, Tucson, AZ, December 4, 2012. http://www.cts.virginia.edu/wp-content/uploads/2014/05/Task2.3._CONOPS_6_Final_Revised.pdf.)

has the intelligence to apply sophisticated strategies rather than "first called, first served" for conflicting requests. The RSE continues to monitor changes in signal status and transit vehicle location so that adjustments can be made as needed. Modifications occur for transit vehicles at nearside bus stops, left turns with protected signals, rail crossings, transit rail vehicles, and certain transit routes.

13.8.3 Dilemma zone protection

Recent approaches to dilemma zone control utilize two fixed-location sensors to protect vehicles [4]. The first sensor starts an extension (gap) timer based on the travel time from the first sensor location to the second. If the vehicle arrives at the second sensor before the extension timer reaches 0.0, then the second sensor reinitializes the extension timer. The sensor locations and extension timer values are designed so that the vehicle is allowed to cross the stop bar safely. If the vehicle is not traveling fast enough to reach the second sensor before the extension timer reaches 0.0, then the vehicle should be able to stop in time. The exception to the process is when the phase reaches the maximum time and is forced to terminate (advance to yellow) regardless of the status of the approaching vehicle(s). Advance warning flashers can be installed at a sufficient stopping distance upstream of the signal. These warning flashers start at a predefined interval before the signal reaches the maximum time (or termination point).

In a connected vehicle environment, the basic signal actuation scenarios (single and multiple vehicles) manage the dilemma zone situation. Although the same condition exists when the phase reaches the maximum green time, now the controller can decide to terminate the phase early (rather than start the first extension timer) since it can track the approaching vehicle over a sufficiently long distance. If one or more equipped vehicles are approaching the signal and the controller has decided to extend the green interval for these vehicles and a new vehicle approaches that will not reach the stop bar before the start of the yellow interval, this new vehicle can be in the same dilemma zone situation. In addition, the equipped vehicle characteristics (type, length, weight, etc.) can be used to determine the safe stopping distance and evaluate the extension or termination decision. For example, a large truck could have more difficulty stopping than a small passenger vehicle. The infrastructure-based warning flashers can be used to warn the vehicle, and a warning message can be sent to the specific vehicle at risk.

The active nodes for dilemma zone protection for equipped vehicles are indicated in Figure 13.6 by their darker color. Dilemma zone protection functions as an extension of basic signal actuation by incorporating special considerations such as (1) a pair of dilemma zone sensors on the approach to the intersection spaced such that vehicles between the first and second sensor could stop if the signal changed to yellow and (2) adjusting the extension timer in the traffic signal controller to be long enough to allow a vehicle to safely cross the stop bar after exiting the second (downstream) sensor.

Dilemma zone protection for equipped vehicles operates as follows:

1. It begins when any one of the equipped vehicles enters the radio range of the RSE.
2. These nine steps occur for each vehicle that approaches the intersection:
   a. The OBE receives MAP and SPaT messages from the RSE.
   b. The RSE receives BSMs from the OBE.
   c. The RSE tracks the OBE to estimate when the vehicle will arrive and want to cross the intersection stop bar (route information is not assumed to be available).
   d. The RSE estimates the required stopping time or distance based on the vehicle characteristics.

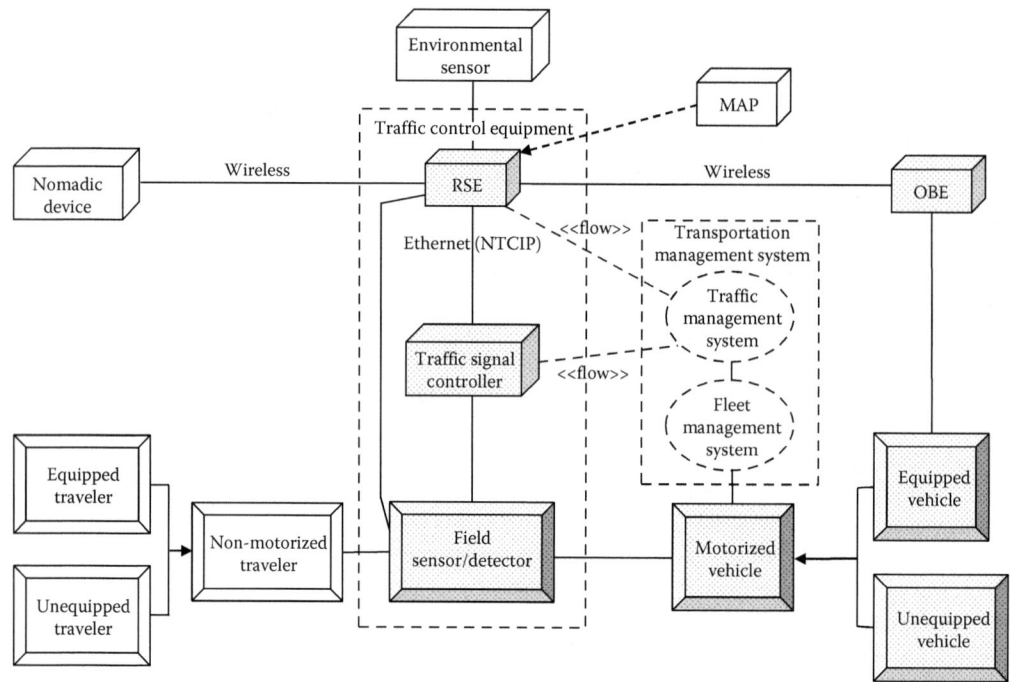

Figure 13.6 Active architecture nodes (darker color) for dilemma zone protection. (Adapted from *MMITSS Final Concept of Operations*, University of Arizona (Lead), University of California PATH Program, Savari Networks, Inc., SCSC, Econolite, Volvo Technology, Version 3.1, University of Arizona, Tucson, AZ, December 4, 2012. http://www.cts.virginia.edu/wp-content/uploads/2014/05/Task2.3._CONOPS_6_Final_Revised.pdf.)

    e. The RSE determines the appropriate traffic signal phase to serve the vehicle by translating BSM data into a phase request.

    f. The RSE matches the sensor call to the OBE location and prevents the gap timer from timing the detection event.

    g. If the service phase is not timing (not active), the RSE places a call for the phase based on when the vehicle will arrive and the phase max time.

    h. If the service phase is timing, the RSE holds the phase green until the vehicle crosses the stop bar or the phase max time or coordination force-off point is reached, unless the vehicle will not be able to reach the stop bar before the maximum time is reached, in which case the phase will be allowed to terminate early to maintain intersection efficiency.

    i. If the vehicle will not reach the stop bar before the maximum time occurs, the infrastructure-based warning flashers are set to an on-state and a warning message is transmitted to the vehicle.

3. The RSE updates the performance measures of the vehicles served.

4. This instance of dilemma zone protection ends.

13.9 PERFORMANCE MEASURES

Performance measures allow the system managers to evaluate if the multimodal traffic signal system is working as desired and if it is providing the expected benefits. Different measures of performance (MOPs) or MOEs are applied to monitor different functions. Thus, appropriate MOEs must be selected for each function in an intelligent transportation system.

The performance measures shown in Tables 13.2 through 13.7 apply to a multimodal traffic signal system that supports intelligent traffic signal operations, transit signal priority, pedestrian mobility, freight signal priority, and emergency vehicle signal priority [4]. There are also crosscutting measures that apply across all user functions. MOEs other than these are used to evaluate limited-access highway performance [6–8].

Table 13.2 Performance measures to evaluate intelligent traffic signal operations

Performance measures	
Overall Vehicle Delay (All Day)	Extent of Congestion (Peak Period)
Overall Vehicle Delay (Peak Period)	Temporal Duration of Congestion (All Day)
Number of Stops (All Day)	Temporal Duration of Congestion (Peak Period)
Number of Stops (Peak Period)	Arterial Total Travel Time (All Day)
Throughput (All Day)	Arterial Total Travel Time (Peak Period)
Throughput (Peak Period)	Arterial Travel Time Variability (All Day)
Maximum Queue Length (All Day)	Arterial Travel Time Variability (Peak Period)
Maximum Queue Length (Peak Period)	Availability of Signal System Health Monitoring State (All Day)
Extent (spatial range) of Congestion (All Day)	Extent of Congestion (Peak Period)

Table 13.3 Performance measures to evaluate transit signal priority

Performance measures
Average Transit Delay (All Day)
Average Transit Delay (Peak Period)
Transit Delay Variability (All Day)
Transit Delay Variability (Peak Period)

Table 13.4 Performance measures to evaluate pedestrian mobility

Performance measures
Overall Pedestrian Delay (All Day)
Overall Pedestrian Delay (Peak Period)

Table 13.5 Performance measures to evaluate freight signal priority

Performance measures
Overall Truck Delay (All Day)
Overall Truck Delay (Peak Period)
Freight/Goods Reliability (Peak Period)
Freight-Intersection Accident Rates
Dilemma Zone Incursions by Trucks
Truck Stops at Signalized Intersections (All Day)
Truck Stops at Signalized Intersections (Peak Period)

Table 13.6 Performance measures to evaluate EVP

Performance measures
Overall EV Delay (All Day)
Overall EV Delay (Peak Period)
EV Delay Variability (All Day)
EV Delay Variability (Peak Period)
EV Response Time (All Day)
EV Response Time (Peak Period)
EV Accidents/Incidents Ratio at Intersections

Note: EV, emergency vehicle.

Table 13.7 Crosscutting performance measures

Performance measures
Data Availability and Usability
System Data Security and Information Assurance
System Reliability
System Availability
System Interoperability
System Mean Time to Repair
Synchronized Time Source Availability
Availability of System Performance Measures

13.10 SUMMARY

Systems engineering encompasses the entire life cycle of a system. It focuses on understanding and defining stakeholder needs and functionality early in the development cycle, documenting requirements and system design options, obtaining stakeholder buy-in for the proposed system design through creation of a ConOps, and only then proceeding with the subsystem and component-level design and build, and system validation. Systems engineering processes and techniques support systems thinking (e.g., consideration of all aspects of a system's design including requirements; hardware specifications; technology selection; software coding, testing, and verification; interface control; and life-cycle costs), which is critical on all ITS projects. An important area of system design is interface and standards specification and definition to ensure that components and subsystems integrate properly and data and information get transmitted over all the required communications channels. As systems engineering becomes incorporated into transportation project development, it provides another set of tools to improve the effectiveness and lower the cost of developing transportation facilities. As a result, expertise in ITS development and management is broadened through the creation of a pool of human and physical resources that can support future ITS projects.

Systems engineering is applicable to all sizes and complexities of projects. The degree of formality and rigor applied to the systems engineering process can be tailored to the complexity of the project. Projects can include more formality when the projects are complex or less when they are simpler.

REFERENCES

1. *Systems Engineering Guidebook for Intelligent Transportation Systems*, Ver. 3.0, U.S. Department of Transportation, Federal Highway Administration, California Division and California Department of Transportation, November 2009. http://www.fhwa.dot.gov/cadiv/segb/. Accessed November 9, 2015.
2. *National Transportation Communications for Intelligent Transportation System Protocol (NTCIP) Standards*, http://www.standards.its.dot.gov/DevelopmentActivities/Published Standards. Accessed April 28, 2015.
3. J.N. Spiller, N. Compin, A. Reshadi, B. Umfleet, T. Westhuis, K. Miller, and A. Sadegh, *Advances in Strategies for Implementing Integrated Corridor Management (ICM)*, Scan Team Report, NCHRP Project 20-68A, Scan 12-02, October 2014. http://www.domesticscan.org/wp-content/uploads/NCHRP20-68A_12-02.pdf. Accessed November 17, 2015.
4. *MMITSS Final Concept of Operations*, University of Arizona (Lead), University of California PATH Program, Savari Networks, Inc., SCSC, Econolite, Volvo Technology, Version 3.1, University of Arizona, Tucson, AZ, December 4, 2012. http://www.cts.virginia.edu/wp-content/uploads/2014/05/Task2.3._CONOPS_6_Final_Revised.pdf. Accessed November 9, 2015.
5. *Message Access Profile (MAP) Bluetooth Specification*, Document No 2009-06-04, Revision V10r00, prepared by Car Working Group, Car-feedback@bluetooth.org.
6. L.A. Klein, *Sensor Technologies and Data Requirements for ITS*, Artech House, Norwood, MA, June 2001.
7. T. Shaw, K. Fisher, Lisa Klein, M.C. Larson, H. Lieu, G. Morgan, V. Pearce, J.L. Powell, J. Skinner, and W. Walsek, *Performance Measures of Operational Effectiveness for Highway Segments and Systems, A Synthesis of Highway Practice, NCHRP Synthesis 311*, Transportation Research Board, Washington, DC, 2003. www.trb.org/publications/nchrp/nchrp_syn_311.pdf. Accessed August 8, 2016.
8. R.E. Brydia, W.H. Schneider, S.P. Mattingly, M.L. Sattler, and A. Upayokin, *Operations-Oriented Performance Measures for Freeway Management Systems: Year 1 Report*, FHWA/TX-07/0-5292-1, Texas Transportation Institute, The Texas A&M University System, College Station, TX 77843-3135, 45–54, April 2007. http://tti.tamu.edu/documents/0-5292-1.pdf. Accessed August 8, 2016.

Chapter 14

National ITS architectures

An architecture is a structure of components, their relationships, and the principles and guidelines governing their design and evolution over time [1,2]. An architecture has the following attributes:

- Identifies a focused purpose.
- Facilitates user understanding and communication.
- Permits comparison and integration.
- Promotes expandability, modularity, and reusability.
- Achieves most useful results with least development costs.
- Applies to the required range of situations.

This chapter explores National Intelligent Transportation System (ITS) Architectures in several countries. The first is the U.S. National ITS Architecture, which provides a framework for planning, programming, and implementing intelligent transportation systems over an extended time period in urban, interurban, and rural environments across the United States. The Architecture facilitates the ability of local, regional, state or provincial, and interstate jurisdictions to operate collaboratively and to harness the benefits of a regional approach to transportation challenges. Its structure offers flexibility at the local level and supports the evolution and incorporation of technological improvements and changing user needs as, for example, brought about by connected vehicle requirements.

A corresponding architecture exists in the European Union (EU) referred to as the FRamework Architecture Made for Europe (FRAME). FRAME encourages a common approach to ITS planning and project development from country to country. It provides functional, physical, communications, organizational, and information viewpoints, but allows each country to configure ITS functionality in its own way without a physical guidance mechanism to influence their individual approaches. Like the U.S. National ITS Architecture, the FRAME Architecture includes options needed to integrate cooperative vehicles into the architecture planning process. The salient features of the Japanese and the Canadian National ITS Architectures are discussed in the concluding sections of the chapter.

14.1 U.S. NATIONAL ITS ARCHITECTURE STRUCTURE

Figure 14.1 illustrates the three-layer architecture structure composed of the Institutional Layer and two technical layers, namely, the Transportation Layer and the Communication

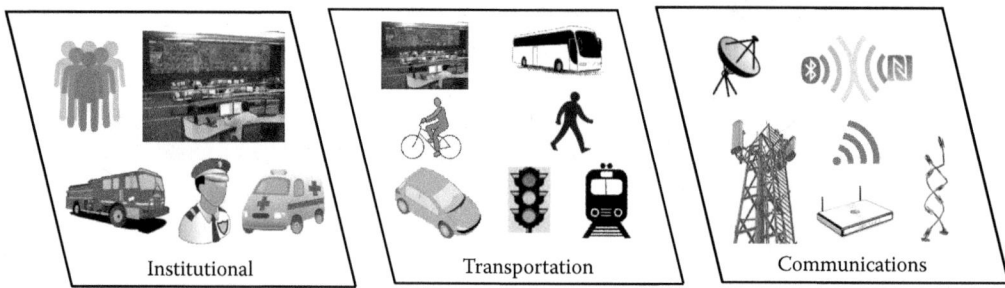

Figure 14.1 Three-layer U.S. National ITS Architecture.

Layer. The technical layers operate according to the policies and constraints imposed by the Institutional Layer [3,4].

- The Institutional Layer includes the institutions, policies, funding mechanisms, and processes required for effective implementation, operation, and maintenance of an intelligent transportation system. These may exist in the form of governmental agencies and departments, volunteer industry consortiums, and professional societies that are relied upon to add their support and expertise to the development of the architecture and the elements it includes. The collaborative discussions that occur at this layer lead to the establishment of the objectives and requirements for the architecture.
- The Transportation Layer defines the architecture's transportation components in terms of the underlying functionality or required services, physical entities or subsystems that implement the desired functions, interface standards that connect the physical subsystems and ensure data transfer across them, and data required to provide each transportation service. This layer is the heart of the National ITS Architecture. An agency or other entity may appear in more than one layer as illustrated by the traffic management center shown in the Institutional and Transportation Layers. In the former, its function is one of contributing culture and policy, while in the latter it is operations.
- The Communications Layer provides the means for the accurate and timely exchange of information among the mobile and fixed infrastructure constituents of the architecture.

14.2 REGIONAL ARCHITECTURES

One of the uses of the U.S. National ITS Architecture is the development of regional ITS architectures for integrating transportation services in a specific state, metropolitan area, or other regions of interest. The National ITS Architecture includes a broad menu of options and other features that can be selectively tailored and applied to an area. Basing each regional architecture on the National ITS Architecture also aids the private sector as any expertise developed in utilizing it in one region can be transferred to another region.

A regional ITS architecture creates benefits on many levels. It is a tool that aids in visualizing and articulating the overall ITS system for the region so that stakeholders can allocate their resources in a compatible instead of a competitive manner. For strategic planning, a regional ITS architecture can function as a bridge between an integrated surface transportation system and the ITS projects that support the strategic vision. The regional architecture also is useful in linking the transportation planning process and the initial phases of project development.

At the metropolitan and statewide transportation planning level, a regional ITS architecture affords peer agencies the opportunity to jointly define their vision for ITS development based on regional goals and objectives. A regional ITS architecture allows the area it serves to plan for the introduction and integration of new technologies to support more effective operations. In this regard, the physical architecture acts as a guide for consistency across the state.

At the individual project level, the regional ITS architecture assists each project in applying systems engineering principles to the development of a well-designed and maintained architecture. Application of these principles supports the definition of stakeholder roles, responsibilities, and desired benefits, and the creation of agreements and operating procedures needed to develop the concept of operations. Systems engineering also highlights the need for the high-level functional requirements and the interfaces and ITS standards that support project design.

The U.S. National Architecture provides a six-step process to assess the completeness and quality of a regional ITS architecture, namely,

1. *Collect materials*: Gather the published documentation and the underlying Turbo Architecture database as discussed in Section 14.6. In most cases, the assessment will require access to an architecture document, the Turbo database, Web pages if they are used to convey the architecture content, and any additional documents, for example, an ITS strategic plan associated with the architecture.
2. *Assign reviewers*: Normally, an assessment will be conducted by a small team of two to three reviewers. This team ideally possesses local knowledge of the transportation agencies and systems in the region and the transportation planning and project development processes that are used. The team should also have some background and expertise in ITS architectures.
3. *Read the documentation*: Before assessing the architecture in detail, a team member should skim through the available documentation and ancillary files and identify the location and descriptions of the key architecture components to be reviewed.
4. *Assess the architecture*: All architecture components are assessed based on any national requirements and the regional ITS architecture guidance document. The architecture team should use an assessment checklist to ensure that all components are reviewed in the same manner.
5. *Document assessment results*: The architecture team should prepare a feedback report for each assessment that covers all architecture components and provides an overview of the strengths and recommended improvements in each area. The feedback reports should be candid and shared only with the National Architecture office and the local agency that is responsible for the regional ITS architecture.
6. *Discuss findings*: A face-to-face meeting is not required, but it is helpful to discuss the assessment results with the agency that is responsible for the architecture.

14.3 PRINCIPAL COMPONENTS OF THE U.S. NATIONAL ITS ARCHITECTURE

The U.S. National ITS Architecture defines the functions (e.g., gather traffic information or request a route) required to implement the system architecture, the physical entities or subsystems where these functions reside (e.g., the field or the vehicle), the data flows and architecture flows that connect the functions and physical subsystems together, respectively, into an integrated system (e.g., receive and process data from sensors at the roadway), and the

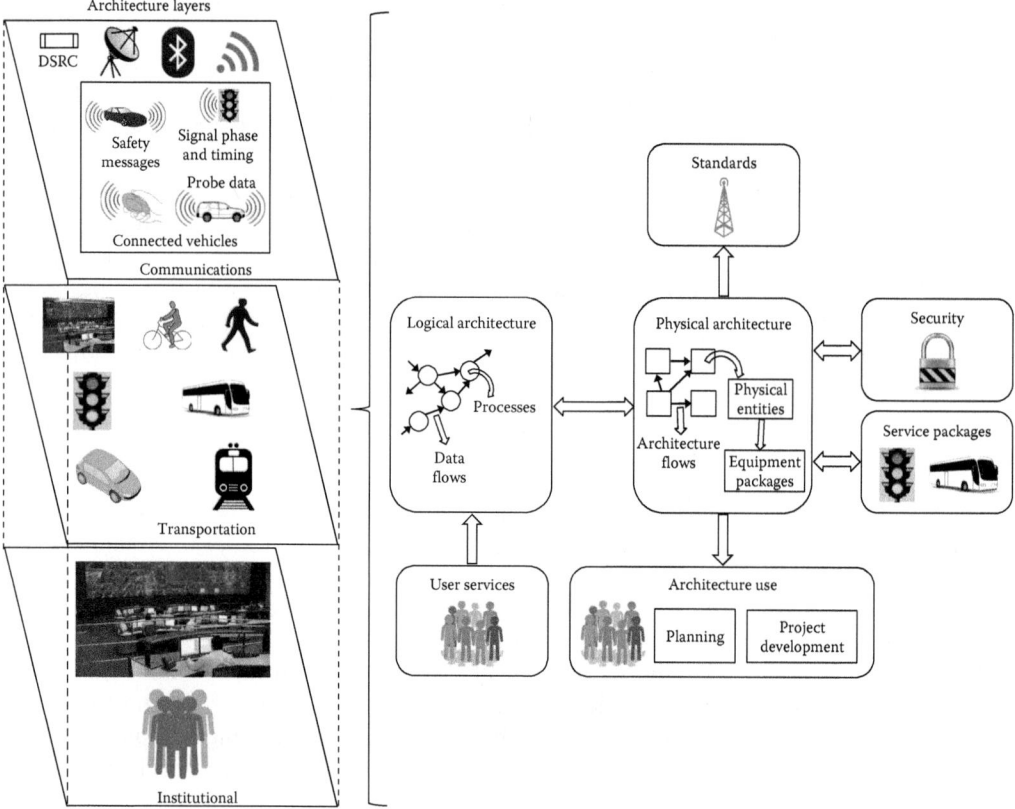

Figure 14.2 U.S. National ITS Architecture framework showing the architecture layers and its sub-elements.

interface standards that connect the physical elements and support data exchanges between them.

This is accomplished through the principal elements of the architecture shown in Figure 14.2, namely,

- User services and service bundles.
- Logical architecture.
- Physical architecture.
- Equipment packages.
- Service packages.

14.3.1 User services and service bundles

User services represent what the system offers from the perspective of the user. They facilitate system definition by requiring the system architect to consider what services will be provided to address the identified problems and needs. The user services available in the National Architecture were jointly defined by a collaborative process involving the USDOT and the Intelligent Transportation Society of America (ITS America) with significant stakeholder input. A user might be the public, a system operator, transit agency, commercial vehicle company, or an emergency response provider.

Table 14.1 User service bundles and services in the U.S. National ITS Architecture

User service bundle	User service
Travel and Traffic Management	En-Route Driver Information
	Route Guidance
	Ride Matching and Reservation
	Traveler Services Information
	Traffic Control
	Incident Management
	Travel Demand Management
	Emissions Testing and Mitigation
	Highway-Rail Intersection
Public Transportation Management	Public Transportation Management
	En-Route Transit Information
	Personalized Public Transit
	Public Travel Security
Electronic Payment	Electronic Payment Services
Commercial Vehicle Operations	Commercial Vehicle Electronic Clearance
	Automated Roadside Safety Inspection
	Onboard Safety and Security Monitoring
	Commercial Vehicle Administrative Processes
	Hazardous Material Security and Incident Response
	Freight Mobility
Emergency Management	Emergency Notification and Personal Security
	Emergency Vehicle Management
	Disaster Response and Evacuation
Advanced Vehicle Safety Systems	Longitudinal Collision Avoidance
	Lateral Collision Avoidance
	Intersection Collision Avoidance
	Vision Enhancement for Crash Avoidance
	Safety Readiness
	Pre-Crash Restraint Deployment
	Automated Vehicle Operation
Information Management	Archived Data Function
Maintenance and Construction Management	Maintenance and Construction Operations

Table 14.1 displays the 33 user services in the U.S. National ITS Architecture, grouped into eight bundles. A user service bundle is a logical grouping of user services that provides a convenient way to discuss the range of requirements in a broad stakeholder area of interest. The eight bundles are Travel and Traffic Management, Public Transportation Management, Electronic Payment, Commercial Vehicle Operations, Emergency Management, Advanced Vehicle Safety Systems, Information Management, and Maintenance and Construction Management.

14.3.2 Logical architecture

The logical architecture defines the functions or processes and data flows (also called information flows) of a system and guides development of functional requirements for new systems and improvements to existing systems. The logical architecture consists of processes, data flows, terminators, and data stores. The processes are the tasks, that is, the logical functions, to be performed by the system. Data flows identify the information that is shared by the processes. The entry and exit points for the logical architecture are the sensors, computers, and human operators (called terminators) of the intelligent transportation system. The terminators, which reside at the boundaries of the system, appear in the physical architecture as well. Data stores are repositories of information maintained by the processes.

Different users of the National Architecture will use the logical architecture in distinctive ways. For example, most public sector agency employees are not required to deal with the logical architecture directly. However, it is essential they review software operation to verify that it meets their requirements. They may use the logical architecture to write their own system and interface requirements specifications.

Consultants who are often the developers of an architecture, on the other hand, must often create logical and physical architectures that are traceable to users' requirements. Hence, the linking and organization of the physical and the logical architectures is of paramount interest to them. The additional detail provided by the logical architecture may also assist developers as they begin to implement a project.

A logical architecture should be independent of institutions and technology; that is, it should not define where or by whom functions are performed in the system, nor identify how functions are to be implemented. The functions and information flows are specified by the selected user services.

Processes and data flows, also referred to as logical data flows, are grouped to form particular transportation management functions (e.g., Manage Traffic) and are represented graphically by data flow diagrams (DFDs) or bubble charts, which decompose into several levels of detail. In the DFDs, processes are represented as bubbles and data flows as arrows. The lowest level of detail in the functional hierarchy is the process specification, referred to as a PSpec. They represent the elemental functions that are executed to satisfy the user service requirements and are not broken down any further. The information exchanges between processes and between PSpecs are shown in the DFD.

Figure 14.3 is a simplified DFD of the Manage Traffic process, which interacts with eight other processes. The Manage Traffic process itself is decomposed into six subprocesses as illustrated on the left side of Figure 14.4. Each of these subprocesses is decomposed still

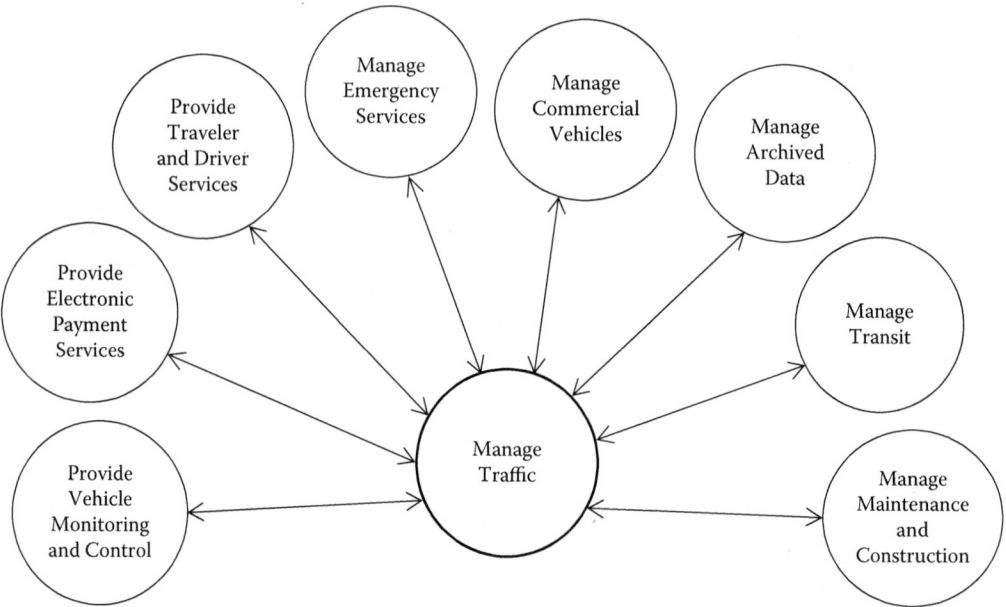

Figure 14.3 Manage Traffic process interacts with eight other processes.

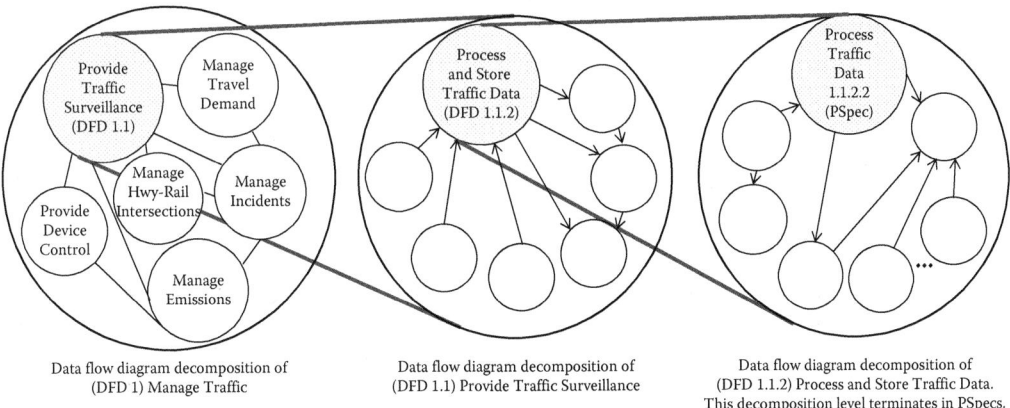

Data flow diagram decomposition of
(DFD 1) Manage Traffic

Data flow diagram decomposition of
(DFD 1.1) Provide Traffic Surveillance

Data flow diagram decomposition of
(DFD 1.1.2) Process and Store Traffic Data.
This decomposition level terminates in PSpecs.

Figure 14.4 Manage Traffic process decomposition into six subprocesses (DFD 1) and eventually into PSpecs at the DFD 1.1.2 level.

further until a complete functional view of the system emerges. For example, Figure 14.4 shows the decomposition of the Provide Traffic Surveillance subprocess into seven additional subprocesses, each of which can be decomposed still further until a complete functional view of the system emerges at the PSpec level [5].

For each PSpec, the architecture provides the associated subsystem and equipment packages. The PSpec level is shown in the last large bubble on the right of Figure 14.4 for Process and Store Traffic Data. One of its 10 PSpecs is Process Traffic Data (PSpec 1.1.2.2), which is defined as

> This process shall receive and process data from sensors at the roadway. The data include sensor and video data coming from traffic sensors and inputs for pedestrians, multimodal crossings, parking facilities, highway-rail intersections, high-occupancy vehicle (HOV) and high-occupancy toll (HOT) lanes, and reversible lanes. The process distributes data to Provide Device Control processes that manage freeway, highway-rail intersections, parking facilities, and surface streets. It also sends the data to another Provide Traffic Surveillance process for loading into the stores of current and long-term data.

14.3.3 Physical architecture

A physical architecture supplies agencies with a tangible representation (though not a detailed design) of how the system provides the required functionality as it is structured around the processes and data flows in the logical architecture. A physical architecture takes the processes (or PSpecs) identified in the logical architecture and assigns them to physical entities (called subsystems in the U.S. National ITS Architecture). The subsystems generally provide a rich set of capabilities, sometimes more than are required to be implemented at any one place or time. The data flows from the logical architecture that originate from one subsystem and end at another are grouped together into physical architecture flows.

An architecture flow may contain one or more detailed data flows. The architecture flows and their communication requirements define the interfaces between subsystems. The interfaces, in turn, are specified through ITS standards. The physical architecture also identifies the desired communications and interactions between different transportation management organizations.

Equipment packages decompose the subsystems into deployment-sized pieces. The designers of an ITS architecture often traverse between the physical architecture structure and the related process and data flow requirements in the logical architecture.

Figure 14.5 illustrates the relation between the logical and physical architectures by showing, in broad terms, the purpose of the logical architecture, namely, defining what has to be done in terms of functions or processes, and the purpose of the physical architecture, namely, how functions are grouped together for implementation in subsystems. The figure depicts the data flows that occur in the logical architecture and the architecture flows that occur in the physical architecture.

Figure 14.6 represents the highest-level characterization of the Transportation and Communications Layers of the physical architecture. There are 22 transportation subsystems (smaller rectangles) that can exchange information at any time, depending on the active application. The four general communication links (narrow oblong rectangles) are used to exchange information between subsystems. The subsystems roughly correspond to physical elements of transportation management systems and are grouped into four classes (larger rectangles): Centers, Field, Vehicles, and Travelers.

In addition to the 22 subsystems, the physical architecture defines interfaces to the terminators, which represent systems that are on the boundary of the architecture. The architecture does not define functionality for the terminators, just interfaces to them.

The representation of a basic traffic signal control system in Figure 14.7 is an example of how the physical architecture is applied to create ITS implementations. This traffic control system is embodied by functions within 2 of the 22 subsystems, namely, the Traffic Management Subsystem and the Roadway Subsystem, which were selected from among the subsystems in Figure 14.6. The Traffic Management Subsystem contains the traffic management center, while the Roadway Subsystem includes the sensors, traffic signals, and controller.

The Traffic Management and Roadway Subsystems together with the necessary communications (shown by the curved arrows in the figure) exchange control and surveillance information and provide capabilities typically associated with traffic signal control systems, namely, area-wide signal coordination, arterial network traffic condition monitoring, and a range of adaptive control strategies.

14.3.4 Equipment packages

An equipment package represents a set of capabilities that exist in a subsystem. Equipment packages group like functions (PSpecs) of a particular subsystem together into an implementable

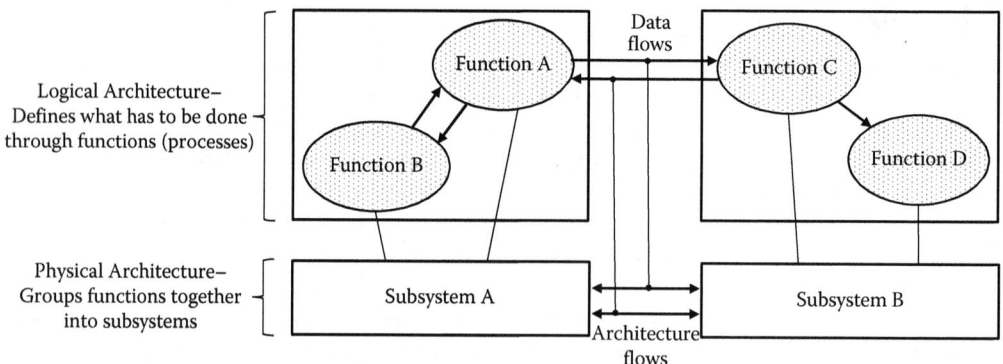

Figure 14.5 Relation between logical and physical architectures.

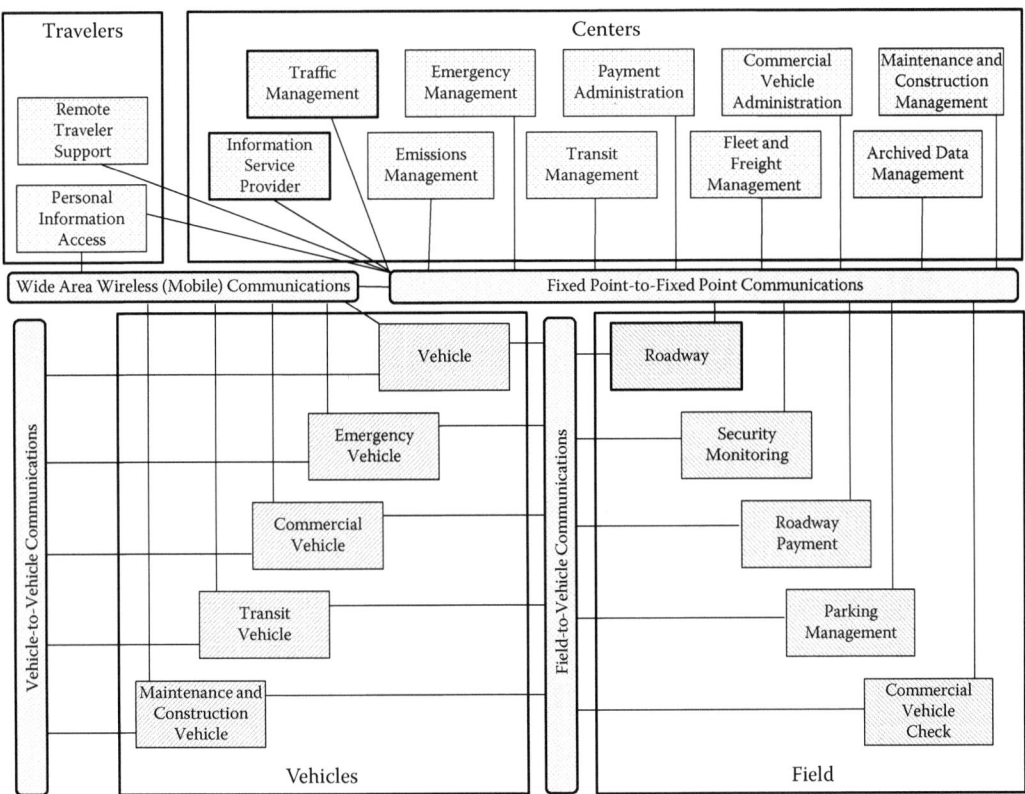

Figure 14.6 Highest-level representation of the Transportation and Communications Layers of the physical architecture.

set of hardware and software capabilities. The isolated portion of service package capabilities and elements allocated to each subsystem is defined as an equipment package. The collected functions take into account the user services and the need to accommodate various levels of functionality within them. There may be more than one set of equipment packages available to represent a combination of functions within a subsystem. The U.S. National ITS Architecture defines 233 equipment packages, that is, groups of PSpecs [6].

Since equipment packages are the most detailed elements of the physical architecture and are associated with specific service packages, there is clear traceability between the interface-oriented architecture framework and the deployment-oriented service packages as illustrated in Figure 14.8 [7]. As an example of the information the U.S. National Architecture makes available for each equipment package, consider the Basic Information Broadcast equipment package of Figure 14.8 (second row, second column).

Upon entering the equipment package Internet Web link [6] and scrolling down to the Information Service Provider subsystem, one finds a listing of the equipment packages that are part of this subsystem. Clicking on the Basic Information Broadcast package opens a new page where there are three tabs. The first identifies the service packages that use the equipment package; the second, functional requirements of the equipment package and the applicable PSpecs; and the third, architecture flow source and destination pairs along with any applicable standards.

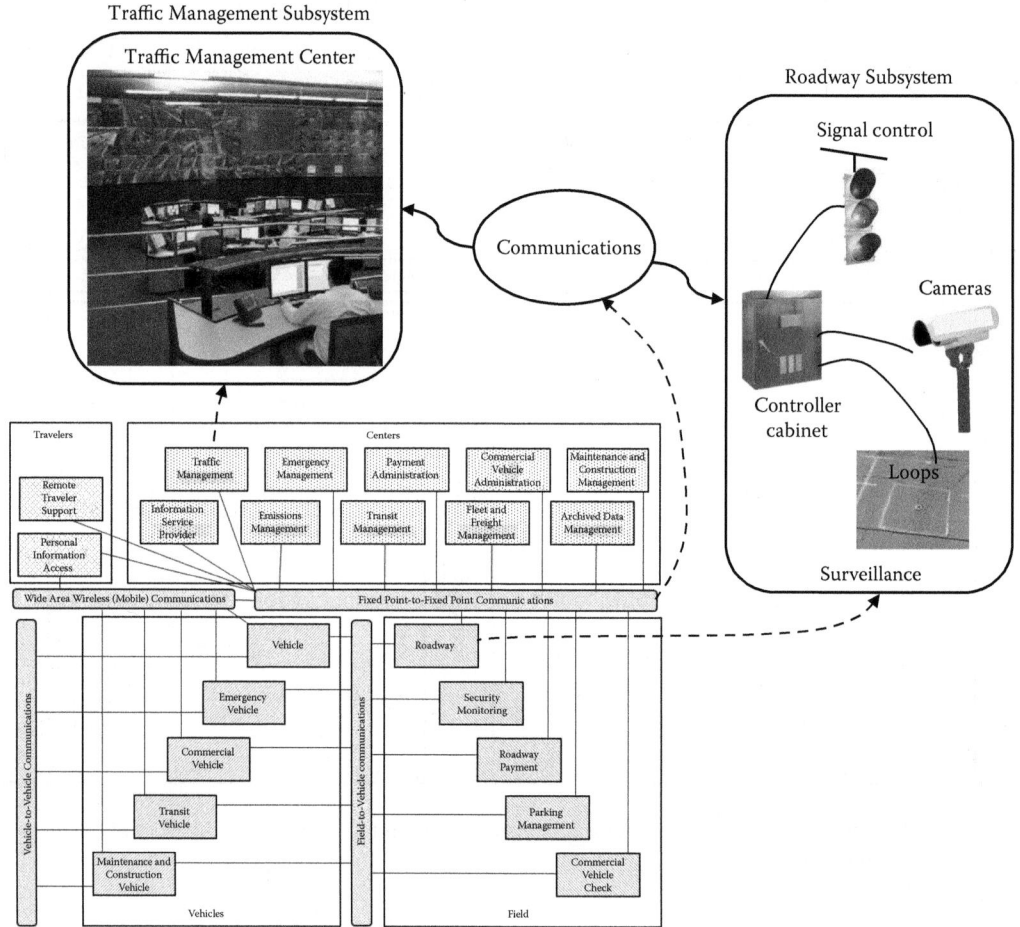

Figure 14.7 Basic traffic signal control system physical architecture.

14.3.5 Service packages

Service packages identify the architecture components required to implement a service that focuses on specific transportation problems and needs [8]. Their necessity became apparent as developers realized that many user services were too broad in scope to be convenient in planning actual deployments. Additionally, user services often do not fit easily into existing institutional environments and do not distinguish between major levels of functionality. In order to address these concerns and to support the creation of service-based regional ITS architectures, a finer grained set of deployment-oriented ITS service building blocks or service packages were defined from the original user services.

A service package is the part of the physical architecture that relates to a specific service such as traffic signal control. A service package gathers together equipment packages from several different subsystems (usually two or more), terminators, and architecture flows that provide the desired service. They also support the major architecture flows between the equipment packages and other external systems.

Table 14.2 lists the service package groups that contain the individual service packages required to implement a particular service or application. Service packages can be linked back to the user services and their more detailed requirements. For instance, the service

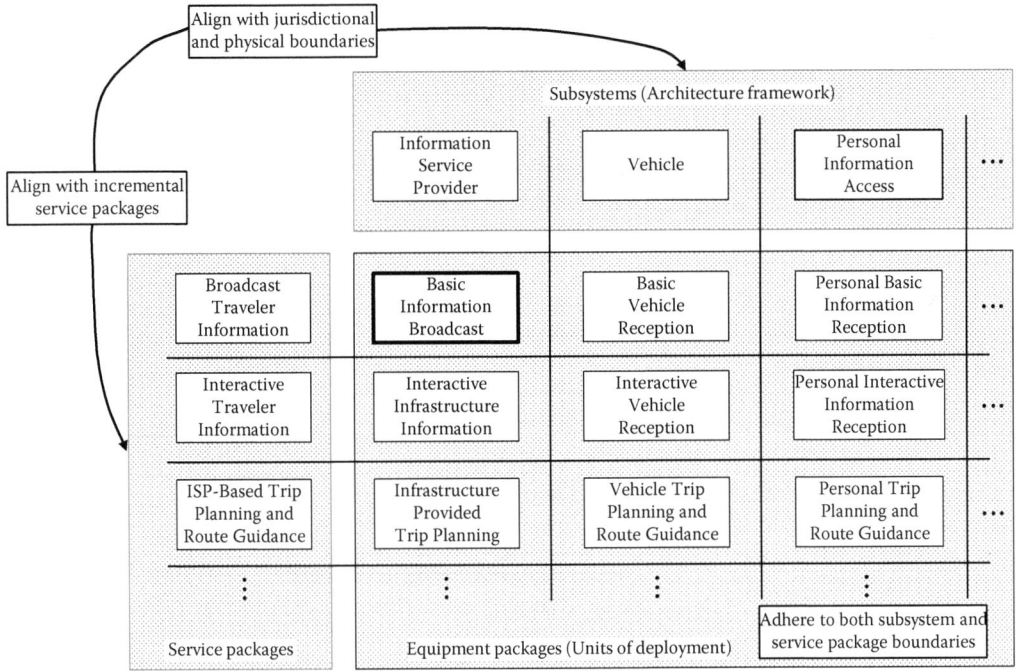

Figure 14.8 Physical architecture subsystem service-package equipment-package relationship.

package Traffic Signal Control that is part of the Traffic Management service package group supports the Highway-Rail Intersection, Traffic Control, and Incident Management user services. Traffic Control, for example, provides for the integration and adaptive control of freeway and surface street systems to improve the flow of traffic, give preference to transit and other high-occupancy vehicles, and minimize congestion while maximizing the movement of people and goods. This is accomplished through four high-level functions, namely, traffic flow optimization, traffic surveillance, control, and provide information. Each of these functions is decomposed into a number of PSpecs, which are found on the Traffic Control user service Web page [9].

Continuing with the Traffic Control user service as an example, its functionality is divided into several service packages to allow for explicit consideration of the following:

- Basic functions such as surveillance as represented by the Network Surveillance and Probe Surveillance service packages.
- Institutional settings by separating control functions typically performed by different agencies into separate service packages such as Traffic Signal Control and Traffic Metering.
- Functional levels of service by including a Regional Traffic Management service package that provides for coordination of control strategies across jurisdictions.

Other service packages that relate to Traffic Control are found in other service package groups, namely, the Transit Signal Priority service package in the Public Transportation group, which contains the functionality for transit vehicle priority at traffic signals, and the Emergency Routing service package in the Emergency Management group that includes the functionality for emergency vehicle preemption at traffic signals.

Table 14.2 Service packages in the U.S. National ITS Architecture

Service package group	Service package
Traffic Management	Network Surveillance
	Probe Surveillance
	Traffic Signal Control
	Traffic Metering
	HOV Lane Management
	Traffic Information Dissemination
	Regional Traffic Management
	Traffic Incident Management System
	Traffic Decision Support and Demand Management
	Electronic Toll Collection
	Emissions Monitoring and Management
	Roadside Lighting System Control
	Standard Railroad Grade Crossing
	Advanced Railroad Grade Crossing
	Railroad Operations Coordination
	Parking Facility Management
	Regional Parking Management
	Reversible Lane Management
	Speed Warning and Enforcement
	Drawbridge Management
	Roadway Closure Management
	Variable Speed Limits
	Dynamic Lane Management and Shoulder Use
	Dynamic Roadway Warning
	VMT Road User Payment
	Mixed Use Warning Systems
Public Transportation	Transit Vehicle Tracking
	Transit Fixed-Route Operations
	Demand Response Transit Operations
	Transit Fare Collection Management
	Transit Security
	Transit Fleet Management
	Multimodal Coordination
	Transit Traveler Information
	Transit Signal Priority
	Transit Passenger Counting
	Multimodal Connection Protection
Traveler Information	Broadcast Traveler Information
	Interactive Traveler Information
	Autonomous Route Guidance
	Dynamic Route Guidance
	ISP Based Trip Planning and Route Guidance
	Transportation Operations Data Sharing
	Travel Services Information and Reservation
	Dynamic Ridesharing
	In-Vehicle Signing
	Short Range Communications Traveler Information
Vehicle Safety	Vehicle Safety Monitoring
	Driver Safety Monitoring
	Longitudinal Safety Warning
	Lateral Safety Warning
	Intersection Safety Warning
	Pre-Crash Restraint Deployment
	Driver Visibility Improvement

(Continued)

Table 14.2 (Continued) Service packages in the U.S. National ITS Architecture

Service package group	Service package
	Advanced Vehicle Longitudinal Control
	Advanced Vehicle Lateral Control
	Intersection Collision Avoidance
	Automated Vehicle Operations
	Cooperative Vehicle Safety Systems
Commercial Vehicle Operations	Carrier Operations and Fleet Management
	Freight Administration
	Electronic Clearance
	CV Administrative Processes
	International Border Electronic Clearance
	Weigh-In-Motion
	Roadside CVO Safety
	Onboard CVO Safety
	CVO Fleet Maintenance
	HAZMAT Management
	Roadside HAZMAT Security Detection and Mitigation
	CV Driver Security Authentication
	Freight Assignment Tracking
Emergency Management	Emergency Call-Taking and Dispatch
	Emergency Routing
	Mayday and Alarms Support
	Roadway Service Patrols
	Transportation Infrastructure Protection
	Wide-Area Alert
	Early Warning System
	Disaster Response and Recovery
	Evacuation and Reentry Management
	Disaster Traveler Information
Archived Data Management	ITS Data Mart
	ITS Data Warehouse
	ITS Virtual Data Warehouse
Maintenance & Construction Operations	Maintenance & Construction Vehicle & Equipment Tracking
	Maintenance & Construction Vehicle Maintenance
	Maintenance & Construction Activity Coordination
	Road Weather Data Collection
	Weather Information Processing and Distribution
	Roadway Automated Treatment
	Winter Maintenance
	Roadway Maintenance and Construction
	Work Zone Management
	Work Zone Safety Monitoring
	Environmental Probe Surveillance
	Infrastructure Monitoring

14.4 U.S. ITS ARCHITECTURE SECURITY

Security and protection ensure the reliability and availability of ITS information system applications. Security is represented in the U.S. National ITS Architecture in two ways, through ITS Security Services and ITS Security Areas as illustrated in Figure 14.9 [10]. ITS security services are the foundation of security as they apply to and support all subsystems and architecture flows in the National ITS Architecture. Its systems must be secure in their

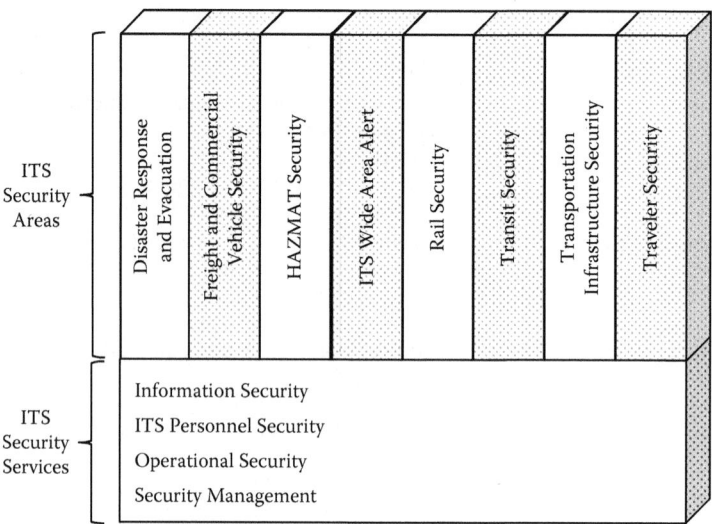

Figure 14.9 Security layers in the U.S. National ITS Architecture. (From Security, http://www.iteris.com/itsarch/html/security/securityhome.htm.)

own right before they can reliably and safely be utilized to improve the security of the rest of the surface transportation system. The eight ITS security areas at the top of the figure define the ways in which an intelligent transportation system detects, responds, and recovers from threats against the surface transportation system. Specific subsystems, architecture flows, service packages, and supporting physical and logical architecture definitions are defined for each ITS security area as part of the Architecture.

14.5 U.S. NATIONAL ARCHITECTURE MAINTENANCE AND UPDATES

Version 7.1 of the U.S. National ITS Architecture provides the following:

- Updated mappings to the Moving Ahead for Progress in the 21st Century Act (MAP-21) goals, objectives, and performance measures.
- Updated mapping to the most current ITS and connected vehicle standards.
- Functionality and interfaces to align with the connected vehicle environment, advanced travel demand management strategies, electronic freight manifest, integrated corridor management, and ITS standards.
- Physical and logical architecture enhancements to support expanded Commercial Vehicle Information Systems and Networks applications that use wireless roadside inspection and transportation planning features.
- New linkages that connect service packages with the connected vehicle applications defined in the Connected Vehicle Reference Implementation Architecture (CVRIA).

14.6 TURBO ARCHITECTURE

Turbo Architecture is a software application that supports development of regional and project ITS architectures using the U.S. National ITS Architecture as a guide [11]. This tool allows a region's ITS architecture to be matched to its transportation planning objectives, strategies, and needs, thereby aligning the regional ITS architecture with the planning process. In addition, it assists the user in integrating multiple project architectures with each other and with a regional architecture. To properly use the software, the Turbo user must be familiar with the National ITS Architecture. The architectures are saved in Microsoft Access–compatible data files. Each data file may contain one regional architecture and multiple project architectures. The software can also be used to produce user-friendly documentation and Web pages.

Information can be entered into Turbo Architecture using tabular forms. As the architecture definition process progresses, the user identifies stakeholders, inventory, services, functional requirements, interfaces, standards, and agreements. Once this initial data input is complete, Turbo Architecture provides tools to customize the architecture to its specific requirements. Outputs are available for display, print, or publication of the results. The user can extend the National ITS Architecture by adding their own information flows and transportation elements for those areas not included in the National ITS Architecture.

Turbo Architecture Version 7.1 contains these features:

- *View service package diagrams*: View the service package diagrams from the National ITS Architecture with the search capability that locates and reviews the service packages for the architecture.
- *Publish project architecture documentation*: Click a button to publish a project architecture document in Microsoft Word that includes chapters for stakeholders, inventory, operational concept, services, interfaces, and standards.
- *Autoselect services*: The planning tab allows transportation goals and objectives to be defined and related to performance measures and service packages. An autoselect button under the services tab transfers the service package selections from the planning tab to the services tab.
- *Context menus*: Make your text entries quickly and accurately using context menus. Right clicking allows the user to undo, cut, copy, paste, select all, or spell check text entries.
- *Conversion facility*: Supports quick and easy conversion of existing Turbo databases, providing a migration path for existing Turbo users.

Future tasks include development of software to support V2X Cooperative Systems, complete development of a CVRIA software tool (SET-IT) to assist project implementers and planners identify and define connected vehicle interfaces for their projects, and creation of outreach and guidance materials to promote deployment of integrated, secure V2X cooperative systems.

14.7 EU FRAME ARCHITECTURE

Following the recommendation of the High Level Group on Telematics and a resolution of the Transport Council, the European ITS Framework Architecture, known as the FRAME Architecture, was created by the European Commission's Keystone Architecture Required for European Networks (KAREN) project (1998–2000). The architecture has been maintained and enhanced continuously since then, with cooperative systems being added by project E-FRAME (2008–2011). Its purpose is to encourage a common approach to ITS planning and project development from country to country [12,13].

Because the FRAME Architecture is intended for application within the EU, it conforms to the precepts of subsidiarity, and thus does not mandate any physical or organizational structures on its users. Hence, the architecture makes no assumptions about the way that subsystems and systems are implemented. It does, however, provide a common approach for exploitation throughout the EU so that the implementation of integrated and interoperable ITS can be planned and operated.

14.7.1 The FRAME Architecture as part of the ITS planning process

The FRAME Architecture is an integral part of an ITS Action Plan and is utilized in a top-down approach to conceive, develop, design, and deploy integrated intelligent transportation systems as indicated in Figure 14.10 [13]. The overall concept and system structure shown in the upper two-thirds of the figure are technology independent so that, as technology evolves, all the higher-level requirements remain unchanged. The information contained within the system structure enables the ITS industry to produce the equipment and systems that provide the desired stakeholder services, each with its own distinctive features, but conforming to the purposes expressed in the overall concept and system structure. Accordingly, integrated and interoperable ITS services can be provided across the EU. Only the bottom third, system design, is technology dependent.

The FRAME Architecture that defines the system structure contains logically consistent subsets of user needs and associated functions as depicted in Figure 14.11. This methodology begins with understanding the wishes or aspirations of the stakeholders to fulfill needs through ITS functions and is supported by computer-based tools. The needs and functions are then identified within the FRAME Architecture and a subset is selected. The subset of functions is next customized to fit the region in which they are to be deployed.

14.7.2 FRAME Architecture scope

The FRAME Architecture addresses the following ITS areas:

- Electronic Fee Collection.
- Emergency Notification and Response—Roadside and In-Vehicle Notification.

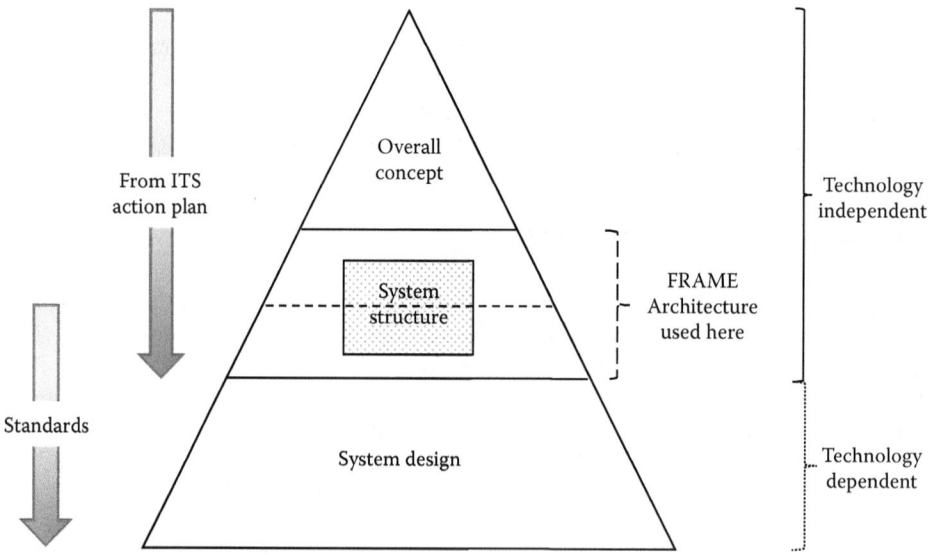

Figure 14.10 Use of the FRAME Architecture in the ITS planning and design process. (From Turbo architecture, http://www.iteris.com/itsarch/html/turbo/turbooverview.htm.)

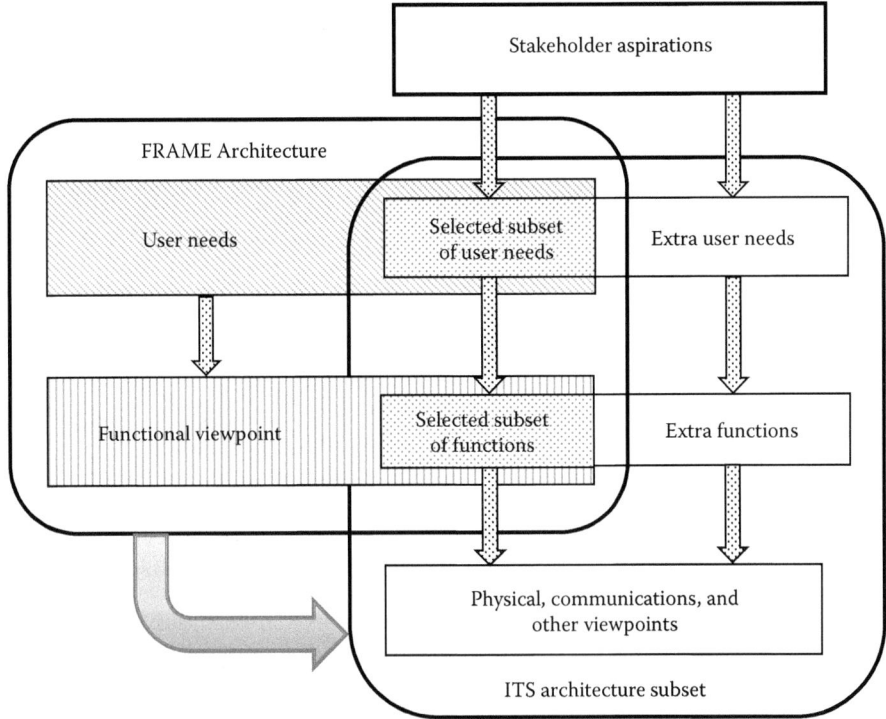

Figure 14.11 Creating an architecture subset from the FRAME Architecture.

- Traffic Management—Urban, Interurban, Simulation, Parking, Tunnels and Bridges, Maintenance, together with the Management of Incidents, Road Vehicle Based Pollution, and Road Use Demand.
- Public Transport Management—Schedules, Fares, On-Demand Services, Fleet and Driver Management.
- In-Vehicle Systems—Various Levels of Automation, Cooperative Systems.
- Traveler Assistance—Pre-Journey and On-Trip Planning, Travel Information.
- Law Enforcement Support.
- Freight and Fleet Management—Pre-clearance, Safety, and Administration.
- Cooperative Systems Support—Specific services not included elsewhere such as bus lane use and freight vehicle parking.
- Multimodal Interfaces—Links to other modes when required, for example, travel information and multimodal crossing management.

These are refined further into 10 functional areas that contain the identified user needs shown in Table 14.3 [14].

14.7.3 Architecture viewpoints

The architecture structure contains a number of viewpoints. The functionality needed to implement ITS services is provided by the Functional Viewpoint, which does not impose any specific technical solutions on its users. The Physical Viewpoint shows the stakeholder choices for the components required to support the services and the links between components.

Table 14.3 User needs addressed by the FRAME Architecture

Functional area	User need	Function
1. General	Properties that either the FRAME Architecture should possess, or that systems built in conformance to the FRAME Architecture should possess	1.1 Architectural Properties 1.2 Data Exchange 1.3 Adaptability 1.4 Constraints 1.5 Continuity 1.6 Cost/Benefit Ratio 1.7 Expandability 1.8 Maintainability 1.9 Quality of Data Content 1.10 Robustness 1.11 Safety 1.12 Security 1.13 User Friendliness 1.14 Special Needs 1.15 Privacy 1.16 Communications
2. Infrastructure Planning and Maintenance	Activities associated with long-term planning, modeling, reporting, and maintenance of the infrastructure[a]	2.1 Transport Planning Support 2.2 Infrastructure Maintenance Management
3. Law Enforcement	Activities associated with the enforcement of traffic laws and regulations, and the collection of evidence[a]	3.1 Policing/Enforcing Traffic Regulations
4. Financial Transactions	Activities associated with the payment for traffic or travel services, and includes the manner of the transaction, its enforcement, and the sharing of revenues[a]	4.1 Electronic Financial Transactions
5. Emergency Services	eCall (a system that calls emergency services either automatically or at the request of a vehicle occupant) and stolen vehicle management; prioritizing of emergency vehicles, hazardous goods (i.e., goods that need to be tracked), and incident management[a]	5.1 Emergency Notification and Personal Security 5.2 Emergency Vehicle Management 5.3 Hazardous Materials and Incident Notification
6. Travel Information and Guidance	Activities concerned with the handling of pre-trip and on-trip information, including mode choice and change, and route guidance	6.1 Pre-trip Information 6.2 On-trip Driver Information 6.3 Personal Information Services 6.4 Route Guidance and Navigation
7. Traffic, Incidents, Demand Management, and Cooperative Systems	Activities linked with monitoring, planning, flow control, exceptions management, speed management, lane and parking management, HOV, road pricing and zoning, and vulnerable road users	7.1 Traffic Control 7.2 Incident Management 7.3 Demand Management 7.4 Cooperative Systems—Traffic Safety 7.5 Cooperative Systems—Traffic Efficiency 7.6 Cooperative Systems—Value Added and Other Services
8. Intelligent Vehicle Systems	Functions found within a vehicle, including vision enhancement, longitudinal and lateral collision avoidance, lane keeping, platooning, speed control, driver alertness, and eCall initiation	8.1 Vision Enhancement 8.2 Automated Vehicle Operation 8.3 Longitudinal Collision Avoidance 8.4 Lateral Collision Avoidance 8.5 Safety Readiness 8.6 Pre-crash Restraint Deployment

(Continued)

Table 14.3 (Continued) User needs addressed by the FRAME Architecture

Functional area	User need	Function
9. Freight and Fleet Management	Activities associated with freight and fleet management, including statutory data collection and reporting; orders and document management; planning, scheduling, monitoring, reporting, and operations management; vehicle and cargo safety; and management of intermodal interfaces	9.1 Commercial Vehicle Pre-clearance 9.2 Commercial Vehicle Administrative Processes 9.3 Automated Roadside Safety Inspection 9.4 Commercial Vehicle Onboard Safety Monitoring 9.5 Commercial Fleet Management
10. Public Transport Management	Management, scheduling, monitoring, information handling, communications and priority activities associated with public transport, demand responsive and shared public transport, on-trip public transport information, and traveler security	10.1 Public Transport Management 10.2 Demand Responsive Public Transport 10.3 Shared Transport Management 10.4 On-trip Public Transport Information 10.5 Public Travel Security

Source: *FRAME User Needs V4.1.* http://frame-online.eu/wp-content/uploads/2014/10/FRAME-User-Needs-V4.1-01.pdf.

[a]   These user needs have links with Groups 6–10.

Further analysis, also based on specific choices or decisions, can then specify the Communications Viewpoint that contains the requirements for communications between the components; the Organizational Viewpoint that stipulates who owns, manages, and operates each component, the management structure, and rules and regulations for providing the services; and the Information Viewpoint that identifies the information that is needed and its attributes and relationships.

14.7.4 Supporting the ITS Action Plan throughout the EU

Once a European Specification for each ITS application and service has been adopted, a corresponding ITS architecture can be created using a subset of the FRAME Architecture. This enables the required standards to be identified and, if necessary, their creation initiated. The creation of the European Specification is usually performed by a team of experts in the topic under consideration, with the addition of a small ITS architecture team who assist in imparting a common format to the result. This purpose of this process is the creation of a Physical and possibly other viewpoints for use throughout the EU. These can then be used directly by, for example, application developers to respond to a quickly changing market, but preserving the links to the overall structure. Thus, over time, the need for separate customized ITS architectures within member states or their subdivisions may diminish.

The advantages of using the FRAME Architecture in EU ITS project development are the following:

- Common language—Each resulting ITS architecture will be based on the FRAME Architecture, and thus use the same terminology.
- Common elements will be easy to identify, as will be the merging of two or more ITS architectures. This will be important as member states with their own ITS architectures need to include those that result from the ITS Action Plan or ITS Directive.
- Efficient—The FRAME Architecture already exists and contains about 80% of the information and work that will be needed to create operational ITS architectures.

14.7.5 International harmonization efforts concerning connected and cooperative vehicles

The emergence of the Connected Vehicle (CV) Program in the United States and Cooperative ITS (C-ITS) in the EU has driven expansion of both the U.S. and EU ITS architectures. In the United States, the CVRIA was developed to identify connected vehicle standards development needs. CVRIA contains Functional, Physical, Communications, and Enterprise Viewpoints and uses ITS entities from the U.S. National ITS Architecture where ITS interfaces are needed. The USDOT is currently integrating the CVRIA with the National ITS Architecture. The integrated product will be available at the beginning of 2017 [14]. The FRAME Architecture has also added functionality to incorporate C-ITS.

The United States, EU, and Australia along with other countries are working together to harmonize standards in the CV/C-ITS environment. These activities are taking place through working groups called Harmonization Task Groups (HTG). In particular, standards and security are being addressed in HTG 6 producing a policy framework for security solutions. HTG 7 is identifying needed standards and performing gap analysis for C-ITS. These working groups have used the CVRIA along with information from FRAME's C-ITS components to discuss and analyze the international standards that need to be developed, adopted, or adapted for CV/C-ITS (Steve Sill, personal communication).

14.8 JAPANESE ITS ARCHITECTURE

The Japanese ITS Architecture was completed in 1999 through the joint efforts of five government ministries involved in ITS, and in cooperation with the Vehicle, Road, and Traffic Intelligence Society (VERTIS), now ITS Japan. The objectives of the architecture are to promote the following [15]:

- Efficient construction of an integrated intelligent transportation system.
- A maintainable and expandable intelligent transport system.
- The development of domestic and international ITS standards.

The development of the Japanese ITS architecture was guided by two principles:

- Assure that the architecture is able to flexibly meet changing social needs and evolving technology.
- Assure that the architecture leads to an ITS that is interoperable and interconnectible with other parts of Japan's advanced information and telecommunications environment.

Like other major national system architectures, the Japanese ITS Architecture includes an enumeration of user services as presented in Table 14.4, a logical architecture, a physical architecture, and areas for creating ITS standards [15,16]. The Japanese ITS physical architecture is similar to that developed for the U.S. National Architecture shown in Figure 14.6. However, the Japanese physical architecture contains the four subsystem classes of Centers, Roadside, Vehicles, and Humans as illustrated in Figure 14.12 rather than Centers, Field, Vehicles, and Travelers found in the U.S. Architecture. The names and number of subsystems also are different as is the inclusion of the External Elements subsystem, whose components receive data through wide-area wireless communications networks.

Table 14.4 Japanese ITS Architecture development areas and user services

Development areas	User services
1. Advances in navigation systems	1. Provision of route guidance traffic information 2. Provision of destination-related information
2. Electronic toll collection systems	3. Electronic toll collection
3. Assistance for safe driving	4. Provision of driving and road conditions information 5. Danger warnings 6. Assistance for driving 7. Automated highway systems
4. Optimization of traffic management	8. Optimization of traffic flow 9. Provision of traffic restriction information in case of incident
5. Increasing efficiency in road management	10. Improvement of maintenance operations 11. Management of specially permitted commercial vehicles 12. Provision of roadway hazard information
6. Support for public transport	13. Provision for public transport information 14. Assistance for public transport operations management
7. Increasing efficiency in commercial vehicle operations	15. Assistance for commercial vehicle operations management 16. Automated platooning of commercial vehicles
8. Support for pedestrians	17. Pedestrian route guidance 18. Vehicle–pedestrian accident avoidance
9. Support for emergency vehicle operations	19. Automated emergency notification 20. Route guidance for emergency vehicles and support for relief activities
10. General	21. Utilization of information in the advanced information and telecommunications society

14.9 CANADIAN ITS ARCHITECTURE

The ITS Architecture for Canada, developed by Transport Canada, is a common framework for planning, defining, and integrating intelligent transportation systems. Version 2.0 of the Canadian architecture (issued in 2010) is the direct result of inputs from the ITS community and a re-alignment with Version 6.1 of the U.S. National ITS Architecture. Changes were incorporated into the physical and logical architectures, and several other key architecture definition documents for this version. The architecture defines the interactions among physical components of the transportation systems including travelers, vehicles, roadside devices, and control centers. It also describes the information and communications system requirements, data uses and sharing mechanisms, and the standards required to facilitate information sharing. Its structure supports ITS implementations in urban, interurban, and rural environments and is the foundation for ongoing ITS standards work [17]. Version 2 contains 37 user services grouped into nine development areas or bundles as illustrated in Table 14.5.

14.10 SUMMARY

The U.S. National ITS Architecture provides a common structure for the design of intelligent transportation systems in the United States. It defines

- Functions that must be performed by components or subsystems. These are contained in the logical architecture.

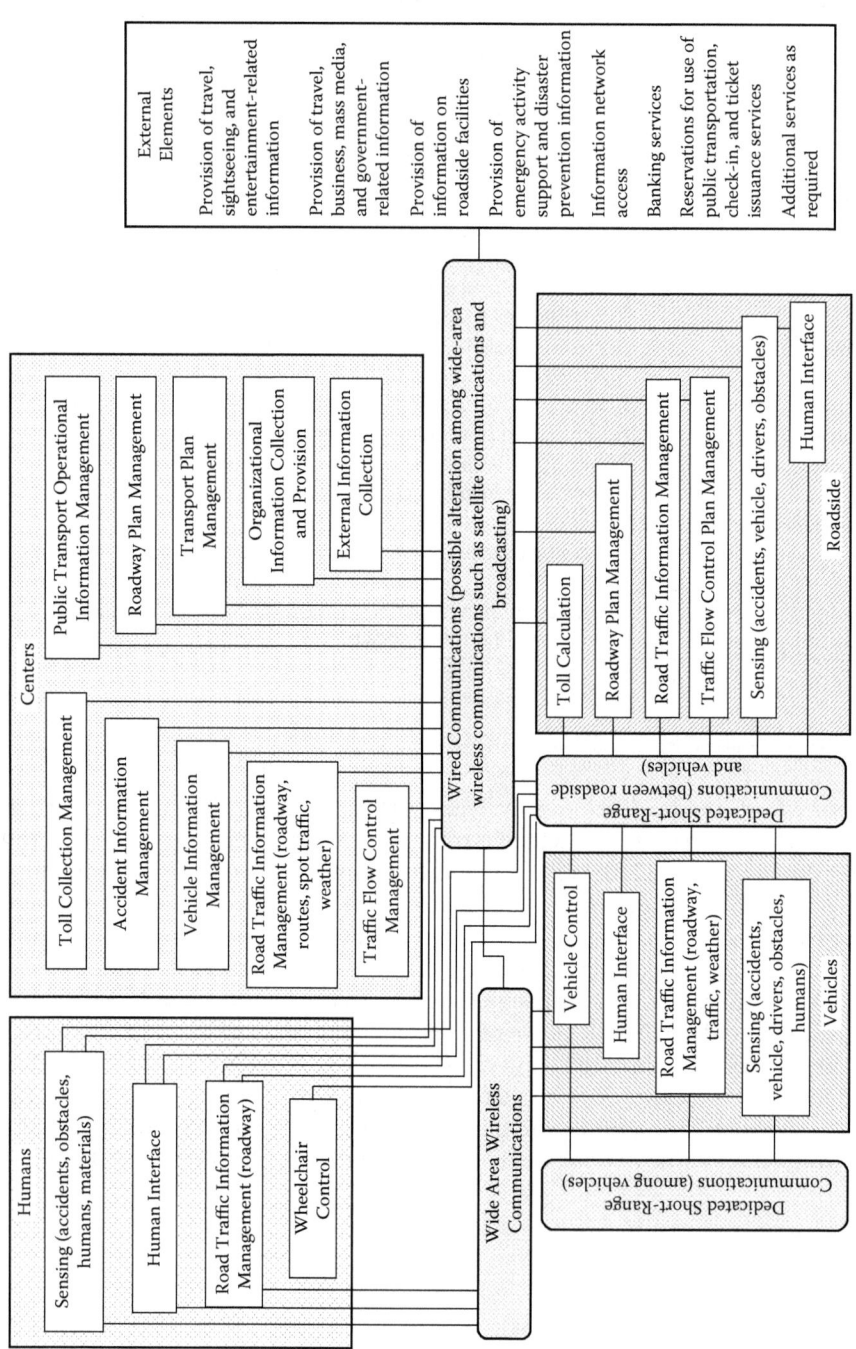

Figure 14.12 Japanese physical ITS architecture. *Author's note:* The Road Traffic Information Management subsystem that appears in the Humans subsystem class does not show any connections with communications media in the reference documents. However, the author believes that this was an oversight and has indicated what he thinks are the intended connections. (From T. Yokota and R.J. Weiland, *ITS System Architectures for Developing Countries, Technical Note 5*, Transport and Urban Development Department, World Bank, Jul. 22, 2004. https://www.google.com/url?q=http:// siteresources.worldbank.org/EXTROADSHIGHWAYS/Resources/ITSNote5.pdf&sa=U&ved=0ahUKEwjojqPQnt7OAhUOgx4KHcphC9sQFgg FMAA&client=internal-uds-cse&usg=AFQjCNFjRhLEKbN_LoRlguRqs-zcf7ktkQ.)

Table 14.5 Canadian ITS Architecture development areas (user bundles) and corresponding services

Development areas	User services
1. Traveler information	1. Pre-trip travel information 2. En-route driver information 3. Route guidance and navigation 4. Ride matching and reservation 5. Traveler services information
2. Traffic management	6. Traffic control 7. Incident management 8. Travel demand management 9. Emissions testing and mitigation 10. Highway-rail intersection 11. Automated dynamic warning and enforcement 12. Nonvehicular road user safety
3. Public transportation management	13. Public transportation management 14. En-route transit information 15. Demand responsive transit 16. Public travel security
4. Electronic payment	17. Electronic payment services
5. Commercial vehicle operations	18. Commercial vehicle electronic clearance 19. Automated roadside safety inspection 20. Onboard safety and security monitoring 21. Commercial vehicle administrative processes 22. Hazardous materials planning and incident response 23. Freight mobility 24. Intermodal freight management 25. International border transportation management
6. Emergency management	26. Emergency notification and personal security 27. Emergency vehicle management 28. Disaster response and evacuation
7. Advanced vehicle safety systems	29. Longitudinal collision avoidance 30. Lateral collision avoidance 31. Intersection collision avoidance 32. Vision enhancement for crash avoidance 33. Safety readiness 34. Pre-crash restraint deployment 35. Automated vehicle operation
8. Information management	36. Archived data
9. Maintenance and construction management	37. Maintenance and construction operations

- Where these functions reside (e.g., field, traffic management center, in-vehicle, or on pedestrian). These are shown in the physical architecture.
- Data flows between functions, architecture flows between subsystems, and their required interfaces.
- Communications requirements for the data flows in order to address the underlying user service requirements.

Interface and information exchange requirements established by the architecture today will likely facilitate or ease the transition to incorporating ITS standards-compliant interfaces in connected and cooperative vehicles and architectures of the future.

Programs similar to the U.S. effort are found in the EU, Japan, and other countries. In the EU, the FRAME Architecture offers a common language, terminology, and tool that can be utilized to develop ITS architectures. The FRAME Architecture structure contains a number of viewpoints. The functionality needed to implement ITS services is provided by the Functional Viewpoint, types of subsystems and components and the links between them are contained in the Physical Viewpoint, requirements for communications between the components are provided by the Communications Viewpoint, stipulations concerning who owns, manages, and operates the components and other organizational issues are found in the Organizational Viewpoint, and the information that is needed and its attributes and relationships are given in the Information Viewpoint. Because the FRAME Architecture is intended for application within the EU, it conforms to the precepts of subsidiarity, and thus does not mandate any physical or organizational structures on its users.

The Japanese ITS Architecture appears similar in design to that of the United States in that it contains a number of user services and a physical architecture structure analogous to those found in the U.S. architecture.

Other countries are also pursuing national ITS architectures. The ITS Architecture for Canada, developed by Transport Canada, is aligned with the U.S. National ITS Architecture. In Australia, assessment of many international ITS architectures identified the European ITS Framework Architecture as the best prototype for the Australian National ITS Architecture [18]. Mexico's ITS architecture and planning incorporate many of the concepts found in U.S. ITS strategies that apply to rural areas, and to a lesser extent, border crossings [19,20].

REFERENCES

1. L.A. Klein, N.A. Rantowich, C.C. Jacoby, and J. Mingrone, IVHS architecture development and evaluation process, *IVHS Journal*, 1(1):13–34, 1993.
2. C4ISR, *ITF Integrated Architecture Panel. C4ISR Architecture Framework, Version 1.0*, Report CISA-0000-104-96, Federation of American Scientists, Washington, DC, 1996.
3. National ITS Architecture, http://www.iteris.com/itsarch/index.htm. Accessed November 6, 2015.
4. *Architecture layers*, http://www.iteris.com/itsarch/html/archlayers/archlayers.htm. Accessed November 5, 2015.
5. *Key Concepts of the National ITS Architecture*, http://local.iteris.com/itsarch/documents/key-concepts/keyconcepts.pdf. Accessed November 16, 2016.
6. *Equipment packages*, http://www.iteris.com/itsarch/html/ep/epindex.htm. Accessed November 10, 2015.
7. Architecture Development Team, *National ITS Architecture Service Packages*, Prepared for Research and Innovation Technology Administration (RITA) U.S. Department of Transportation, Washington, DC, Jan. 2012. http://www.iteris.com/itsarch/documents/mp/sp.pdf. Accessed November 9, 2015.
8. *Service packages*, http://www.iteris.com/itsarch/html/mp/mpindex.htm. Accessed November 6, 2015.
9. *Traffic control user service*, http://www.iteris.com/itsarch/html/user/usr16.htm. Accessed November 6, 2015.
10. *Security*, http://www.iteris.com/itsarch/html/security/securityhome.htm. Accessed November 6, 2015.
11. *Turbo architecture*, http://www.iteris.com/itsarch/html/turbo/turbooverview.htm. Accessed November 6, 2015.
12. Home Page of the European Intelligent Transport Systems (ITS) Framework Architecture. frame-online.eu. Accessed August 26, 2016.
13. *The FRAME Architecture and the ITS Action Plan*, Booklet of the E-FRAME Project, Jun. 2011. http://frame-online.eu/wp-content/uploads/2014/10/FRAME-ITS-Action-Plan.pdf. Accessed August 26, 2016.

14. *FRAME User Needs V4.1.* http://frame-online.eu/wp-content/uploads/2014/10/FRAME-User-Needs-V4.1-01.pdf. Accessed August 26, 2016.
15. T. Yokota and R.J. Weiland, *ITS System Architectures for Developing Countries, Technical Note 5*, Transport and Urban Development Department, World Bank, Jul. 22, 2004. https://www.google.com/url?q=http://siteresources.worldbank.org/EXTROADSHIGHWAYS/Resources/ITSNote5.pdf&sa=U&ved=0ahUKEwjojqPQnt7OAhUOgx4KHcphC9sQFgg FMAA&client=internal-uds-cse&usg=AFQjCNFjRhLEKbN_LoRlguRqs-zcf7ktkQ. Accessed Aug. 26, 2016.
16. T. Hasegawa, Chapter 5: Intelligent transport systems, In *Traffic and Safety Sciences: Interdisciplinary Wisdom of IATSS*, First Edition, International Association of Traffic and Safety Sciences, Tokyo, Japan, Mar. 31, 2015. http://www.iatss.or.jp/common/pdf/en/publication/commemorative-publication/iatss40_theory_05.pdf. Accessed September 5, 2016.
17. *ITS Architecture for Canada*, https://www.itscanada.ca/about/architecture/. Accessed August 29, 2016.
18. http://www.austroads.com.au/road-operations/network-operations/national-its-architecture. Accessed August 29, 2016.
19. M. Schiemer and J. Labaco, *Development of Mexico's National Intelligent Transportation (ITS) Strategic Plan.* https://www.google.com/url?q=http://nationalruralitsconference.org/downloads/11documents/Abstracts/A1abstract_Schiemer.pdf&sa=U&ved=0ahUKEwiNrcTZlejOAhVFGR4KHV6tDFAQFggEMAA&client=internal-uds-cse&usg=AFQjCNH_jnT4z-WJ2F_uCiWEMG4FfGP9asQ. Accessed August 29, 2016.
20. R. Rajbhandari, J. Villa, R. Macias, and W. Tate, Chapter 4. Transportation operations and traffic management and enforcement, *Border-Wide Assessment of Intelligent Transportation System (ITS) Technology—Current and Future Concepts Final Report*, FHWA-HOP-12-015, U.S. Department of Transportation, Federal Highway Administration, Washington, DC, July 2012. http://ops.fhwa.dot.gov/publications/fhwahop12015/fhwahop12015.pdf. Accessed August 29, 2016.

Chapter 15

Connected vehicle architectures and applications

It is appropriate after discussing National ITS Architectures to explore a derivative architecture and several additional applications that were developed to benefit from information available from connected vehicles. To this end, we describe the Connected Vehicle Reference Implementation Architecture (CVRIA) first used in Southeast Michigan, a simulation of a traffic signal control system for connected vehicles that was evaluated with a calibrated model of a test network of four intersections along Route 50 in Chantilly, Virginia, and a lane management system for connected and conventional vehicles that alerts drivers to when it is productive and safe to change lanes on a controlled-access highway.

15.1 CONNECTED VEHICLE REFERENCE IMPLEMENTATION ARCHITECTURE

Continuing connected vehicle research recognizes that a framework is needed from which potential vehicle, handheld device, and infrastructure interfaces can be identified and analyzed for standardization. Because there are many types of connected vehicle applications and underlying system definitions, the subsequent system architecture was based on the fundamentals of the International Organization for Standardization/International Electrotechnical Commission/Institute of Electrical and Electronic Engineers (ISO/IEC/ IEEE) 42010:2011 standard "Systems and software engineering—Architecture description" [1]. The architecture development process includes steps to define not just data and messages, but the full environment in which stakeholders have concerns. This includes understanding the functionality, the high-level physical partitioning into subsystems or alternative configurations, the enterprise or institutional relationships that govern how those systems are deployed and operated, and the communications protocols needed for the interfaces to function properly. This class of architecture becomes a framework for developers, standards organizations, and implementers to utilize as a common reference structure for developing the eventual systems.

Once a multifaceted architecture is available, interfaces can be defined and analyzed to determine what areas need to be standardized. Some interfaces may already be standardized, others may just need modification to accommodate a new design or concept, or some may not be standardized at all. A standards development plan with input from the U.S. Department of Transportation or the appropriate transportation agency and the stakeholder community may be required to establish the priorities for harmonizing these interfaces. Ultimately, reference implementation architectures based on the new and modified standards can be established to inform policy makers, implementers, and the international community of the outcome.

The CVRIA is one such architecture that was established to identify connected vehicle standards development needs. It is defined from four points of view, namely, Enterprise, Functional, Physical, and Communications and uses ITS entities from the U.S. National ITS Architecture where needed [2]. Eventually the CVRIA will be integrated into the National ITS Architecture. The figures and tables in the sections below were selected to be representative of the various types of tools and exhibits available to depict the four viewpoints in the CVRIA.

15.1.1 Enterprise Viewpoint

The Enterprise Viewpoint addresses the relationships between organizations and the roles of those organizations in the delivery of services in the connected vehicle environment. It also deals with the personnel (including operators, users, and support staff) that are part of those organizations.

In the Enterprise Viewpoint, the CVRIA is depicted as a set of enterprise objects that interact to exchange information, and manage and operate systems beyond the scope of one organization. The Enterprise View describes the organizations that are involved and the roles they play in installing, operating, maintaining, and certifying all of the components in the connected vehicle environment. The relationships between enterprise objects are largely determined by roles, responsibilities, policies, and goals of the enterprises, not by CVRIA policies or goals, although there are exceptions for all-encompassing supporting functionality such as the management of digital certificates. Enterprise objects include Traffic Manager, Traffic Information or Management Center, Support Personnel and Services, Field Devices, Vehicles, Travelers, Pedestrians, and Other Entities. Enterprise View diagrams are utilized to identify the relationships between the institutions and people involved in the four phases of an application's life cycle: development and installation, operations, maintenance, and certification. Figure 15.1 provides an example of the Enterprise View for an intelligent traffic signal system [3].

15.1.2 Functional Viewpoint

The Functional View addresses the analysis of abstract functional elements and their behavior, structure, and logical interactions rather than engineering concerns of how functions are implemented, where they are allocated, how they transfer information, which protocols are used, and what method is used to implement them. Functional Views define processes that control and manage system behavior such as monitoring. The behavior of a function (also referred to as a process) is the set of actions performed by the process to achieve an application objective or to support actions of another process. This may involve data collection, data transformation, data generation, data processing, data stores, and the logical flows of data and information among these processes [4]. The Functional View also identifies other active control elements that determine the functional behavior of the system.

An example of a Functional View was shown in Figure 14.4 where the data flows for the Manage Traffic process were decomposed into its subprocesses. The Functional View may also be displayed as a table of process specifications and corresponding data flows between source and destination objects.

15.1.3 Physical Viewpoint

The Physical View consists of a set of integrated physical objects that interact and exchange information to support a particular connected vehicle application. Physical objects are

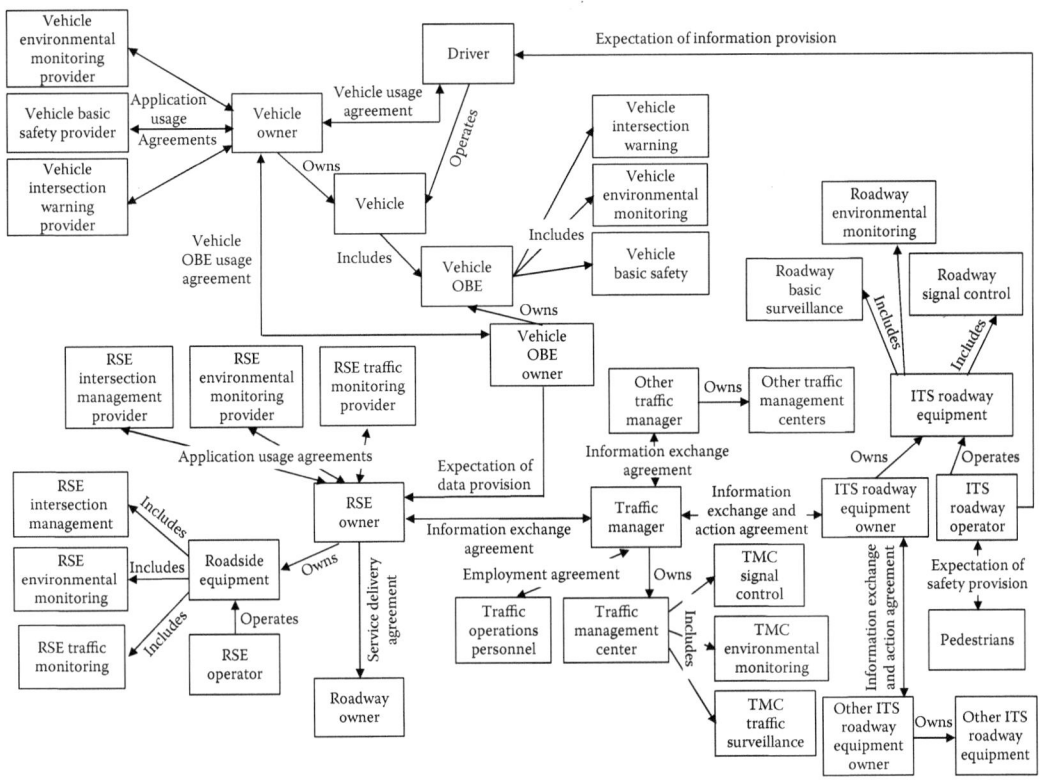

Figure 15.1 Enterprise View of an intelligent traffic signal system. (Adapted from Enterprise view of an Intelligent Traffic Signal System, http://www.iteris.com/cvria/html/applications/app43.html#tab-1, November 2015.)

linked to application objects that provide the specific functionality and interfaces required to implement the service. The objects represent physical elements that operate in the mobile environment, the field, and the back office where information from connections between elements and interactions with the external environment is processed. They include vehicle onboard equipment, traffic information or management centers, field devices, vehicles, travelers of all types, drivers, application servers, data stores, network components, mobile and nonmobile transportation elements, wired and wireless links, their physical connections and interactions, and the allocation of connected vehicle functionality to those elements. Information flows portray the exchange of information that occurs between physical objects and application objects.

The Physical View is interrelated to the other CVRIA views. Physical objects and application objects are linked to the Enterprise View, while application objects are also linked to the Functional View. In the Communications View, information exchanges defined in the Physical View are identified by triples that specify the source and destination physical objects and the information flow that is exchanged.

Figure 15.2 illustrates an application of the CVRIA to the highway management network in Southeast Michigan. This depiction of the physical layer shows the principal data flows. Other examples of physical views were illustrated in Figures 13.2 and 13.4 through 13.6 for the Multimodal Intelligent Traffic Signal System. Examples of physical and application objects for the Traffic management center (TMC) object are shown in Table 15.1 [5].

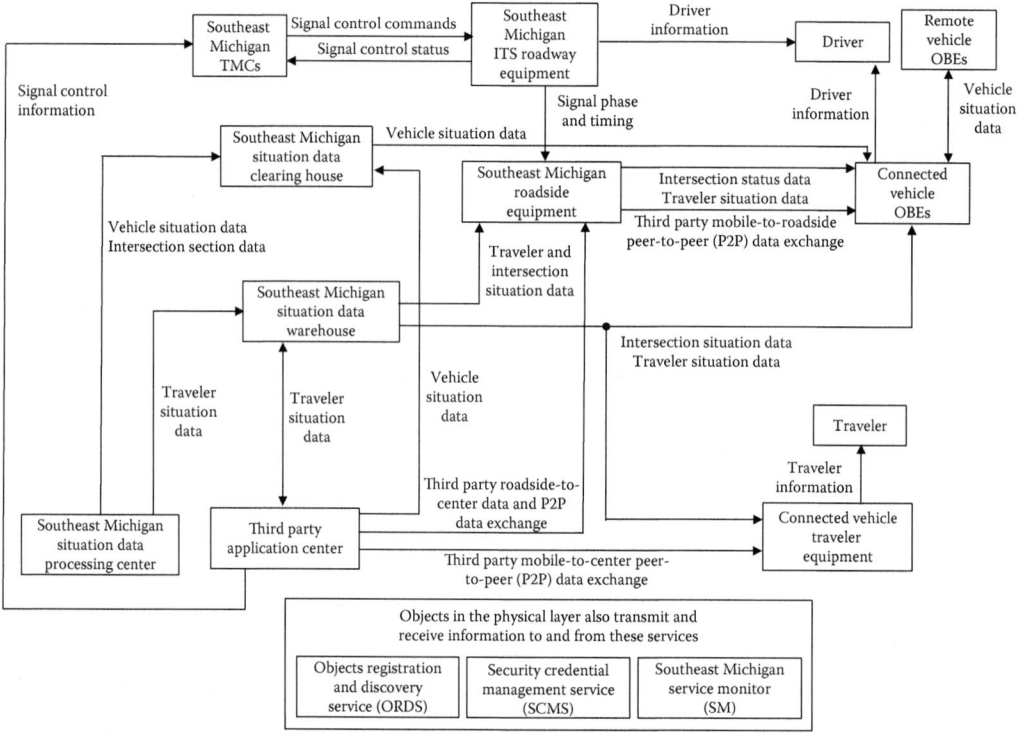

Figure 15.2 Southeast Michigan CVRIA physical layer. The major data flows are shown.

15.1.4 Communications Viewpoint

The Communications View describes the design and implementation of communications protocols and standards that provide interoperability between physical objects in the Physical View. It depicts the implementation choices and the specification and allocation of communications functionality to the components of the system.

The CVRIA communications model and Communications View diagrams are based on the open system interconnection (OSI) model, the National Transportation Communications for Intelligent Transportation System Protocol (NTCIP) framework, and DSRC/WAVE implementation guide as illustrated in Figure 15.3. Each Communications View diagram shows the information flow triple (the source, the transmitted data, and the destination) at the top followed by the source and destination communications protocols used for deployment in a layered stack. Each triple from the Physical View is mapped into one or more data dictionary standards, also referred to as an information layer standard, and one or more profiles defined by 16 standards that identify the communications protocols necessary to transport the data described by an information flow. In Figure 15.3, the information flow triple consists of the source physical object, here the vehicle OBE, the destination physical object, here the remote vehicle OBE, and the information flow, here the vehicle location and motion.

The names and functions of the seven layers in the CVRIA communications model are as follows:

1. *Process information layer*: The process information layer standards specify the structure, meaning, and control the exchange of information between two end points.

Table 15.1 TMC physical object with its corresponding application objects

Physical object	Application objects
Traffic Management Center	TMC Automated Vehicle Operations
	TMC Dynamic Lane Management and Shoulder Use
	TMC Environmental Monitoring
	TMC Evacuation Support
	TMC Incident Dispatch Coordination/ Communication
	TMC Infrastructure Restriction Warning
	TMC Intersection Safety
	TMC In-Vehicle Signing Management
	TMC Lighting System Control
	TMC Multimodal Coordination
	TMC Rail Crossing Management
	TMC Regional Traffic Management
	TMC Restricted Lanes CV Application
	TMC Roadway Warning
	TMC Signal Control
	TMC Speed Warning
	TMC Traffic Gap Assist
	TMC Traffic Information Dissemination
	TMC Traffic Metering
	TMC Traffic Surveillance
	TMC Variable Speed Limits
	TMC Work Zone Traffic Management

2. *Facility layer*: The facility layer standards define rules and procedures for exchanging encoded data.
3. *Encoding layer*: The encoding layer standards define the rules for representing the bits and bytes of information content to be transferred.
4. *Session layer*: The session layer provides the mechanism for opening, closing, and managing a dialogue between application processes. Sessions may be asynchronous as in paired requests and responses (information exchanges), asynchronous as in an unsolicited publication of information, and may require acknowledgement or receipt or not.
5. *Transport layer*: The transport layer standards define the rules and procedures for exchanging application data between endpoints on a network, including any necessary routing, message disassembly and re-assembly, and network management functions.
6. *Link layer*: The link layer standards define the rules and procedures for exchanging data between two adjacent devices over some communications media. These standards are roughly equivalent to the Data Link Layer of the OSI model.
7. *Physical layer*: The physical layer is a general term that describes the numerous signaling standards within this layer, typically developed for specific communications media and industry needs. With the exception of IEEE 802.11p (air interface to the 5.9 GHz spectrum) developed to address the needs of WAVE/DSRC, these standards are largely governed by the telephony industry.

The security plane identifies standards that specify system-to-system policies and authentication protocols, and encryption of data across one or more layers of the communications stack.

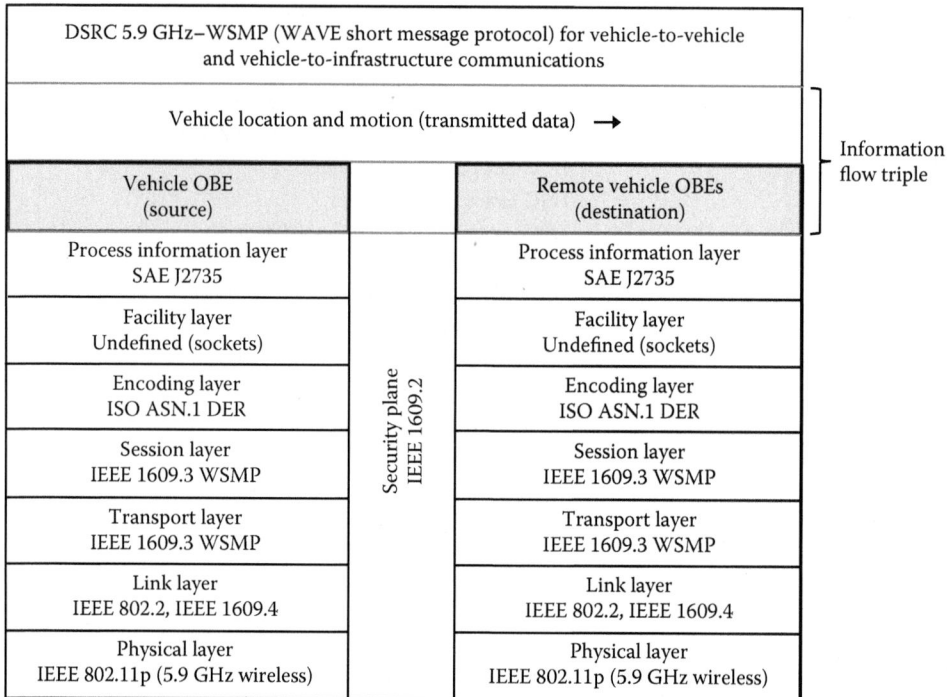

Figure 15.3 CVRIA V2V and V2I dedicated short-range communications based on the OSI seven-layer model.

Table 15.2 contains illustrations of information flow triples as seen through the Communications View of an intelligent traffic signal system [6].

15.1.5 Alternate connected vehicle viewpoint of the CVRIA

Another way to view the CVRIA is from the perspective of connected vehicle safety, mobility, environment, and support impact areas. Each of these is split into groups and the groups into applications as listed in Table 15.3 [7]. Safety contains transit safety, V2I safety, and V2V safety groupings of applications. Mobility contains border, commercial vehicle fleet operations, commercial vehicle roadside operations, freight advanced traveler information systems, planning and performance monitoring, public safety, traffic network, traffic signals, transit, and traveler information applications. Environmental applications are grouped into AERIS and sustainable travel, and road and weather systems. Support consists of core services, security, and signal phase and timing. An interactive Internet-based link is available for every application where descriptions, source references, and the CVRIA Enterprise, Functional, Physical, and Communications Views are found [7].

15.2 TRAFFIC SIGNAL CONTROL SYSTEM FOR CONNECTED VEHICLES

Traffic signal operation is currently dependent on data available from traditional point sensors. Point sensors, frequently in-ground inductive loops, may provide only limited vehicle

Table 15.2 Information flow triples for an intelligent traffic signal system

Source	Destination	Information flow
Driver	Vehicle OBE	Driver input
ITS roadway equipment	Roadside equipment	Conflict monitor status
ITS roadway equipment	Pedestrians	Crossing permission
ITS roadway equipment	Driver	Driver information
ITS roadway equipment	Traffic management center	Environmental sensor data
ITS roadway equipment	Roadside equipment	Intersection control status
ITS roadway equipment	Other ITS roadway equipment	Signal control data
ITS roadway equipment	Traffic management center	Signal control status
ITS roadway equipment	Traffic management center	Traffic flow
Other ITS roadway equipment	ITS roadway equipment	Signal control data
Other traffic management centers	Traffic management center	Device data
Other traffic management centers	Traffic management center	Device status
Other traffic management centers	Traffic management center	Road network conditions
Pedestrians	ITS roadway equipment	Pedestrian detection
Roadside equipment	ITS roadway equipment	Environmental situation data
Roadside equipment	Traffic management center	Environmental situation data
Roadside equipment	Traffic management center	Intersection management application status
Roadside equipment	Vehicle OBE	Intersection status
Roadside equipment	ITS roadway equipment	Intersection status monitoring
Roadside equipment	ITS roadway equipment	Signal service request
Roadside equipment	Traffic management center	Traffic situation data
Roadside equipment	ITS roadway equipment	Traffic situation data
Roadside equipment	Vehicle OBE	Vehicle situation data parameters
Traffic management center	Other traffic management centers	Device data
Traffic management center	Other traffic management centers	Device status
Traffic management center	ITS roadway equipment	Environmental sensors control
Traffic management center	Roadside equipment	Intersection management application info
Traffic management center	Other traffic management centers	Road network conditions
Traffic management center	ITS roadway equipment	Signal control commands
Traffic management center	ITS roadway equipment	Signal control device configuration
Traffic management center	ITS roadway equipment	Signal control plans
Traffic management center	ITS roadway equipment	Signal system configuration
Traffic management center	Traffic operations personnel	Traffic operator data
Traffic management center	ITS roadway equipment	Traffic sensor control
Traffic operations personnel	Traffic management center	Traffic operator input
Vehicle data bus	Vehicle OBE	Driver input information
Vehicle data bus	Vehicle OBE	Host vehicle status
Vehicle OBE	Vehicle data bus	Driver update information
Vehicle OBE	Driver	Driver updates
Vehicle OBE	Roadside equipment	Vehicle environmental data
Vehicle OBE	Roadside equipment	Vehicle location and motion for surveillance
Vehicle OBE	Roadside equipment	Vehicle situation data

Table 15.3 CVRIA applications by impact area and group

Impact area	Group	Application name
Safety	Transit Safety	Transit Pedestrian Indication
		Transit Vehicle at Station/Stop Warnings
		Vehicle Turning Right in Front of a Transit Vehicle
	V2I Safety	Curve Speed Warning
		In-Vehicle Signage
		Oversize Vehicle Warning
		Pedestrian in Signalized Crosswalk Warning
		Railroad Crossing Violation Warning
		Red Light Violation Warning
		Reduced Speed Zone Warning/Lane Closure
		Restricted Lane Warnings
		Spot Weather Impact Warning
		Stop Sign Gap Assist
		Stop Sign Violation Warning
		Warnings about Hazards in a Work Zone
		Warnings about Upcoming Work Zone
	V2V Safety	Blind Spot Warning + Lane Change Warning
		Control Loss Warning
		Do Not Pass Warning
		Emergency Electronic Brake Light Warning
		Emergency Vehicle Alert
		Forward Collision Warning
		Intersection Movement Assist
		Motorcycle Approaching Indication ⊕
		Pre-crash Actions
		Situational Awareness
		Slow Vehicle Warning ⊕
		Stationary Vehicle Warning ⊕
		Tailgating Advisory
		Vehicle Emergency Response
Mobility	Border	Border Management Systems
	Commercial Vehicle Fleet Operations	Container Security
		Container/Chassis Operating Data
		Electronic Work Diaries ⊕
		Intelligent Access Program ⊕
		Intelligent Access Program—Mass Monitoring ⊕
	Commercial Vehicle Roadside Operations	Intelligent Speed Compliance ⊕
		Smart Roadside Initiative
	Electronic Payment	Electronic Toll Collection
		Road Use Charging
	Freight Advanced Traveler Information Systems	Freight Drayage Optimization
		Freight-Specific Dynamic Travel Planning
	Planning and Performance Monitoring	Performance Monitoring and Planning
	Public Safety	Advanced Automatic Crash Notification Relay
		Emergency Communications and Evacuation
		Incident Scene Pre-Arrival Staging Guidance for Emergency Responders
		Incident Scene Work Zone Alerts for Drivers and Workers
	Traffic Network	Cooperative Adaptive Cruise Control
		Queue Warning
		Speed Harmonization
		Vehicle Data for Traffic Operations

(Continued)

Table 15.3 (Continued) CVRIA applications by impact area and group

Impact area	Group	Application name
	Traffic Signals	Emergency Vehicle Preemption Freight Signal Priority Intelligent Traffic Signal System Pedestrian Mobility Transit Signal Priority
	Transit	Dynamic Ridesharing Dynamic Transit Operations Integrated Multimodal Electronic Payment Intermittent Bus Lanes Route ID for the Visually Impaired Smart Park and Ride System Transit Connection Protection Transit Stop Request
	Traveler Information	Advanced Traveler Information Systems Receive Parking Space Availability and Service Information Traveler Information—Smart Parking
Environmental	AERIS and Sustainable Travel	Connected Eco-Driving Dynamic Eco-Routing Eco-Approach and Departure at Signalized Intersections Eco-Cooperative Adaptive Cruise Control Eco-Freight Signal Priority Eco-Integrated Corridor Management Decision Support System Eco-Lanes Management Eco-Multimodal Real-Time Traveler Information Eco-Ramp Metering Eco-Smart Parking Eco-Speed Harmonization Eco-Traffic Signal Timing Eco-Transit Signal Priority Electric Charging Stations Management Low Emissions Zone Management Roadside Lighting
	Road and Weather Systems	Enhanced Maintenance Decision Support System Road Weather Information and Routing Support for Emergency Responders Road Weather Information for Freight Carriers Road Weather Information for Maintenance and Fleet Management Systems Road Weather Motorist Alert and Warning Variable Speed Limits for Weather-Responsive Traffic Management
Support	Core Services	Connected Vehicle Map Management Core Authorization Data Distribution Infrastructure Management Location and Time Object Registration and Discovery Privacy Protection System Monitoring
	Security	Security and Credentials Management
	Signal Phase & Timing	Signal Phase & Timing

Note: European Union and Australian applications are designated with the international icon ⊕.

information at a fixed location. Furthermore, advanced adaptive traffic signal control strategies are often not implemented in the field due to their operational complexity and their need for many detection areas, especially on facilities that serve multilane intersections. However, connected vehicles would allow for the wireless transmission of vehicles' positions, headings, and speeds for use by the traffic controller. The predictive microscopic simulation algorithm (PMSA) was developed to utilize these new, more robust data [8,9]. This decentralized, fully adaptive traffic signal control algorithm uses a rolling horizon strategy, whose phasing minimizes an objective function over a 15-s period in the future [10]. Its objective function utilizes either delay-only, or a combination of delay, stops, and decelerations. To measure the objective function, the algorithm employs a microscopic simulation driven by present vehicle positions, headings, and speeds. Unlike most adaptive control strategies, the algorithm is relatively simple, does not require point sensors or traffic-signal to traffic-signal communication, and is responsive to immediate vehicle demands. To ensure drivers' privacy, the algorithm does not store individual or aggregate vehicle locations. Simulation results show that the algorithm maintains or improves performance compared to a state-of-practice coordinated-actuated timing plan optimized by Synchro at low- and mid-level volumes. However, performance worsens during saturated and oversaturated conditions. Testing also showed improved performance during periods of unexpected high demand and the ability to automatically respond to year-to-year growth without retiming.

15.2.1 Predictive microscopic simulation algorithm

The PMSA has three objectives:

1. Match or significantly improve the performance of a state-of-practice actuated-coordinated traffic signal system.
2. Respond to real-time demands only, thereby eliminating the need for manual timing plan updates to adjust for traffic growth or fluctuations.
3. Never re-identify, track, or store any records of individual or aggregate vehicle movements for any length of time, thereby protecting driver privacy.

The objectives are fulfilled through a rolling horizon approach whereby the traffic signal controller attempts to minimize an objective function (in this case, total delay) over a short time period in the future. Total delay is the combination of delay due to slower than normal speed and delay due to stops.

A microscopic traffic model simulates vehicles over the horizon period and calculates the objective function delay directly from the vehicle's simulated behavior. An intersection's movement is defined as a single controlled vehicle path, for example, westbound left, whereas a phase is defined as two noncontradictory movements, for example, westbound left and eastbound left. When the algorithm recalculates the signal's phase, it first collects a snapshot of the position, heading, and speed of every equipped vehicle within 300 m (984 ft) of the intersection. This is the distance a vehicle travels during the 15-s horizon at 45 mi/h (72 km/h), which is the speed of this particular corridor.

Figure 15.4 displays the information that is utilized to populate a model of the vehicles in the vicinity of the intersection. The blue rectangular objects represent vehicle positions predicted by the simulation model, while the red and green colors show the signal phase of the intersections.

The algorithm operates completely without inductive loop or video detection, with no knowledge of expected demand or memory of past demand, and is decentralized. There is no communication with any other signal on the corridor, either ad hoc or through synchronized

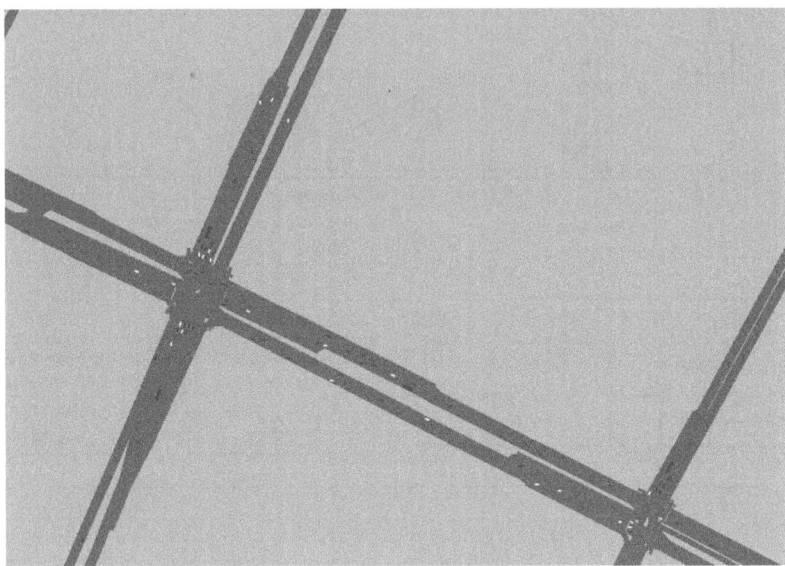

Figure 15.4 Intersection model as populated by the PMSA with the positions and speed of equipped vehicles that represent real-world intersection data. (From Goodall, N. J., B. L. Smith, and B. Park. *Traffic Signal Control with Connected Vehicles. Transportation Research Record: Journal of the Transportation Research Board*, No. 2381, Figure, p. 67, 2013. Reproduced with permission of the Transportation Research Board.)

timing. The algorithm was designed to be compatible with the SAE J2735 standard for V2V and V2I DSRC communications. It requires only the information broadcast in the basic safety message no more frequently than once per second, whereas the message is sent 10 times per second according to the standard. Further, the algorithm is able to protect driver privacy by clearing any vehicle data seconds after it is recorded. The algorithm does not store any vehicle location data, neither aggregated volumes nor individual vehicle trajectories, once the next phase is determined.

15.2.2 Simulation parameters

Once the model has been populated with the new vehicles, the vehicles are simulated 15 s into the future. Because the turn lanes in the test network were between 75 and 300 m (246 and 984 ft) in length, the turning movement of many vehicles can be assumed based on their current lane. For vehicles upstream of the turning lane, it was assumed that 50% of those in the lane nearest a turning lane would use the turning lane. This is repeated once for each possible new phase configuration, and for the possibility of maintaining the current phasing. Four-second amber phases and two-second red phases are simulated as well. The phase with the optimal objective function over the 15-s horizon is selected as the next phase.

The new phase's green time is determined from the horizon simulation as the time required to clear all simulated vehicles from a single movement. The green time has a 5-s minimum and a 15-s maximum before recalculation.

Figure 15.5 illustrates the PMSA's decision process. To ensure smooth operation of the signal, several restrictions are put in place. Because the algorithm is acyclic and allows phase skipping, each movement has a maximum red time of 120 s. This was considered reasonable since the Synchro-recommended timing plan for the corridor was 120 s. Also, to take

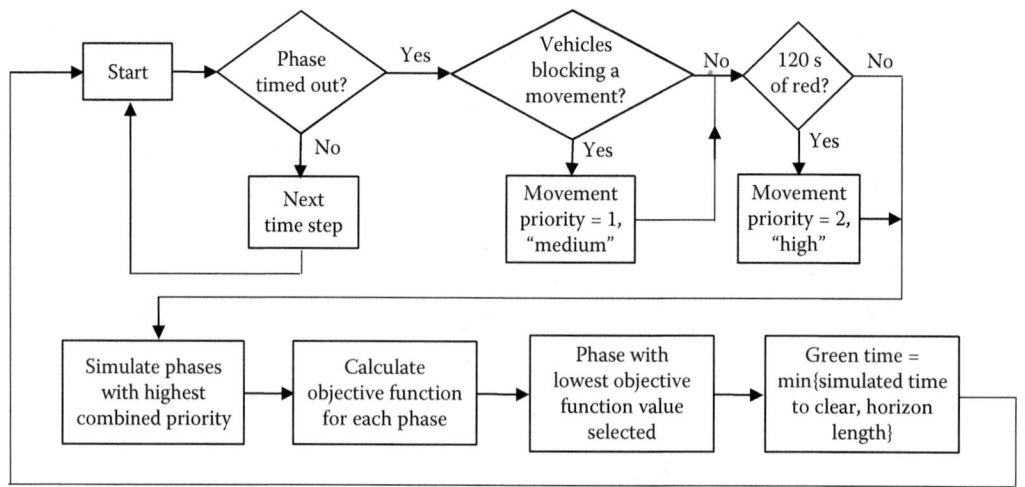

Figure 15.5 PMSA decision flow chart. (Reprinted with permission of N.J. Goodall, *Traffic Signal Control with Connected Vehicles*, A dissertation presented to the Faculty of the School of Engineering and Applied Science, University of Virginia, May 2013.)

advantage of the queue detection capabilities of connected vehicles, the algorithm will not allow queues to block a turning lane or a through lane. When a vehicle is detected within 40 ft (12 m) of blocking a movement, the vehicle's movement is given priority at the next phase recalculation.

VISSIM was used to simulate the position and speed of the connected vehicles as it allows users to easily access individual vehicle information via a COM interface, and also sequences a second "future" simulation parallel to the primary simulation. The test network is a calibrated model of four intersections along Route 50 in Chantilly, Virginia. Actual vehicle volumes and turning movements were collected in 2003 between 3:00 and 4:00 p.m. on weekdays [11]. Pedestrian movements, which were very low at these intersections, were eliminated for this analysis as the minimum pedestrian crossing time often exceeded 60 s, well beyond the algorithm's 15-s horizon.

Synchro was applied off-line to create an optimized coordinated-actuated timing plan with a 120-s cycle length as a base case for comparison with the PMSA. Synchro's recommended timing plans were programmed into and tested in the VISSIM network. Vehicle volumes were converted to approximate intersection saturation rates using Synchro's intersection capacity utilization (ICU) metric, and measured at an average of 0.75. To investigate the sensitivity of the algorithm to various equipped vehicle penetration rates, the algorithm was tested at 10%, 25%, 50%, and 100% vehicle participation, using total delay over the horizon as the sole element of the objective function. Each scenario was evaluated for 30 min after 400 s of simulation initialization. Each scenario was assessed 10 times at different random seeds, and all produced statistically similar results at a 95% confidence level.

15.2.3 Test results

Improvements in delay and speed were only experienced at penetration rates of 50% and higher. There were fewer stops at higher penetration rates, but always more stops compared to a coordinated-actuated system. Stopped delay improves at 25% penetration and higher, with a 34% improvement experienced with 50% of vehicles participating [8]. These improvements are experienced without assumed knowledge of historical demand volumes, nor of any coordination or communication with neighboring signals. The algorithm has the advantage of responding to unexpected demands due to incidents with minimal transition time compared to a time-of-day plan because the PMSA requires no knowledge of historical traffic demands. This can be important because 25% of congestion is caused by incidents according to a Federal Highway Administration (FHWA) estimate [12].

To evaluate the PMSA's ability to cope with large unexpected variations in flow, a simulation was run where volumes entering the mainline heading east increased by 30%. This represents a realistic scenario for vehicles rerouting to avoid an incident on a parallel freeway or arterial. The PMSA, operating with 100% equipped vehicle penetration rate, is able to respond instantly to the increased demand, with no outside input from operators or communication with roadside infrastructure or nearby signals. With the unexpected volume increase, the PMSA produces greater benefits as compared to the improvements provided by a correctly timed coordinated-actuated system.

Another common cause of congestion is poor signal timing, estimated by FHWA to be responsible for 5% of all congestion. The PMSA, because it responds only to immediate traffic demand, can accommodate annual volume increases without adjustments. To evaluate this benefit of the PMSA, the algorithm was tested at 100% market penetration against a coordinated-actuated timing plan that was optimized for the much lower volumes from 10 years in the past, assuming a 3% annual growth rate for all approach volumes. This equates to a 34% volume increase in all directions, with no change to the timing plan. The PMSA showed significant benefits across all metrics.

15.2.4 Other objective functions

Other objective functions incorporating acceleration and stops were added to the delay function for evaluation. A multivariable objective function was defined as

$$f = \alpha \times d + \beta \times a + \gamma s, \tag{15.1}$$

where α, β, and γ are adjustable factors, d is delay per second per vehicle, a is negative acceleration per second per vehicle, and s is stops per vehicle.

The delay, acceleration, and stops are defined as follows:

$$d = \min\left\{1, \frac{\sum_{i=1}^{n}\sum_{j=1}^{t} d_{ij}}{ntd_{\max}}\right\}, \tag{15.2}$$

$$a = \min\left\{1, \frac{\sum_{i=1}^{n}\sum_{j=1}^{t} \max\{a_{ij}, 0\}}{nta_{\max}}\right\}, \tag{15.3}$$

and

$$s = \min\left\{1, \frac{\sum_{i=1}^{n}\sum_{j=1}^{t} s_{ij}}{ns}\right\}, \tag{15.4}$$

where i represents an individual vehicle, j represents a single time interval, n represents the total number of vehicles, t represents the total time, and d_{ij} represents the delay of vehicle i over time j.

The addition of acceleration and stops to the objective function was motivated by the following. Delay and stops are commonly used measures of effectiveness for signal timing. Negative acceleration was selected because of its relationship with emissions (although positive acceleration is correlated with emissions, it is an unrealistic metric in practice as it discourages phase changes under all circumstances).

A range of α, β, and γ between 0 and 1 was tested in increments of 0.1, and where the sum of α, β, and γ was always equal to 1. The maximum delay per second per vehicle d_{max}, maximum acceleration a_{max}, and maximum stops per vehicle s_{max} were set to 1 s/s/veh, 3 m/s³/veh, and 2 stops/veh, respectively, to cap and normalize the observed values. All scenarios assumed an equipped vehicle penetration rate of 100%.

Different objective functions were unable to significantly improve on the delay-only function, either in average delay or stops. A high acceleration factor in particular produces poor performance when compared to a delay-only function.

While considerable progress has been shown in optimizing the performance of the PMSA, additional development is needed to improve its performance at low connected vehicle penetration rates. Research suggests that the behavior of a few connected vehicles can estimate positions of unequipped vehicles in real-time on freeways and delay on arterials [13–15]. These techniques may be adapted for signal control where they can provide real-time estimates of individual vehicle locations, thereby artificially augmenting the equipped penetration rate.

15.3 LANE MANAGEMENT SYSTEM FOR CONNECTED AND CONVENTIONAL VEHICLES

When a motorist enters the freeway, it becomes necessary to select a lane. Furthermore, drivers of conventional vehicles often switch freeway lanes expecting to go faster. Though performed more or less subconsciously, the choice of lanes is often suboptimal or poor because the driver does not receive a detailed picture of traffic conditions at some downstream distance in the driver's current lane and in other lanes. The result is often lane switching that does not accomplish the driver's objective. This unnecessary lane switching results in nonproductive maneuvers, disturbance to traffic flow, waste of fuel, and driving workloads that are greater than necessary. It also results in more accidents (4% of the crashes result from lane changes). The situation is even more complex for automated vehicles.

15.3.1 Advanced Lane Management Assist concept

The proposed Advanced Lane Management Assist (ALMA) concept (patented) functions as a decision support system that provides lane changing advice to drivers of both conventional vehicles and connected vehicles containing automated driving features [16]. The system informs the driver if a lane change is appropriate, assists in selecting the most appropriate travel lane, and gives the target speed for the selected lane. It also determines if there is a suitable gap in traffic for the lane change to occur safely. With conventional vehicles, ALMA serves as an extension of the navigation system to the lane guidance level. However, the motorist makes the decision and performs the corresponding maneuvers by manually utilizing the vehicle's lateral and longitudinal control systems. For connected vehicles incorporating automated features, ALMA enables the driver to select a lane, while the vehicle's lateral and longitudinal controls automatically perform the actual lane changing maneuvers and associated safety functions.

As shown in Figure 15.6, ALMA bridges the gap between traditional vehicle navigation systems and the vehicle's lateral and longitudinal control systems. ALMA obtains data from freeway TMCs operated by states and other agencies. It processes the data to calculate lane speed, volume (vehicles/hour/lane), average vehicle time headway (hours/vehicle/lane), traffic density (vehicles/mile/lane), average vehicle length, gap between vehicles, and passenger car equivalent volume per hour per lane. The lane-specific time- and space-based information supports timely decision-making by the driver.

Figure 15.7 displays the principal data flow relationships among ALMA modules, the TMC, and the vehicle. ALMA components are represented as rectangles enclosed by solid lines. The dash–dot rectangles represent components that allow the vehicle to respond to information provided by the computer-based ALMA Management Center (ALMAMC). The dashed rectangles are other components of the system.

The ALMAMC combines the data it receives from freeway TMCs and other sources with information in the ALMAMC portion of the ALMA static database. ALMA database information includes dynamic, static, and user-selected lane use requirements and restrictions. TMC data consists of mandatory lane controls, lane speeds and closures, variable speed limits, hard shoulder running permissibility, weight restrictions, HOV and toll-tag requirements, and entry and exit ramp closures.

The ALMA data formatter transforms the TMC data into ALMA data structures. This information is transmitted to the ALMA Static Database Module in the vehicle using any suitable means such as satellite radio, cellular networks, and cloud communication. ALMA can also employ DSRC connected vehicle communications technology as available. The Static Database Module combines the TMC data with information from the vehicle manufacturer relating to vehicle handling and acceleration characteristics and

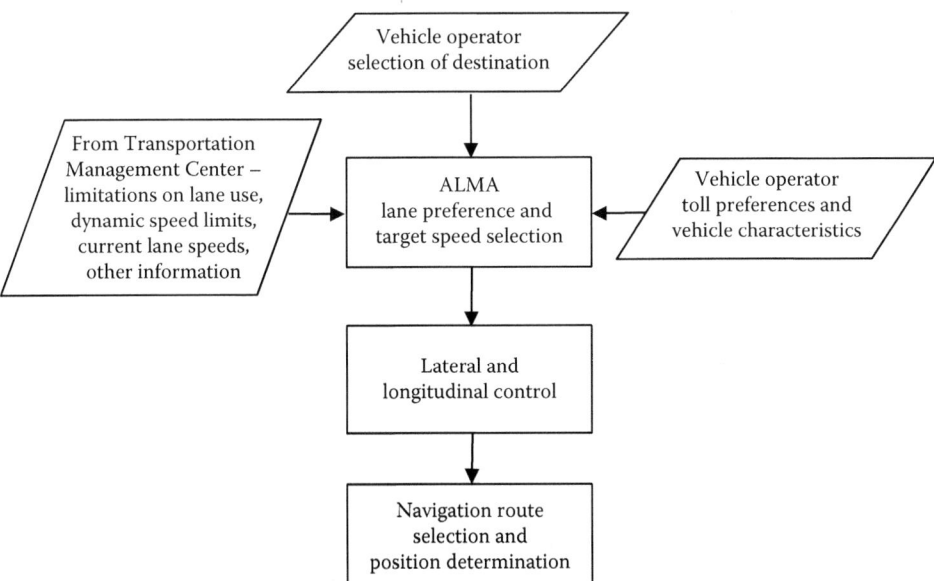

Figure 15.6 ALMA concept. (Adapted from R.L. Gordon, ALMA ends freeway lane lottery, *ITS International*, 20(1):NA6–NA7, 2014.)

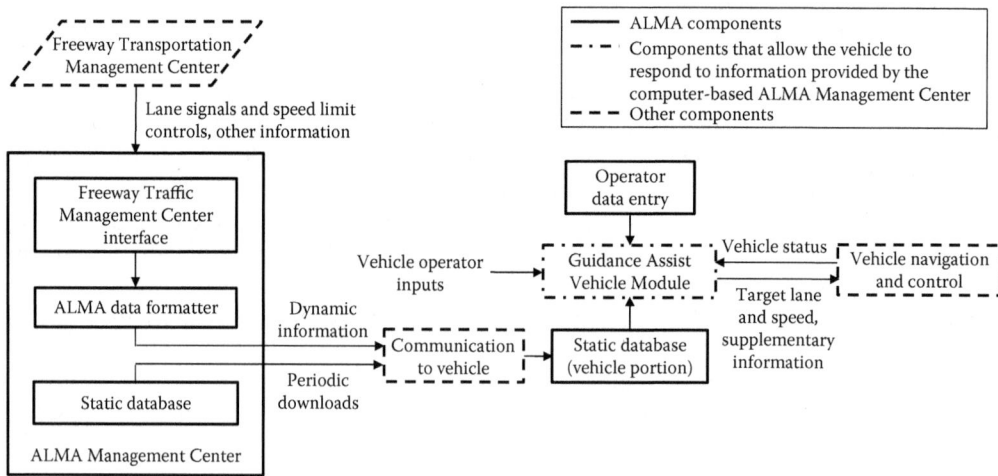

Figure 15.7 ALMAMC interfaces and data flows with TMCs and the vehicle. (Adapted from R.L. Gordon, ALMA ends freeway lane lottery, *ITS International*, 20(1):NA6–NA7, 2014.)

driver and passenger occupancy information. Then the Guidance Assist Vehicle Module (GAVM) computes and recommends lane-change strategies and target speeds for the selected lane. This strategy takes into account the motorist's driving speed preferences, motorist aggressiveness, class of vehicle, number of passengers, exit proximity, tolling preferences, and the likely availability of a suitable gap in alternative lane traffic in the event of a merge. The ALMA static database is periodically updated as new information becomes available.

Information presented to the driver must enhance the following assessments:

1. Lane choice based on traffic regulations, toll preferences, type of vehicle, number of vehicle occupants, automatic speed enforcement considerations, and freeway exiting requirements.
2. Motorist satisfaction with current lane speed.
3. Motorist satisfaction with speed in an alternative available lane.
4. Presence of a gap in a selected lane that is sufficiently wide for a safe merge.

15.3.2 ALMA lane-change strategy

Before ALMA will recommend, or in the case of a connected or automated vehicle, initiate a lane change, the GAVM *requires a degree of long-term speed differentiation between the current and other lanes* for the next few kilometers and the calculation of the likelihood of a *suitable gap* in the traffic in the alternative lane. For example, if the vehicle is currently in Lane 2 and the average speed in Lane 3 is 3 km/h (2 mi/h) higher for the next 5 km (3 mi), then a lane change would not be advised.

Lane-change indications are limited to where they will work meaningfully and the lack of any lane-change advice provides drivers with the assurance that they will not progress more slowly if they remain in their current lane.

Where the ALMA system detects little difference in relative lane speeds, such information could also be conveyed to all drivers by variable message signs. By avoiding over-corrections as multiple drivers swap lanes, traffic flow will become more stable, trip quality improved, and the potential for accidents during lane changing reduced.

15.3.3 Alternative lane-change strategy

The lane-change strategy described above provides ample opportunities for motorists to implement their driving preferences. An alternative strategy, ALMA Truck (ALMATR), recommends a lane change when the change will improve travel time and safety as may be applicable to operators of commercial vehicles (Robert Gordon, personal communication). By not recommending changes when lane speeds indicate that another lane will not move significantly faster for the next few miles, unnecessary lane changes are avoided, with resulting safety benefits. Lane changes are recommended when lane speeds indicate that time savings may be significant. ALMATR can also be used for cars, although individual motorists often have their own preferred driving styles.

Figure 15.8 illustrates the benefits of ALMATR. The figure was constructed from data provided by the California Department of Transportation's Performance Measurement System (PeMS) along a 4-mi section of I-880 in Fremont, CA. Trucks travel almost exclusively in Lane 4 (lane next to the right shoulder) and Lane 3 (the adjacent lane to the left). The plot shows the difference between the average speeds and travel times in these lanes during 5-min time periods on a weekday. The data incorporate two assumptions: (1) observed local lane speed differences of 4 mi/h (6 km/h) will induce many drivers to change lanes, and (2) a 1-min time saving over the 4-mi section is significant, since this savings of time when added to similar savings over other roadway sections results in a meaningful travel-time reduction. Many of the data points within the dashed rectangle show that significant speed differences do not result in meaningful time savings (because they occur at higher speeds) and may induce unnecessary lane changes and excess fuel consumption from the acceleration and deceleration maneuvers. These unnecessary lane changes also increase crash exposure. The solid rectangle shows the 5-min time periods that result in favorable lane-change opportunities. The settings and thresholds in ALMATR described above are adjustable by the system manager or the truck operator.

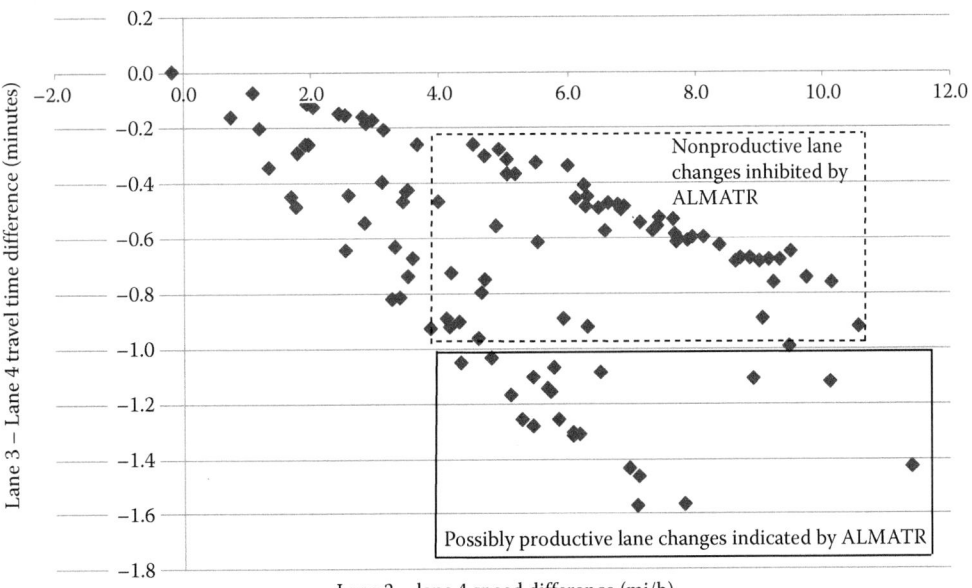

Figure 15.8 Travel-time difference as a function of speed difference for a 4-mi section of I-880 in Fremont, CA. (From Robert Gordon, personal communication.)

15.4 SUMMARY

The introduction and use of connected vehicles and other devices makes available new sources of data that are ideally suited for enhancing the mobility and safety of travelers. Three applications that seek to exploit this information were described. The first is the CVRIA developed to assist in the definition of interfaces and data flows from four viewpoints: Enterprise, Functional, Physical, and Communications. The Enterprise Viewpoint addresses the relationships between organizations and their personnel and the roles the organizations serve to deliver services in the connected vehicle environment. The Functional View addresses the analysis of abstract functional elements, their logical interactions, and the flows of data and information among them. The Physical View consists of a set of integrated physical objects that interact and exchange information to support a particular connected vehicle application. The Communications View describes the design and implementation of communications protocols and standards that provide interoperability between physical objects in the Physical View.

The second application is the PMSA that utilizes wireless transmission of connected vehicle positions, headings, and speeds to optimize the timing of traffic signals over a section of roadway. The rolling horizon strategy minimizes an objective function, in this example total delay, over a 15-s period in the future to meet the objectives of the adaptive traffic signal control system. Test results show that the algorithm was able to reduce stopped delay when connected vehicles were at least 25% of the vehicle mix.

The third concept is the ALMA decision support system that provides lane changing advice to drivers of conventional and connected vehicles containing automated driving features. The system informs the driver if a lane change is appropriate, assists in selecting the most appropriate travel lane, and gives the target speed for the selected lane. The system also determines if there is a suitable gap in traffic for the lane change to occur safely.

REFERENCES

1. International Standard "Systems and Software Engineering—Architecture Description," ISO/IEC/IEEE 42010:2011(E), First Edition 2011-12-01, International Organization for Standardization, ISO Central Secretariat, BIBC II, Chemin de Blandonnet 8, CP 401, 1214 Vernier, Geneva, Switzerland. http://cabibbo.dia.uniroma3.it/asw/altrui/iso-iec-ieee-42010-2011.pdf, Authorized licensed use limited to Biblioteca D'Area Scientifico Tecnologica Roma 3. Accessed November 5, 2015.
2. Connected Vehicle Reference Implementation Architecture (CVRIA) Website, http://www.iteris.com/cvria/index.html
3. Enterprise view of an Intelligent Traffic Signal System, http://www.iteris.com/cvria/html/applications/app43.html#tab-1
4. Functional processes in an Intelligent Traffic Signal System, http://www.iteris.com/cvria/html/applications/app43.html#tab-2
5. Application objects found in the physical view, http://www.iteris.com/cvria/html/appobjects/appobjects.html
6. Application triples for an Intelligent Traffic Signal System, http://www.iteris.com/cvria/html/applications/app43.html#tab-4
7. Connected Vehicle Reference Implementation Architecture Applications, http://www.iteris.com/cvria/html/applications/applications.html
8. N.J. Goodall, B.L. Smith, and B. Park, Traffic signal control with connected vehicles, Transportation Research Record: Journal of the Transportation Research Board, No. 2381, Washington, DC, 65–72, 2013. http://dx.doi.org/10.3141/2381-08.

9. N.J. Goodall, *Traffic Signal Control with Connected Vehicles*, A Dissertation Presented to the Faculty of the School of Engineering and Applied Science, University of Virginia, May 2013.

10. I. Porche and S. Lafortune. Adaptive look-ahead optimization of traffic signals, *Journal of Intelligent Transportation Systems*, 4(3):209–254, 1999.

11. K.L. Head, Event-based short-term traffic flow prediction model, *Transportation Research Record: Journal of the Transportation Research Board*, No. 1510, Washington, DC, 45–52, 1995.

12. *Describing the congestion problem*, www.fhwa.dot.gov/congestion/describing_problem.htm. Accessed July 20, 2012.

13. National Transportation Operations Coalition, *National Traffic Signal Report Card*, Institute of Transportation Engineers, 2012.

14. S. Lämmer and D. Helbing, Self-control of traffic lights and vehicle flows in urban road networks, *Journal of Statistical Mechanics: Theory and Experiment*, 4, April 2008.

15. J.C. Herrera and A.M. Bayen, Incorporation of Lagrangian measurements in freeway traffic state estimation, *Transportation Research Part B: Methodological*, 44(4):460–481, May 2010.

16. R.L. Gordon, ALMA ends freeway lane lottery, *ITS International*, 20(1):NA6–NA7, January/February 2014. http://www.itsinternational.com/sections/nafta/features/proposed-system-to-take-guesswork-out-of-choosing-a-freeway-lane/. Accessed June 6, 2016.

Chapter 16

Sensor and data fusion in traffic management

Multisensor data fusion offers many benefits to traffic management. It aids in the interpretation of information gathered from a complex environment characterized by the presence of different types of vehicles, unexpected objects such as pedestrians darting across a roadway, inclement weather, vehicles changing lanes, and roadside structures or weather effects that interfere with the normal observation of traffic patterns and the gathering of needed data [1–3]. Data fusion processes and algorithms that combine information from infrastructure-based sensors and other sources offer the potential for overcoming these impediments to data acquisition. Furthermore, sensor and data fusion can assist traffic management personnel in gaining insight into what created the situation causing the observed events and its impact on the occurrence of future events that could affect traffic flow [1,3,4].

This chapter introduces the application of sensor and data fusion to traffic management. It reviews the definitions of sensor and data fusion, their role in enhancing the effectiveness of traffic management strategies, and examines factors that influence the selection of a fusion architecture. The U.S. Department of Defense Joint Directors of Laboratories six-level data fusion model is discussed as it is frequently applied to describe the processing and inferences that are typical of data fusion practices [2,5]. A taxonomy for object detection, classification, and identification algorithms is presented, followed by one for the state estimation algorithms utilized to track objects. The specific types of algorithms identified in these taxonomies are then briefly discussed. Klein [2], Waltz and Llinas [5], Khaleghi et al. [6], and Castanedo [7] provide additional information concerning these procedures. The choice of which data fusion algorithm to use in a given application is often made by considering which technique makes correct inferences and the availability of the required computer resources and algorithm input parameters.

16.1 WHAT IS MEANT BY SENSOR AND DATA FUSION?

Data fusion is concerned with the following:

1. The representation of information within a computational database, particularly the information gained through data fusion.
2. The presentation of this information in a manner that supports the required decision processes when a human operator or decision-maker is involved.

Data fusion should not be the goal or end result of a transportation management strategy. Rather, the goal is to provide a control system, in the form of a machine or a human, the information necessary to support automated or semiautomated decision-making, such as in

ITS applications where vehicle systems, drivers, or traffic management personnel may have to take corrective actions to ensure traveler safety and the smooth flow of traffic.

Several definitions of sensor and data fusion are found in the literature. The Joint Directors of Laboratories (JDL) model, perhaps the most widely cited, defines data fusion as "a multilevel, multifaceted process dealing with the automatic detection, association, correlation, estimation, and combination of data and information from single and multiple sources to achieve refined position and identity estimates, and complete and timely assessments of situations and threats and their significance" [5,8]. The Institute of Electrical and Electronics Engineers (IEEE) Geoscience and Remote Sensing Society's definition is "the process of combining spatially and temporally-indexed data provided by different instruments and sources in order to improve the processing and interpretation of these data." The University of Skövde provides a definition in terms of information fusion as "the study of efficient methods for automatically or semiautomatically transforming information from different sensors and different points in time into a representation that provides effective support for human or automated decision making" [9]. These definitions offer different insights into the role of sensor and data fusion. Their existence is a reflection of the diverse applications for sensor and data fusion.

The terms *data fusion* and *sensor fusion* are often used interchangeably. Strictly speaking, data fusion is defined as in the preceding paragraph. Sensor fusion, then, describes the use of more than one sensor or information source in a configuration that enables more accurate or additional data to be gathered about events or objects that occur in the observation space of the sensors. More than one sensor may be needed to completely and continually monitor the observation space for a number of reasons. For instance, some objects may be detected by one sensor but not another because each sensor may respond to a different signature-generation phenomenology. The signature of an object may be masked or otherwise hidden with respect to one sensor but not another; or one sensor may be blocked from viewing objects because of the geometric relation of the sensor to the objects in the observation space, but another sensor located elsewhere in space may have an unimpeded view of the object. In this case, the data or tracks from the sensor with the unimpeded view may be combined with past information (i.e., data or tracks) from the other sensor to update the state estimate of the object [2].

16.2 APPLICATION OF SENSOR AND DATA FUSION TO TRAFFIC MANAGEMENT

Figure 16.1 illustrates the benefits that multisensor fusion can bring to traffic management. It assists in the interpretation of information gathered from a multifaceted environment that contains a variety of information sources and vehicles, some of which may be difficult to detect because of their location or missing structural elements such as license plates, appearance of unexpected objects such as a crowd of pedestrians exiting a venue and crossing a street or the chaos that ensues from an incident, inclement weather, multiple vehicles changing lanes, inoperable roadway sensors, and high clutter as indicated on the left of the figure. Data fusion processes and algorithms that combine information from multiple sensors and other resources offer the potential to enhance traveler safety and mobility, provide higher-confidence vehicle detection and tracking, extend vehicle tracking spatially and temporally, and afford insight into what created the situation causing the observed events and its impact on the occurrence of future events.

The need for real-time and accurate data for implementation of ITS traffic management strategies, including automatic incident detection, active transportation and demand

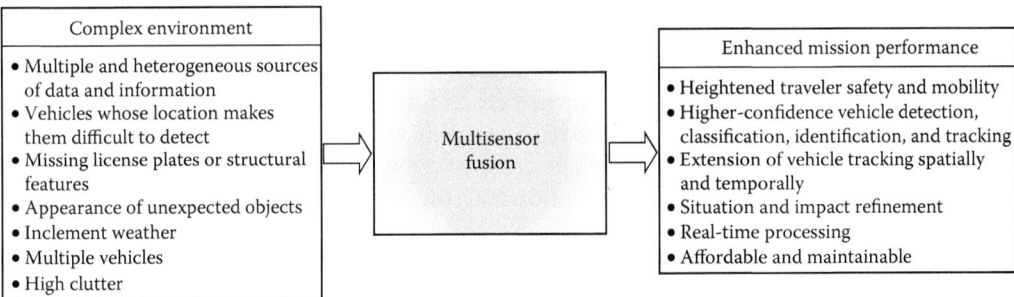

Figure 16.1 Enhancing mission performance in a complex environment with multisensor fusion.

management, route guidance, and safety warnings found in connected vehicle applications, underscores the importance of complementary sources of data, that is, those other than traffic flow sensors, for traffic flow parameter estimation. Since a wide spectrum of data and heterogeneous and independent sources of information can potentially be utilized in a given traffic management situation, many traffic engineering challenges become a typical data fusion problem. Accordingly, data fusion can be applied to produce an improved model or estimate of system parameters or events, where the desired model is the state vector describing the traffic phenomena of interest. These estimates may include current or future vehicular speeds, mean speeds, travel times, vehicle classification, red light running assessment, road surface state (i.e., dry, wet, snow or ice covered), and similar parameters of significance to travelers and traffic management personnel.

16.3 DATA AND INFORMATION OPTIONS

Advances in road telematics have expanded and led to improved methods of traffic data collection. Examples are the availability of novel sensor and communications technologies as exemplified by enhanced roadside-mounted sensors that provide innovative data types or improved spatial resolution, dedicated short-range communications (DSRC) associated with V2V and V2I connected and cooperative vehicle applications [10,11], in-car sensors that determine driving conditions related to the vehicle (e.g., braking, velocity, acceleration, steering wheel position, traction loss, and lane departure warning), floating car data that provide emissions information in addition to normal traffic flow parameters, and crowd sourcing applications and personal device monitoring that measure travel behavior of pedestrians and bicyclists. Improvement and expansion of wireless communications networks and the growth of cellular phone utilization have enabled cellular positioning services to exploit cell phone–equipped drivers as traffic probes [12,13]. Automatic vehicle identification (AVI) systems and technologies, including toll-tag readers, automatic license plate recognition (ALPR), and GPS and Bluetooth® MAC address readers, allow vehicle tracking and re-identification in support of route travel-time and speed measurement and determination of origin–destination pairs.

Basic traffic flow data (volume, occupancy, and speed) needed by traffic operations personnel are typically obtained from sensors embedded in the pavement or mounted on roadside poles or structures. The predominant sensor of this type is the inductive loop detector (ILD) that measures temporal traffic flow characteristics at a given location. Other point sensor technologies such as acoustic, ultrasonic, passive infrared, magnetometer, and microwave Doppler can be used to gather roadway network data. While these devices provide point data, they fail in measuring the spatial behavior of traffic flow [14,15]. In addition, their deployment and maintenance costs may become prohibitive when large-area coverage of a roadway network or a multilane intersection is required. Roadside sensors with improved spatial coverage have been developed and deployed to supplement loop detector data. These consist of visible and infrared spectrum video detection systems and surveillance cameras and multilane presence-detecting microwave radar sensors. Table 16.1 summarizes the advantages and limitations of the narrow and wider field-of-view sensors.

Every data source has its own accuracy, detection area and resolution, latency, and data refresh rate. No one data source may be capable of providing a complete kinematic description of all vehicles and pedestrians at every location. Therefore, it is often necessary to gather needed data and information from a variety of devices, which requires that the data be temporally and spatially correlated. To this end, it is valuable to time stamp data since data sources may report information at different times. Spatial correlation is needed since different data source modalities may report from different locations or provide information from different sized detection areas. In addition, a method must be developed to address how overlapping data and missing data will be treated.

16.4 SENSOR AND DATA FUSION ARCHITECTURES FOR ITS APPLICATIONS

The selection of a data fusion architecture requires an overall system perspective that simultaneously considers the viewpoints of four major participants [2]:

1. System stakeholders, whose concerns include system requirements, user constraints, and operations.
2. Numerical or statistical specialists, whose knowledge includes numerical techniques, statistical methods, and algorithm design.
3. Operations analysts concerned with the man–machine interface (MMI), transaction analysis, and operational concepts.
4. Systems engineers concerned with performance, interoperability with other systems, and system integrity. Traffic management personnel at a traffic management center frequently assume this role in traffic management applications.

There are several ways to classify sensor and data fusion architectures. For example, architectures may be organized by where the majority of data processing occurs, for example, in the individual sensors or in a central fusion processor, or by the type of data or information that is combined, for example, raw data, feature-based information, or decision-type information. Dasarathy [16] discusses other frameworks such as those based on the category of the entities at both the input and output of the fusion system (e.g., data, features, or decisions), and the notion of the fusion architecture being constructed to provide feedback to the individual sensors. Durrant-Whyte and Henderson [17] generalize multisensor fusion architectures and classify them according to selections made from among four independent design dimensions: (1) centralized–decentralized, (2) local–global interaction of components, (3) modular–monolithic, and (4) heterarchical–hierarchical. The most prevalent combinations are centralized, global interaction, and hierarchical; decentralized, global interaction, and heterarchical; decentralized, local interaction, and hierarchical; and decentralized, local interaction, and heterarchical. The reader is referred to [2,3,16,17] for a detailed discussion of these schemes.

Table 16.1 Sensor options for traffic management applications

Sensor	Advantages	Limitations
Inductive loop detector	Standardization of loop electronics units (detectors) Mature, well-understood technology Excellent counting accuracy with properly installed and maintained loop Presence and occupancy data Some models provide classification data	Not suitable for bridges, over passes, viaducts, poor roadbeds Reliability and useful life are dependent on installation procedures and practices Installation and maintenance require lane closure Decreases pavement life Susceptible to damage by heavy vehicles, road repair, and utilities Multiple sensors usually required at a site
Magnetometer	Less susceptible than loops to stresses of traffic Detects stopped and moving vehicles Some models transmit data over wireless RF link Count, presence, occupancy data	Small detection zone Installation requires intrusion into pavement Installation and maintenance require lane closure Decreases pavement life Multiple sensors usually required at a site
Magnetic detector	Can be used where loops are not feasible (e.g., bridge decks) Some models installed under roadway without need for pavement cuts Less susceptible than loops to stresses of traffic Count and passage data	Small detection zone Installation requires pavement cut or tunneling under roadway Cannot detect stopped vehicles (exception for 1 model using multiple sensors and application-specific software from vendor)
Visible spectrum video detection system	Best-resolution images Passive—detects reflected light Multilane data collection Speed, count, occupancy, limited classification by vehicle length, and other data	Night operation may require street lights Affected by clouds, heavy rain and snow, fog, haze, dust, smoke, sun glint and glare May be affected by shadows (false or missed calls), reflections from wet pavement (false calls) Vehicle occlusion in distant lanes when camera is side mounted Tall vehicles can project into adjacent lanes (false calls) and headlights can project past stop bar (dropped calls) Vehicle-to-road contrast, day/night transitions, camera vibration, and debris on camera lens may affect performance No range data
Microwave/millimeter-wave radar	All weather Lower frequencies penetrate foliage Multilane data collection Day/night operation Range, speed, count, occupancy, tracking data, limited classification by vehicle length	Vehicle occlusion in distant lanes when side mounted Offset mounting distance must be accommodated May require multiple detection zones per lane or other measures to ensure 100% vehicle detection at a stopline
Microwave Doppler sensor	All weather Day/night operation Count, occupancy, and speed of moving vehicles	Most only detect vehicles traveling greater than some minimum speed—does not detect stopped vehicles or provide presence indication

(Continued)

Table 16.1 (Continued) Sensor options for traffic management applications

Sensor	Advantages	Limitations
Infrared video detection system	Fine spatial and spectral resolution imagery Passive—detects heat emissions Day/night operation Does not require external illumination Potential for same data as visible spectrum Less affected by rain, fog, haze, dust, smoke, snow, sun glint and glare, reflections	Poor foliage and cloud penetration Requires cooled focal plane to maximize signal-to-noise ratio Cost may be an issue
Passive infrared	Ease of pole-mounted installation Day/night operation	Performance possibly degraded by heavy rain, fog, overcast skies, or snow
Lidar (laser radar)	Fine spatial and spectral resolution imagery Range and reflectance data Velocity, track, count, occupancy, classification data Day/night operation	Affected by rain, fog, haze, dust, smoke, heavy snow Poor foliage penetration Most effective when used to monitor a relatively small area
Acoustic	Passive—detects engine and road noise Side mounted Day/night operation Multilane data collection Count, presence, occupancy speed data	May undercount in congested flow Performance may degrade in rain
Ultrasonic	Ease of pole-mounted installation Day/night operation Count, presence, occupancy data are typical	Traffic interrupted with overhead installation and repair Performance affected by variations in temperature and extreme air turbulence One per lane required Low PRF may degrade occupancy measurement on freeways with moderate- to high-speed vehicles

16.5 DETECTION, CLASSIFICATION, AND IDENTIFICATION OF A VEHICLE

An observer's ability to interpret and extract information from an image is a function of the contrast and brightness of the object (in traffic management applications, the object is a vehicle or conveyance of some type, a pedestrian, or a fixed object such as a lamp or signal support, sign bridge, building, or tree), properties of the device used to enhance the image or the algorithm used to process the data, and the physiological response characteristic of the human eye–brain interface when a human interprets the imagery. Contrast is determined by the image signal-to-noise ratio, while brightness is controlled by the luminance of the object of interest. A review of the literature concerning human interpretation of images may be found in [18].

The first row in Figure 16.2 depicts the detection process, whose purpose is to determine if an object of interest is present. This involves differentiating between images or data that

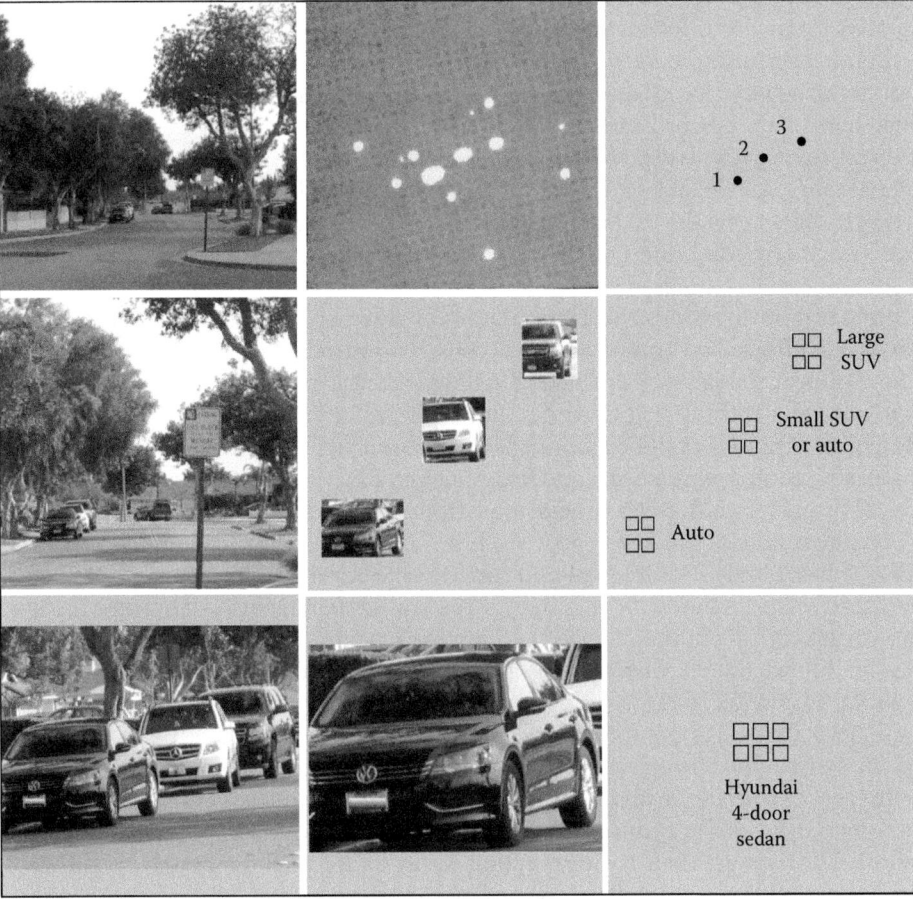

Figure 16.2 Detection, classification, and identification of a vehicle. The squares in the last column of rows two and three indicate the number of pixels needed across the minimum dimension of an object to achieve a 50% probability of classifying or identifying the object, respectively.

belong to relevant objects and those that belong to clutter or objects not of interest such as trees, rocks, fences, or perhaps puddles of water. The second row shows the classification process, namely, discerning the class to which an object belongs. For traffic management, classes of interest include sedans, buses and other transit vehicles, emergency vehicles, sport utility vehicles, pickup trucks, 18-wheelers, pedestrians, and bicycles, for example. The third row shows the identification process, which describes the objects to the limit of the observer's or data processing algorithm's knowledge, for example, identifying the manufacturer and model of a vehicle.

16.6 THE JDL DATA FUSION MODEL

The JDL data fusion model has five levels, with a potential sixth one to address the human–computer interface (HCI). Level 0 of the model deals with the preprocessing of data from the contributing source. Typical operations that occur at this level are data or image registration, noise removal, normalizing, formatting, ordering, batching, and compressing input data [5,19–25]. It may even identify sub-objects or features in the data that are used later in

Level 1 processing. Level 1 data processing is where the identity and state estimation data fusion algorithms are implemented. For traffic management, Level 1 processing concerns the combining or fusing of data from all appropriate sources, including real-time point and wide-area traffic flow sensors, Bluetooth and toll-tag roadside readers, ALPR information, transit system operators, toll data, cellular telephone calls, emergency call box reports, probe vehicle and roving tow truck messages, commercial vehicle transmissions, roadway-based weather sensors, and V2V and V2I connected vehicle DSRC devices as these become available [1–3,15,26]. Level 2 processing identifies the probable situation causing the observed data and events by combining the results of the Level 1 processing with information from other sources and databases. These sources may include police reports and databases, roadway configuration drawings, local and national weather reports, anticipated traffic mix, time-of-day and seasonal traffic patterns, construction schedules, and special event schedules. Level 3 processing assesses the traffic flow patterns and other data with respect to the likely impact of a traffic event on traffic flow (e.g., duration of traffic congestion, incident, fire, or police action). Level 4 processing seeks to improve the entire data fusion process by continuously refining predictions and assessments, and evaluating the need for additional sources of information. Level 5 processing is concerned with enabling a human to interpret and apply the results of the fusion process. Level 1 fusion is expanded upon in Sections 16.7 and 16.8, while Levels 2–5 are explored further in Section 16.10.

The data fusion methods investigated in the traffic management literature involve basic functions such as temporal and spatial alignment of input data, data association, and data mining for knowledge extraction. The latter is also one of the objectives of multisource information fusion [22,27,28].

Figure 16.3 illustrates the interrelationships in the Level 1 through Level 3 fusion processes. In some applications, object detection, classification, and state estimation occur simultaneously rather than in separate paths as displayed in the figure. As part of Level 1 fusion, each sensor or data source provides data or imagery concerning objects in the scene of interest. The data are then associated, that is, combined to enhance and update detection, classification, and tracking of the objects of interest. Level 2 and 3 fusion processes perform detection of behavior patterns, association of entities and events, classification of the situation causing the observed events, and prediction of future behavior.

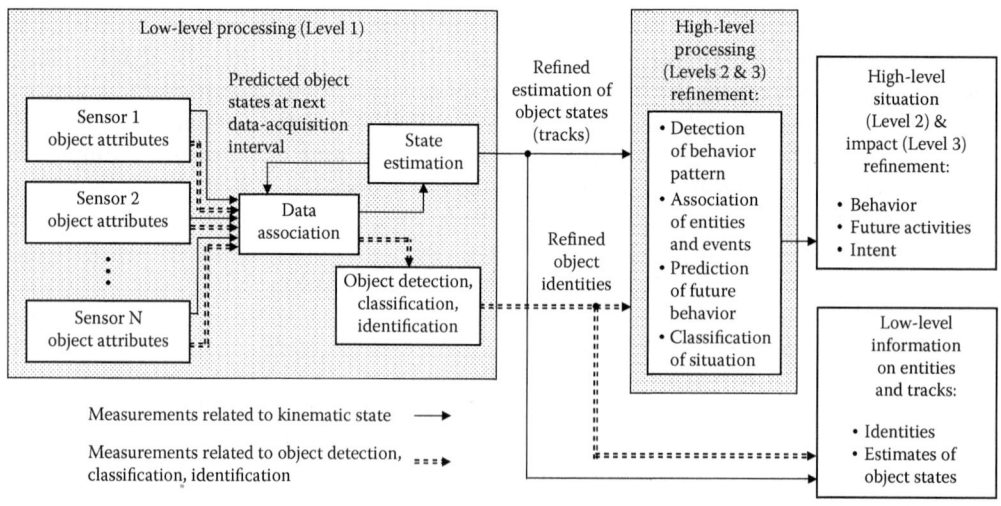

Figure 16.3 Data fusion processing Levels 1, 2, and 3 of the JDL model.

Figure 16.4 shows the six-level JDL data fusion model [23,29]. Data gathered from all appropriate sources, including real-time sensor information, highway and freeway service patrols, maps, weather reports, planned work zone and sports events, transit agencies, Bluetooth and toll-tag roadside readers, DSRC devices, time-of-day and seasonal traffic flow predictions, and information from other databases, are input to the fusion domain as illustrated on the left of the figure. The data may be subject to preprocessing or pass directly into one of the other fusion levels. A significant amount of information from external databases is usually needed to support the Level 2 and 3 fusion processes. The final interpretation of the data is typically performed by traffic management agency personnel with the assistance of data and information processing devices.

An important caveat in the design of data fusion systems is to realize that data fusion levels are intended only as a convenient categorization of data fusion functions. Data fusion levels were never intended to be, nor should they be taken as a prescription for designing systems: do Level 0 fusion first, then Level 1, then Level 2, and so on. Processing should be partitioned in terms of the individual system requirements [30].

16.7 LEVEL I FUSION: DETECTION, CLASSIFICATION, AND IDENTIFICATION ALGORITHMS

Figure 16.5 contains a taxonomy of Level 1 detection, classification, and identification algorithms [5,20,22,31,32]. The major algorithm categories are physical models, feature-based inference techniques, and cognitive-based models. Other mathematical concepts, not shown in the figure, are also utilized for data fusion. These include random set theory, conditional algebra, and relational-event algebra. Random set theory deals with random variables that are sets rather than points. Goodman et al. use random set theory to reformulate multisensor, multi-object estimation problems into single-sensor, single-object problems [33]. They also apply the theory to incorporate ambiguous evidence (e.g., natural language reports and rules) into multisensor, multi-object estimation, and to incorporate various expert system methods (e.g., fuzzy logic and rule-based inference) into multisensor, multi-object estimation. Conditional-event algebra is a type of probabilistic calculus suited for contingency

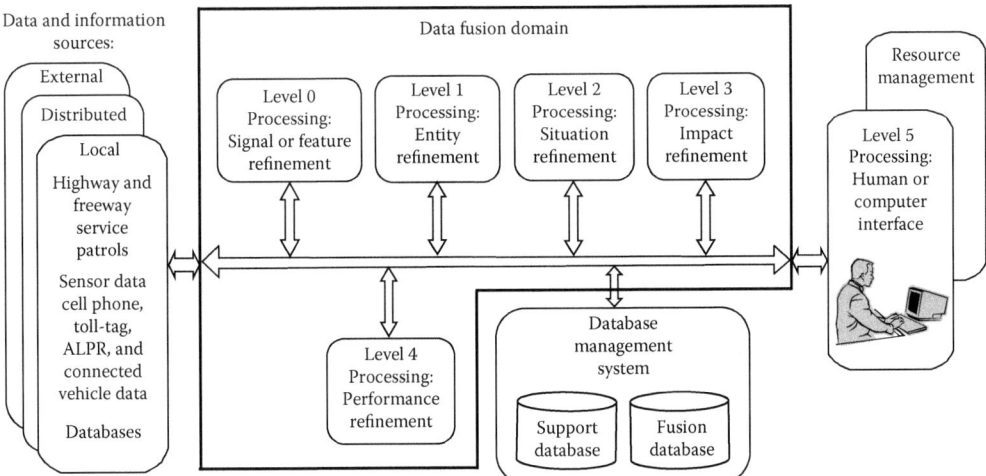

Figure 16.4 Six-level JDL data fusion processing model. Data fusion levels are not a prescription for designing systems. Instead, they are meant to be a convenient way to categorize the purposes of data fusion.

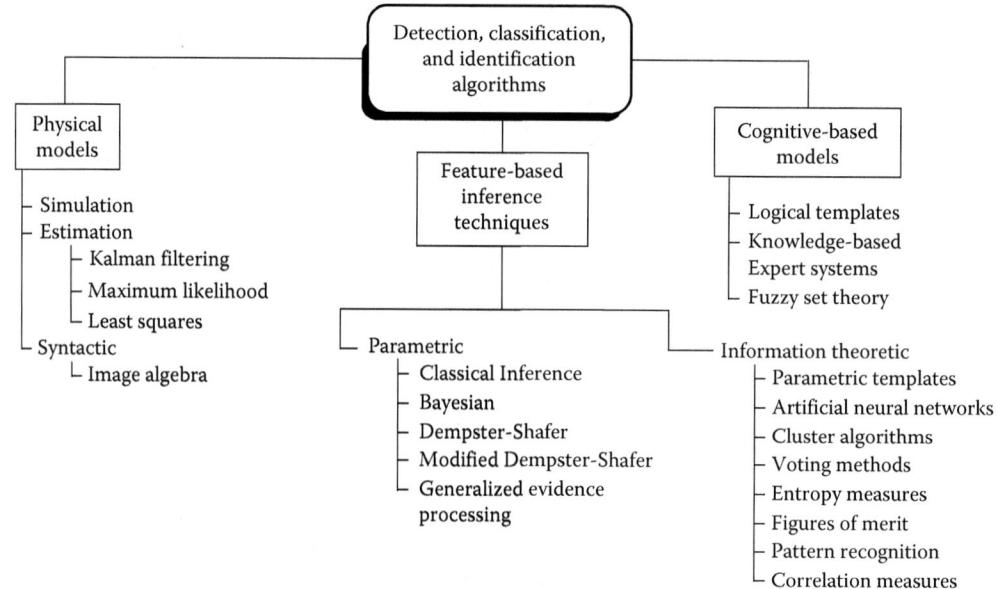

Figure 16.5 Detection, classification, and identification algorithm taxonomy for Level I processing.

problems such as knowledge-based rules and contingent decision-making. Relational-event algebra is a generalization of conditional-event algebra that provides a systematic basis for solving problems involving pooling of evidence. Still other data fusion approaches combine several of the illustrated methods, such as combinations of Bayesian with fuzzy logic, Dempster–Shafer with fuzzy logic, artificial neural networks with fuzzy logic, and artificial neural networks with Dempster–Shafer.

16.7.1 Physical models

Physical models replicate object discriminators that are easily and accurately observable or calculable. Examples of discriminators are radar cross section as a function of aspect angle; infrared emissions as a function of vehicle type, engine temperature, or surface characteristics such as roughness, emissivity, and temperature; multispectral signatures; and height profile images.

This approach classifies and identifies objects by matching the prestored or modeled signatures to observed data as illustrated in Figure 16.6. The signature or imagery gathered by a sensor is analyzed for pre-identified physical characteristics or attributes, which are input into an identity declaration process. Here, the characteristics identified by the analysis of the observed data are compared with stored physical models or signatures of the objects of interest and other items that may be present, such as trees or buildings. The stored model or signature having the closest match to the real-time sensor data is declared to be the correct identity of the object in the sensor's field of view.

Physical modeling techniques include simulation, estimation, and syntactic methods. Simulation is used when the physical characteristics to be measured can be accurately and predictably modeled. Estimation processes include Kalman filtering, maximum likelihood, and least squares approximation. The Kalman filter provides a general solution to the recursive, minimum mean-square state estimation problem within the class of *linear* estimators.

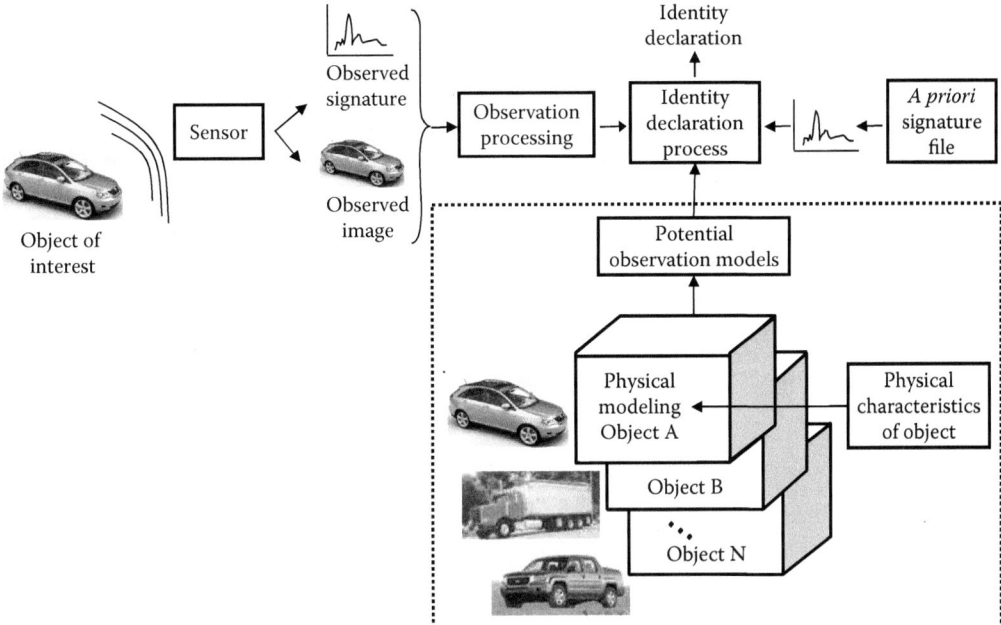

Figure 16.6 Physical modeling concept.

It minimizes the mean-squared error as long as the tracked object's dynamics and measurement noise are accurately modeled. The Kalman filter and its nonlinear motion counterparts are examples of physical models since the kinematics of the objects being tracked are modeled.

The extended Kalman filter (EKF) is applied when nonlinearities are present in the observation matrix or covariance matrices of the process and measurement noise sources. The EKF linearizes about the current mean and covariance of the state using first-order Taylor approximations to the time-varying transition and observation matrices. There are alternatives to the EKF when the system of interest is highly nonlinear [3,6,7]. The first of these employs an unscented Kalman filter (UKF), which operates on the premise that it is easier to approximate a Gaussian distribution than it is to approximate an arbitrary nonlinear function. Instead of linearizing using Jacobian matrices as in the EKF, the UKF uses a deterministic sampling approach to capture estimates of the mean and covariance with a minimal set of carefully chosen sample points. Application of the UKF allows the posterior mean and covariance to be accurately represented to the third order (Taylor series expansion) for any nonlinearity. The EKF, in contrast, only achieves first-order accuracy. The computational complexity of the UKF is the same order as that of the EKF.

Monte Carlo techniques are appropriate for problems where state transition models and measurement models are highly nonlinear. They are flexible as they do not make any assumptions regarding the probability densities to be approximated. These methods describe probability distributions as a set of weighted samples of an underlying state space. The samples simulate probabilistic inference usually through Bayes' rule. Many simulations are performed and by analyzing the statistics of the simulations, a probabilistic picture of the process can be discerned. Sequential Monte Carlo filtering is a simulation of the recursive Bayes' update equations using sample support values and weights to describe the underlying probability distributions.

Particle filters extend the sequential Monte Carlo algorithm by utilizing a weighted ensemble of randomly drawn samples called particles as an approximation of the probability density of interest. When applied within a Bayesian framework, particle filters approximate the posterior probability of the system state as a weighted sum of random samples. The random samples are usually drawn (predicted) from the prior density (transition model) with their weights updated according to the likelihood of the given measurement (sensing model).

The syntactic methods, although listed here under physical models, appear again as part of pattern recognition, a subset of information theoretic techniques. Syntactic pattern recognition is applied when the significant information in a pattern is not merely the presence or absence of numerical values, but rather the interconnections of features that yield its structure. Pattern similarity is assessed by quantifying and extracting structural information utilizing, for example, the syntax of a formally defined language. Typically, syntactic approaches formulate hierarchical descriptions of complex patterns from simpler subpatterns or primitives.

16.7.2 Feature-based inference techniques

Feature-based inference techniques perform classification or identification by mapping data, such as statistical knowledge about an object or recognition of object features, into a declaration of identity. Feature-based algorithms may be further divided into parametric and information theoretic techniques (i.e., algorithms that have some commonality with information theory).

16.7.2.1 Parametric techniques

Parametric classification directly maps parametric data (e.g., features) into a declaration of identity. Stochastic properties of features may be modeled although physical models are not used. Parametric techniques include classical inference, Bayesian inference, Dempster–Shafer evidential theory, modified Dempster–Shafer methods, and generalized evidence processing.

Classical inference gives the probability that an observation can be attributed to the presence of an object or event, given an assumed hypothesis. Its major disadvantages are (1) difficulty in obtaining the density function that describes the observable used to classify the object, (2) complexities that arise when multivariate data are encountered, (3) its capability to assess only two hypotheses at a time, and (4) its inability to take direct advantage of *a priori* and likelihood probabilities.

Bayesian inference resolves some of the difficulties with classical inference. It updates the *a priori* probability of a hypothesis given a previous likelihood estimate and additional observations and is applicable when more than two hypotheses are to be assessed [22,34]. The limitations of Bayesian inference include (1) difficulty in defining the prior probabilities and likelihood functions, (2) complexities that arise when multiple potential hypotheses and multiple conditionally dependent events are evaluated, (3) mutual exclusivity required of competing hypotheses, and (4) inability to account for general uncertainty. When multiple sensors collect information and apply Bayesian inference to determine the identity of an object, each sensor provides an identity declaration D or hypothesis about the object's identity based on the observations and a sensor-specific algorithm. The previously established performance characteristics of each sensor's classification algorithm (developed either theoretically or experimentally) provide estimates of the likelihood function, that is, the probability $P(D|O_i)$, that the sensor will declare the object to be a certain type, given that the object is in fact type i. The sensor declarations are then combined using a generalization of Bayes' rule to produce an updated, joint probability for each entity O_i.

Dempster–Shafer evidential reasoning generalizes Bayesian inference to allow for uncertainty by distributing support for a proposition (e.g., that an object is of a particular type) not only to the proposition itself, but also to the union of propositions (disjunctions) that include it and to the negation of a proposition. Any support that cannot be directly assigned to a proposition or its negation is assigned to the set of all propositions in the hypothesis space (i.e., uncertainty). Support provided by multiple sensors for a proposition is combined using Dempster's rule. Bayesian and Dempster–Shafer produce identical results when all propositions are singleton, that is, consist of only one object, are mutually exclusive, and there is no support assigned to uncertainty. The Dempster–Shafer method requires definition of processes in each sensor that assign the degree of support for a proposition. Limitations of the method include the inability to make direct use of prior probabilities when they are known and the counterintuitive output sometimes produced when support for conflicting propositions is large. Several methods have been proposed to modify Dempster's rule through the use of probability transformations that better accommodate conflicting beliefs [35] and, in some cases, through the use of prior knowledge and spatial information [36–42].

Generalized evidence processing (GEP) allows a Bayesian decision process to be extended into a multiple-hypothesis space (called the frame of discernment in Dempster–Shafer evidential theory). Evidence that supports nonmutually exclusive propositions can be combined to arrive at a decision by minimizing a Bayesian risk function tying probability masses to likelihood ratios, or equivalently, by maximizing a detection probability for fixed *a priori* miss and false-alarm probabilities [43–46].

16.7.2.2 Information theoretic techniques

Information theoretic techniques transform or map parametric data into an identity declaration. They relate a similarity in identity to a similarity in the observable parameters. No attempt is made to directly model the stochastic aspects of the observables. The techniques included under this category are parametric templates, artificial neural networks, cluster algorithms, voting methods, entropy-measuring techniques, figures of merit, pattern recognition, and correlation measures.

In *parametric templating*, multisensor or multispectral data acquired over time and multisource information are matched with preselected conditions to determine if the observations contain evidence to identify an entity. Templating can be applied to event detection, situation assessment, and single-object identification [5,24]. Figure 16.7 shows an application of parametric templating to the identification of an emitter of electromagnetic radiation, whose pulse repetition frequency and pulse width are measured by a sensor. In this example, the emitted energy arises from radio frequency transmissions from the vehicles. The measured parameters are overlaid on a template such as the one depicted in the lower right portion of the figure. Identification is made when the parameters lie in a region that corresponds to the characteristics of a known device. If the parameters fall into a region that has not been previously mapped, then the object is declared to be unknown. In this example, Emitter 1 would be identified as a Class A emitter and Emitter 2 as unknown.

Artificial neural networks are hardware or software systems that are trained to map input data into selected output categories. The transformation of the input data into output classifications is performed by artificial neurons that attempt to emulate the complex, nonlinear, and massively parallel computing processes that occur in biological nervous systems [47].

The particular artificial neural network in Figure 16.8 is an example of a fully connected, three-layer feedforward network [2]. In a fully connected network, each processing element receives inputs from every output in the preceding layer. In general, a feedforward network

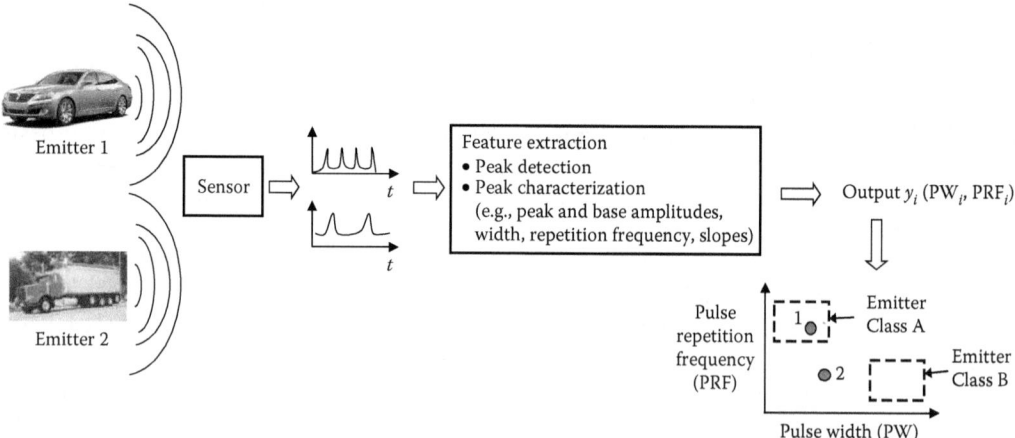

Figure 16.7 Parametric templating concept.

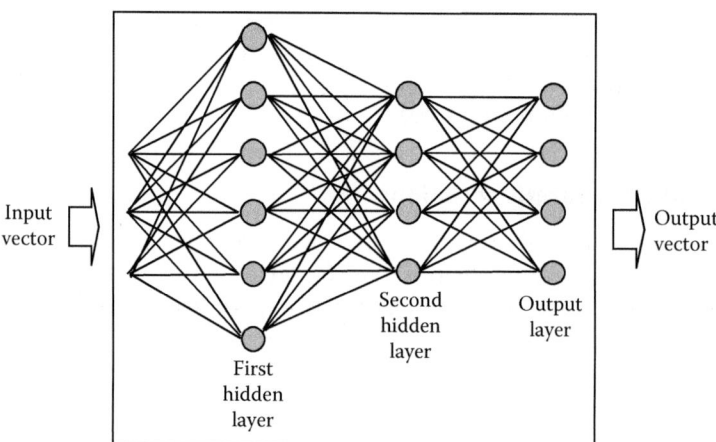

Figure 16.8 Fully connected, feedforward artificial neural network.

contains a hierarchy of processing elements that are organized as a series of two or more mutually exclusive layers. The input elements are a holding place for the values applied to the network. These elements do not implement a separate mapping or conversion of input data, and their weights are insignificant. The last, or output layer, permits the final state of the network to be read. Between these two extremes are zero or more layers of hidden elements. The hidden layers remap the inputs and results of previous layers' analyses and, thereby, produce a more separable or more easily classifiable representation of the data.

Links or weights connect each element in one layer to only those in the next higher layer. An implied directionality exists in these connections, whereby the output of one element, scaled by the connecting weight, is fed forward to provide a portion of the activation for the elements in the next higher layer. There are other forms of feedforward networks, such as one where the processing elements receive signals directly from each input component and from the output of each preceding processing element.

Voting methods combine detection and classification declarations from multiple sensors by treating each sensor's declaration as a vote in which majority, plurality, or decision-tree rules are used often with the aid of Boolean algebra. Additional discrimination can be introduced via weighting of the sensor's declaration.

Pattern recognition concerns the description or classification of data. The three major approaches to pattern recognition are statistical (or decision theoretic), syntactic (or structural), and artificial neural networks. In statistical pattern recognition, a set of characteristic measurements or features is extracted from the input data and used to assign the feature vector to one of c classes. Assuming features are generated by a state of nature, the underlying statistical model represents a state of nature, set of probabilities, or probability density functions that correspond to a particular class [48]. Syntactic pattern recognition is applied when the significant information in a pattern is not merely the presence or absence of numerical values, but rather the interconnections of features that yield structural information. The structural similarity of patterns is assessed by quantifying and extracting structural information using, for example, the syntax of a formally defined language. Typically, syntactic approaches formulate hierarchical descriptions of complex patterns from simpler subpatterns or primitives. Neural computing was described earlier in this section.

Descriptions of the other algorithms in the information theoretic category are found in Klein [2] and Waltz and Llinas [5].

16.7.3 Cognitive-based models

Cognitive-based models, including logical templates, knowledge-based systems, and fuzzy set theory, attempt to emulate and automate the decision-making processes employed by human analysts.

16.7.3.1 Logical templates

Templating matches predetermined and stored patterns against observed data to infer the identity of an object or to assess a situation. Parametric templates that compare real-time patterns with stored ones can be combined with logical templates derived, for example, from Boolean relationships [2,5]. Fuzzy logic may also be applied to the pattern-matching technique to account for uncertainty in either the observed data or the logical relationships that define a pattern.

16.7.3.2 Knowledge-based expert systems

Knowledge-based systems incorporate rules and other knowledge from known experts to automate the object identification process. They retain the expert knowledge for use at a time when the human inference source is no longer available. Computer-based expert systems frequently consist of four components: (1) a knowledge base that contains facts, algorithms, and a representation of heuristic rules; (2) a global database that contains dynamic input data or imagery; (3) a control structure or inference engine; and (4) a human–machine interface. The inference engine processes the data by searching the knowledge base and applying the facts, algorithms, and rules to the input data. The output of the process is a set of suggested actions that is presented to the end user [49].

The knowledge-based system in Figure 16.9 depicts processed sensor data or imagery as the source of the features that identifies the object or situation. Three types of rules are listed to assist in correlating information contained in the real-time feature vector with information in the stored knowledge base. Syntactic rules are expressed as IF–THEN statements. The IF or antecedent clause states the conditions that must be present for the action specified in the THEN or conditional clause to occur. Expert systems typically rely on binary

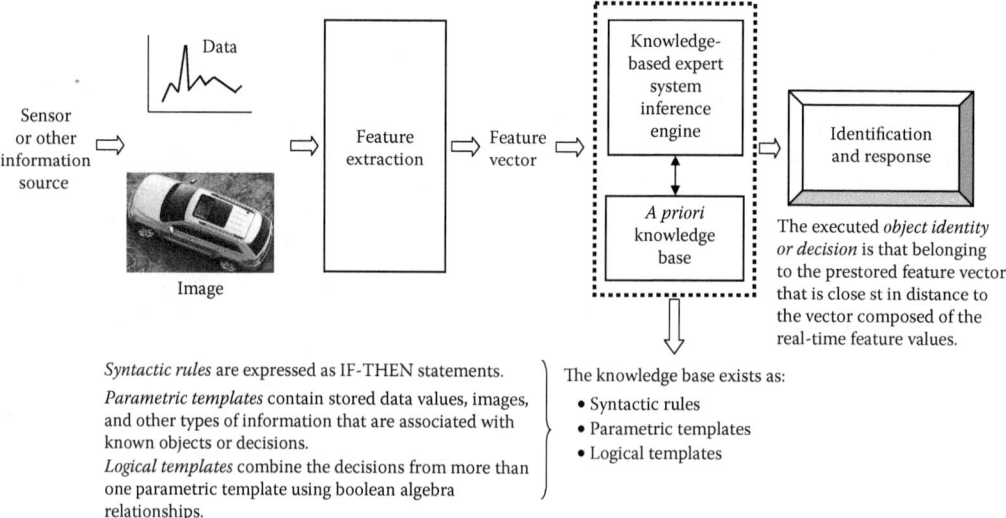

Syntactic rules are expressed as IF-THEN statements.

Parametric templates contain stored data values, images, and other types of information that are associated with known objects or decisions.

Logical templates combine the decisions from more than one parametric template using boolean algebra relationships.

The knowledge base exists as:

• Syntactic rules
• Parametric templates
• Logical templates

Figure 16.9 Knowledge-based expert system concept.

on–off logic and probability to develop the inferences found in the IF–THEN statements. Parametric templates contain stored data values, images, and other types of information that are associated with known objects or decisions. Logical templates combine the decisions from more than one parametric template using Boolean-algebra relationships. The executed object identity or decision is the one belonging to the prestored feature vector closest in distance to the vector composed of the real-time feature values.

16.7.3.3 Fuzzy set theory

Fuzzy set theory opens the world of imprecise knowledge or indistinct boundary definition to mathematical treatment. It facilitates the mapping of system state–variable data into control, classification, or other outputs [50–52]. There are four essentials to a fuzzy system, namely, fuzzy sets, membership functions, production rules, and a defuzzification mechanism.

Fuzzy sets are the state variables defined in imprecise terms. In Figure 16.10, the sets represent height categories of people. In conventional set theory, the set boundaries leave no doubt as to the person's height class. However, in fuzzy set theory, the set boundaries are vague, that is they overlap into neighboring sets. Therefore, fuzzy set theory allows a person to belong to more than one fuzzy set simultaneously. Membership functions are the graphical representation of the boundary between fuzzy sets. Membership can range from 0 (definitely not a member) to 1 (definitely a member). Production rules (also known as fuzzy associative memory) are the constructs that specify the membership value of a state variable in a given fuzzy set. The production rules, which govern the behavior of the system, are in the form of IF–THEN statements. An expert specifies the production rules and fuzzy set boundary shapes that represent the characteristics of each input and output variable. Defuzzification is the process that converts the result of the application of the production rules into a crisp output value, which is used to control the system.

Fuzzy set theory is intuitively appealing in that it permits uncertainties in knowledge or identity boundaries to be applied to such diverse applications as identification of battlefield threats, tracking of objects, and control of industrial and automotive processes. Unlike neural networks that sum throughputs, fuzzy systems sum outputs.

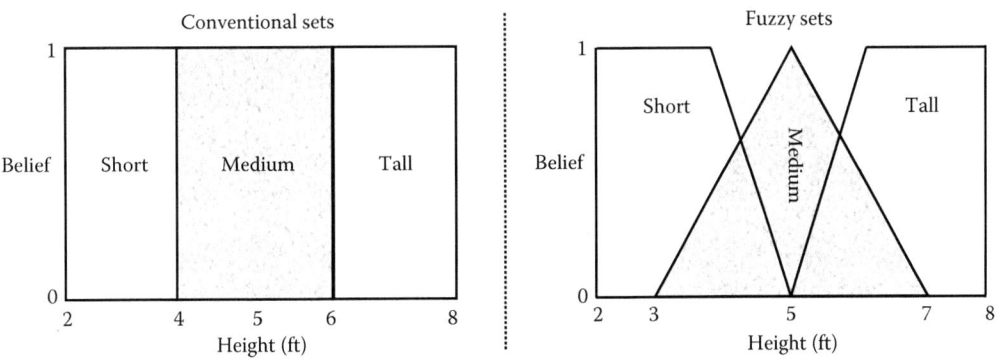

Figure 16.10 Conventional (left) and fuzzy (right) sets representing a person's height.

16.8 LEVEL I FUSION: STATE ESTIMATION AND TRACKING ALGORITHMS

Figure 16.11 displays a taxonomy for state estimation and tracking algorithms used in Level 1 processing [5,20,22,31]. At the top level, it shows the options available for (1) conducting data-driven or track-driven searches to update tracks and (2) correlating and associating data and tracks. Correlation and association are further separated into data alignment; data and track association; and position, kinematic, and attribute estimation. The majority of the algorithms are concerned with data and track association techniques.

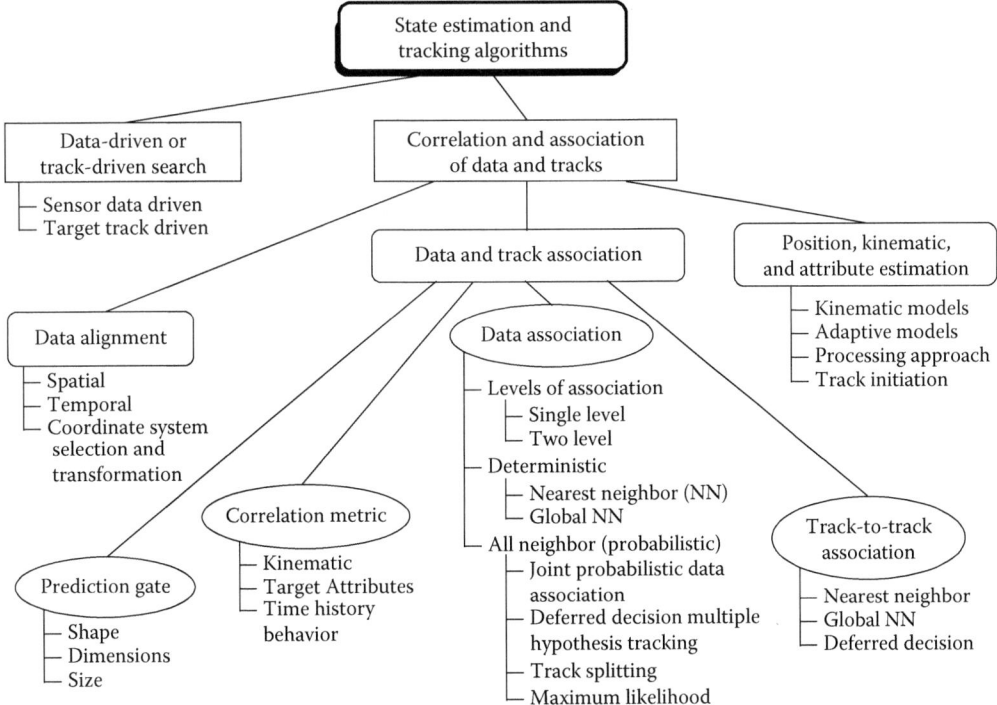

Figure 16.11 State estimation and tracking algorithm taxonomy for Level 1 processing.

In sensor data-driven systems, sensor data (also referred to as target reports) consisting of combinations of range, azimuth, elevation, and range-rate sensor measurements initiate a search through the known track file for *tracks that can be associated with the target reports*. Target track-driven systems use a primary sensor for tracking and use the target track to direct other sensors to acquire data or search databases for *reports that can be associated with particular tracks*.

The proper correlation and association of measurement data and tracks from multisensor inputs ultimately generate optimal central track files. Each file ideally represents a unique physical object or entity. Correlation and association require algorithms that define data alignment, prediction gates, correlation metrics, data and track association methods, and position, kinematic, and attribute estimation. A review of all of these procedures is found in Bar-Shalom and Fortmann [53], Klein [2], and the references they contain. However, it is instructive here to establish uniform definitions of data and track correlation and association and to briefly describe several of the data association methods.

16.8.1 Prediction gates, correlation metrics, and data association

Prediction gates are the mechanism that controls the association of data sets into one of two categories, namely, candidates for track update or initial observations for forming a new tentative track. The size of the gates reflects the calculated or otherwise anticipated object position and velocity errors associated with their calculation, sensor measurement errors, and desired probability of correct association.

Correlation metrics quantify the closeness of measurement data to existing tracks. They are also used in track-to-track association to assist in associating tracks produced by different sensors. Metrics are evaluated using the kinematic parameters (e.g., range, range rate, angle, and position) and object attributes (e.g., temperature, size, shape, and edge structure) that are observed and measured. The metric can be based on spatial distance (e.g., Euclidean distance) or statistical measures of correlation between observations and predictions (e.g., Mahalanobis distance), heuristic functions such as figures-of-merit that use the kinematic and object attribute information, and measures that quantify the realism of an observation or track based on prior assumptions such as track lengths, object densities, or track behavior.

In a multiple object and multiple sensor scenario, *data association* refers to the statistical decision process that associates sets of measurement data from overlapping gates, multiple returns in a gate, clutter in a gate, and new objects that appear in a gate on successive scans for the purpose of updating existing tracks or initiating new tracks. Thus, data association partitions the measurements into sets that could have originated from the same objects [54].

16.8.2 Single- and two-level data and track association

Association techniques that merge data and tracks from several sources into a single track usually employ either single-level tracking systems or two-level tracking systems. In a single-level tracking system, measurement data from several sensors are transmitted to a single processing node. Here, the data are correlated and associated to initiate new tracks and update estimates of existing tracks in the central track file.

Two-level tracking systems have four variants: (1) track-to-track association at the sensors and at a central node; (2) sensor data and track association at a central node; (3) sensor data association to form tracks at a central node; and (4) sensor track association at a central node. The first type of two-level tracking system maintains separate sensor-level and central-level trackers. Each sensor-level tracker initiates and updates tracks based on its own data. The tracks generated by each sensor are sent to a central site where track-to-track association is performed and redundant tracks are eliminated, producing a central track file. The second two-level system performs tracking with local measurement data only. The resulting tracks are reported to a designated track management center for distribution to the users. The third uses either sensor measurement data or sensor tracks to initiate and maintain a central track file. Track-to-track association of sensor tracks is initiated at a central node to form a central track file. The fourth two-level system distributes all correlated measurement data to all tracking subsystems for association with new or existing tracks. This approach forms tracks with all available data processed identically at all sensor subsystems, creating a common representation at each site.

16.8.3 Deterministic and probabilistic (all-neighbor) association

In general, there are two distinct approaches to the data association problem, deterministic and probabilistic or all neighbor [2]. The simpler deterministic method includes nearest neighbor and global nearest neighbor data association. It takes the most likely of several possible associations, and completely ignores the possibility that this selected association may be inappropriate. The alternatives are probabilistic approaches based on a Bayesian framework (probabilistic data association and joint probabilistic data association), multiple hypothesis tracking, and maximum likelihood. The probabilistic data association methods are applied in situations where a measurement may fall inside the intersection of two or more validation gates of several different targets and, therefore, could have originated from any of these targets or from clutter, and also when multiple returns from a large target or a closely spaced group of targets occur. Multiple hypothesis tracking allows the association of data to more than one track until a definitive assignment can be made at a later time. A variation of multiple hypothesis tracking, called track splitting, associates each report in the gate with a track, but does not specifically generate new tracks, nor compute the probability of correct association. Maximum likelihood selects the most likely single set of measurement data for association with a track.

16.9 DATA FUSION ALGORITHM SELECTION

How does one know which data fusion algorithm or technique to use in a given application? A starting point is to evaluate the choice of algorithm and its performance based on the degree to which the technique makes correct inferences and the availability of required computer resources and algorithm input parameters [2,3]. The selection process also seeks to identify algorithms that meet the following goals:

1. *Maximum effectiveness*: Algorithms are sought that make inferences with maximum specificity in the presence of uncertain or missing data. Required *a priori* data such as likelihood functions and probability masses are often unavailable for a particular scenario and must be estimated within time and budget constraints.
2. *Operational constraints*: The selection process should consider the constraints and perspectives of both automatic data processing and the analyst's desire for tools and useful products that are executable within the time constraints posed by the application. If the output products are examined by more than one decision-maker, then multiple sets of user expectations must be addressed.
3. *Resource efficiency*: Algorithm operation should minimize the use of computer resources (when they are scarce or in demand by other processes), for example, CPU time and required input and output devices.

4. *Operational flexibility*: Evaluation of algorithms should include the potential for different operational needs or system applications, particularly for data-driven algorithms versus alternative logic approaches. The ability to accommodate different sensors or sensor types may also be a requirement in some systems.

5. *Functional growth*: Data flow, interfaces, and algorithms must accommodate increased functionality as the system evolves.

Many of the Level 1 entity refinement data fusion algorithms are mature in the context of mathematical development. They encompass a broad range from numerical techniques to heuristic approaches such as knowledge-based expert systems. Practical real-world implementations of specific procedures (e.g., Kalman filters, Bayesian inference, and Dempster–Shafer evidential theory) exist. Algorithm selection criteria and the requisite *a priori* data are still major challenges however. Applying Bayesian inference, Dempster–Shafer evidential theory, artificial neural networks, fuzzy logic, and Kalman filtering data fusion algorithms to vehicle and event detection, classification, identification, and state estimation requires expert knowledge, probabilities, or other information from the analyst or data fusion specialist in the form of

- *A priori* probabilities and likelihood functions (Bayesian inference).
- Probability mass (Dempster–Shafer).
- Neural-network type, numbers of hidden layers and weights, and training data sets (artificial neural networks).
- Membership functions, production rules, and defuzzification method (fuzzy logic).
- Object kinematic and measurement models, process noise, and model transition probabilities when multiple state models are utilized (Kalman filtering).

More detailed examples of the prerequisite information typically required to utilize Bayesian inference, Dempster–Shafer evidential theory, artificial neural network, fuzzy logic, and Kalman filtering algorithms are found in [2].

Table 16.2 shows the data fusion algorithms that are currently embedded in ITS applications [1–4,55–57]. These include Bayesian inference, Dempster–Shafer evidential theory and some of its modifications, artificial neural networks, fuzzy logic, knowledge-based expert systems, and vehicle and pedestrian tracking based on the Kalman filter, EKF, UKF, and Monte Carlo techniques including the particle filter.

16.10 LEVEL 2, 3, 4, AND 5 FUSION

Level 2 processing identifies the probable situation causing the observed data and events by combining the results of the Level 1 processing with information from other sources and databases. These sources may include highway patrol or local police reports and databases, roadway configuration drawings, local and national weather reports, anticipated traffic mix, time-of-day and seasonal traffic patterns, construction schedules, and special event schedules.

In terms of a traffic management application, Level 2 fusion includes the following:

- Object aggregation—temporal, geometric, communications, and functional dependence relations among pedestrians, vehicles, transit riders, bicyclists, work zone personnel, incident responders, and roadside objects.
- Event and activity aggregation—temporal relations among drivers, other travelers, vehicles, work zone personnel, fleet operators, emergency service providers, and other entities to identify meaningful events.

Table 16.2 Data fusion algorithms and architectures currently found in ITS applications

Application	Data fusion algorithm	Architecture
Ramp metering	Fuzzy logic	Sensor level[a]
Pedestrian crossing time	Fuzzy logic	Central level[b]
Automatic incident detection	Artificial neural network	Sensor level
Automatic incident detection	Bayesian inference	Sensor level
Automatic incident detection	Dempster–Shafer	Sensor level or decision level[a]
Travel time estimation	Inference rules	Sensor level
Travel time estimation	Dempster–Shafer	Sensor level
Travel time estimation	Weighted mean of several travel-time estimators. Weights are a function of the variance or covariance of the estimators.	Sensor level
Travel time estimation	Weighted mean where weights are a function of the data source reliability	Sensor level
Travel time estimation	Fuzzy logic	Sensor level
Vehicle and object tracking	Kalman filter	Central level
Lane departure warning	Image processing using edge detection and extraction of other features	Pixel level[c]
Traffic state estimation	Extended Kalman filter	Central level
Crash analysis and prevention	k-means algorithm	Sensor level or decision level
Traffic forecasting and monitoring	Bayesian inference	Sensor level
Traffic forecasting and monitoring	Artificial neural network	Sensor level
Traffic forecasting and monitoring	Kalman filter	Central level
Traffic forecasting and monitoring	Extended Kalman filter	Central level
Traffic forecasting and monitoring	Kernel estimator	Central level
Traffic forecasting and monitoring	Particle filter	Central level
Vehicle position estimation	Unscented Kalman filter	Central level
Vehicle position estimation	Artificial neural network	Central level

[a] In sensor-level and decision-level fusion, each sensor detects, classifies, identifies, and provides state estimates of the objects of interest before data entry into the fusion processor, which combines the individual sensor information to improve the classification, identification, or state estimate of the objects. Sensor- and decision-level fusion are optimal for detecting and classifying objects only if the sensors involved rely on independent signature-generation phenomena to develop information about the identity of objects in the field of regard, that is, they derive object signatures from different physical processes and generally do not report a false alarm on the same artifacts.

[b] In central-level fusion (also called centralized fusion), minimally processed sensor data are sent to a fusion processor that analyzes the data for object features or attributes that aid in identifying and tracking the objects.

[c] In pixel-level fusion, minimally processed data from each sensor are combined at the pixel or resolution-cell level of the sensors using a central-level fusion architecture. Little, if any, preprocessing of data occurs before reaching the fusion processor.

- Contextual interpretation—data analysis with respect to the context of the evolving situation, including weather, terrain, road and bike lane configurations, local driving habits, local traffic management strategies, availability of first responders, line-of-sight restrictions, and so on. Contextual analysis requires large databases. This in turn mandates the need for a balance between fast data insertion and fast data retrieval. Placing a time tag on database entries aids in determining the relevance of inferences drawn from the database.

Level 3 processing develops a traffic management impact-oriented perspective to estimate vehicle and driver capabilities, incident opportunities, vehicle and driver (also pedestrian and bicycle as applicable) intent, and danger levels. It assesses the traffic flow patterns and other data to determine the likelihood and time of occurrence of a traffic event (e.g., traffic congestion, incident, construction or other preplanned special event, fire, or police action) that impacts traffic flow. Information is analyzed to categorize the event as one that is imminent (i.e., seconds away), one that may occur within several minutes, or one that may occur when a longer time period has passed.

In terms of a traffic management application, Level 3 fusion includes the following:

- Capability estimation—prediction of size, location, and capabilities of vehicles, drivers, other travelers, incident responders, and fixed roadside and randomly appearing objects.
- Prediction of vehicle and traveler intent—based on actions, driving culture, road configuration (e.g., presence of medians, multiple lanes, roundabouts, hills, curves, and elevation changes), entrance and exit ramp locations, congestion level, location of crosswalks and bicycle lanes (if any), traffic signal timing, and so on.
- Incident identification—based on prediction of driver and other traveler actions, operational readiness of incident responders, weather.
- Multiperspective assessment—data analysis with respect to first responder, commercial vehicle fleet operators, transit system operators, traffic management agency, traveler, and other perspectives.
- Offensive and defensive analysis—prediction of results of hypothesized scenarios involving drivers, other travelers, and levels of congestion.

Level 4 processing seeks to improve the entire data fusion process by continuously monitoring and evaluating the ongoing fusion process to enhance the effectiveness of the process itself and to regulate the acquisition of data for optimum results. It refines predictions and assessments, and evaluates the need for additional sources of information.

Key functions of Level 4 fusion are as follows:

- Evaluations—assess performance and effectiveness of fusion process.
- Fusion control—identify changes to processing functions that will improve performance.
- Source requirements processing—determine source-specific data requirements to improve multilevel fusion products.
- Mission management—recommend allocation of resources to achieve overall mission goals. Resources include roadside sensors, Bluetooth and toll-tag readers, DSRC transmitters and receivers such as those found in connected vehicles, and ALPR.

Level 5 processing addresses issues concerning human interpretation of the results of the data fusion process and their application. Although not officially incorporated into the JDL fusion model, the broader impacts of human–computer interactions in terms of cognitive science and information fusion systems are widely discussed in the literature [58,59]. Research into cognitive science has focused not only on a single individual's internal thought processes, but also on the interactions with the surroundings, including other individuals and groups, artifacts, and additional types of information systems. Thus, cognition can be considered as distributed in a threefold sense: (1) across individuals in a group or organization; (2) between internal mechanisms in the human organism (e.g., memory and perception) and external mechanisms that include computer systems, material, and social and cultural environments; and (3) over time.

HCI functions provide the mechanisms through which the results of fusion processing are conveyed to one or more human operators or analysts, and the means by which an operator controls and guides the fusion inference process. Data must be presented to a user, and often multiple users, in a timely fashion without overwhelming the user with constant interruptions from incoming data or extraneous information.

Fundamental design questions for HCI systems are, what does the user need to know, and when does it need to be known? Another complicating factor for HCI in data fusion is due to the magnitude and variety of data that can be displayed, including fixed and free-text message formats under multiple protocols, and asynchronous, out-of-sequence, and false sensor reports.

In addition to these issues, operators of data fusion systems should be aware of human performance costs that have been linked to particular forms of automation. The costs are reduced situation awareness, complacency, and skill degradation [60]. The first of these occurs when automated decision-making reduces the operator's awareness of the system and of specific dynamic features of the work environment. If a decision aid, expert system, or other type of decision automation repeatedly selects and executes decision choices in a dynamic environment, human operators may not be able to sustain an acceptable understanding of the information sources because they are not actively engaged in evaluating the sources leading to a decision. The second cost occurs if automation is highly, but not perfectly reliable in executing decisions. This may cause operators to be complacent in monitoring the automation and its information sources and, hence, bring about a failure to detect the occasional times when the automation fails. Complacency is greatest when the operator is engaged in multiple tasks. It also happens if the algorithms used for data and information analysis are reliable, but not perfectly so. The third detriment takes place when the decision-making function is consistently executed by automation, leading to a time when the human operator will not be as skilled in performing that function.

Other challenging issues arise concerning HCI design for traffic management applications. Since these fusion systems may operate in a stressful environment where people's lives may be at stake, such as when a traffic accident occurs, they should guide traffic management personnel through an effective decision-making paradigm in the face of stress. Where several agencies are involved in resolving an incident or managing traffic flow, shared situation awareness is vital and it is necessary to achieve a common state of understanding within all agencies through the exchange of data and information [61]. This requires that the lead agency's intent be accessible and understandable, and the understanding that shared situation awareness can only be developed over time and with practice [58]. There are also different decision-making styles employed by different agencies that affect the way they search for relevant data and information and perform analysis procedures.

These and other concerns that information fusion research attempts to address are summarized in Table 16.3 [59]. It contains an overview of categories that can influence user interactions, specific factors associated with each category, and the constraints often imposed when attempting to implement the functions contained in an information fusion system. The table also indicates the flow of information between categories. For example, the external environment, comprising sensors, databases, and the organization's functional relationships, affects the users in terms of their cognitive abilities and the activities they can perform. The users' cognitive abilities, in turn, often limit the possible tasks they can execute. The trust factor relates to the acceptance level on the part of the user to the automated output of the particular tool. The user exploits the interface to assist in completing various activities and, consequently, the interface is required to access the functions supported by the information fusion system. Lastly, the information fusion system itself captures various aspects of the environment.

Table 16.3 Human–computer interaction issues in an information fusion context

Category	Factor	Constraint
External environment *affects*	Organizational demands	Enable different levels of information availability to facilitate access for individuals and groups with different authorizations and job descriptions. Provide option of protecting sensitive data. Capture organizational information that guides interaction to inform users. Encourage role-based systems. Integrate information fusion system into those currently operational within the organization.
	Multiple decision-makers	Provide overlapping information to facilitate communication among team members. Use similar language to facilitate team communication. Introduce standard and advanced functions to meet varying user needs.
	Risk	Introduce thresholds to facilitate similar user decisions. Provide guidelines on how to respond to probabilities and other information provided.
	Temporal aspects	Clearly indicate temporal data, e.g., time and date, on displays to aid users.
	Dynamism	Provide flexibility in system for evolving requirements and tasks.
	Environment	Indicate if and how sensors are affected by environmental factors.
User's cognitive abilities	Cognitive issues	Allow interface personalization. Direct user's attention toward specific areas of interest. Restrict distracting clutter so as not to overload users. Focus on a subset of information to reduce cognitive workload. Support user's mental model for the system. Limit amount of data that needs to be processed simultaneously.
determines	Situation awareness	Provide alternative views of the situation at hand. Enable switching between detailed or local view and a global view. Show your own situation in relation to that of others.
	Trust	Present uncertainty in information provided by automated process. Provide transparency to enable understanding of recommendations and predictions. Direct user training toward confidence building (in the outputs of the tool) rather than training as such, that is, trust builds up over time.
User activities	User tasks	Provide interaction opportunities for users. Filter information but keep it available for users with flexibility. Do not allow information fusion system design to interfere with user tasks.
utilize	Decision-making	Provide a fit between decision-makers and decision-making process at IF system output. Incorporate explanatory capabilities, feature matching strategies, and story generation or exploration according to decision at hand. Enable filtering options to extract relevant information according to decision at hand without hindering access to nonfiltered (original) data. Provide access to both fused data and original data. Facilitate fast decisions through easy access to certain information without a requirement for interaction.

(Continued)

Table 16.3 (Continued) Human–computer interaction issues in an information fusion context

Category	Factor	Constraint
Interface access	Input/output devices	Use multiple modalities to support simultaneous processing of information. Present data in visual form when possible.
	Visualization	Visualize uncertainty, information reliability, and quality of information. Display past, present, and future (predicted) information. Present different levels of abstraction or granularity in time and space.
Information fusion system captures	Multiple information sources	Indicate type of source when using multiple information sources to aid interpretations. Provide access to original data and fused data.
	Uncertainty	Convey uncertainty (when it exists) in the information provided to others.
External environment and completes the cycle	Information flow	Provide flexibility to support both a top-down and bottom-up approach when required.
	Automation	Automate tasks that computers do best.

16.11 SUMMARY

Sensor fusion is a term that describes a configuration of sensors or other sources of information utilized to gather more accurate or additional data about events or objects that are present in the area of interest. According to the JDL data fusion model, data fusion is a multilevel, multifaceted process dealing with the automatic detection, association, correlation, estimation, and combination of data and information from single and multiple sources to achieve refined position and identity estimates, and complete and timely assessments of situations and impacts and their significance. The JDL model references six levels of fusion.

Level 0 preprocesses data with techniques such as normalizing, formatting, ordering, batching, and compressing. Level 1 data processing identifies the objects or events and their state by combining or fusing data from all available sources. Level 2 processing identifies the probable situation causing the observed data and events by combining the results of the Level 1 processing with information from other sources and databases such as police reports, roadway configuration drawings, local and national weather reports, anticipated traffic mix, time-of-day and seasonal traffic patterns, construction schedules, and special event schedules. Level 3 processing assesses the traffic flow patterns and other data with respect to the likely impact of a traffic event on traffic flow (e.g., duration of traffic congestion, incident, fire, or police action). Level 4 processing seeks to improve the entire data fusion process by continuously refining predictions and assessments, and evaluating the need for additional sources of information or redeployment of existing sources. Level 5 processing is concerned with enabling a human to interpret and apply the results of the fusion process. The taxonomy for object detection, classification, and identification algorithms includes physical models, feature-based inference techniques, and cognitive-based models. The taxonomy for state estimation algorithms identifies two general methods for searching and updating existing track files, namely, data-driven or track-driven, and identifies the techniques available to correlate and associate data and tracks.

The JDL data fusion levels are intended only as a convenient categorization of data fusion functions. Data fusion levels are not intended to be, nor should they be taken as a prescription for designing systems: do Level 0 fusion first, then Level 1, then Level 2, and so on. Processing should be partitioned in terms of the individual system requirements.

Data fusion applications to transportation management have been ongoing for at least two decades. Yet, it is a still maturing resource. Level 1 data fusion algorithms that have been successfully applied to traffic management include fuzzy logic, artificial neural networks, Bayesian inference, Dempster–Shafer evidential reasoning, inference rules, weighted mean of several travel-time estimators, Kalman filtering and its variants, image processing using edge detection and extraction of other features, k-means algorithm, kernel estimator, and Monte Carlo techniques including the particle filter.

Level 2 and 3 fusion applications are also appearing. For example, in England, the National Traffic Information Service is combining inductive loop speed and flow information with automatic license plate reader travel-time data and accident, inclement weather, and work zone locations to generate predictive forecasts concerning the duration of an event [62].

For the applications reported, data fusion techniques appear promising. However, these encouraging results should not conceal the challenges that still remain. These include obtaining data with the necessary accuracy to create effective applications, dynamic and real-time issues associated with data quality as traffic flow changes, processing of data in real time, and the development of methods to combine sensor or hard data with human-generated or soft data [3,6].

REFERENCES

1. N.-E. El Faouzi, H. Leung, and A. Kurian, Data fusion in intelligent transportation systems: Progress and challenges–A survey, *Information Fusion*, 12:4–10, Elsevier, 2011.
2. L.A. Klein, *Sensor and Data Fusion: A Tool for Information Assessment and Decision Making*, Second Edition, Press Monograph 222, SPIE, Bellingham, WA, 2012.
3. N.-E. El Faouzi and L.A. Klein, Data fusion in intelligent traffic and transportation engineering—Recent advances and challenges, In *Multisensor Data Fusion: From Algorithm and Architecture Design to Applications*, H. Fourati (ed.), Chapter 32, CRC Press/Taylor and Francis, Boca Raton, FL, 2015.
4. N.-E. El Faouzi and L.A. Klein, Data fusion for ITS: Techniques and research needs, ISEHP 2016 International Symposium, *Transportation Research Procedia*, 15:495–512, 2016.
5. E. Waltz and J. Llinas, *Multisensor Data Fusion*, Artech House, Norwood, MA, 1990.
6. B. Khaleghi, A. Khamis, F.O. Karray, and S.N. Razav, Multisensor data fusion: A review of the state-of-the-art, *Information Fusion*, 14:28–44, 2013.
7. F. Castanedo, A review of data fusion techniques, Article ID 704504, *The Scientific World Journal*, 2013. http://dx.doi.org/10.1155/2013/704504. Accessed June 22, 2016.
8. O. Kessler, K. Askin, N. Beck, J. Lynch, F. White, D. Buede, D. Hall, and J. Llinas, *Functional Description of the Data Fusion Process*, Office of Naval Technology, Office of Naval Technology, Naval Air Development Center, Warminster, PA, 1991.
9. H. Boström, S.F. Andler, M. Brohede, R. Johansson, A. Karlsson, J. van Laere, L. Niklasson, M. Nilsson, A. Persson, and T. Ziemke, *On the Definition of Information Fusion as a Field of Research*, Univ. of Skövde Tech Rpt. HS-IKI-TR-006, Skövde, SW, 2007.
10. *Connected Vehicle Pilot Deployment Program*, ITS Joint Program Office, U.S. Department of Transportation, Washington, DC, 2015. http://www.its.dot.gov/pilots/index.htm. Accessed November 18, 2015.
11. P.E. Ross, Europe's smart highway will shepherd cars from Rotterdam to Vienna, *IEEE Spectrum*, posted Dec. 30, 2014. http://spectrum.ieee.org/transportation/advanced-cars/europes-smart-highway-will-shepherd-cars-from-rotterdam-to-vienna. Accessed November 12, 2015.

12. J.L. Ygnace, *Travel Time/Speed Estimates on the French Rhone Corridor Network using Cellular Phones as Probes: STRIP Project*, European Community Research Program SERTI, INRETS-LESCOT, Lyon, France, 2001.

13. Y. Youngbin and R. Cayford, *Investigation of Vehicles as Probes using Global Positioning System and Cellular Phone Tracking*, University of California PATH, Berkeley, CA, 2000.

14. L.A. Klein and M.R. Kelley, *Detection Technology for IVHS, Volume I: FHWA-RD-95-100*, Final Report, U.S. Department of Transportation, Federal Highway Administration, Washington, DC, 1996.

15. L.A. Klein, *Sensor Technologies and Data Requirements for ITS*, Artech House, Boston, MA, 2001.

16. B.V. Dasarathy, Sensor fusion, potential exploitation: innovative architectures and illustrative applications, *Proceedings of IEEE*, 85:24–38, 1997.

17. H. Durrant-Whyte and T. Henderson, Multisensor data fusion, In *Handbook of Robotics, Part C—Sensing and Perception*, O. Khatib and B. Siciliano (eds.), Chapter 25, Springer, New York, NY, 2008.

18. L.A. Klein, *Millimeter-Wave and Infrared Multisensor Design and Signal Processing*, Artech House, Norwood, MA, August 1997.

19. F.E. White Jr., Joint directors of laboratories data fusion subpanel report: SIGINT session, Tech. Proc. Joint Service Data Fusion Symposium, Vol. I, DFS-90, 469–484, 1990.

20. 1989 data fusion survey, *Proc. 1990 Data Fusion Symposium*, Johns Hopkins University Applied Physics Laboratory, Laurel, MD, May 1990.

21. Data Fusion Development Strategy Panel, *Functional Description of the Data Fusion Process*, Office of Naval Technology, Warminster, PA, November 1991.

22. D.L. Hall, *Mathematical Techniques in Multisensor Data Fusion*, Artech House, Norwood, MA, 1992.

23. A.N. Steinberg, C.L. Bowman, and F.E. White Jr., Revisions to the JDL data fusion model, *Proceedings of SPIE 3719*, 430–441, 1999. Also in A.N. Steinberg, C.L. Bowman, and F.E. WhiteJr., Revisions to the JDL model, *Joint NATO/IRIS Conference Proceedings*, Quebec, October 1998.

24. J. Llinas, *Data Fusion Overview*, Univ of Buffalo, Buffalo, NY, October 2002.

25. J. Llinas, C. Bowman, G. Rogova, A. Steinberg, E. Waltz, and F. White, Revisiting the JDL data fusion model II, In *Proc. 7th International Conf. on Information Fusion*, Vol. II, P. Svensson and J. Schubert (eds.), International Society of Information Fusion, Mountain View, CA, 1218–1230, June 2004.

26. N.-E. El Faouzi, Multiform traffic data collection and data fusion in road traffic engineering, In *Proceedings of the Multisource Data Fusion in Traffic Engineering Workshop*, N.-E. El Faouzi (ed.), INRETS, Bron, France, 2003.

27. B.V. Dasarathy, Information fusion, data mining, and knowledge discovery, *Information Fusion*, 85(1):1, Elsevier, 2003.

28. E.P. Blasch and S. Plano, JDL Level 5 fusion model 'user refinement' issues and application in group tracking, Aerosense, *Proceedings of SPIE* 4729:270–279, 2002.

29. A.N. Steinberg and C.L. Bowman, *Rethinking the JDL Data Fusion Levels*, NSSDF JHAPL, June 2004.

30. A.N. Steinberg, C.L. Bowman, and F.E. White, *Revisions to the JDL Data Fusion Model*, ERIM International, Arlington, VA, 1999.

31. D.L. Hall and R.J. Linn, Algorithm selection for data fusion systems, *1987 Tri-Service Data Fusion Symposium Technical Proceedings*, Vol. I, DFS-87, Naval Air Development Center, Warminster, PA, June 1987.

32. D.L. Hall and R.J. Linn, A taxonomy of algorithms for multi-sensor data fusion, *Tech. Proc. Joint Service Data Fusion Symposium*, Vol. I, DFS-90, 594–610, Naval Air Development Center, Warminster, PA, 1990.

33. I.R. Goodman, R.P.S. Mahler, and H.T. Nguyen, *Mathematics of Data Fusion*, Kluwer Academic Publishers, Norwell, MA, 1997.

34. J. Pearl, *Probabilistic Reasoning in Intelligent Systems: Networks of Plausible Inference*, Morgan Kaufmann Publishers, San Mateo, CA, 1988.

35. J. Dezert, Foundations for a new theory of plausible and paradoxical reasoning, *Information and Security*, 9:1–45, 2002.

36. C.K. Murphy, Combining belief functions when evidence conflicts, *Decision Support Systems*, 29:1–9, Elsevier Science, 2000.

37. P. Smets and R. Kennes, The transferable belief model, *Artificial Intelligence*, 66:191–234, 1994.

38. B.R. Cobb and P.P. Shenoy, A comparison of methods for transforming belief function models to probability models, In *Symbolic and Quantitative Approaches to Reasoning with Uncertainty, Lecture Notes in Artificial Intelligence*, T.D. Nielsen and N. L. Zhang (eds.), Springer-Verlag, Berlin, Germany, 255–266, 2003.

39. A. Jøsang, The consensus operator for combining beliefs, *AI Journal*, 14(1–2):157–170, 2002.

40. R. Haenni and N. Lehmann, Probabilistic augmentation systems: a new perspective on Dempster-Shafer theory, *International Journal of Intelligent Systems*, 18(1):93–106, 2003.

41. D. Fixsen and R. P. S. Mahler, The modified Dempster-Shafer approach to classification, *IEEE Transactions on Systems, Man, and Cybernetics—Part A: Systems and Humans*, SMC-27(1):96–104, January 1997.

42. A.-S. Capelle, C. Fernandez-Maloigne, and O. Colot, Introduction of spatial information within the context of evidence theory, IEEE Int. Conf. on Acoustics, Speech, and Sig. Proc. (ICASSP), 785–788, 2003.

43. S.C.A. Thomopoulos, R. Viswanathan, and D. C. Bougoulias, Optimal decision fusion in multiple sensor systems, *IEEE Transactions on Aerospace and Electronic Systems*, AES-23(5):644–653, September 1987.

44. S.C.A. Thomopoulos, Theories in distributed decision fusion: Comparison and generalization, *Sensor Fusion III: 3-D Perception and Recognition, Proceedings of SPIE*, 1383:623–634, 1990 [doi: 10.1117/12.25302].

45. S.C.A. Thomopoulos, Sensor integration and data fusion, *Journal of Robotic Systems*, 7(3):337–372, June 1990.

46. P. Rohan, *Surveillance Radar Performance Prediction*, Peter Peregrinus, Ltd., London, UK, 1983.

47. S. Haykin, *Neural Networks: A Comprehensive Foundation*, 2nd Edition, Prentice Hall PTR, Upper Saddle River, NJ, 1998.

48. R. Schalkoff, *Pattern Recognition: Statistical, Structural, and Neural Approaches*, John Wiley, New York, NY, 1992.

49. J. Roy, Combining elements of information fusion and knowledge-based systems to support situation analysis, *Proceedings of SPIE 6242*, Multisensor, Multisource Information Fusion: Architectures, Algorithms, and Applications 2006, Paper 6242-02, 2006.

50. L.A. Zadeh, *Fuzzy Sets and Systems*, North-Holland Press, Amsterdam, the Netherlands, 1978.

51. L.A. Zadeh, Fuzzy logic, *IEEE Computer Magazine*, 21(4):83–93, April 1988.

52. B. Kosko, *Neural Networks and Fuzzy Systems: A Dynamical Systems Approach to Machine Intelligence*, Prentice-Hall, Englewood Cliffs, NJ, 1992.

53. Y. Bar-Shalom and T.E. Fortmann, *Tracking and Data Association*, Academic Press, Orlando, FL, 1988.

54. M.E. Liggins II, C.-Y. Chong, I. Kadar, M.G. Alford, V. Vannicola, and S. Thomopoulos, Distributed fusion architectures and algorithms for target tracking, *Proceedings of IEEE*, 85(1):95–107, January 1997.

55. L.A. Klein, Dempster-Shafer data fusion at the traffic management center, *Presented at Transportation Research Board 79th Annual Meeting*, Transportation Research Board, Paper 00-1211, Washington, DC, 2000.

56. L.A. Klein, P. Yi, and H. Teng, Decision support system for advanced traffic management through data fusion, *Transportation Research Record*, 1804:173–178, 2002.

57. N.-E. El Faouzi, L.A. Klein, and O. De Mouzon, Improving travel time estimates from inductive loop and toll collection data with Dempster-Shafer data fusion, *Transportation Research Record*, 2129:73–80, 2009.

58. M. Nilsson and T. Ziemke, Rethinking Level 5: Distributed cognition and information fusion, *Proc. 9th International Conf. on Information Fusion*, Florence, IT, July 10–13, 2006.

59. M. Nilsson and T. Ziemke, *Investigating Human-Computer Interaction Issues in Information-Fusion-Based Decision Support*, Tech. Rpt. HS-IKI-TR-08-002, School of Humanities and Informatics, University of Skövde, Sweden, July 2008.

60. R. Parasuraman, T.B. Sheridan, and C.D. Wickens, A model for types and levels of human interaction with automation, *IEEE Transactions on Systems, Man and Cybernetics—Part A: Systems and Humans*, 30(3):286–297, May 2000.

61. S. Snell, The dissemination and fusion of geographical data to provide distributed decision making in a network-centric environment, 9th International Command and Control Research and Technical Symp., Washington, DC, 2004.

62. I. Patey, Use of big data for managing England's National Network, International Symposium on Enhancing Highway Performance, Berlin, Germany, June 2016.

Chapter 17

Bayesian inference and Dempster–Shafer evidential reasoning and their application to traffic management

Bayesian inference and Dempster–Shafer evidential reasoning are widely applied for detection, classification, and identification of objects and events that occur in conjunction with traffic management strategies. The applications include travel time estimation, automatic incident detection, and decision support. This chapter describes the underlying principles of these techniques followed by several examples that illustrate their effectiveness for traffic management.

17.1 BAYESIAN INFERENCE

Bayesian inference is a probability-based reasoning discipline grounded in Bayes' rule. When used to support data fusion, Bayesian inference belongs to the class of parametric algorithms that use *a priori* knowledge about events or objects in an observation space to make inferences about the identity of events or objects in that space. Bayesian inference provides a method for calculating the conditional *a posteriori* (posterior) probability of a hypothesis being true given supporting evidence. Thus, Bayes' rule offers a technique for updating beliefs in response to information or evidence that would cause the belief to change [1].

Another interpretation of Bayesian inference is given by this scenario. Suppose it is necessary to determine the likelihoods of different values of an unknown state H_i. There may be prior beliefs about what H_i might be expected. These are encoded in the form of relative likelihoods in the prior probability $P(H_i)$. An observation E is made to obtain additional information about the particular H_i. The observations are modeled as a conditional probability $P(E|H_i)$ that describes, for each H_i, the probability of E given that a particular H_i is true. The new likelihoods associated with the H_i are computed from the product of the original prior information and the information gained by observation (the evidence). This is encoded in the posterior probability $P(H_i|E)$ that describes the likelihoods associated with H_i given the evidence E. In a fusion process, the factor $P(E)$ that appears in the denominator of Bayes' rule is used to normalize the posterior and is not generally evaluated using its formal definition as the sum of the products of the likelihood functions and the prior probabilities [2].

17.1.1 Derivation of Bayes' rule

Bayes' rule may be derived by evaluating the probability of occurrence of an arbitrary event E assuming that another event H has occurred. The probability is given by

$$P(E \mid H) = \frac{P(EH)}{P(H)}, \tag{17.1}$$

where H is an event with positive probability. The quantity $P(E|H)$ is the probability of E conditioned on the occurrence of H. The conditional probability is not defined when H has zero probability. The factor $P(EH)$ represents the probability of the intersection of events E and H.

To illustrate the meaning of Equation 17.1, consider a population of N people that includes N_E left-handed people and N_H females as shown in the Venn diagram of Figure 17.1. Let E and H represent the events that a person chosen at random is left-handed or female, respectively. Then,

$$P(E) = \frac{N_E}{N} \tag{17.2}$$

and

$$P(H) = \frac{N_H}{N}. \tag{17.3}$$

The probability that a female chosen at random is left-handed is N_{EH}/N_H, where N_{EH} is the number of left-handed females. In this example, $P(E|H)$ denotes the probability of selecting a left-handed person at random assuming the person is female. In terms of population parameters, $P(E|H)$ is

$$P(E \mid H) = \frac{N_{EH}}{N_H} = \frac{P(EH)}{P(H)}. \tag{17.4}$$

Returning to the derivation of Bayes' rule, Equation 17.1 may be rewritten as

$$P(EH) = P(E \mid H)P(H), \tag{17.5}$$

which is referred to as the theorem on compound probabilities.

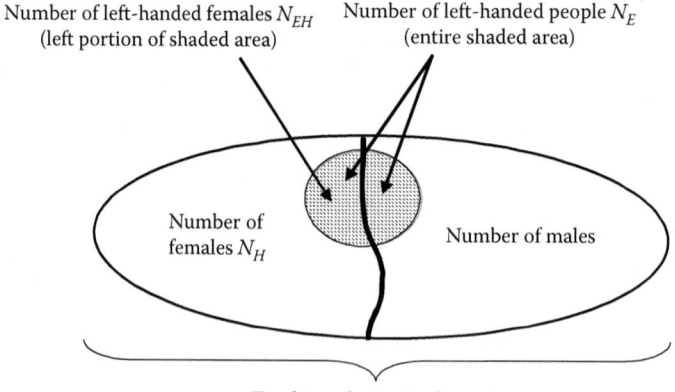

Figure 17.1 Venn diagram illustrating intersection of events E (person chosen at random is left-handed) and H (person chosen at random is female).

When H consists of a set of mutually exclusive and exhaustive hypotheses $H_1, ..., H_n$, conditional probabilities, which may be easier to evaluate than unconditional probabilities, can be substituted for $P(EH)$ as follows. The mutually exhaustive property implies that one hypothesis necessarily is true, that is, the union of $H_1, ..., H_n$ is the entire sample space. Under these conditions, any event E can occur only in conjunction with some H_j such that

$$E = EH_1 \cup EH_2 \cup ... \cup EH_n. \tag{17.6}$$

Since the EH_j are mutually exclusive, their probabilities add as

$$P(E) = \sum_{i=1}^{n} P(EH_i). \tag{17.7}$$

Upon substituting H_j for H and summing over i, Equation 17.5 becomes

$$P(E) = \sum_{i} [P(E \mid H_i) P(H_i)], \tag{17.8}$$

when the identity in Equation 17.7 is applied.

Equation 17.8 states that the belief in any event E is a weighted sum over all the distinct ways that E can be realized.

In Bayesian inference, we are interested in the probability that hypothesis H_i is true given the existence of evidence E. This statement is expressed as

$$P(H_i \mid E) = \frac{P(EH_i)}{P(E)}. \tag{17.9}$$

If Equations 17.5 and 17.8 are introduced into Equation 17.9, Equation 17.9 takes the form of Bayes' rule as

$$P(H_i \mid E) = \frac{P(E \mid H_i) P(H_i)}{P(E)} = \frac{P(E \mid H_i) P(H_i)}{\sum_{i} [P(E \mid H_i) P(H_i)]}, \tag{17.10}$$

where $P(H_i|E)$ is the *a posteriori* (posterior) probability that hypothesis H_i is true given evidence E, $P(E|H_i)$ is the probability of observing evidence E given that H_i is true (sometimes referred to as the likelihood function), $P(H_i)$ is the *a priori* (prior) probability that hypothesis H_i is true,

$$\sum_{i} P(H_i) = 1, \tag{17.11}$$

and $\sum_i P(E \mid H_i) P(H_i)$ is the preposterior or probability of observing evidence E given that hypothesis H_i is true, summed over all hypotheses i.

To summarize, Bayes' rule simply states that the posterior probability is equal to the product of the likelihood function and the prior probabilities divided by the evidence. The likelihood functions represent the extent to which the posterior probability is subject to change. These functions are evaluated through offline experiments or by analyzing the available information for the problem at hand. A general method of estimating the parameter(s) that maximize the likelihood function given the data is to find the maximum likelihood estimate. This procedure selects the parameter value that makes the data actually observed as likely as possible [3–5]. The preposterior is the sum of the products of the likelihood functions and the *a priori* probabilities. It can also be calculated from its role as a normalizing constant [6]. The approach used to compute its numerical value depends on the information available in a particular problem.

17.1.2 Monty Hall problem

The classical Monty Hall problem from the game show Let's Make a Deal® describes "gifts" hidden behind three doors. Only one of the doors hides a valuable gift, such as an automobile, while the other two hide less desirable gifts such as goats. In the first formulation of the problem, Monty knows what's behind each door. This is critical information, as shown later. Monty asks the contestant to select the door that he thinks is hiding the valuable gift. Suppose the contestant chooses Door 1 initially. Monty then reveals the goat located behind Door 2 or Door 3 as depicted in Figure 17.2. The contestant is then asked if he wants to switch his door selection. Is it to the advantage of the contestant to switch or not? The problem can be solved using any of three methods.

17.1.2.1 Case-by-case analysis solution

In the first approach, a case-by-case analysis, a truth table confirms that the odds of winning the automobile if the contestant does not switch are 1:3 as only one of the three doors hides the automobile. Table 17.1 shows where the two goats (Billy and Millie) may be hidden

Figure 17.2 Monty Hall problem showing a goat behind one of three doors. In this example, there is a second goat behind either Door 1 or Door 2.

Table 17.1 Monty Hall problem with car behind one door and goats Billy or Millie behind the other two

Door 1	Door 2	Door 3
Car	Billy	Millie
Car	Millie	Billy
Billy	Car	Millie
Millie	Car	Billy
Billy	Millie	Car
Millie	Billy	Car

Table 17.2 Case-by-case analysis of Monty Hall problem

Case	Explanation
1	Monty Hall reveals a goat behind either Door 2 or Door 3. It is *not* to the contestant's advantage to switch. Record *No.*
2	Similar to case 1. It is *not* to the contestant's advantage to switch. Record *No.*
3	Monty Hall reveals a goat behind Door 3 and it *is* to the contestant's advantage to switch. Record *Yes.*
4	Similar to Case 3 and it *is* to the contestant's advantage to switch. Record *Yes.*
5	Monty Hall reveals a goat behind Door 2 and it *is* to the contestant's advantage to switch. Record *Yes.*
6	Similar to Case 5 and it *is* to the contestant's advantage to switch. Record *Yes.*

by any two of the three doors. Monty will always reveal a goat, never the more valuable automobile.

The odds that the contestant will win the automobile by switching doors are described in Table 17.2. The tally of the case-by-case analysis is four *Yes* and two *No*. Thus, odds are 2:1 in favor of switching after Monty Hall reveals the goat.

17.1.2.2 Conditional probability solution

The second way of solving the Monty Hall problem is by applying conditional probabilities. Let the three doors be denoted by 1, 2, and 3 and let C_1 be the event that the car is behind Door 1, C_2 be the event that the car is behind Door 2, and C_3 be the event that the car is behind Door 3.

Let H_1 represent the event that Monty Hall opens Door 1, H_2 represent the event that Monty Hall opens Door 2, and H_3 represent the event that Monty Hall opens Door 3.

Assuming the game show host knows what item is behind each door and, furthermore, that the contestant chooses Door 1 to begin the game, the probability that the contestant wins a car if he then switches his choice is given by

$$P(H_3 \cap C_2) + P(H_2 \cap C_3) = P(C_2)P(H_3 \mid C_2) + P(C_3)P(H_2 \mid C_3) = \left(\frac{1}{3}\right)(1) + \left(\frac{1}{3}\right)(1) = \frac{2}{3}. \quad (17.12)$$

The values of $P(H_3|C_2)$ and $P(H_2|C_3)$ are unity since Monty will always reveal a goat as he knows what item is behind each door.

If the game show host does *not* know what item is behind each door or he forgets, then the probability of the contestant winning the car does not change if he switches his choice of doors. Mathematically, the probabilities become

$$P(H_3 \cap C_2) + P(H_2 \cap C_3) = P(C_2)\, P(H_3 \mid C_2) + P(C_3)\, P(H_2 \mid C_3) = \left(\frac{1}{3}\right)\left(\frac{1}{2}\right) + \left(\frac{1}{3}\right)\left(\frac{1}{2}\right) = \frac{1}{3}.$$
$$(17.13)$$

In this case, Monty makes a random choice between the two remaining doors and $P(H_3|C_2)$ and $P(H_2|C_3)$ are each equal to 1/2.

A Web site demonstration of the Monty Hall problem for both situations is found at www.math.ucsd.edu/~crypto/Monty/monty.html and further discussions of the problem are offered at www.cut-the-knot.org/hall.shtml.

17.1.2.3 Bayesian inference solution

In Bayesian terms, a probability $P(A|I)$ is a number in $|0, 1|$ associated with a proposition A. The number expresses a degree of belief in the truth of A, subject to whatever *background* information I happens to be known.

For this problem, the background is provided by the rules of the game. The propositions of interest are as follows:

C_i: The car is behind Door i, for i equal to 1, 2, or 3.
H_{ij}: The host opens Door j after the player has picked Door i, for i and j equal to 1, 2, or 3.

For example, C_1 denotes the proposition *the car is behind Door 1*, and H_{12} denotes the proposition *the host opens Door 2 after the player has picked Door 1*.

The assumptions underlying the common interpretation of the Monty Hall puzzle are then formally stated as follows.

First, the car can be behind any door, and all doors are *a priori* equally likely to hide the car. In this context *a priori* means *before the game is played*, or *before seeing the goat*. Hence, the prior probability of a proposition C_i is

$$P(C_i \mid I) = \frac{1}{3}. \tag{17.14}$$

Second, the host will always open a door that has no car behind it chosen from among the two not picked by the player. If two such doors are available, each one is equally likely to be opened. This rule determines the conditional probability of a proposition H_{ij} subject to where the car is, that is, *conditioned* on a proposition C_k according to

$$P(H_{ij} \mid C_k, I) = \begin{cases} 0 & \text{if } i = j \quad \text{(the host cannot open the door picked by the player)} \\ 0 & \text{if } j = k \quad \text{(the host cannot open a door with a car behind it)} \\ \frac{1}{2} & \text{if } i = k \quad \text{(the two doors with no car are equally likely to be opened)} \\ 1 & \text{if } i \neq k \text{ and } j \neq k \quad \text{(there is only one door available to open)} \end{cases} \tag{17.15}$$

The problem can now be solved by scoring each strategy with its associated posterior probability of winning, that is, with its probability subject to the host's opening of one of the doors. Without loss of generality, assume, by re-numbering the doors if necessary, that the player picks Door 1 and that the host then opens Door 3 revealing a goat. In other words, the host *makes* proposition H_{13} true.

The posterior probability of winning by *not* switching doors, subject to the game rules and H_{13}, is then $P(C_1|H_{13}, I)$. Bayes' theorem expresses this probability as

$$P(C_1 \mid H_{13}, I) = \frac{P(H_{13} \mid C_1, I)\, P(C_1 \mid I)}{P(H_{13} \mid I)}. \tag{17.16}$$

Using the above assumptions, the numerator of the right-hand side becomes

$$P(H_{13} \mid C_1, I)\, P(C_1 \mid I) = \frac{1}{2} \times \frac{1}{3} = \frac{1}{6}. \tag{17.17}$$

The normalizing constant in the denominator is evaluated by expanding it using the definitions of marginal probability and conditional probability. Thus,

$$P(H_{13} \mid I) = \sum_i P(H_{13}, C_i \mid I) P(C_i \mid I)$$

$$= P(H_{13}, C_1 \mid I) P(C_1 \mid I) + P(H_{13}, C_2 \mid I) P(C_2 \mid I) + P(H_{13}, C_3 \mid I) P(C_3 \mid I) \quad (17.18)$$

$$= \frac{1}{2} \times \frac{1}{3} + 1 \times \frac{1}{3} + 0 \times \frac{1}{3} = \frac{1}{2}.$$

Dividing the numerator by the normalizing constant yields

$$P(C_1 \mid H_{13}, I) = \frac{1}{6} \div \frac{1}{2} = \frac{1}{3}.$$

This is equal to the prior probability of the car being behind the initially chosen door, meaning that the host's action has not contributed any novel information with regard to this eventuality. In fact, the following argument shows that the effect of the host's action consists entirely of redistributing the probabilities for the car being behind either of the *other* two doors.

The probability of winning by switching the selection to Door 2, $P(C_2 \mid H_{13}, I)$, is evaluated by requiring that the posterior probabilities of all the C_i propositions add to 1. That is,

$$1 = P(C_1 \mid H_{13}, I) + P(C_2 \mid H_{13}, I) + P(C_3 \mid H_{13}, I). \quad (17.19)$$

There is no car behind Door 3, since the host opened it, so the last term must be zero. This is proven using Bayes' theorem and the previous results as

$$P(C_3 \mid H_{13}, I) = \frac{P(H_{13} \mid C_3, I) P(C_3 \mid I)}{P(H_{13} \mid I)} = \left(0 \times \frac{1}{3}\right) \div \frac{1}{2} = 0. \quad (17.20)$$

Hence,

$$P(C_2 \mid H_{13}, I) = 1 - \frac{1}{3} - 0 = \frac{2}{3}. \quad (17.21)$$

This shows that the winning strategy is to switch the selection to Door 2. It also makes clear that the host's showing of the goat behind Door 3 has the effect of *transferring* the 1/3 of winning probability, *a priori* associated with that door, to the remaining unselected and unopened one, thus making it the most likely winning choice.

17.1.3 Application of Bayes' rule to cancer screening

Suppose a patient visits a physician who proceeds to administer a low-cost screening test for cancer. The test has an accuracy of 95% (i.e., the test will indicate positive 95% of the time if the patient has the disease) with a 4% false-alarm probability. Furthermore, suppose that cancer occurs in 5 out of every 1000 people in the general population. If the patient is informed that he has tested positively for cancer, what is the probability the patient actually has cancer?

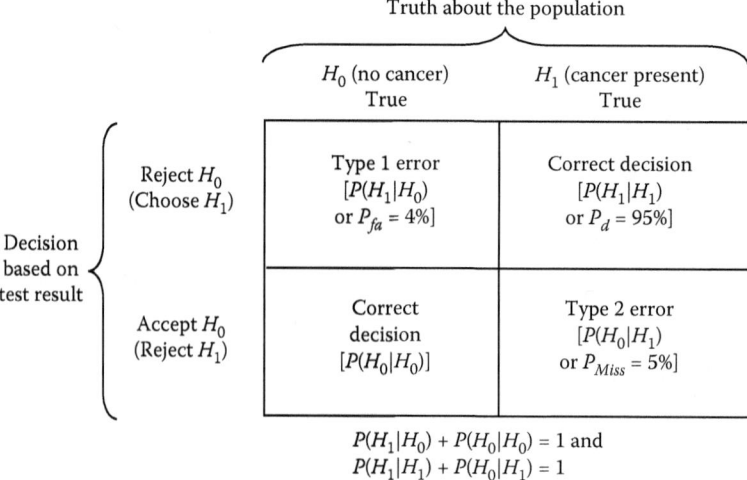

Figure 17.3 Hypotheses and errors in cancer screening example.

The truth diagram in Figure 17.3 summarizes the statistics for this example in terms of H_0 (patient does not have cancer) and H_1 (patient has cancer). The Bayesian formulation of Equation 17.10 predicts the required probability as

$$P(\text{patient has cancer} \mid \text{test positive}) = \frac{P(\text{test positive} \mid \text{cancer})P(\text{cancer})}{P(\text{test positive})}, \qquad (17.22)$$

where

$$P(\text{test positive}) = P(\text{test positive} \mid \text{cancer})P(\text{cancer}) + P(\text{test positive} \mid \text{no cancer})P(\text{no cancer}). \qquad (17.23)$$

The probability $P(\text{test positive} \mid \text{no cancer})$ is the false-alarm probability or Type 1 error. The Type 2 error is the probability of missing the detection of cancer in a patient with the disease.

Upon substituting the statistics from Figure 17.3 into Equation 17.22, we find

$$P(\text{patient has cancer} \mid \text{test positive}) = \frac{(0.95)(0.005)}{(0.95)(0.005) + (0.04)(0.995)} = 0.107 \quad \text{or} \quad 10.7\%. \qquad (17.24)$$

Intuitively, this result may appear smaller than expected. It asserts that in only 10.7% of the cases in which the test gives a positive result and declares cancer to be present, is it actually true that cancer is present. Further testing is thus required when this type of initial test is administered. The screening test may be said to be reliable because it will detect cancer in 95% of the cases in which cancer is present. However, the critical Type 2 error is 0.05, implying that the test will not diagnose 1 in 20 cancers.

To increase the probability of the patient actually having cancer, given a positive test, and concurrently reduce the Type 2 error requires a test with a greater accuracy. A more-effective method of increasing the *a posteriori* probability is to reduce the false-alarm probability. If, for example, the test accuracy is increased to 99.9% and the false-alarm probability reduced to 1%, the probability of the patient actually having cancer, given a positive test, is increased to 33.4%. The Type 2 error now implies a missed diagnosis in only 1 out of 1000 patients. Increasing the test accuracy to 99.99% has a minor effect on the *a posteriori* probability, but it reduces the Type 2 error by another order of magnitude.

In other situations, the Type 1 error may be the more serious error. Such a case occurs if an innocent man is tried for a crime and his freedom relied on the outcome of a certain experiment. If a hypothesis corresponding to his innocence was constructed and was rejected by the experiment, then an innocent man would be convicted and a Type 1 error would result. On the other hand, if the man was guilty and the experiment accepted the hypothesis corresponding to innocence, the guilty man would be freed and a Type 2 error would result.

Another breast cancer screening test is based on exposing single strands of hair to concentrated X-ray beams and examining the diffraction pattern formed. It is known from test trial data that

$$P(\text{test positive} \mid \text{cancer present}) = P(H_1 \mid H_1) = 0.82 \qquad (17.25)$$

and

$$P(\text{test negative} \mid \text{cancer not present}) = P(H_0 \mid H_0) = 0.77. \qquad (17.26)$$

The problem is to find the probability that a patient has cancer given a positive test result when the probability of cancer in the general population is 0.005.

From $P(H_1 \mid H_1) + P(H_0 \mid H_1) = 1$, calculate

$$P(H_0 \mid H_1) = 0.18 = P(\text{test negative} \mid \text{cancer present}). \qquad (17.27)$$

From $P(H_1 \mid H_0) + P(H_0 \mid H_0) = 1$, calculate

$$P(H_1 \mid H_0) = 0.23 = P(\text{test positive} \mid \text{cancer not present}), \qquad (17.28)$$

where H_1 represents the presence of cancer and H_0 represents the absence of cancer.

The required probability is then found as

$$
\begin{aligned}
&P(\text{patient has cancer} \mid \text{test positive}) \\
&= \frac{P(\text{test positive} \mid \text{cancer})\, P(\text{cancer})}{P(\text{test positive})} \\
&= \frac{P(\text{test positive} \mid \text{cancer})\, P(\text{cancer})}{P(\text{test positive} \mid \text{cancer present})\, P(\text{cancer}) + P(\text{test positive} \mid \text{no cancer})\, P(\text{no cancer})} \\
&= \frac{(0.82)(0.005)}{(0.82)(0.005) + (0.23)(0.995)} = 0.0176 \approx 1.8\%
\end{aligned}
$$

$$(17.29)$$

17.1.4 Bayesian inference in support of data fusion

Computing probabilities from subjective information allows the Bayesian inference process to be applied to multisensor fusion since probability density functions are not

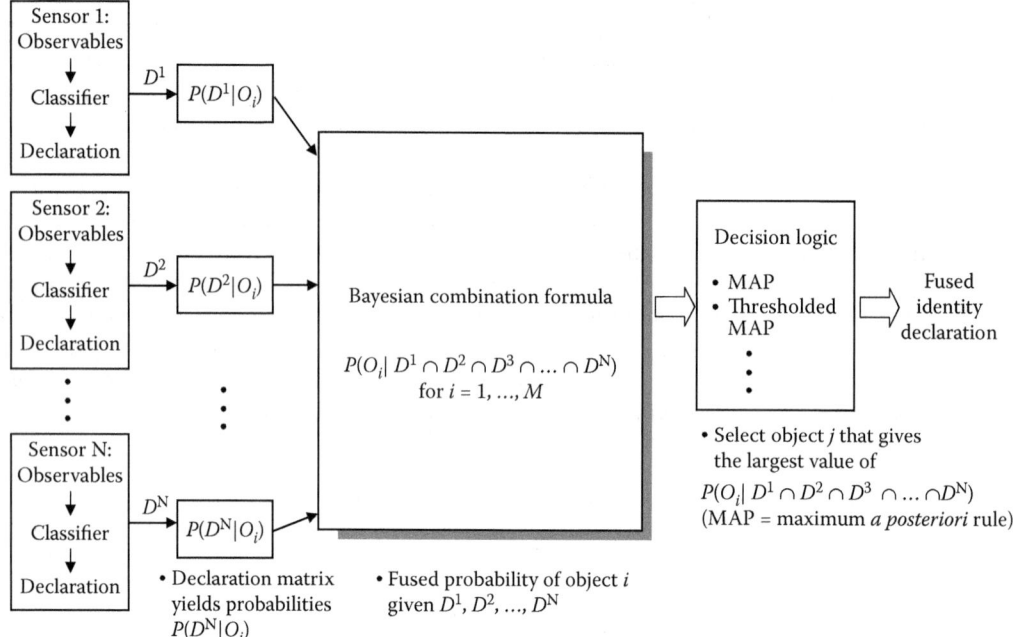

Figure 17.4 Bayesian inference fusion process. (Adapted from E. Waltz and J. Llinas, *Multisensor Data Fusion*, Artech House, Norwood, MA, 1990.)

required. A cautionary note—the output of such a process is only as good as the input *a priori* data.

Figure 17.4 illustrates the Bayesian inference process as applied to the fusion of multisensor identity information [7]. In this example, multiple sensors observe parametric data (e.g., visible spectrum or infrared video images and data, radar cross section, pulse repetition interval, rise and fall times of pulses, or frequency-spectrum signal characteristics) about an entity whose identity is unknown. Each of the sensors provides an identity declaration D or hypothesis about the object's identity based on the observations and a sensor-specific algorithm. The previously established performance characteristics of each sensor's classification algorithm (developed either theoretically or experimentally) provide estimates of the likelihood function, that is, the probability $P(D|O_i)$ that the sensor will declare the object to be a certain type, given that the object is in fact type i. These declarations are then combined using a generalization of Bayes' rule (Equation 17.10) to produce an updated, joint probability for each entity O_i founded on the multisensor declarations.

Thus, the probability of having observed object i from the set of M objects given declaration (evidence) D^1 from Sensor 1, declaration D^2 from Sensor 2, and so on is given by the Bayesian combination formula:

$$P(O_i \mid D^1 \cap D^2 \cap D^3 \cap ... \cap D^N), \quad i = 1,...,M. \tag{17.30}$$

By applying a decision logic, a joint declaration of identity can be selected by choosing the object whose joint probability given by Equation 17.30 is greatest. The choice of the maximum value of Equation 17.30 is referred to as the maximum *a posteriori* probability (MAP) decision rule. Other decision rules exist as indicated in the figure. The Bayesian formulation, therefore, provides a method to combine identity declarations from multiple sensors to obtain a new and hopefully improved joint identity declaration. Required inputs for the Bayes method are the ability to compute or model $P(D|O_i)$ for each sensor and entity and the *a priori* probabilities that the hypotheses $P(O_i)$ are true. When *a priori* information is lacking concerning the relative likelihood of O_i, the principle of indifference may be invoked in which $P(O_i)$ for all i are initially set equal to one another.

The application of Bayes' rule is often contrasted in modern probability theory with the application of confidence intervals. While Bayes' rule provides an inference approach suitable for some data fusion applications, the theory of confidence intervals is better suited when it is desired to assert, with some specified probability, that the true value of a certain parameter (e.g., mean and variance) that characterizes a known distribution is situated between two limits.

The Bayesian approach to recursive updating of the posterior probability occurs by utilizing the previous posteriors as the new values for the prior probabilities. In Equation 17.31, H_i denotes a hypothesis as before. The vector $\mathbf{E}^N = E^1, E^2,..., E^N$ represents a sequence of data observed from N sources in the past, while E represents a new fact (or new datum). If once we have calculated $P(H_i|\mathbf{E}^N)$ and we can discard past data, the impact of the new datum E is expressed as [6–8]

$$P(H_i \mid \mathbf{E}^N, E) = \frac{P(E \mid \mathbf{E}^N, H_i)\, P(H_i \mid \mathbf{E}^N)}{P(E \mid \mathbf{E}^N)} = \frac{P(E \mid \mathbf{E}^N, H_i)\, P(H_i \mid \mathbf{E}^N)}{\sum_i [P(E \mid \mathbf{E}^N, H_i)\, P(H_i)]}, \tag{17.31}$$

where $P(H_i \mid \mathbf{E}^N, E)$ is the *a posteriori* (posterior) probability of H_i for the current period, given the evidence or data $\mathbf{E}^N$, E available at the current period; $P(E \mid \mathbf{E}^N, H_i)$ is the probability of observing evidence E given H_i and the evidence $\mathbf{E}^N$ from past observations (i.e., the likelihood function); $P(H_i \mid \mathbf{E}^N)$ is the *a priori* (prior) probability of H_i, set equal to the posterior probability calculated using the evidence $\mathbf{E}^N$ from past observations; and $\sum_i P(E \mid \mathbf{E}^N, H_i)P(H_i)$ is the preposterior or probability of the evidence E occurring given the evidence $\mathbf{E}^N$ from past observations, conditioned on all possible outcomes H_i.

In this updating of the *a posteriori* probabilities, the old belief $P(H_i \mid \mathbf{E}^N)$ assumes the role of the prior probability when computing the new posterior. It completely summarizes past experience. Accordingly, updating of the posterior is accomplished by multiplying the old belief by the likelihood function $P(E \mid \mathbf{E}^N, H_i)$, which is equal to the probability of the new datum E given the hypothesis and the past observations.

A simplification of Equation 17.31 occurs under two conditions. The first is allowing the discarding of past evidence or data once the posterior probability $P(H_i \mid \mathbf{E}^N)$ is calculated. Then the likelihood function is independent of the past data and involves only E and H_i such that

$$P(E \mid \mathbf{E}^N, H_i) = P(E \mid H_i), \tag{17.32}$$

where $\mathbf{E}^N$ is the sequence of data observed from N sensors or sources in the past.

The second condition is fulfilled when conditionally independent sensors exist (i.e., a sensor's response is independent of that of the other sensors) such that

$$P(E^1, E^2,..., E^N \mid H) = \prod_{k=1}^{N} P(E^k) \mid H). \tag{17.33}$$

Under these conditions, Bayes' rule becomes (same as Equation 17.10)

$$P(H_i \mid E) = \frac{P(E \mid H_i)P(H_i)}{P(E)} = \frac{P(E \mid H_i)P(H_i)}{\sum_i [P(E \mid H_i)P(H_i)]}. \tag{17.34}$$

17.1.5 Posterior calculation with multivalued hypotheses and recursive updating

The following discussion follows a method found in Pearl [6]. Suppose several hypotheses $\mathbf{H} = \{H_1, H_2, H_3, H_4\}$ exist where each represents one of four possible conditions, such as

$H_1 =$ passenger vehicle on a highway
$H_2 =$ bus on a highway
$H_3 =$ 18-wheeler on a highway
$H_4 =$ motorcycle on a highway

Assume that the evidence variable $\mathbf{E}^k$ produced by a sensor has one of several output states in response to the detection of a vehicle. For example, when a multispectral sensor is used, three types of outputs may be available as represented by

$E_1^k =$ evidence from detected emission in radiance spectral band 1
$E_2^k =$ evidence from detected emission in radiance spectral band 2
$E_3^k =$ evidence from detected emission in radiance spectral band 3

The causal relations between $\mathbf{H}$ and $\mathbf{E}^k$ are quantified by a $q \times r$ matrix $\mathbf{M}^k$, where q is the number of hypotheses under consideration and r is the number of output states or output values of the sensor. The (i, j)th matrix element of $\mathbf{M}^k$ represents

$$M_{ij}^k = P(E_j^k \mid H_i). \tag{17.35}$$

For example, the sensitivity of the kth sensor having $r = 3$ output states to $\mathbf{H}$ containing $q = 4$ hypotheses is represented by the 4×3 evidence matrix in Table 17.3.

On the basis of the given evidence, the overall belief in the ith hypothesis H_i is (from Equation 17.10)

$$P\left(H_i \mid E_1, \ldots, E_r\right) = \alpha P(E_1, \ldots, E_r \mid H_i) P(H_i), \tag{17.36}$$

where α is a normalizing constant computed by requiring Equation 17.36 to sum to unity over i. Accordingly, α is equal to the inverse of the sum of the elements of $P(E_1, \ldots, E_r \mid H_i)$. In contrast, the likelihood functions $P(E^k \mid H_i)$ are not required to sum to unity over i.

When a sensor's response is conditionally independent, that is, each $P(E_j \mid H_i)$ is independent of that of the other sensors, Equation 17.33 can be applied to give

$$P(H_i \mid E_1, \ldots, E_r) = \alpha P(H_i) \prod_{k=1}^{N} P(E^k \mid H_i). \tag{17.37}$$

Table 17.3 Likelihood functions $P(E^k|H_i)$ corresponding to evidence produced by kth sensor with three output states in support of four hypotheses

	E_1^k: detection of emission in spectral band 1	E_2^k: detection of emission in spectral band 2	E_3^k: detection of emission in spectral band 3
H_1	0.35	0.40	0.10
H_2	0.26	0.50	0.44
H_3	0.35	0.10	0.40
H_4	0.70	0	0

A likelihood vector λ^k is defined to describe the evidence produced by each sensor E^k as

$$\lambda^k = (\lambda_1^k, \lambda_2^k, ..., \lambda_q^k), \tag{17.38}$$

where

$$\lambda_i^k = P(E^k \mid H_i). \tag{17.39}$$

Equation 17.37 can be evaluated using a two-step vector-product process as follows:

1. The individual likelihood vectors from each sensor are multiplied together, term by term, to obtain an overall likelihood vector $\Lambda = \lambda_1, ..., \lambda_q$ given by

$$\Lambda_i = \prod_{k=1}^{N} P(E^k \mid H_i). \tag{17.40}$$

2. The overall belief vector $P(H_i|E^1, ..., E^N)$ is computed from the product

$$P(H_i \mid E^1, ..., E^N) = \alpha P(H_i)\Lambda_i. \tag{17.41}$$

Only estimates for the relative magnitudes of the conditional probabilities in Equation 17.39 are required. Absolute magnitudes do not affect the outcome because α can be found later from the requirement

$$\sum_i P(H_i \mid E^1, ..., E^N) = 1. \tag{17.42}$$

To model the behavior of a multisensor system, let us assume that two sensors are deployed, each having the identical evidence matrix shown in Table 17.3. Furthermore, the prior probabilities for the hypotheses $\mathbf{H} = \{H_1, H_2, H_3, H_4\}$ are assigned as

$$P(H_i) = (0.42, 0.25, 0.28, 0.05), \tag{17.43}$$

where Equation 17.11 is satisfied by this distribution of prior probabilities.

If Sensor 1 detects emission in spectral band 3 and Sensor 2 detects emission in spectral band 1, the elements of the likelihood vector are

$$\lambda^1 = P(E^1 \mid H_i) = (0.10, 0.44, 0.40, 0) \tag{17.44}$$

and

$$\lambda^2 = P(E^2 \mid H_i) = (0.35, 0.26, 0.35, 0.70). \tag{17.45}$$

Therefore, the overall likelihood vector is

$$\Lambda = \lambda^1 \lambda^2 = (0.035, 0.1144, 0.140, 0) \tag{17.46}$$

and from Equation 17.41,

$$P(H_i \mid E^1, E^2) = \alpha P(H_i)\Lambda = \alpha(0.42,\ 0.25,\ 0.28,\ 0.05) \cdot (0.035,\ 0.1144,\ 0.140,\ 0)$$
$$= \alpha(0.0147,\ 0.0286,\ 0.0392,\ 0) = (0.178,\ 0.347,\ 0.475,\ 0), \tag{17.47}$$

where α is found from the requirement of Equation 17.42 as the inverse of the sum of $0.0147 + 0.0286 + 0.0392 + 0$, which is equal to 12.1212.

From Equation 17.47, we conclude that the probability of a large vehicle on the highway, H_2 or H_3, is $0.347 + 0.475 = 0.822$ or 82.2% and the probability of a passenger vehicle being present is 17.8%. The combined probability for some vehicle being present is 100%.

Updating of the posterior belief does not have to be delayed until all the evidence is collected, but can be implemented incrementally. For example, if it is first observed that Sensor 1 detects emission in spectral band 3, the belief in **H** becomes

$$P(H_i \mid E^1) = \alpha(0.42,\ 0.25,\ 0.28,\ 0.05) \cdot (0.10,\ 0.44,\ 0.40,\ 0)$$
$$= \alpha(0.042,\ 0.110,\ 0.112,\ 0) = (0.1591,\ 0.4167,\ 0.4242,\ 0), \tag{17.48}$$

with $\alpha = 3.7879$.

These values of the posterior are now utilized as the new values of the prior probabilities when the next datum arrives, namely, evidence from Sensor 2, which detects emission in spectral band 1. Upon incorporating this evidence, the posterior updates to

$$P(H_i \mid E^1, E^2) = \alpha' \lambda_i^2 \cdot P(H_i \mid E^1)$$
$$= \alpha'(0.35, 0.26, 0.35, 0.70) \cdot (0.1591, 0.4167, 0.4242, 0) \tag{17.49}$$
$$= \alpha'(0.0557, 0.1083, 0.1485, 0) = (0.178, 0.347, 0.475, 0),$$

where $\alpha' = 3.2003$. This is the same result given by Equation 17.47 for $P(H_i \mid E^1, E^2)$.

Thus, the evidence from Sensor 2 lowers the probability of a large vehicle being present slightly from 84.1% to 82.2%, but increases the probability of a passenger vehicle being present by the same amount from 15.9% to 17.8%. The result specified by Equation 17.47 or 17.49 is unaffected by which sensor's evidence arrives first and is subsequently used to update the priors for incorporation of the evidence from the next datum.

Reference 1 contains an incident detection application of Bayesian inference that utilizes information from roadway sensors, cell phones, and commercial truck radio transmissions. In that example, the likelihood functions are dependent on weather and lighting conditions.

17.1.6 Bayesian inference summary

Bayesian inference determines the likelihoods that unknown states H_i exist, where the new likelihoods associated with the H_i are computed from the product of the original prior information and the information gained by observation (the evidence). This is encoded in the posterior probability $P(H_i \mid E)$ that describes the likelihoods associated with H_i given the evidence E. Updating of the posterior belief does not have to be delayed until all the evidence is collected, but can be implemented incrementally.

In a data fusion application of Bayesian inference, each sensor provides an identity declaration D or hypothesis about the object's identity based on observations and a sensor-specific algorithm. The classification algorithm provides estimates of the probability that the sensor will declare the object to be a certain type, given the object is in fact type i.

This probability is denoted by $P(D|O_i)$. Sensor declarations are combined using a generalization of Bayes' rule to produce an updated, joint probability for each entity O_i.

The method requires the following:

- Ability to compute or model the likelihood functions $P(E \mid H_i)$ for each sensor and object.
- A *priori* probabilities that hypotheses $P(H_i)$ are true.
- When *a priori* information about $P(H_i)$ is lacking, invoke principle of indifference, that is, initial values for $P(H_i)$ for all i are set equal to each other.
- Discarding of past evidence or data once posterior probability $P(H_i \mid E^N)$ is calculated, where E^N is sequence of data observed from N sources in the past.
- Conditionally independent sensors.

Caution: Bayesian inference does not have a convenient representation for ignorance or uncertainty. Therefore, it equates concepts of ignorance and belief when direct knowledge is lacking. This limitation is overcome with Dempster–Shafer evidential reasoning, which is discussed in the next section.

17.2 DEMPSTER–SHAFER EVIDENTIAL REASONING

Dempster–Shafer evidential theory, a probability-based data fusion classification algorithm, is useful when the sensors (or more generally, the information sources) contributing information cannot associate a 100% probability of certainty to their output decisions. The algorithm captures and combines whatever certainty exists in the object-discrimination capability of the sensors. Knowledge from multiple sensors about events (called propositions) is combined using Dempster's rule to find the intersection or conjunction of the propositions and their associated probabilities. When the intersection of the propositions reported by the sensors is an empty set, Dempster's rule redistributes the conflicting probability to the nonempty set elements. When the conflicting probability becomes large, application of Dempster's rule can lead to counterintuitive conclusions. Several modifications to the original Dempster–Shafer theory have been proposed to accommodate these situations.

17.2.1 Overview of the process

Figure 17.5 contains an overview of the Dempster–Shafer data fusion process as might be configured to identify objects [7]. Each sensor has a set of observables corresponding to the phenomena that generate information received about objects and their surroundings. In this illustration, a sensor operates on the observables with its particular set of object classification algorithms. The knowledge gathered by each Sensor k, where $k = 1,...,$ N, associates a declaration of object type (referred to in the figure by object o_i where $i = 1,...,n$) with a probability mass or basic probability assignment $m_k(o_i)$ between 0 and 1. The probability mass expresses the certainty of the declaration or hypothesis, that is, the amount of support or belief attributed directly to the declaration. Probability masses closer to unity characterize decisions made with more definite knowledge or less uncertainty about the nature of the object. The probability masses for the decisions made by each sensor are then combined using Dempster's rules of combination. The hypothesis favored by the largest accumulation of evidence from all contributing sensors is selected as the most probable outcome of the fusion process. A computer stores the relevant information from each sensor concerning

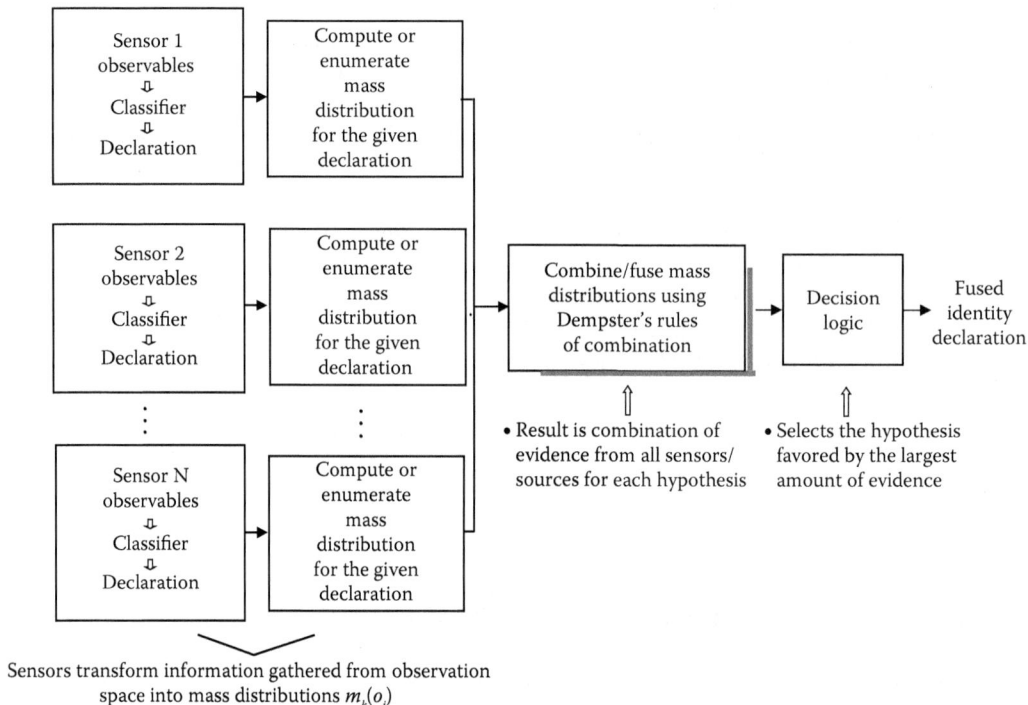

Sensors transform information gathered from observation space into mass distributions $m_k(o_i)$

Figure 17.5 Dempster–Shafer fusion process. (Adapted from E. Waltz and J. Llinas, *Multisensor Data Fusion*, Artech House, Norwood, MA, 1990.)

an object's identity and probability mass. The converse is also true, namely, objects not supported by evidence from any sensor are not stored.

In addition to real-time sensor data, other information or rules can be stored in the information base to improve the overall decision or object-discrimination capability. Examples of such rules are "Ships detected in known shipping lanes are cargo vessels" and "Objects in previously charted Earth orbits are weather or reconnaissance satellites."

17.2.2 Implementation of the method

Assume a set of n mutually exclusive and exhaustive propositions exists, for example, an object is of type $a_1, a_2,...,$ or a_n. This is the set of all propositions making up the hypothesis space, called the frame of discernment, and is denoted by θ. A probability mass $m(a_i)$ is assigned to any of the original propositions or to the union of the propositions based on available sensor information. Thus, the union or disjunction that the object is of type a_1 or a_2 (denoted $a_1 \cup a_2$) can be assigned probability mass $m(a_1 \cup a_2)$ by a sensor. A proposition is called a focal element if its probability mass is greater than zero. The number of combinations of propositions that exists (including all possible unions and θ itself, but excluding the null set) is equal to $2^n - 1$. For example if $n = 3$, there are $2^3 - 1 = 7$ propositions given by $a_1, a_2, a_3, a_1 \cup a_2, a_1 \cup a_3, a_2 \cup a_3,$ and $a_1 \cup a_2 \cup a_3$ where θ equals $a_1 \cup a_2 \cup a_3$. When the frame of discernment contains n focal elements, the power set consists of 2^n elements including the null set.

In the event that all of the probability mass cannot be directly assigned by the sensor to any of the propositions or their unions, the remaining mass is assigned to the frame of discernment θ (representing uncertainty as to further definitive assignment) as $m(\theta) = m(a_1 \cup a_2 \cup \ldots \cup a_n)$ or to the negation of a proposition such as $m(\bar{a}_1) = m(a_2 \cup a_3 \ldots \cup a_n)$. A raised bar is used to denote the negation of a proposition. The mass assigned to θ represents the uncertainty the sensor has concerning the accuracy and interpretation of the evidence [9]. The sum of probability masses over all propositions, uncertainty, and negation equals unity.

To illustrate these concepts, suppose that two sensors observe a scene in which there are three objects. Sensor A identifies the object as belonging to one of the three possible types: a_1, a_2, or a_3. Sensor B declares the object to be of type a_1 with a certainty of 80%. The intersection of the data from the two sensors is written as

$$(a_1 \text{ or } a_2 \text{ or } a_3) \text{ and } (a_1) = (a_1), \tag{17.50}$$

or upon rewriting as

$$(a_1 \cup a_2 \cup a_3) \cap (a_1) = (a_1). \tag{17.51}$$

Only a probability of 0.8 can be assigned to the intersection of the sensor data based on the 80% confidence associated with the output from Sensor B. The remaining probability of 0.2 is assigned to uncertainty represented by the union (disjunction) of $(a_1 \text{ or } a_2 \text{ or } a_3)$ [10].

17.2.3 Support, plausibility, and uncertainty interval

According to Shafer, "an adequate summary of the impact of the evidence on a particular proposition a_i must include at least two items of information: a report on how well a_i is supported and a report on how well its negation $\bar{a}_i$ is supported" [11]. These two items of information are conveyed by the proposition's degree of support and its degree of plausibility.

Support for a given proposition is defined as "the sum of all masses assigned *directly* by the sensor to that proposition or its subsets" [11,12]. A subset is called a focal subset if it contains elements of θ with mass greater than zero. Thus, the support for object type a_1, denoted by $S(a_1)$, contributed by a sensor is equal to

$$S(a_1) = m(a_1). \tag{17.52}$$

Support for the proposition that the object is either type a_1, a_2, or a_3 is

$$S(a_1 \cup a_2 \cup a_3) = m(a_1) + m(a_2) + m(a_3) + m(a_1 \cup a_2) + m(a_1 \cup a_3) + m(a_2 \cup a_3) + m(a_1 \cup a_2 \cup a_3). \tag{17.53}$$

Plausibility of a given proposition is defined as "the sum of all mass not assigned to its negation." Consequently, plausibility defines the mass free to move to the support of a proposition. The plausibility of a_i, denoted by $Pl(a_i)$, is written as

$$Pl(a_i) = 1 - S(\bar{a}_i), \tag{17.54}$$

where $S(\bar{a}_i)$ is called the dubiety and represents the degree to which the evidence impugns a proposition, that is, supports the negation of the proposition.

Plausibility can also be computed as the sum of all masses belonging to subsets a_j that have a non-null intersection with a_i. Accordingly,

$$Pl(a_i) = \sum_{a_j \cap a_i \neq 0} m(a_j) \tag{17.55}$$

Thus, when $\theta = \{a_1, a_2, a_3\}$, the plausibility of a_1 is computed as the sum of all masses compatible with a_1, which includes all unions containing a_1 and θ, such that

$$Pl(a_1) = m(a_1) + m(a_1 \cup a_2) + m(a_1 \cup a_3) + m(a_1 \cup a_2 \cup a_3). \tag{17.56}$$

The more general equation for calculating plausibility is Equation 17.54 as Equation 17.55 is only valid when values for the probability masses of all the subsets are known.

An uncertainty interval is defined by $[S(a_i), Pl(a_i)]$, where

$$S(a_i) \leq Pl(a_i). \tag{17.57}$$

The Dempster–Shafer uncertainty interval shown in Figure 17.6 illustrates the concepts just discussed [13,14]. The lower bound or support for a proposition is equal to the minimal commitment for the proposition based on direct sensor evidence. The upper bound or plausibility is equal to the support plus any potential commitment. Therefore, these bounds show what proportion of evidence is truly in support of a proposition and what proportion results merely from ignorance, or the requirement to normalize the sum of the probability masses to unity.

Support and probability mass obtained from a sensor (knowledge source) represent different concepts. Support is calculated as the sum of the probability masses that directly support the proposition and its unions. Probability mass is determined from the sensor's ability to assign some certainty to a proposition based on the evidence.

Table 17.4 provides further interpretations of uncertainty intervals. For example, the uncertainty interval [0, 1] represents total ignorance about proposition a_i since there is no direct support for a_i, but also no refuting evidence. The plausible range is equal to unity, as is the uncertainty interval. The uncertainty interval denoted by [0.6, 0.6] contains equal support and plausibility values. It indicates a definite probability of 0.6 for proposition a_i since both the direct support and plausibility are 0.6. In this case, the uncertainty interval equals zero. Support and plausibility values represented by [0, 0] indicate that the proposition a_i is false as all the probability mass is assigned to the negation of a_i. Therefore, the support for a_i is zero and the plausibility, $1 - S(\overline{a_i})$, is also zero since $S(\overline{a_i}) = 1$. When a_i is known to be

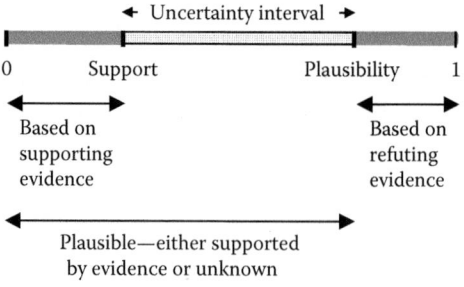

Figure 17.6 Dempster–Shafer uncertainty interval for a proposition.

Table 17.4 Interpretation of uncertainty intervals for proposition a_i

Uncertainty interval $[S(a_i), Pl(a_i)]$	Interpretation
[0, 1]	Total ignorance about proposition a_i
[0.6, 0.6]	A definite probability of 0.6 for proposition a_i
[0, 0]	Proposition a_i is false
[1, 1]	Proposition a_i is true
[0.25, 1]	Evidence provides partial support for proposition a_i
[0, 0.85]	Evidence provides partial support for $\bar{a}_i$
[0.25, 0.85]	Probability of a_i is between 0.25 and 0.85, that is, the evidence simultaneously provides support for both a_i and $\bar{a}_i$

true, [1, 1] represents the support and plausibility values. The uncertainty interval is zero since all the probability mass is assigned to the proposition a_i. Therefore, the support for a_i is 1 and the plausibility, $1 - S(\bar{a}_i)$, is also 1 since $S(\bar{a}_i) = 0$.

The support and plausibility values [0.25, 1] imply evidence that partially supports proposition a_i with a support value of 0.25. A plausibility of one indicates there is not any direct evidence to refute a_i. All the probability mass in the uncertainty interval of length 0.75 is free to move to the support of a_i. The interval [0, 0.85] implies partial support for the negation of a_i since there is no direct evidence to support a_i while there is partial evidence to support $\bar{a}_i$, that is, $S(\bar{a}_i) = 0.15$. The support and plausibility represented by [0.25, 0.85] show partial direct support for a_i and partial direct support for its negation. In this case, the uncertainty interval represents probability mass that is available to move to support a_i or $\bar{a}_i$.

As an example of how the uncertainty interval is computed from the knowledge a sensor provides, consider once more a single sensor denoted as Sensor A that gathers information concerning three propositions a_1, a_2, and a_3. The frame of discernment θ is given by

$$\theta = \{a_1, a_2, a_3\}. \tag{17.58}$$

The negation of proposition a_1 is represented by

$$\bar{a}_1 = \{a_2, a_3\}. \tag{17.59}$$

Assume probability masses are contributed by Sensor A to the propositions a_1, $\bar{a}_1$, $a_1 \cup a_2$, and θ as

$$m_A(a_1, \bar{a}_1, a_1 \cup a_2, \theta) = (0.4, 0.2, 0.3, 0.1). \tag{17.60}$$

Table 17.5 shows the uncertainty intervals for a_1, $\bar{a}_1$, $a_1 \cup a_2$, and θ calculated using these numerical values. The uncertainty interval computations for a_1 and $\bar{a}_1$ are straightforward since they are based on direct sensor evidence. The uncertainty interval for proposition $a_1 \cup a_2$ is found using the direct evidence from Sensor A that supports a_1 and $a_1 \cup a_2$. The probability mass $m_1(\theta)$, that is, the mass not assignable to a smaller set of propositions, is not included in any of the supporting or refuting evidence for $a_1 \cup a_2$ because $m_1(\theta)$ represents the residual uncertainty of the sensor in distributing the remaining probability mass directly to any other propositions or unions in θ based on the evidence. That is, the evidence has allowed the sensor to assign direct probability mass only to propositions a_1, $\bar{a}_1$, and $a_1 \cup a_2$.

Table 17.5 Uncertainty interval calculation for propositions a_1, $\bar{a}_1$, $a_1 \cup a_2$, and θ

Proposition	Support $S(a_i)$	Plausibility $1 - S(\bar{a}_i)$	Uncertainty interval
a_1	0.4 (given)	$1 - S(\bar{a}_1) = 1 - 0.2 = 0.8$	$[0.4, 0.8]$
$\bar{a}_1$	0.2 (given)	$1 - S(a_1) = 1 - 0.4 = 0.6$	$[0.2, 0.6]$
$a_1 \cup a_2$	$S(a_1) + S(a_1 \cup a_2) = 0.4 + 0.3 = 0.7$	$1 - S(\overline{a_1 \cup a_2}) = 1 - S(\bar{a}_1 \cap \bar{a}_2)^a$ $= 1 - 0 = 1$	$[0.7, 1]$
θ	$S(\theta) = 1$	$1 - S(\bar{\theta}) = 1 - 0 = 1$	$[1, 1]$

[a] Only probability mass assigned directly by Sensor A to $\bar{a}_1 \cap \bar{a}_2$ is used in the calculation. Because Sensor A has not assigned any probability mass directly to $\bar{a}_1 \cap \bar{a}_2$, the support for $\bar{a}_1 \cap \bar{a}_2$ is zero. Thus, the plausibility of $a_1 \cup a_2$ is unity.

Table 17.6 Uncertainty interval calculation for propositions a_1, $a_2 \cup a_3$, a_3, $\bar{a}_2$, and a_2

Proposition	Support $S(a_i)$	Plausibility $1 - S(\bar{a}_i)$	Uncertainty interval
a_1	0.4 (given)	$1 - S(\bar{a}_1) = 1 - 0 = 1$	$[0.4, 1]$
$a_2 \cup a_3$	$m_A(a_2) + m_A(a_3) + m_A(a_2 \cup a_3)$ $= 0 + 0.1 + 0.2 = 0.3$	$1 - S(\overline{a_2 \cup a_3}) = 1 - S(\bar{a}_2 \cap \bar{a}_3) = 1 - 0 = 1$	$[0.3, 1]$
a_3	$S(a_3) = 0.1$ (given)	$1 - S(\bar{a}_3) = 1 - 0 = 1$	$[0.1, 1]$
$\bar{a}_2$	$S(\bar{a}_2) = 0.3$ (given)	$1 - S(a_2) = 1 - 0 = 1$	$[0.3, 1]$
a_2	$S(a_2) = 0$	$1 - S(\bar{a}_2) = 1 - 0.3 = 0.7$	$[0, 0.7]$

The remaining mass is assigned to $m_1(\theta)$, implying that it is distributed in some unknown manner among the totality of all propositions. The uncertainty interval for the proposition θ is found as follows: support for θ is equal to unity because θ is the totality of all propositions; plausibility for θ is also unity because support is not assigned outside of θ; therefore, $m_1(\bar{\theta}) = 0$ and $Pl(\theta) = 1 - S(\bar{\theta}) = 1 - 0 = 1$.

As a final example of uncertainty interval calculation, consider a sensor that provides information about the identification of propositions a_1, a_2, a_3 such that $\theta = \{a_1, a_2, a_3\}$ and $\bar{a}_2$ is represented by the set $\{a_1, a_3\}$. The probability mass assignments supplied by Sensor A are $m_A(a_1, a_2 \cup a_3, a_3, \bar{a}_2) = (0.4, 0.2, 0.1, 0.3)$. Table 17.6 presents the results of the uncertainty interval calculation for all propositions that have been assigned probability mass by Information Source A and for a_2.

17.2.4 Dempster's rule for combination of multiple-sensor data

Dempster's rule supplies the formalism to combine the probability masses provided by multiple sensors or information sources for compatible propositions. The output of the fusion process is given by the intersection of the propositions having the largest probability mass. Propositions are compatible when their intersection exists. Dempster's rule also treats intersections that form a null set, that is, incompatible propositions. In this case, the rule adjusts the probability masses associated with null intersections to zero and increases the probability masses of the nonempty set intersections by a normalization factor K such that their sum is unity.

The general form of Dempster's rule for the total probability mass committed to an event c defined by the combination of evidence $m_A(a_i)$ and $m_B(b_j)$ from Sensors A and B is given by

$$m(c) = K \sum_{a_i \cap b_j = c} [m_A(a_i) m_B(b_j)], \qquad (17.61)$$

where $m_A(a_i)$ and $m_B(b_j)$ are probability mass assignments on θ,

$$K^{-1} = 1 - \sum_{a_i \cap b_j = \phi} [m_A(a_i) m_B(b_j)],$$ (17.62)

and ϕ is defined as the empty set. If K^{-1} is zero, then m_A and m_B are completely contradictory and the sum defined by Dempster's rule does not exist. The probability mass calculated in Equation 17.61 is termed the orthogonal sum and is denoted by $m_A(a_i) \oplus m_B(b_j)$.

Application of Dempster's rule is illustrated with the following four-object, two-sensor example.

Suppose that two vehicle types and locations are possible, namely,

$$\begin{pmatrix} a_1 = \text{commercial truck in Lane 1} & a_3 = \text{commercial truck in Lane 2} \\ a_2 = \text{passenger car in Lane 1} & a_4 = \text{passenger car in Lane 2} \end{pmatrix}.$$

Furthermore, Sensor A provides truck classification data according to the probability mass matrix given by

$$m_A = \begin{bmatrix} m_A(a_1 \cup a_3) = 0.6 \\ m_A(\theta) = 0.4 \end{bmatrix},$$ (17.63)

where $m_A(\theta)$ represents the uncertainty associated with the rules that determine the vehicle is a truck.

Sensor B is able to classify Lane 2 vehicles with the probabilities

$$m_B = \begin{bmatrix} m_B(a_3 \cup a_4) = 0.7 \\ m_B(\theta) = 0.3 \end{bmatrix},$$ (17.64)

where $m_B(\theta)$ represents the uncertainty associated with the rules that determine the vehicle is in Lane 2.

Dempster's rule is implemented by forming a matrix with the probability masses that are to be combined entered along the first column and last row as illustrated in Table 17.7. Inner matrix (row, column) elements are computed as the product of the probability mass in the same row of the first column and the same column of the last row. The proposition corresponding to an inner matrix element is equal to the intersection of the propositions that are multiplied. Accordingly, matrix element $(1, 2)$ represents the proposition formed by the intersection of uncertainty θ from Sensor A and $(a_3 \cup a_4)$ from Sensor B, namely, that a vehicle is in Lane 2. The probability mass $m(a_3 \cup a_4)$ associated with the intersection of these propositions is

$$m(a_3 \cup a_4) = m_A(\theta) m_B(a_3 \cup a_4) = (0.4)(0.7) = 0.28.$$ (17.65)

Table 17.7 Application of Dempster's rule

$m_A(\theta) = 0.4$	$m(a_3 \cup a_4) = 0.28$	$m(\theta) = 0.12$
$m_A(a_1 \cup a_3) = 0.6$	$m(a_3) = 0.42$	$m(a_1 \cup a_3) = 0.18$
	$m_B(a_3 \cup a_4) = 0.7$	$m_B(\theta) = 0.3$

Matrix element (1, 3) represents the intersection of the uncertainty propositions from Sensors A and B. The probability mass $m(\theta)$ associated with the uncertainty intersection is

$$m(\theta) = m_A(\theta)m_B(\theta) = (0.4)(0.3) = 0.12. \qquad (17.66)$$

Matrix element (2, 2) represents the proposition formed by the intersection of $(a_1 \cup a_3)$ from Sensor A and $(a_3 \cup a_4)$ from Sensor B, namely, that the vehicle is a commercial truck in Lane 2. The probability mass $m(a_3)$ associated with the intersection of these propositions is

$$m(a_3) = m_A(a_1 \cup a_3)m_B(a_3 \cup a_4) = (0.6)(0.7) = 0.42. \qquad (17.67)$$

Matrix element (2, 3) represents the proposition formed by the intersection of $(a_1 \cup a_3)$ from Sensor A and θ from Sensor B. Accordingly, the probability mass associated with this element is

$$m(a_1 \cup a_3) = m_A(a_1 \cup a_3)m_B(\theta) = (0.6)(0.3) = 0.18 \qquad (17.68)$$

and corresponds to the proposition that the vehicle is a commercial truck, either in Lane 1 or in Lane 2.

The proposition represented by $m(a_3)$, namely, a commercial truck in Lane 2, has the highest probability mass in the matrix. Thus, it is typically the one selected as the output to represent the fusion of the evidence from Sensors A and B. Note that the inner matrix element values add to unity.

When three or more sensors contribute information, the application of Dempster's rule is repeated using the inner elements calculated from the first application of the rule as the new first column and the probability masses from the next sensor as the entries for the last row (or vice versa).

17.2.5 Dempster's rule with empty set elements

When the intersections of the propositions that define the inner matrix elements form an empty set, the probability mass of the empty set elements is set equal to zero and the probability mass assigned to the nonempty set elements is increased by the factor K. To illustrate this process, suppose that Sensor B had identified Objects 2 and 4 instead of Objects 3 and 4, with probability mass assignments given by m_B' as

$$m_B' = \begin{bmatrix} m_B'(a_2 \cup a_4) = 0.5 \\ m_B'(\theta) = 0.5 \end{bmatrix} \qquad (17.69)$$

Application of Dempster's rule gives the results presented in Table 17.8, where element (2, 2) now belongs to the empty set. Since mass is assigned to ϕ, we find the value K that redistributes this mass to the nonempty set members by calculating

$$K^{-1} = 1 - 0.30 = 0.70, \qquad (17.70)$$

Table 17.8 Application of Dempster's rule with an empty set

$m_A(\theta) = 0.4$	$m(a_2 \cup a_4) = 0.20$	$m(\theta) = 0.20$
$m_A(a_1 \cup a_3) = 0.6$	$m(\phi) = 0.30$	$m(a_1 \cup a_3) = 0.30$
	$m_B'(a_2 \cup a_4) = 0.5$	$m_B'(\theta) = 0.5$

Table 17.9 Probability masses of nonempty set elements increased by K

$m_A(\theta) = 0.4$	$m(a_2 \cup a_4) = 0.286$	$m(\theta) = 0.286$
$m_A(a_1 \cup a_3) = 0.6$	0	$m(a_1 \cup a_3) = 0.429$
	$m'_B(a_2 \cup a_4) = 0.5$	$m'_B(\theta) = 0.5$

and its inverse K as

$$K = 1.429. \tag{17.71}$$

Table 17.9 shows the probability mass corresponding to the null set element is set equal to zero and the probability masses of the nonempty set elements are multiplied by K such that their sum is unity. In this example, a commercial truck is declared present, but its location by lane is undetermined.

17.2.6 Dempster's rule with singleton propositions

When probability mass assignments are provided by sensors that report unique singleton events (i.e., probability mass is not assigned to the union of propositions or the uncertainty class), the problem becomes Bayesian and the number of empty set elements increases as shown in the following example. Assume two possible vehicle types and locations are present as before, namely,

$$\begin{pmatrix} a_1 = \text{commercial truck in lane 1} & a_3 = \text{commercial truck in lane 2} \\ a_2 = \text{passenger car in lane 1} & a_4 = \text{passenger car in lane 2} \end{pmatrix}.$$

Only now Sensor A's probability mass matrix is given by

$$m_A = \begin{bmatrix} m_A(a_1) = 0.35 \\ m_A(a_2) = 0.06 \\ m_A(a_3) = 0.35 \\ m_A(a_4) = 0.24 \end{bmatrix} \tag{17.72}$$

and Sensor B's probability mass matrix by

$$m_B = \begin{bmatrix} m_B(a_1) = 0.10 \\ m_B(a_2) = 0.44 \\ m_B(a_3) = 0.40 \\ m_B(a_4) = 0.06 \end{bmatrix}. \tag{17.73}$$

Next apply Dempster's rule to compute the orthogonal sum by applying the m_A and m_B values as indicated in Table 17.10. The only commensurate elements are those along the diagonal. All others are empty set members. The value of K that redistributes the probability mass of empty set members to nonempty set propositions is found from

$$\begin{aligned} K^{-1} &= 1 - 0.006 - 0.035 - 0.024 - 0.154 - 0.154 - 0.1056 - 0.140 - 0.024 - 0.096 - 0.021 \\ &\quad - 0.0036 - 0.021 = 1 - 0.7842 = 0.2158 \end{aligned}$$

$$\tag{17.74}$$

Table 17.10 Application of Dempster's rule with singleton events

$m_A(a_1) = 0.35$	$m(a_1) = 0.035$	$m(\phi) = 0.154$	$m(\phi) = 0.140$	$m(\phi) = 0.021$
$m_A(a_2) = 0.06$	$m(\phi) = 0.006$	$m(a_2) = 0.0264$	$m(\phi) = 0.024$	$m(\phi) = 0.0036$
$m_A(a_3) = 0.35$	$m(\phi) = 0.035$	$m(\phi) = 0.154$	$m(a_3) = 0.140$	$m(\phi) = 0.021$
$m_A(a_4) = 0.24$	$m(\phi) = 0.024$	$m(\phi) = 0.1056$	$m(\phi) = 0.096$	$m(a_4) = 0.0144$
	$m_B(a_1) = 0.10$	$m_B(a_2) = 0.44$	$m_B(a_3) = 0.40$	$m_B(a_4) = 0.06$

Table 17.11 Redistribution of probability mass to nonempty set elements

$m_A(a_1) = 0.35$	$m(a_1) = 0.1622$	0	0	0
$m_A(a_2) = 0.06$	0	$m(a_2) = 0.1223$	0	0
$m_A(a_3) = 0.35$	0	0	$m(a_3) = 0.6487$	0
$m_A(a_4) = 0.24$	0	0	0	$m(a_4) = 0.0667$
	$m_B(a_1) = 0.10$	$m_B(a_2) = 0.44$	$m_B(a_3) = 0.40$	$m_B(a_4) = 0.06$

as

$$K = 4.6339. \tag{17.75}$$

Table 17.11 contains the resulting probability mass matrix. In this case, the most likely event is the presence of a commercial truck in Lane 2.

17.2.7 Singleton proposition problem solved with Bayesian inference

Since the probability mass assignment to the uncertainty interval and the union of propositions is zero, we can reformulate the last four-object example in Bayesian terms. The likelihood vector Λ is computed as the scalar product of the likelihood functions λ^1 and λ^2 (in this example, these are the probability mass assignments from each sensor). Accordingly,

$$\lambda^1 = P(E^A \mid H_i) = (0.35,\ 0.06,\ 0.35,\ 0.24), \tag{17.76}$$

$$\lambda^2 = P(E^B \mid H_i) = (0.10,\ 0.44,\ 0.40,\ 0.06), \tag{17.77}$$

and

$$\Lambda = \lambda^1 \cdot \lambda^2 = (0.035,\ 0.0264,\ 0.140,\ 0.0144). \tag{17.78}$$

From Equation 17.41,

$$P(H_i \mid E^A, E^B) = \alpha(0.035,\ 0.0264,\ 0.140,\ 0.0144) = (0.1622,\ 0.1223,\ 0.6487,\ 0.0667), \tag{17.79}$$

where $\alpha = 1/(0.035 + 0.0264 + 0.140 + 0.0144) = 4.6339$, the same value calculated for K with the Dempster–Shafer method. The result in Equation 17.79 is identical to that obtained with the Dempster–Shafer method in Table 17.11, namely, that the most likely event is a_3 (commercial truck in Lane 2).

In computing $P(H_i|E^A, E^B)$ in Equation 17.79, the values for $P(H_i)$ drop out as they are set equal to each other for all i by the principle of indifference. For example, if $P(H_i)$ equal to 0.25 for all i were included, α would be 18.5357 (four times larger), but the final values for $P(H_i|E^A, E^B)$ would be the same.

17.2.8 Comparison with Bayesian inference

Dempster–Shafer evidential theory accepts an incomplete probabilistic model. Bayesian inference does not. Thus, Dempster–Shafer can be applied when the prior probabilities and likelihood functions or ratios are unknown. The available probabilistic information is interpreted as phenomena that impose truth values to various propositions for a certain time period, rather than as likelihood functions. Dempster–Shafer theory estimates how close the evidence is to forcing the truth of a hypothesis, rather than estimating how close the hypothesis is to being true [6,15].

Dempster–Shafer allows sensor classification error to be represented by a probability assignment directly to an uncertainty class θ. Furthermore, Dempster–Shafer permits probabilities that express certainty or confidence to be assigned directly to an uncertain event, namely, any of the propositions in the frame of discernment θ or their unions. Bayesian theory permits probabilities to be assigned only to the original propositions themselves. This is expressed mathematically in Bayesian inference as

$$P(a + b) = P(a) + P(b) \tag{17.80}$$

under the assumption that a and b are disjoint propositions. In Dempster–Shafer,

$$P(a + b) = P(a) + P(b) + P(a \cup b). \tag{17.81}$$

Shafer expresses the limitation of Bayesian theory in a more general way: "Bayesian theory cannot distinguish between lack of belief and disbelief. It does not allow one to withhold belief from a proposition without according that belief to the negation of the proposition" [11].

Bayesian theory does not have a convenient representation for ignorance or uncertainty. Prior distributions have to be known or assumed with Bayesian. A Bayesian support function ties all of its probability mass to single points in θ. There is no freedom of motion, that is, no uncertainty interval [16]. The user of a Bayesian support function must somehow divide the support among singleton propositions. This may be easy in some situations such as an experiment with a fair die. If we believe a fair die shows an even number, we can divide the support into three parts, namely, 2, 4, and 6. If the die is not fair, then Bayesian theory does not provide a solution.

Thus, the difficulty with Bayesian theory is in representing what we actually know without being forced to overcommit when we are ignorant. With Dempster–Shafer, we use information from the sensors (information sources) to find the support available for each proposition. For the fair-die example, Dempster–Shafer gives the probability mass m_k(even). If the die were not fair, Dempster–Shafer would still give the appropriate probability mass.

Therefore, there is no inherent difficulty in using Bayesian statistics when the required information is available. However, Dempster–Shafer offers an alternative approach when knowledge is not complete; that is, ignorance exists about the prior probabilities associated with the propositions in the frame of discernment. The Dempster–Shafer formulation of a problem collapses into the Bayesian when the uncertainty interval is zero for all propositions and the probability mass assigned to unions of propositions is zero. On the other hand, any discriminating proposition information that may have been available from prior probabilities is ignored when Dempster–Shafer in its original formulation is applied.

17.2.9 Modifications to the original Dempster–Shafer method

Criticism of Dempster–Shafer has been expressed concerning the way it reassigns probability mass originally allocated to conflicting propositions and the effect of the redistribution on the proposition selected as the output of the fusion process [17,18]. This concern is of particular consternation when there is a large amount of conflict that produces counterintuitive results. Several alternatives have been proposed to modify Dempster's rule to better accommodate conflicting beliefs [12,19,20]. Several of these are discussed in [1] including the transferable belief model due to Smets [21–23] that modifies the basic probability assignment in proportion to the number of elements it contains, the plausibility transformation due to Cobb and Shenoy [24] that utilizes the concept of plausibility to decide among propositions, the modified Dempster–Shafer approach of Fixsen and Mahler [25] that utilizes *a priori* probability measures as weighting functions on the probability masses, and plausible and paradoxical reasoning due to Dezert [20] that allows evidence from the conjunction (AND) operator ∩ as well as the disjunction (OR) operator ∪ to be admitted.

The effect of conflicting propositions is illustrated by the following example. Consider two physicians who examine a patient and agree the patient suffers from either meningitis (M), concussion (C), or brain tumor (T). Thus $\theta = \{M, C, T\}$. Furthermore, the doctors agree in their low expectation of a tumor, but disagree in the other likely cause and provide diagnoses as follows:

$$m_1(M) = 0.99 \quad m_1(T) = 0.01 \quad \text{and} \quad m_2(C) = 0.99 \quad m_2(T) = 0.01,$$

where the subscript 1 represents the diagnosis of Physician 1 and subscript 2 the diagnosis of Physician 2.

Combining the physicians' belief functions using Dempster's rule to form the orthogonal sum gives

$$m(T) = \frac{0.001}{1 - 0.0099 - 0.0099 - 0.9801} = 1. \tag{17.82}$$

This is certainly an unexpected result that arises from the bodies of evidence (i.e., the physicians' diagnoses) agreeing that the patient does not suffer from a tumor, but being in almost full contradiction for the other causes of the disease. The result conveys a negative implication for the practical use of Dempster–Shafer in automated reasoning or data analysis where a counterintuitive result may not be uncovered. The reader can verify the result in Equation 17.82 through an orthogonal sum calculation such as that shown in Tables 17.8 and 17.9.

The issue highlighted with the above example is not unique to Dempster–Shafer. In general, the inherent uncertain nature of information sources, either due to sensor imprecision or due to changes in operator priorities, causes conditions to exist in which the information fusion algorithms are inappropriate for those conditions [26]. Nevertheless, information acquisition and automated fusion of information may still be retained at a relatively high level as long as the operator has access to the raw data and the operator is aware of the level of unreliability, such that some attention will be allocated to the original information.

17.2.10 Constructing probability mass functions

Perhaps, the most difficult part of applying Dempster–Shafer theory in its original or modified forms is obtaining probability mass functions. Three methods for developing these probabilities are noted in no particular preference order. The first utilizes knowledge of the characteristics of the data gathered by the sensors for different objects to create plots that relate probability mass to the value of a parameter measured by the sensor. The second uses confusion matrices derived from a comparison of real-time sensor data with reliable ground-truth data. A third method defines probability masses based on the ability of features extracted from incoming sensor signals to match expected object traits. Detailed discussions and additional references to these methods are found in Klein [1].

17.2.11 Dempster–Shafer evidential reasoning summary

The Dempster–Shafer approach to object detection, classification, and identification allows each sensor or information source to contribute information to the extent of its knowledge. Incomplete knowledge about propositions that corresponds to objects in a sensor's field of view is accounted for by assigning a portion of the sensor's probability mass to the uncertainty class. Dempster–Shafer can also assign probability mass to the union of propositions if the evidence supports it. It is in these regards that Dempster–Shafer differs from Bayesian inference, as Bayesian theory does not have a representation for uncertainty and permits probabilities to be assigned only to the original propositions themselves.

The uncertainty interval is bounded on the lower end by the support for a proposition and on the upper end by the plausibility of the proposition. Support is the sum of *direct* sensor evidence for the proposition. Plausibility is the sum of all probability mass not directly assigned by the sensor to the negation of the proposition. Thus, the uncertainty interval depicts what proportion of evidence is truly in support of a proposition and what proportion results merely from ignorance.

Dempster's rule provides the formalism to combine probability masses from different sensors or information sources. The intersection of propositions with the largest probability mass is selected as the output of the Dempster–Shafer fusion process. If the intersections of the propositions form an empty set, the probability masses of the empty set elements are redistributed among the nonempty set members.

Several alternative methods have been proposed to render the output of the Dempster–Shafer fusion process more intuitively appealing by preventing assignment of probability mass to highly conflicting propositions. These approaches involve transformations of the belief functions into probability functions that are used to make a decision based on the available information. Perhaps, the most difficult part of applying Dempster–Shafer theory in its original or modified forms is obtaining the probability mass functions.

REFERENCES

1. L.A. Klein, *Sensor and Data Fusion: A Tool for Information Assessment and Decision Making*, Second Edition, Press Monograph 222, SPIE, Bellingham, WA, 2012.
2. H. Durrant-Whyte and T. Henderson, Multisensor data fusion, In *Handbook of Robotics, Part C—Sensing and Perception*, O. Khatib and B. Siciliano (eds.), Chapter 25, Springer, New York, NY, 2008.

3. P.G. Hoel, *Introduction to Mathematical Statistics*, John Wiley and Sons, New York, NY, 1947.

4. H. Cramér, *Mathematical Methods of Statistics*, Ninth Printing, Princeton Univ. Press, Princeton, NJ, 1961.

5. L.L. Chao, *Statistics: Methods and Analyses*, McGraw-Hill Book Company, New York, NY, 1969.

6. J. Pearl, *Probabilistic Reasoning in Intelligent Systems: Networks of Plausible Inference*, Morgan Kaufmann Publishers, San Mateo, CA, 1988.

7. E. Waltz and J. Llinas, *Multisensor Data Fusion*, Artech House, Norwood, MA, 1990.

8. D.L. Hall, *Mathematical Techniques in Multisensor Data Fusion*, Artech House, Norwood, MA, 1992.

9. R.A. Dillard, *Computing Confidences in Tactical Rule-Based Systems by Using Dempster-Shafer Theory*, NOSC TD 649 (AD A 137274), Naval Ocean Systems Center, San Diego, CA, September 14, 1983.

10. S.S. Blackman, *Multiple Target Tracking with Radar Applications*, Artech House, Norwood, MA, 1986.

11. G. Shafer, *A Mathematical Theory of Evidence*, Princeton Univ. Press, Princeton, NJ, 1976.

12. C.K. Murphy, Combining belief functions when evidence conflicts, *Decision Support Systems*, 29:1–9, Elsevier Science, 2000.

13. P.L. Bogler, Shafer-Dempster reasoning with applications to multisensor target identification systems, *IEEE Transactions on Systems, Man and Cybenetics*, SMC-17(6):968–977, 1987.

14. T.D. Garvey, J.D. Lowrance, and M.A. Fischler, An inference technique for integrating knowledge from disparate sources, *Proc. Seventh International Joint Conference on Artificial Intelligence*, Vol. I, IJCAI-81, 319–325, 1981.

15. D. Koks and S. Challa, *An introduction to Bayesian and Dempster-Shafer data fusion*, DSTO-TR-1436, DSTO Systems Sciences Laboratory, Australian Government Department of Defence, Nov. 2005. www.dsto.defence.gov.au/publications/2563/DSTO-TR-1436.pdf.

16. J.A. Barnett, Computational methods for a mathematical theory of evidence, *Proc. Seventh International Joint Conference on Artificial Intelligence*, Vol. II, IJCAI-81, 868–875, 1981.

17. L.A. Zadeh, A theory of approximate reasoning, In *Machine Intelligence*, J. Hayes, D. Michie, and L. Mikulich (eds.), Vol. 9, 149–194, 1979.

18. L.A. Zadeh, Review of Shafer's a mathematical theory of evidence, *AI Magazine*, 5(3):81–83, 1984.

19. L. Valet, G. Mauris, and Ph. Bolon, A statistical overview of recent literature in information fusion, *IEEE AESS Systems Magazine*, 7–14, Mar. 2001.

20. J. Dezert, Foundations for a new theory of plausible and paradoxical reasoning, *Information and Security*, 9:1–45, 2002.

21. P. Smets, Constructing the pignistic probability function in a context of uncertainty, *Artificial Intelligence*, 5:29–39, 1990.

22. P. Smets, The transferable belief model and random sets, *International Journal of Intelligent Systems*, 7:37–46, 1992.

23. P. Smets and R. Kennes, The transferable belief model, *Artificial Intelligence*, 66:191–234, 1994.

24. B.R. Cobb and P.P. Shenoy, A comparison of methods for transforming belief function models to probability models, Lecture Notes in Artificial Intelligence, In *Symbolic and Quantitative Approaches to Reasoning with Uncertainty*, T.D. Nielsen and N.L. Zhang (eds.), Springer-Verlag, Berlin, Germany, 255–266, 2003. www.business.ku.edu/home/pshenoy/ECSQARU03.pdf. Accessed Jul. 2003.

25. D. Fixsen and R.P.S. Mahler, The modified Dempster-Shafer approach to classification, *IEEE Transactions on Systems, Man and Cybenetics—Part A: Systems and Humans*, SMC-27(1):96–104, January 1997.

26. R. Parasuraman, T.B. Sheridan, and C.D. Wickens, A model for types and levels of human interaction with automation, *IEEE Transactions on Systems, Man and Cybernetics—Part A: Systems and Humans*, 30(3):286–297, May 2000.

Index